Inorganic Chemistry: Toward the 21st Century

Malcolm H. Chisholm, EDITOR
Indiana University

Based on a symposium cosponsored by the Divisions of Inorganic Chemistry of both the American Chemical Society and the Chemical Institute of Canada, and jointly sponsored by the Dalton Division of the Royal Society of Chemistry, Indiana University, Bloomington, Indiana, May 16–19, 1982

ACS SYMPOSIUM SERIES 211

AMERICAN CHEMICAL SOCIETY
WASHINGTON, D.C. 1983

Library of Congress Cataloging in Publication Data

Inorganic chemistry.

(ACS symposium series, ISSN 0097-6156; 211)

"Based on a symposium cosponsored by the Divisions of Inorganic Chemistry of both the American Chemical Society and the Chemical Institute of Canada, and jointly sponsored by the Dalton Division of the Royal Society of Chemistry, [held at] Indiana University, Bloomington, Indiana, May 16–19, 1982."

Includes bibliographies and index.

1. Chemistry, Inorganic—Congresses.

I. Chisholm, Malcolm H. II. American Chemical Society. Division of Inorganic Chemistry. III Chemical Institute of Canada. Inorganic Chemistry Division. IV. Royal Society of Chemistry (Great Britain). Dalton Division. V. Series.

QD146.I55 546 82-24505
ISBN 0-8412-0763-1 ACSMC8 211 1-567
1983

PRINTED IN THE UNITED STATES OF AMERICA

ACS Symposium Series

M. Joan Comstock, *Series Editor*

FOREWORD

The ACS SYMPOSIUM SERIES was founded in 1974 to provide a medium for publishing symposia quickly in book form. The format of the Series parallels that of the continuing ADVANCES IN CHEMISTRY SERIES except that in order to save time the papers are not typeset but are reproduced as they are submitted by the authors in camera-ready form. Papers are reviewed under the supervision of the Editors with the assistance of the Series Advisory Board and are selected to maintain the integrity of the symposia; however, verbatim reproductions of previously published papers are not accepted. Both reviews and reports of research are acceptable since symposia may embrace both types of presentation.

CONTENTS

NEW ANALYTICAL AND SPECTROSCOPIC TECHNIQUES

PREFACE

INORGANIC CHEMISTRY continues to provide the spawning ground for the evolution of vast areas of chemistry. It reaches into solid state chemistry, polymer chemistry, biochemistry, organic synthesis via organometallics, homogeneous and heterogeneous catalysis, energy storage and energy-related technology, analytical, physical and theoretical chemistry. By definition, inorganic chemistry is concerned with the chemistry of the elements and their compounds, other than those of carbon which fall into the field of organic chemistry. The 1950s were considered the renaissance of inorganic chemistry. Since then, the field has grown rapidly in many new directions. The challenges that the world presents to inorganic chemists have never been greater; nor have the contributions that inorganic chemists make been more than at this time.

This symposium brings together a large number of leading chemists to discuss some of the rapidly developing areas. Emphasis was placed on looking toward the future, and some speculation from the speakers was encouraged as to what may arise in the future in their fields of research. I have also encouraged discussion of these ideas by allotting a greater amount of time for questions, answers, and debate after each lecture. This volume is the proceedings of this historic symposium, the first symposium to be jointly sponsored by the inorganic divisions of the American Chemical Society and The Chemical Institute of Canada, and the Dalton Division of the Royal Society of Chemistry.

I should like to thank the Office of Naval Research, the Petroleum Research Fund administered by the American Chemical Society, E. I. du Pont de Nemours & Company, Exxon Research & Engineering Company, The Chemical Institute of Canada, the ACS Division of Inorganic Chemistry, Monsanto Company, Imperial Chemical Industries (London), Union Carbide Corporation, Strem Chemicals, Indiana University's Friends of Chemistry, and the Office of the Vice President of Indiana University for financial support of this symposium. I should also like to thank my wife, the former Cynthia Truax and Symposium Secretary, for all her assistance.

MALCOLM H. CHISHOLM
Indiana University

October 1982

INORGANIC PHOTOCHEMISTRY AND ASPECTS OF SOLAR ENERGY CONVERSION

1

Solar Electricity: Lessons Gained from Photosynthesis

JAMES R. BOLTON

The University of Western Ontario, Department of Chemistry, Photochemistry Unit, London, Ontario N6A 5B7 Canada

Nature has developed a mechanistically very complex yet conceptually very simple process for the conversion and storage of solar energy. In this lecture I shall first examine the reaction and mechanism of photosynthesis deriving insights into how nature has achieved this remarkable process. I shall then go on to describe various attempts to mimic the primary steps of photosynthesis. Finally, I shall speculate on how these insights into the mechanism of photosynthesis might be used to design a new type of solar cell for the conversion of light to electricity.

The Photosynthesis Reaction

The reaction of photosynthesis is clearly the most important chemical reaction since life could not exist for long without it. The overall reaction is

$$CO_2(g) + H_2O(\ell) \xrightarrow{\text{light}} \frac{1}{6} C_6H_{12}O_6(s) + O_2(g)$$

where $C_6H_{12}O_6(s)$ is D-glucose, the major energy-storage product of photosynthesis from which all plant and animal biomass is derived. The thermodynamic parameters for the photosynthesis reaction at 298K (25°C) are $\Delta H = 467\ kJmol^{-1}$; $\Delta G = 496\ kJmol^{-1}$ and $E^\circ = 1.24V$ (1).

It is helpful to think of the photosynthesis reaction as the sum of an oxidation half reaction and a reduction half reaction as shown in Figure 1. In fact, nature does separate these half reactions, in that the reduction of CO_2 to carbohydrates occurs in the stroma of the chloroplast, the organelle in the leaf where the photosynthesis reaction occurs, - whereas, the light-driven oxidation half reaction takes place on the thylakoid membranes which make up the grana stacks within the chloroplast. Reduced nicotinamide adenine dinucleotide phosphate (NADPH) carries the reducing power and most of the energy to the stroma to drive the fixation of CO_2 with the help of some additional energy provided

0097-6156/83/0211-0003$06.00/0

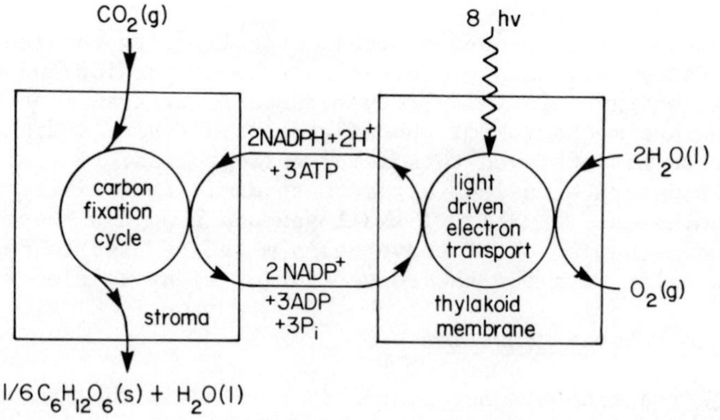

Figure 1. The separation of the half reaction in the chloroplast of the photosynthetic plant cell. The dark reaction (left) and the light-driven reactions (right) are shown. Key: NADP⁺, oxidized form of nicotinamide adenine dinucleotide phosphate; ATP, adenosine triphosphate; and P_i, inorganic phosphate.

by adenosine triphosphate (ATP) which is also generated on the thylakoid membrane. [Readers interested in the structure and mechanism of photosynthesis should refer to references (2-4).] Although the mechanism of the carbon-fixation cycle is very interesting biochemistry, it is the mechanism of the oxidation half reaction which will concern us most, because it is here that the conversion of sunlight to chemical energy takes place. We shall now explore some of the details of this remarkable process.

Mechanism of the Primary Photochemical Reaction of Photosynthesis

The primary photochemistry of photosynthesis takes place within very specialized reaction-center proteins situated in the thylakoid membrane (see Figure 2) of the chloroplast (2). Most (>99%) of the chlorophyll and other pigments act as an antenna system to gather light photons and channel them to one of the two reaction centers. In essence the antenna chlorophyll system acts as a photon concentrator, concentrating the photon flux by a factor of ∿300 over what it would be without an antenna system. In both Photosystems I and II the photochemically active component (P700 or P680) is thought to be a chlorophyll *a* species. P700 is likely a dimer of chlorophyll *a*. The primary photochemical step then involves the transfer of an electron from the donor (P700 or P680) to an acceptor species. In Photosystem II the acceptor Q_1 is thought to be a molecule of plastoquinone. In Photosystem I

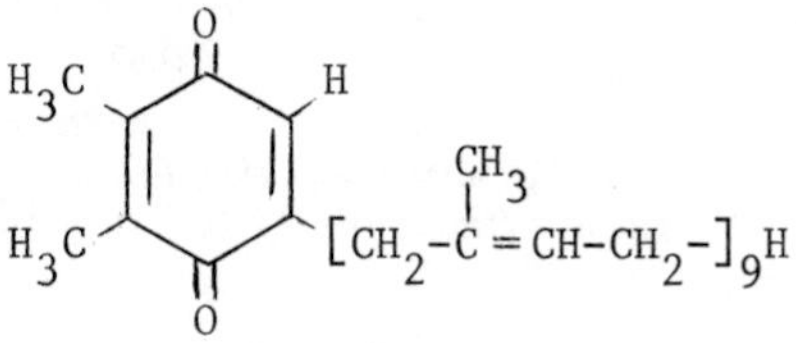

the acceptors are not as well characterized; A_1 may be a molecule of chlorophyll *a* in a special environment and A_2 is thought to be an iron-sulfur center.

Hence, it appears that nature has chosen the fastest and perhaps simplest of all photochemical reactions, namely, photochemical electron transfer, to trap the energy of the elusive sunbeam. In effect, the reaction-center proteins of photosynthesis are solar cells converting light to electricity which is then used to drive the relatively slow biochemical reactions which lead ultimately to D-glucose. These reaction-center solar cells are very effective - the yield for electron transfer is almost unity, and the overall solar energy conversion to electrical energy is ∿16% (5). This is as good as or generally much better than most commercially available silicon solar cells.

Now let us examine more closely the details of how the

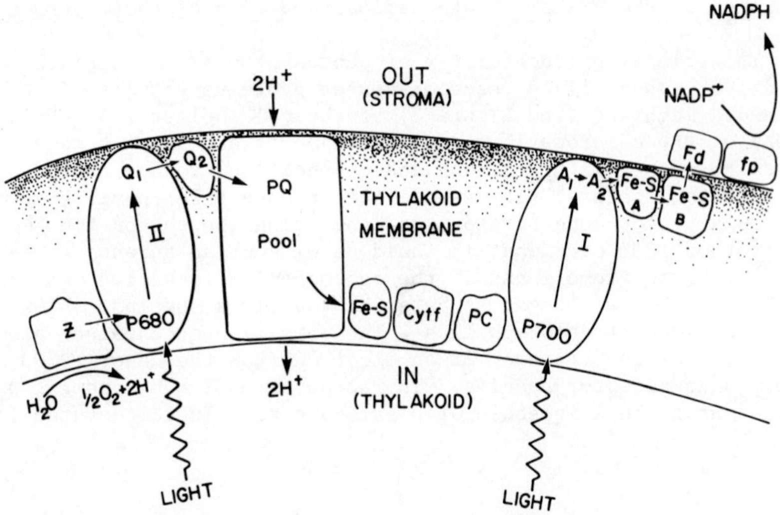

Figure 2. Model of the thylakoid membrane showing the various components involved in electron transport from H_2O to $NADP^+$.

The reaction-center proteins for Photosystems I and II are labeled I and II, respectively. Key: Z, the watersplitting enzyme which contains Mn; P680 and Q_1, the primary donor and acceptor species in the reaction-center protein of Photosystem II; Q_1 and Q_2, probably plastoquinone molecules; PQ, 6–8 plastoquinone molecules that mediate electron and proton transfer across the membrane from outside to inside; Fe–S (an iron–sulfur protein), cytochrome f, and PC (plastocyanin), electron carrier proteins between Photosystems II and I; P700 and A_1, the primary donor and acceptor species of the Photosystem I reaction-center protein; A_2, Fe–S_A and Fe–S_B, membrane-bound secondary acceptors which are probably Fe–S centers; Fd, soluble ferredoxin Fe–S protein; and fp, is the flavoprotein that functions as the enzyme that carries out the reduction of $NADP^+$ to NADPH.

electron is transferred from one side of the thylakoid membrane to the other. Figure 3 illustrates a view of our current knowledge of the composition and structure of the reaction-center protein from photosynthetic bacteria which performs a photochemical electron-transfer reaction analogous to the reactions in the two photosystems of green-plant photosynthesis (2, 7, 8). The bacterial reaction-center protein contains 4 bacteriochlorophyll molecules, 2 bacteriopheophytins, one nonheme iron and a ubiquinone molecule. Two of the bacteriochlorophyll molecules form a special pair called P870 which acts as the photochemical electron donor. The other two bacteriochlorophylls absorb at 800 nm and may act as the last link to the antenna system. Within <6 ps following the absorption of a photon, an electron moves from the P870 to one of the bacteriopheophytin molecules. Subsequently within ∿300 ps the electron moves to the ubiquinone. Thus within ∿300 ps the energy conversion step is over resulting in a charge separation between $P870^+$ and Q^- of 20-30 Å. If transfer to secondary acceptors does not occur, as it does efficiently *in vivo* at physiological temperatures, then the electron returns to $P870^+$ via a tunnelling mechanism with a half-life of ∿20 ms at low temperatures (7, 8). A similar picture, although more complex and less well characterized, has been assembled for the reaction center of Photosystem I in green plants (9).

Now for some speculation - it appears that nature has achieved the very difficult task of assuring rapid and efficient forward electron transfer while at the same time preventing the spontaneous back reaction. In effect, the reaction-center protein is an almost perfect photodiode. However, this facility has a price in that ∿0.8 eV of energy is lost in the primary electron transfer while an electron is moved 20-30 Å away from the primary donor. This process is carried out in several steps rather than one step - this has two advantages - less energy has to be dissipated in each step and it is possible to move the electron away a greater distance and thus minimize the chance of its returning. These seem to be key features in the energy-conversion process.

Spectral and Thermodynamic Considerations

Photosynthesis uses sunlight as its energy source, a source with a broad spectral distribution of photon energies. Thus any device which operates using a sensitizer with a fixed excitation energy will likely be more efficient the lower the excitation energy, because more solar photons may then be utilized. However, it is important to point out that, just as the Second Law of Thermodynamics says that it is impossible to convert heat into work with 100% efficiency, there is a similar thermodynamic restriction on the conversion of light to work or any other form of energy other than heat (6, 10, 11). The plant degrades the energy of all absorbed light photons in the photosynthetically active region (380-720 nm) to the energy of the lowest excited

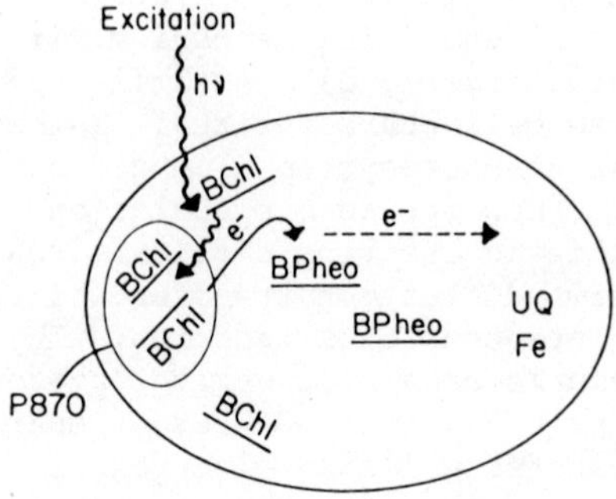

Figure 3. The bacterial reaction-center protein model from Rhodopseudomonas sphaeroides; *the structure and positioning of components are highly speculative.*

state of the chlorophyll molecules in the reaction centers which lie at an energy of ~1.8 eV (174 kJ mol^{-1}, 700 nm) above the ground state. It is a common misconception that in principle all of this energy is available to be converted into work or chemical energy. In fact, the thermodynamic analysis indicates that even in a perfect converter at 700 nm only ~1.4 eV or ~78% of this energy is Gibbs energy and available to do work.

Now let us see what effect these thermodynamic limits have on the mechanism of photosynthesis. The simplest system which nature could have devised would be one in which a single photosystem absorbs light and carries out the necessary reactions of photosynthesis. At most only one electron can be driven through the circuit for every photon absorbed. The thermodynamic limit allows only ~1.4 eV from each photon at 700 nm to operate the overall photosynthesis reaction which requires a minimum of 1.24 eV per electron driven through the circuit. Thus the reaction is allowed thermodynamically; however, to achieve this only ~0.16 eV could be lost in any nonidealities in the process. The problem is analagous to the engineer being required to design an almost frictionless heat engine. The system requires more flexibility and thus we conclude that the photosynthesis reaction cannot be operated with one photosystem. In fact, we have already seen that the energy loss in the primary electron transfer is ~0.8 eV leaving only ~1.0 eV to drive the reactions of photosynthesis. This is why the plant had to develop the much more complex mechanism of two photosystems operating in series (see Figure 2). Thus two photons are absorbed for every electron transferred from water to $NADP^+$ providing ~2.0 eV which is plenty to run the reaction.

Efficiency of Photosynthesis

The efficiency of photosynthesis depends on how the question is asked. Worldwide ~3×10^{21} J of solar energy is stored per year (12) and the annual solar input to the biosphere is ~3×10^{24} J - thus the worldwide efficiency of photosynthesis is ~0.1%. This may sound rather low but then consider that 3×10^{21} J is about 10 times the annual energy consumption by all peoples on the earth! Crops growing under ideal conditions have achieved as much as ~4% efficiency for short terms (13). However, it is possible to analyze the maximum efficiency expected for a plant from a knowledge of the absorption spectrum of a leaf and the experimental quantum yields for O_2 production as a function of wavelength. Such an analysis (12) indicates a maximum gross efficiency of ~9% which drops to ~5% when photorespiration, a light-stimulated back reaction, is accounted for. Hence, it is seen that a considerable loss in efficiency occurs between the primary conversion to electrical potential (~16%) and the end products (~5%). Of course, the end product is usually a very stable and storable fuel.

Lessons Gained from Photosynthesis

Here I shall summarize some of the salient points gained from our examination of the mechanism of photosynthesis.

1. Light is concentrated by a factor of ∿300 by use of an extensive antenna system of pigments. This minimizes the need to duplicate reaction centers.

2. The primary photochemical reaction is an electron transfer reaction which occurs within a highly structured reaction-center protein which spans the thylakoid membrane.

3. Photochemical electron transfer occurs from the excited singlet state of the primary donor.

4. One or more intermediate electron acceptors mediate the electron transfer across the reaction center from the donor to the ultimate acceptor.

5. Relatively long wavelength absorbers are utilized to maximize the capture of the solar spectrum.

6. Two photosystems are required to run the full reaction of photosynthesis.

Artifical Solar Energy Converters Designed to Mimic Photosynthesis

Much has been written about artifical solar-energy converters - the reader is referred to references 10, 12, 14-17 for detailed treatments. Here I shall deal exclusively with those artificial systems designed to mimic various aspects of the photosynthesis reaction.

The primary donor in Photosystem I P700 is thought to be a special pair of chlorophyll *a* molecules. Katz and Hindman (18) have reviewed a number of systems designed to mimic the properties of P700 ranging from chlorophyll *a* in certain solvents under special conditions where dimers form spontaneously (19) to covalently linked chlorophylls (20). Using these models it has been possible to mimic many of the optical, EPR and redox properties of the *in vivo* P700 entity.

Most of the interest in mimicing aspects of photosynthesis has centered on a wide variety of model systems for electron transfer. Among the early studies were experiments involving photoinduced electron transfer in solution from chlorophyll *a* to p-benzoquinone (21, 22) which has been shown to occur via the excited triplet state of chlorophyll *a*. However, these solution studies are not very good models of the *in vivo* reaction center because the *in vivo* reaction occurs from the excited singlet state and the donor and acceptor are held at a fixed relationship to each other in the reaction-center protein.

The next level of sophistication involved studies of systems where separate donors and acceptors interact across an interface. Chlorophyll-quinone photochemical studies have been conducted using liposomes (23-25) and acetate films (26). Calvin and his coworkers (27) have conducted a variety of experiments demonstrating that efficient photochemical electron transfer can under certain conditions be sensitized across artifical vesicles using ruthenium (II) tris(bipyridyl) as a sensitizer (see also 28). In addition micelles (29-32) and monolayers (33-35) have been used to separate donors and acceptors.

The final level of sophistication has involved covalently linking a donor and an acceptor. I shall review these systems extensively as this is where we have concentrated our own work. Kong and Loach (36, 37) were the first to report the synthesis of such an intramolecular donor-acceptor entity - namely (I);

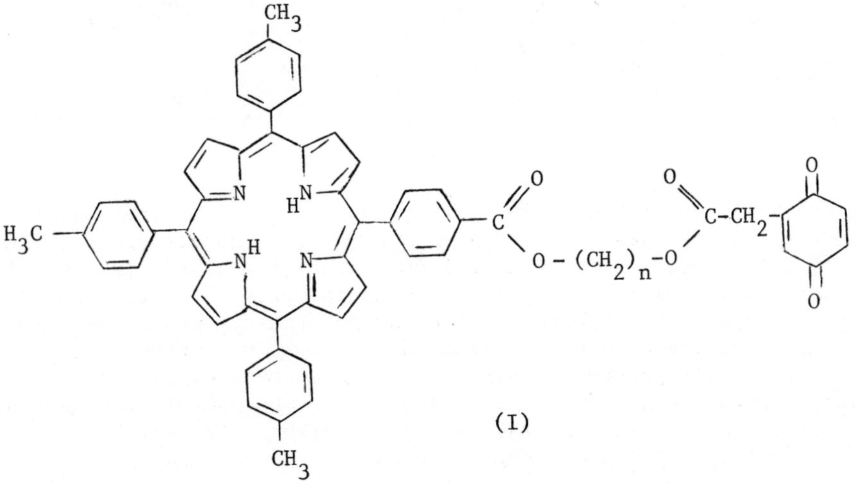

(I)

however, none of the photophysical or photochemical properties of (I) were reported at that time. Tabushi et al (38) reported the synthesis of (II) and found that the fluorescence of (II) was strongly quenched as compared to tetraphenyl porphyrin and

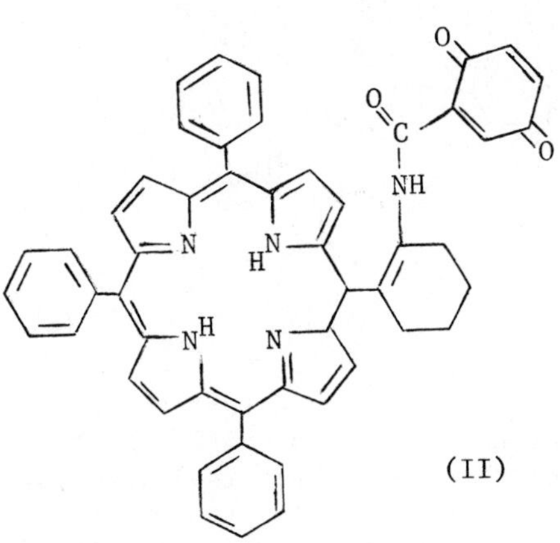

(II)

speculated that the quenching might be due to intramolecular electron transfer. Mataga and his coworkers (39) have synthesized the series of compounds (III) and also observed strong

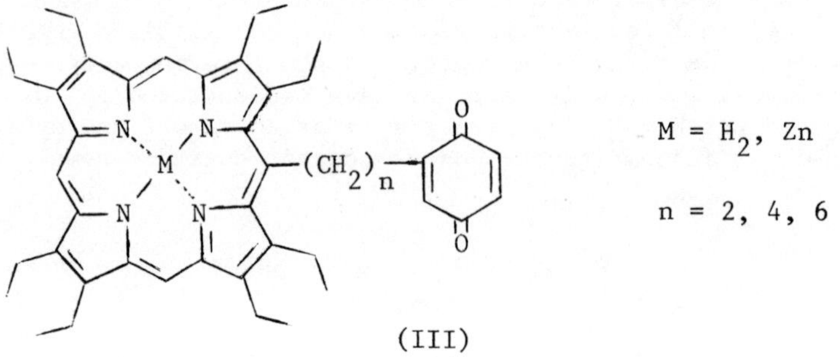

(III)

fluorescence quenching. In another publication (40) they observed optical changes by picosecond laser flash spectroscopy which they interpreted as evidence for formation of a charge transfer state via intramolecular electron transfer from the porphyrin to the quinone. The lifetime of the charge transfer state was found to increase markedly as n increased from 2 to 6. Ganesh and Sanders (41) reported the synthesis of a quinone-capped metalloporphyrin but did not report on any of its photophysical or photochemical properties. Netzel and his coworkers (42, 43) have studied (IV) using picosecond spectroscopy and on the basis of

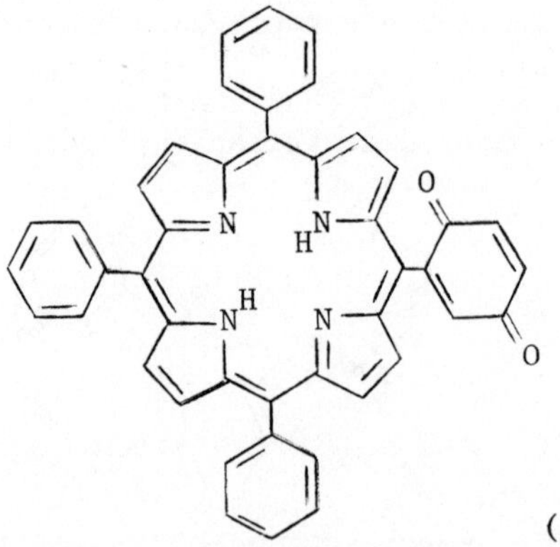

(IV)

a complex kinetic model concluded that the charge transfer state of (IV) comes from the porphyrin triplet state rather than the singlet state. Harriman and Hosie (44) studied a series of porphyrins in which a variety of electron donors and acceptors were attached at the four meso positions. They found strong fluorescence quenching for good electron donors (such as N,N-dimethyl aniline) as well as good electron acceptors (such as p-benzoquinone). Finally Matsuo et al (45) reported the synthesis of ruthenium tris(bipyridyl) covalently linked to viologen units. They found almost complete quenching of the emission from the ruthenium complex and in addition the covalently linked compound considerably enhanced electron transfer to relay systems of aligned viologen units on micelles and polymers.

Although the fluorescence quenching and optical absorption changes found on excitation of many of the above intramolecular donor-acceptor molecules is highly suggestive of intramolecular electron transfer, none of the above studies provided conclusive proof. We have been very interested in the problem of the syntheses of porphyrin-quinone molecules. When Kong and Loach (36) published their synthesis of (I) in 1978, we repeated their synthesis and in 1980 we reported (46) a light-induced electron paramagnetic resonance (EPR) signal which was shown, via a variety of control experiments, to arise from an intramolecular electron transfer from the porphyrin end to the quinone end of (I). Under certain conditions the electron transfer is reversible. Kong and Loach (46) have observed similar EPR signals from intramolecular transfer in the Zn complex of I; however, under all conditions the electron transfer was irreversible. Loach et al (47) have reported a variety of EPR and fluorescence data on (I) and its Zn complex confirming the intramolecular nature of the electron transfer and from the fluorescence data they conclude that the electron transfer most likely occurs out of the singlet state.

Our current work is concentrated on the series of porphyrin-quinone molecules (V). These are very similar to I in that amide linkages have replaced the ester linkages. We chose to study these molecules not only because they provided a definite change

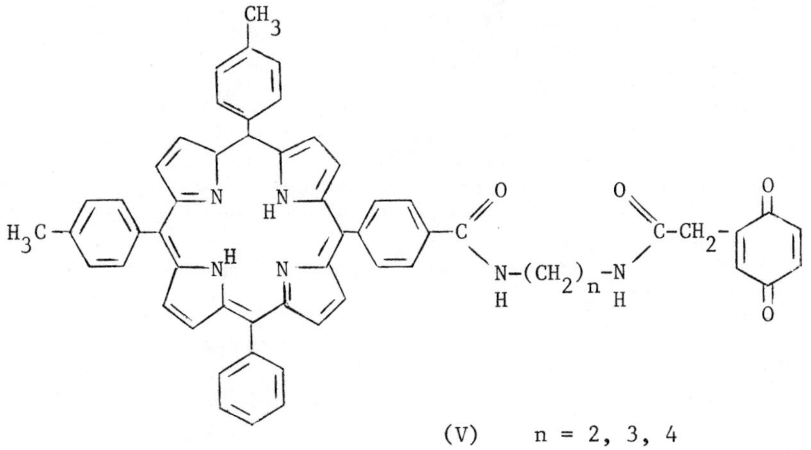

(V) n = 2, 3, 4

in the structure of the linkage but also because they are much less susceptible to decomposition by hydrolysis. The rate constant for forward electron transfer, as inferred from the fluorescence lifetimes, is about three times slower in (V) as compared to (I) (48). We have also observed optical changes using nanosecond optical flash photolysis and low temperature photolysis which show clearly the development of a charge transfer state which parallels the EPR observations (48). Finally, we have observed satellites in the EPR spectrum of (V) with n = 2 which arise from spin exchange between the porphyrin cation radical and the quinone anion radical providing conclusive evidence for intramolecular electron transfer (49). Also the quantum efficiency for electron transfer is found to be ∿1-2% (49).

A Possible Solar Cell

Much fundamental work yet remains in the study of intramolecular donor-acceptor molecules to find out what structural parameters of the donor, acceptor and particularly the linkage enhance the efficiency of forward electron transfer while at the same time inhibiting the rate of reverse electron transfer. Progress so far is very promising.

Assuming that an efficient D-A type of molecule can be synthesized, it should be possible to deposit these molecules as a monolayer onto a glass slide coated with a metal such as aluminum or a wide bandgap semiconductor such as SnO_2. With the acceptor end of the molecule near the conductor and with contact to the other side via an electrolyte solution it should be possible to stimulate electron transfer from D to A and then into the conductor, through an external circuit and finally back to D through the electrolyte. This would form the basis of a new type of solar cell in which the layer of D-A molecules would perform the same function as the p-n junction in a silicon solar cell (50). Only the future will tell whether or not this concept will be feasible but if nature can do it, why can't we?

Literature Cited

1. Bolton, J.R. "Photosynthesis 77: Proceedings of the Fourth International Congress on Photosynthesis" (Hall, D.O.; Coombs, J. and Goodwin, T.W., eds.); The Biochemical Society: London, 1978; pp 621-634.
2. Clayton, R.K. "Photosynthesis: Physical Mechanisms and Chemical Patterns"; Cambridge University Press: Cambridge, England, 1980.
3. Hall, D.L.; Rao, K.K. "Photosynthesis" Third Edition; Edward Arnold Ltd.: London, 1980.
4. Gregory, R.P.F. "Biochemistry of Photosynthesis" Second Edition; John Wiley and Sons Ltd.: New York, 1977.

5. The 16% efficiency figure is obtained as follows: for an air mass 1.5 solar spectra distribution $\sim$36% of the incident solar energy can be converted to excited state energy at 700 nm (6). If an average absorption coefficient of 0.9 is assumed, then 32.2% of the incident irradiance will appear as energy of the excited states of the reaction-center chlorophylls. Out of the 1.8 eV excitation energy at 700 nm $\sim$1.0 eV is generated as electrical energy in the electron transfer reaction. If we assume a fill factor of 0.9 then $\sim$16% of the incident solar power will be available as electrical power following the primary electron-transfer step.
6. Bolton, J.R.; Haught, A.F.; Ross, R.T. "Photochemical Conversion and Storage of Solar Energy" (Connolly, J.S., ed.); Academic Press: New York, 1981; pp 297-399.
7. Bolton, J.R. "The Photosynthetic Bacteria" (Calyton, R.K. and Sistrom, W.R., eds.); Plenum Publishing Corp.: New York, 1978; pp 414-429.
8. Blankenship, R.E.; Parson, W.W. "Photosynthesis in Relation to Model Systems" (Barber, J., ed.); Elsevier/North Holland Biomedical Press: Amsterdam, 1979; pp. 71-114.
9. Bolton, J.R. "Primary Processes of Photosynthesis" (Barber, J., ed.); Elsevier/North Holland Biomedical Press: Amsterdam, 1977; pp 188-201.
10. Bolton, J.R. Science 1978, 202, 705-711.
11. Bolton, J.R.; Haught, A.F.; Ross, R.T. Interamerican Photochemical Society Newsletter 1981, 4, 26-30.
12. Bolton, J.R.; Hall, D.O. Annu. Rev. Energy 1979, 4, 353-401.
13. Hall, D.O. "Solar Power and Fuels" (Bolton, J.R., ed.); Academic Press: New York, 1977; pp 27-52.
14. Bolton, J.R., ed. "Solar Power and Fuels"; Academic Press: New York, 1977.
15. Archer, M.D., ed. Special Issue of J. Photochem. 1979, 10, (1).
16. Hautala, R.R.; King, R.B.; Kutal, C., eds. "Solar Energy Chemical Conversion and Storage"; Humana Press: Clifton, N.J., 1979.
17. Connolly, J.S. "Photochemical Conversion and Storage of Solar Energy"; Academic Press: New York, 1981.
18. Katz, J.J.; Hindman, J.C. "Photochemical Conversion and Storage of Solar Energy" (Connolly, J.S., ed.); Academic Press: New York, 1981; pp 27-78.
19. Shipman, L.L.; Cotton, T.M.; Norris, J.R.; Katz, J.J. Proc. Natl. Acad. Sci. U.S.A. 1976, 73, 1791.
20. Boxer, S.G.; Closs, G.L. J. Am. Chem. Soc. 1976, 98, 5406.
21. Tollin, G. Bioenergetics 1974, 6, 69-87.
22. Tollin, G. J. Phys. Chem. 1976, 80, 2274-2277.
23. Hurley, J.K.; Castelli, F.; Tollin, G. Photochem. Photobiol. 1980, 32, 79-86.
24. Hurley, J.K.; Castelli, F.; Tollin, G. Photochem. Photobiol. 1981, 34, 623-631.

25. Kuriharg, K.; Sukigara, M.; Toyosshima, Y. Biochim. Biophys. Acta 1979, 547, 117-126.
26. Cheddar, G.; Castelli, F.; Tollin, G. Photochem. Photobiol. 1980, 32, 71-78.
27. Calvin, M. "Photochemical Conversion and Storage of Solar Energy" (Connolly, J.S., ed.); Academic Press: New York, 1981; pp 1-26.
28. Tunuli, M.S.; Fendler, J.H. J. Am. Chem. Soc. 1981, 103, 2507-2513.
29. Grätzel, M. "Photochemical Conversion and Storage of Solar Energy" (Connolly, J.S., ed.); Academic Press: New York, 1981; pp 131-160.
30. Thomas, J.K.; Piciulo, P. "Interfacial Photoprocesses"; ACS Symposium Series 1980, 184, 97-111.
31. Schmehl, R.H.; Whitesell, L.G.; Whitten, D.G. J. Am. Chem. Soc. 1981, 103, 3761-3764.
32. Infelta, P.P.; Grätzel, M.; Fendler, J.H. J. Am. Chem. Soc. 1980, 102, 1479-1483.
33. Whitten, D.G.; Mercer-Smith, J.A., Schmehl, R.H.; Worsham, P.R. "Interfacial Photoprocesses"; ACS Symposium Series 1980, 184, 47-67.
34. Möbius, D. Accounts Chem. Res. 1981, 14, 63-68.
35. Kuhn, H. J. Photochem. 1979, 10, 111-132.
36. Kong, J.L.Y.; Loach, P.A. "Frontiers of Biological Energetics: From Electrons to Tissues" Vol. 1 (Dutton, P.L.; Leigh, J.S.; Scarpa, H., eds.); Academic Press: New York, 1978; p 73.
37. Kong, J.L.Y.; Loach, P.A. J. Heterocyclic Chem. 1980, 17, 737-744.
38. Tabushi, I.; Koga, N.; Yanagita, M. Tetrahed. Lett. 1979, 257-260.
39. Nishitani, S.; Kurata, N.; Sakata, Y.; Misumi, S.; Migita, M.; Okada, T.; Mataga, N. Tetrahed. Lett. 1981, 22, 2099-2102.
40. Migita, M.; Okada, T.; Mataga, N.; Nishitani, S.; Kurata, N.; Sakata, Y.; Misumi, S. Chem. Phys. Lett. 1981, 84, 263-266.
41. Ganesh, K.N.; Sanders, J.K.M. J. Chem. Soc. Chem. Commun. 1980, 1129-1131.
42. Netzel, T.L., Bergkamp, M.A., Chang, C-W.; Dalton, J. J. Photochem. 1981, 17, 451-460.
43. Bergkamp, M.A., Dalton, J.; Netzel, T.L. J. Am. Chem. Soc. 1982, 104, 253-259.
44. Harriman, A.; Hosie, R.J. J. Photochem. 1981, 15, 163-167.
45. Matsuo, T.; Sakamoto , T.; Takuma, K.; Sakura, K.; Ohsako, T. J. Phys. Chem. 1981, 85, 1277-1279.
46. Kong, J.L.Y.; Loach, P.A. "Photochemical Conversion and Storage of Solar Energy" (Connolly, J.S., ed.); Academic Press: New York, 1981; p 350.
47. Loach, P.A.; Runquist, J.A.; Kong, J.L.Y.; Dannhauser, T.J.; Spears, K.G. ACS Advances in Chemistry Series (in press).

48. Bolton, J.R.; Ho. T.-F.; McIntosh, A.R.; Siemiarczuk, A.; Weedon, A.C.; Connolly, J.S. "Proceedings of the Solar World Forum Congress of the International Solar Energy Society" Brighton, England; Pergamon Press: New York, (in press).
49. McIntosh, A.R.; Ho, T.-F.; Weedon, A.C.; Siemiarczuk, A.; Stillman, M.J.; Bolton, J.R. (in preparation).
50. Bolton, J.R.; Ho, T.-F.; McIntosh, A.R., Canada, U.S. and Other Country Patents applied for.

RECEIVED August 3, 1982

Discussion

A.W. Adamson, University of Southern California: You mentioned a thermodynamic limitation to the efficiency of conversion of solar energy to useful work. This is an interesting point, and I would like to hear more about it. I can see an entropic effect in terms of the following hypothetical cell for the system A = A*, where A* is the excited state of species A:

Pt/C(s)	/	solution of A with equilibrium concentration of A*	//	solution of A with non-equilibrium concentration of A* produced by irradiation	/	C(s)/M/Pt

Here, C(s) is a solid reduced form of A, and the double line denotes a liquid junction of negligible potential. Also, A* is a thermally equilibrated excited (or thexi) state and, as such, is essentially a different chemical species from A. We suppose, therefore, that it is possible to find an electrode M that is reversible to the reduction of A* to C(s), but completely polarized with respect to the reduction of A to C(s). Operation of this cell under steady state conditions should then give the desired reversible work available from the photoproduction of A*.

J.R. Bolton: Your model is interesting, but I believe it complicates the issue. The system A,A* is a two-component system. In the dark $\mu_A = \mu_{A^*} = \mu_{A^*}^{\circ} + kT\ln X_{A^*}^{eq}$ where the μ's are chemical potentials and $X_{A^*}^{eq}$ is the equilibrium mole fraction of A*. In the light X_{A^*} increases; hence, the chemical potential available to do work is

$$\Delta\mu = kT\ln \frac{X_{A^*}}{X_{A^*}^{eq}}$$

$\Delta\mu$ can never be as large as E_g, the excitation energy.

A.B.P. Lever, York University: Most model photocatalysts undergo electron transfer reactions via their spin triplet states. You point out that there are advantages to using the spin singlet state for electron transfer as exemplified by chlorophyll. What kind of structural or electronic features should be built into model photocatalysts to favour use of their spin singlet states for electron transfer quenching?

J.R. Bolton: In solution most photochemical electron transfer reactions occur from the triplet state because in the collision complex there is a spin inhibition for back electron transfer to the ground state of the dye. Electron transfer from the singlet excited state probably occurs in such systems but the back electron transfer is too effective to allow separation of the electron transfer products from the solvent cage. In our linked compound, the quinone cannot get as close to the porphyrin as in a collision complex, yet it is still close enough for electron transfer to occur from the excited singlet state of the porphyrin Now the back electron transfer is inhibited by the distance and molecular structure between the two ends. Our future work will focus on how to design the linking structure to obtain the most favourable operation as a molecular "photodiode".

A.J. Bard, University of Texas: The mechanism you propose implies that there are spin selection rules operative which affect the relative rates of the electron transfer reactions. Is there any evidence that such spin selection rules are important in these kinds of reactions, especially in the presence of metallic centers?

J.R. Bolton: We have not carried out any experiments as yet on metalloporphyrins linked to quinones. The spin selection rules should be operative in the radical pair. The singlet state of the radical pair should be able to return to the ground state with no spin inhibition; however, the triplet state of the radical pair can return to the ground state only via spin interconversion or via the triplet state of the porphyrin.

T.J. Kemp, University of Warwick: Noting the very low quantum yield for intramolecular electron transfer in low temperatures displayed by your porphyrin-quinone model compound, would it not be possible to 'shock-freeze' a solution undergoing irradiation at a higher temperature (and giving a workable concentration of paramagnetic species) in order to determine a low-temperature spectrum with the particular aim of observing a possible $\Delta m = 2$ transition?

J.R. Bolton: This is a good idea - we will try it.

M.S. Wrighton, M.I.T.: What is the basis for an estimate of 5% efficiency for photosynthesis and the 18% for a comparison with solid state photovoltaics for electricity generation?

J.R. Bolton: The 5% figure refers to the net energy efficiency in the production of d-glucose taking account of photorespiration. The 18% figure refers to the efficiency of conversion of sunlight to electrical energy at the level of the reaction-center protein. The details of the calculation are in my paper. However, as you pointed out, I neglected a fill factor and thus a more realistic value would be ~16%, still an impressive number for the natural photovoltaic cell.

T.J. Meyer, University of North Carolina: You report that the data that you observe suggests a lack of solvent dependence, and therefore that electron transfer occurs through the chemical link rather than through the solvent. However, in either case there is a charge transfer process through distance and there should be a solvent dependence associated with the electron transfer act. Do you have any comments on this point?

J.R. Bolton: If the electron reaches the quinone via the linkage, then the transfer must involve the molecular orbitals of the linking structure and thus the solvent will have only a secondary effect. If the electron transfer occurs through the solvent, then the solvent should have a first-order effect on the rate.

2

Oxidation–Reduction Photochemistry of Polynuclear Complexes in Solution

DANIEL G. NOCERA, ANDREW W. MAVERICK, JAY R. WINKLER, CHI-MING CHE, and HARRY B. GRAY

California Institute of Technology, Arthur Amos Noyes Laboratory, Pasadena, CA 91125

Three classes of polynuclear complexes containing metal-metal bonds possess emissive excited states that undergo oxidation-reduction reactions in solution: the prototypes are $Re_2Cl_8^{2-}$($\underline{d}^4 \cdot \underline{d}^4$), $Pt_2(P_2O_5H_2)_4^{4-}$ ($\underline{d}^8 \cdot \underline{d}^8$), and $Mo_6Cl_{14}^{2-}$ ($\underline{d}^4$)$_6$. Two-electron oxidations of $Re_2Cl_8^{2-}$ and $Pt_2(P_2O_5H_2)_4^{4-}$ have been achieved by one-electron acceptor quenching of the excited complexes in the presence of Cl^-, followed by one-electron oxidation of the Cl^--trapped mixed-valence species. Two-electron photochemical oxidation-reduction reactions also could occur by excited-state atom transfer pathways, and some encouraging preliminary observations along those lines are reported.

Present systems for photochemical energy conversion emphasize one-photon/one-electron-relay/catalyst schemes, as in the famous $Ru(bpy)_3^{2+}$/water splitting experiments of many workers(1-14). All such systems are designed to take advantage of the fact that upon excitation the metal complex sensitizer is a better oxidant and reductant than its ground state, and as a result it can oxidize poor donors and reduce weak acceptors provided it lives long enough in solution to do so (Figure 1). Net storage of photochemical energy is then achieved by employing separate catalysts to convert the photogenerated species into useful products(1).

It has been our goal for some time to run photochemical energy storage reactions without relay molecules or separate catalysts. We have concentrated on the photochemistry of polynuclear metal complexes in homogeneous solutions, because we believe it should be possible to facilitate multielectron transfer processes at the available coordination sites of such cluster species.

One problem we have had to overcome in developing metal-cluster oxidation-reduction photochemistry is the tendency of excited clusters to dissociate into radical fragments (for

0097-6156/83/0211-0021$06.00/0

example, $Mn_2(CO)_{10}$). But after several years of research we have found that (at least) three broad classes of metal-metal bonded systems have attractive excited-state properties:

Electronic Configuration	Example
(1) $d^4 \cdot d^4$	$Re_2Cl_8^{2-}$
(2) $d^8 \cdot d^8$	$Pt_2(P_2O_5H_2)_4^{4-}$
(3) $(d^4)_6$	$Mo_6Cl_{14}^{2-}$

In the first case the metal-metal bond is quite strong in the ground state, and it is weakened only slightly in the lowest singlet ($^1\delta\delta^*$) excited state. In the second case the lowest excited state ($^3d\sigma^*p\sigma$) possesses a relatively strong metal-metal bond. In the third case the nature of the long-lived excited state is not well understood at this time. It may have some relationship to a $\delta \to \delta^*$ or $\delta \to \pi^*$ excited state in a d^4 binuclear complex, however.

Binuclear Complexes

Evidence suggests that the $^1\delta\delta^*$ state of $Re_2Cl_8^{2-}$ is not eclipsed (the δ bond is gone), and that Franck-Condon factors are responsible for the relatively long lifetime of this state (Figure 2)(15,16,17). Various electron acceptors (*e.g.*, TCNE) quench the $Re_2Cl_8^{2-*}$ luminescence in nonaqueous solutions, thereby producing $Re_2Cl_8^-$ and the reduced acceptor(17). A transient signal attributable to $TCNE^-$ was observed in flash kinetic spectroscopic studies of dichloromethane solutions containing TCNE and $(Bu_4N)_2Re_2Cl_8$; the decay of the transient was found to follow second-order kinetics ($k = 3 \times 10^9$ M^{-1} s^{-1}).

The luminescence of $Re_2Cl_8^{2-*}$ also is quenched by secondary and tertiary aromatic amines in acetonitrile solution(17). Neither the electronic absorption nor the emission spectrum of $Re_2Cl_8^{2-}$ changes in the presence of the quenchers, and no evidence for the formation of new chemical species was observed in flash spectroscopic or steady-state emission experiments. The results of these experiments suggest that the products of the quenching reaction form a strongly associated ion pair, $Re_2Cl_8^{3-} \cdot D^+$.

The two reduction potentials involving $Re_2Cl_8^{2-*}$ (-/2-*; 2-*/3-) have been estimated from the results of spectroscopic and electrochemical experiments (-0.51 and 0.90 V *vs.* SCE)(17). The $\delta\delta^*$ singlet provides a facile route to an extremely powerful inorganic oxidant, $Re_2Cl_8^-$ (E° = 1.24 V *vs.* SCE), a species that has not been generated cleanly by other means.

In recent experiments D. G. Nocera has extended this work to include two-electron photochemical oxidation of $Re_2Cl_8^{2-}$. The strategy involved here is to generate $Re_2Cl_8^-$ by acceptor quenching

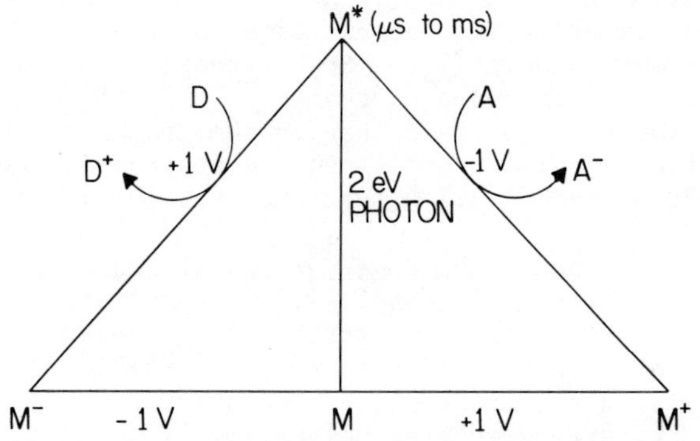

Figure 1. Modified Latimer diagram illustrating the relative reduction potentials of a metal complex (M) and its excited state (M).*

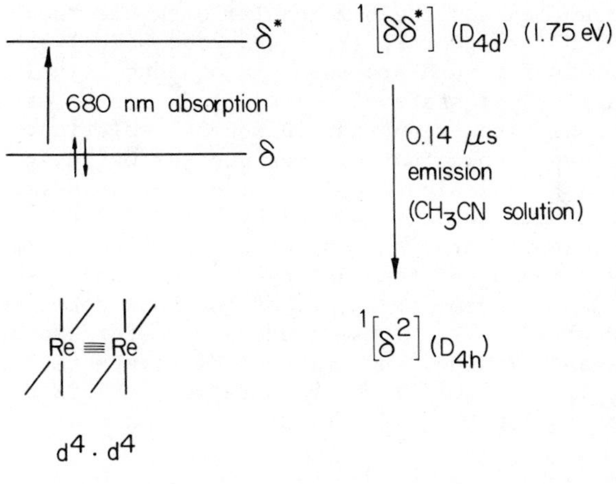

Figure 2. Selected electronic spectroscopic properties of $Re_2Cl_8^{2-}$.

of $^1\delta\delta^*$, and then by adding Cl^- to trap the mixed-valence species, $Re_2Cl_9^{2-}$. Acceptors whose potentials are high enough can then oxidize $Re_2Cl_9^{2-}$ to $Re_2Cl_9^-$ (Figure 3).

Several $d^8 \cdot d^8$ complexes of rhodium(I), iridium(I), and platinum(II) possess relatively long-lived excited states (many are in the μs range)(18,19,20). The excited state is believed to be a $d\sigma^* p\sigma$ triplet (Figure 4). The beautiful green phosphorescence (517 nm; lifetime ~10 μs) exhibited by $Pt_2(P_2O_5H_2)_4^{4-}$ is an interesting case in point(20). Spectroscopic studies on $Ba_2Pt_2(P_2O_5H_2)_4$ at 4 K have revealed that the Pt-Pt bond is much stronger in the excited state than in the ground state (Pt-Pt stretching frequency: 160 cm^{-1} (excited state), 110 cm^{-1} (ground state); Pt-Pt distance: 2.7 Å (excited state), 2.93 Å (ground state)(21).

The phosphorescence of $Pt_2(P_2O_5)_4H_8^{4-}$ in aqueous solution is quenched by 1,1-bis(2-sulfoethyl)-4,4'-bipyridinium inner salt (BSEP). Transient absorption attributable to $BSEP^-$ (λ_{max} ~ 610 nm) is observed in flash kinetic spectroscopic studies of aqueous solutions containing $Pt_2(P_2O_5)_4H_8^{4-}$ and BSEP, thereby establishing an electron transfer quenching mechanism:

$$Pt_2(P_2O_5)_4H_8^{4-*} + BSEP \xrightarrow{k_q} Pt_2(P_2O_5)_4H_8^{3-} + BSEP^-$$

Stern-Volmer analysis of the quenching yields $k_q = 5.5 \times 10^9\ \underline{M}^{-1}\ s^{-1}$ ($[Pt_2(P_2O_5)_4H_8^{4-}] \sim 10^{-4}$ M; 0.1 $\underline{M}$ $NaClO_4$; 25°C). Both the quenching reaction and the bimolecular back electron transfer ($k = 1 \times 10^9\ \underline{M}^{-1}\ s^{-1}$ for $Pt_2(P_2O_5)_4H_8^{3-}$ and $BSEP^-$) are near the diffusion limit for such processes in aqueous solution at 25°C.

The $^3A_{2u}(d\sigma^* p\sigma)$ state of $Pt_2(P_2O_5)_4H_8^{4-}$ is an extremely powerful one-electron reductant in aqueous solution. Preliminary experiments have shown that species such as $Os(NH_3)_5Cl^{2+}$ ($E_{1/2}$ = -1.09 V vs. SCE) and nicotinamide ($E_{1/2}$ = -1.44 V vs. Ag/AgCl; CH_3OH, pH 7.2) are readily reduced by $Pt_2(P_2O_5)_4H_8^{4-*}$. From these and related experiments it is apparent that $Pt_2(P_2O_5)_4H_8^{4-*}$ is a stronger reducing agent than $Ru(bpy)_3^{2+*}$ in aqueous solution.

C.-M. Che has demonstrated that the back reaction of photogenerated $Pt_2(P_2O_5H_2)_4^{3-}$ with a reduced acceptor can be inhibited by axial ligand binding. Addition of Cl^- traps the mixed-valence species, $Pt_2(P_2O_5H_2)_4Cl^{4-}$, which is rapidly oxidized by the acceptor to $Pt_2(P_2O_5H_2)_4Cl_2^{4-}$ (Figure 5).

Hexanuclear Clusters

The $(d^4)_6$ clusters ($M_6X_{14}^{2-}$: M = Mo, W; X = Cl, Br, I) exhibit red luminescence, and the excited state lifetimes are remarkably long:

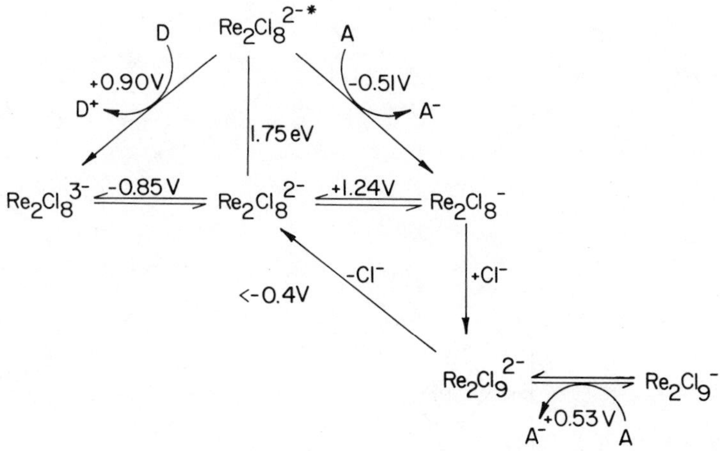

Figure 3. Modified Latimer diagram for the $Re_2Cl_8^{2-}/Re_2Cl_9^-$ system ($E°/V$ vs. SCE).

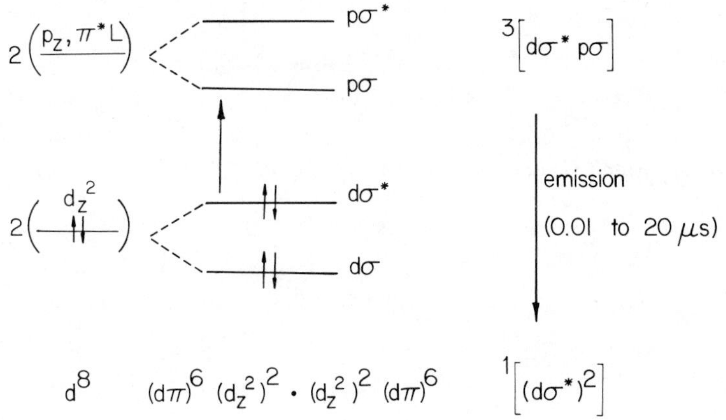

Figure 4. Electronic structure of $d^8 \cdot d^8$ systems that highlights the emissive $d\sigma^* p\sigma$ triplet state.

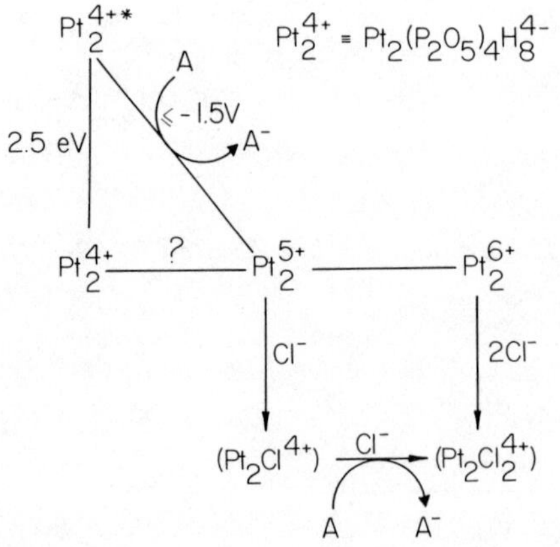

Figure 5. Modified Latimer diagram for the $Pt_2(P_2O_5)_4H_8^{4-}/Pt_2(P_2O_5)H_8Cl_2^{4-}$ system.

M	X	Lifetime (μs)(Bu_4N^+ salt, CH_3CN soln)
Mo	Cl	180
Mo	Br	130
Mo	I	50
W	Cl	2
W	Br	15
W	I	30

In recent studies the diamagnetism of the clusters has been confirmed(22), which accords with the closed-shell ($^1A_{1g}$) ground state obtained in an SCF-Xα-SW calculation(23). The positions of the emission and weak absorption bands at low temperature suggest excited state energies of ∿2 eV for all three cluster ions(22,24). As the energies of these bands are similar for $Mo_6Cl_{14}^{2-}$ and $Mo_6Br_{14}^{2-}$, it appears that both HOMO and LUMO are largely metal-centered. Also, it may be deduced from the symmetry of the EPR spectrum of $Mo_6Cl_{14}^-$ that the oxidized cluster is axially distorted in some way(22). Although there are several possible explanations for the apparent reorganization, an attractive one consistent with the values of the g tensor components is that an e_g cluster orbital is depopulated on oxidation, and that as a result the parent O_h cluster undergoes a tetragonal distortion to give a $^2A_{1g}$ (D_{4h}) or $^2A_1$ (C_{4v}) ground state for $Mo_6Cl_{14}^-$. Determination of the nature of the Mo-Mo and Mo-Cl interactions in the half-occupied cluster orbital will require a full analysis of the g parameters based on structural information for the oxidized ion that is not available at present.

Electrochemical studies have shown that the cluster ions undergo simple one-electron oxidation in aprotic solvents.

Reduction potenitals of $(Bu_4N)_2M_6X_{14}$ complexes in CH_3CN solution at 25°C.

M/X	$E_{1/2}$(V)/Ag/Ag$^+$	$E_{1/2}$(V)/SCE(est.)
Mo/Cl	1.29	1.60
Mo/Br	1.07	1.38
W/Cl	0.83	1.14

The $M_6X_{14}^-$ products are powerful oxidizing agents, rivaled by only a few chemical oxidants. Also, the position of the oxidation wave relative to those of the free halide ions confirms that the halides in $M_6X_{14}^{2-}$ are firmly bound.

The luminescent excited state of $Mo_6Cl_{14}^{2-}$ reacts rapidly with electron acceptors(24). The powerfully oxidizing $Mo_6Cl_{14}^-$is produced in these reactions. Experiments with BSEP as acceptor in particular confirm that the ion's attractive photochemical properties, originally inferred from its photophysical and electrochemical behavior in nonaqueous media, are retained in aqueous solution as well. These properties, namely, the presence of long-lived excited states and the capacity for rapid thermal and photochemical redox reactions without major structural change, are expected to persist in the $Mo_6Br_{14}^{2-}$ and $W_6Cl_{14}^{2-}$ ions also.

Atom Transfer Processes

Excited state atom transfer could be very attractive as a means to photochemical energy storage, because the accompanying structural rearrangements could create significant kinetic barriers to the energy-wasting back reactions. What is more, atom transfers typically are multielectron events, as are most useful energy storage reactions. Recent work in this area has involved systems that are well suited for photochemical O-atom and Cl^+ transfers.

Several luminescent (~650 nm) dioxorhenium(V) systems(25) have been investigated as potential O-atom transfer agents. The emission quantum yields measured with 436 nm excitation are about 0.03 for trans-$ReO_2(pyridine)_4^+$ and its isotopically-substituted derivatives in pyridine solution. The excited state lifetimes of these ions vary from 4 to 17 μs.

Though the electronic spectra of the dioxorhenium(V) ions indicate that their excited states possess activated metal-oxygen bonds(25), it has not been possible to demonstrate any simple oxygen atom transfer photochemistry with these species. The thermodynamic and kinetic barriers to oxygen atom transfer from ReO_2^{+*} must be quite large. As a result, attention has been directed toward a mono-oxo rhenium(V) compound, $ReOCl_3(PPh_3)_2$, in which the oxygen atom acceptor is PPh_3. J. R. Winkler has found that irradiation of this species in noncoordinating solvents leads to a green product whose infrared spectrum exhibits phosphorus-oxygen stretching bands but no rhenium-oxygen stretches. In addition, preliminary laser flash spectra suggest that the formation of this green product is rapid (~7 μs) and obeys first-order kinetics. These data imply a unimolecular photochemical reductive elimination process in which an oxygen atom is transferred from rhenium to coordinated PPh_3.

Evidence obtained in recent experiments by D. G. Nocera suggests that energetic species generated by irradiation of $Re_2Cl_8^{2-}$ may undergo atom transfer reactions (Figure 6). A key observation in this context is that a dichloromethane solution of $Re_2Cl_8^{2-}$ and $PtCl_6^{2-}$ produces $Re_2Cl_9^-$ upon irradiation (>560 nm). Solutions identical with those used for the photochemical experiment do not react in the dark over a period of five days at 50°C. It is reasonable to speculate that the reaction involves that fraction of the higher excited states of $Re_2Cl_8^{2-}$ that are believed to bypass the $\delta\delta^*$ singlet channel, thereby producing a high energy

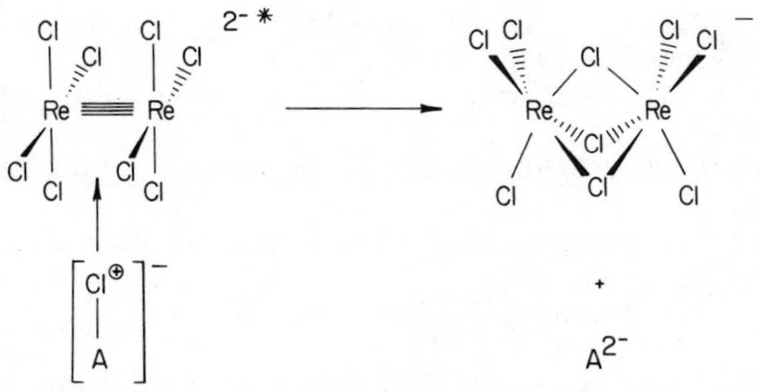

Figure 6. Proposed atom (Cl^+) transfer pathway for the photochemical oxidation of $Re_2Cl_8^{2-}$ to $Re_2Cl_9^-$.

(probably Cl-bridged) intermediate(15) that would be ideally suited for Cl^+ transfer to form $Re_2Cl_9^-$. In any case, these preliminary photochemical experiments are encouraging, because they demonstrate that $Re_2Cl_8^{2-}$ can be converted to $Re_2Cl_9^-$ by Cl^+ transfer, thereby effecting with low energy visible light a net two-electron photoredox reaction in homogeneous solution.

Acknowledgment

D. MacKenzie assisted with the $(Bu_4N)_2M_6X_{14}$ emission lifetime measurements. Research at the California Institute of Technology was supported by National Science Foundation Grants CHE78-10530 and CHE81-20419. This is Contribution No. 6703 from the Arthur Amos Noyes Laboratory.

Literature Cited

1. Gray, H. B.; Maverick, A. W. Science 1981, 214, 1201.
2. Balzani, V.; Bolletta, F.; Gandolfi, M. T.; Maestri, M. Top. Curr. Chem. 1978, 75, 1.
3. Gafney, H. D.; Adamson, A. W. J. Am. Chem. Soc. 1972, 94, 8238.
4. Brugger, P.-A.; Grätzel, M. J. Am. Chem. Soc. 1980, 102, 2461.
5. Tsutsui, Y.; Takauma, K.; Nishijima, T.; Matsuo, T. Chem. Lett. 1979, 617.
6. Lehn, J.-M.; Sauvage, J. P.; Ziessel, R. Nouv. J. Chim. 1981, 5, 291.
7. Brugger, P.-A.; Cuendet, P.; Grätzel, M. J. Am. Chem. Soc. 1981, 102, 2923.
8. Kalyanasundaram, K.; Grätzel, M. Angew. Chem. Int. Ed. Engl. 1980, 19, 646.
9. Borgarello, E.; Kiwi, J.; Pelizzetti, E.; Visca, M.; Grätzel, M. J. Am. Chem. Soc. 1981, 103, 6324.
10. Kirch, M.; Lehn, J.-M.; Sauvage, J.-P. Helv. Chim. Acta 1979, 62, 1345.
11. Chan, S.-F.; Chou, M.; Creutz, C.; Sutin, N. J. Am. Chem. Soc. 1981, 103, 369.
12. Shafirovich, V. Ya.; Khannonov, N. K.; Strelets, V. V. Nouv. J. Chim. 1981, 4, 81.
13. Krishnan, C. V.; Sutin, N. J. Am. Chem. Soc. 1981, 103, 2141.
14. Brown, G. M.; Brunschwig, B. S.; Creutz, C.; Endicott, J. F.; Sutin, N. J. Am. Chem. Soc. 1979, 101, 1298.
15. Fleming, R. H.; Geoffroy, G. L.; Gray, H. B.; Gupta, A.; Hammond, G. S. J. Am. Chem. Soc. 1976, 98, 48.
16. Miskowski, V. M.; Goldbeck, R. A.; Kliger, D. S.; Gray, H. B. Inorg. Chem. 1979, 18, 86.
17. Nocera, D. G.; Gray, H. B. J. Am. Chem. Soc. 1981, 103, 7349.
18. Milder, S. J.; Goldbeck, R. A.; Kliger, D. S.; Gray, H. B. J. Am. Chem. Soc. 1980, 102, 6761.

19. Smith, T. P. Ph.D. Thesis, California Institute of Technology, Pasadena, California, 1982.
20. Che, C.-M.; Butler, L. G.; Gray, H. B. J. Am. Chem. Soc. 1981, 103, 7796.
21. Rice, S. F. Ph.D. Thesis, California Institute of Technology, Pasadena, California, 1982.
22. Maverick, A. W. Ph.D. Thesis, California Institute of Technology, Pasadena, California, 1982.
23. Cotton, F. A.; Stanley, G. G. Chem. Phys. Lett. 1978, 58, 1978.
24. Maverick, A. W.; Gray, H. B. J. Am. Chem. Soc. 1981, 103, 1298.
25. Winkler, J. R.; Gray, H. B. J. Am. Chem. Soc., submitted 1982.

RECEIVED September 24, 1982

Discussion

A.W. Adamson, University of Southern California: You spoke of the bond energy being larger in the excited state, in the case of $Pt_2(P_2O_5H_2)_4^{4-}$. It seems to me that what the i.r. data indicate is that the bond strength, that is, the bond force constant, may be larger. The concept of an excited state bond energy probably requires some care of definition. In the case of ground state molecules, one ordinarily would like the sum of all bond energies to add up to the atomization energy. Depending on the definition used, this might not be the case with an excited state species. Could you comment on this matter?

H. B. Gray: Care of definition is certainly needed. The reference dissociation limit must be specified. Discussion of this point has been published (1).

(1) Rice, S. F.; Gray, H. B. J. Am. Chem. Soc. 1981, 103, 1593.

C.H. Langford, Concordia University: Gray rightly emphasizes the value of the search for long-lived excited states as initiators of photoreaction in homogeneous solution. This is because the reactive excited state must accomplish an encounter with a reaction partner. The long lifetime does impose a disadvantage. It usually implies a fully relaxed excited state which is prevented from relaxing to the ground state by a suitable barrier. This is one of the loss processes discussed by Bolton which limits the efficiency of energy storage. In the context of energy conversion, I suspect that nature uses an alternate strategy. Both chlorophyll and rhodopsin appear to achieve an irreversible step very fast without waiting for full relaxation. Thus, efficiency photochemical synthesis of endergonic products could be argued to require assembly first to avoid a diffusion encounter requirement then the examination of "selection rules" to discover useful irreversible pathways of energy degradation. Hollebone (1) has proposed an approach to such selection rules.

(1) Hollebone, B.R.; Langford, C.H.; Serpone, N. Coord. Chem. Revs. 1981, 39, 101.

H. B. Gray: We must remember that we are still in the initial stages of systematic study of inorganic oxidation-reduction photochemistry. Nature has indeed some slick ways to optimize photochemical energy conversion. I am confident that inorganic chemists will do as well or better, perhaps even before the turn of the century!

A.B.P. Lever, York University: The next few years will see growth in our knowledge of two electron photoredox reagents as suggested by the atom transfer work you have reported. The purposeful design of two electron reagents might lead to sequential two electron redox or concerted two electron redox. For purposes such as the oxidation of water, how do you see such reagents developing?

H. B. Gray: Work on the excited states of soluble metal oxo species that in a sense are the homogeneous-solution analogues of TiO_2-type materials is a promising direction to take, in my view.

D. Cole-Hamilton, Liverpool University: Professor Gray, some of the platinum complexes that you have prepared at the same time as producing hydrogen contain, formally at least, Pt(III). Can these compounds act as oxidising agents, either thermally or photochemically?

H. B. Gray: Yes, certain of the binuclear Pt(III) complexes act as mild thermal oxidizing agents, and they are more potent oxidizers when irradiated.

D.R. McMillin, Purdue University: You referred to a reactive excited state of $Re_2Cl_8^{4-}$ which had a lifetime of the order of a microsecond in CH_2Cl_2 solution and suggested that it could be assigned to the singlet state of the $\delta^1\delta^{*1}$ configuration. That would require intersystem crossing to the associated triplet state to occur with a rate constant of the order of 10^6 sec^{-1} which seems quite low for a transition metal complex, especially one involving a third row ion. Would you comment on this point?

H. B. Gray: Multiconfiguration SCF calculations by P. J. Hay indicate that the $^1\delta\delta^*$-$^3\delta\delta^*$ energy separation is over 1 eV, and there is no evidence for intervening states that could provide a facile intersystem pathway. Thus a relatively small singlet→triplet intersystem crossing rate constant is not all that peculiar.

R.A. Walton, Purdue University: The question as to rotational conformation (eclipsed or staggered) that characterizes the excited state geometries of the complex anions $[Re_2Cl_8]^{2-}$ and $[Mo_2Cl_8]^{4-}$ has been addressed by Professor Gray (1). There is obviously no question that a sterically hindered complex such as $Mo_2Cl_4(P\text{-}n\text{-}Bu_3)_4$ (1) will retain the eclipsed geometry that pertains to the ground state, thereby providing an interesting comparison with $[Mo_2Cl_8]^{4-}$. I should like to mention that we have now prepared the sterically hindered quadruply bonded dirhenium(III) complex $[Re_2Cl_4(PMe_2Ph)_4](PF_6)_2$ (2), a molecule that is isoelectronic with the unhindered octachlorodirhenate(III) anion, $[Re_2Cl_8]^{2-}$. A study of its emission behavior would perhaps be of interest.

(1) Miskowski, V.M.; Goldbeck, R.A.; Kliger, D.S.; Gray, H.B. Inorg. Chem. 1979, 18, 86.

(2) Dunbar, K.; Walton, R.A., unpublished work.

H. B. Gray: Indeed it would. Please send me a sample!

3

Photochemical Intermediates

J. J. TURNER and M. POLIAKOFF

University of Nottingham, Department of Chemistry, Nottingham NG7 2RD England

Few would deny the importance of photochemistry - few would further deny that full exploitation of photochemistry demands an intimate knowledge of the identity, structure and behavior of the appropriate intermediates. In this article we review some of the methods used to obtain this information, and the techniques are illustrated with some detailed examples taken from Organometallic - mostly metal carbonyl - Photochemistry (1). Since this symposium looks towards the 21st century(!) a few speculations on future experiments are included.

Clearly, mechanistic investigations can provide circumstantial evidence for the participation of particular intermediates in a reaction but, here, we are concerned with the definitive observation of these species. If the intermediates are relatively stable then direct spectroscopic observation of the species during a room-temperature reaction may be possible. As a rather extreme example of this, the zero-valent manganese radicals, $Mn(CO)_3L_2$ (L = phosphine) can be photochemically generated from $Mn_2(CO)_8L_2$, and, in the absence of O_2 or other radical scavengers, are stable in hydrocarbon solution for several weeks (2,3). However, we are usually more anxious to probe reactions in which unstable intermediates are postulated. There are, broadly speaking, three approaches - *continuous generation*, *instantaneous methods* and *matrix isolation*.

Continuous generation simply means that the intermediate is continuously replenished by some method and examined under pseudo-equilibrium conditions. For instance, Whyman (4) was able, using a special IR cell working at high pressure and temperature, to monitor the behavior of several species of importance in the thermal hydroformylation catalytic cycle. Similarly, Koerner von Gustorf and colleagues (5) have monitored the photochemical

0097-6156/83/0211-0035$07.00/0

production of substituted iron carbonyls by IR spectroscopy at -60°C.

In the most popular *instantaneous technique*, flash photolysis, a burst of light (or in pulse radiolysis, a pulse of electrons) generates a high concentration of intermediates whose disappearance is monitored by kinetic spectroscopy. There are very many applications of this technique to Organometallic Chemistry but the great problem is that, using UV methods in solution chemistry, it is almost impossible to identify an intermediate positively. Unfortunately, conventional IR spectroscopy, which could provide this identification, is much too slow to monitor really fast reactions. There are, however, likely to be technical developments which will overcome this problem. Fourier-Transform IR with its ability to acquire data quickly will prove increasingly valuable for fairly slow reactions. Thus, using FTIR, Chase and Weigert (6) have monitored the catalyst lifetime in the $Fe(CO)_5$-catalyzed olefin isomerization, and Kazlauskas and Wrighton (7) have characterized the *mono*dentate intermediate in the photochemical formation of tetracarbonyl (4,4'-dialkyl-2,2'bipyridine)-metal from $M(CO)_6$. For genuinely fast reactions the recent experiments of Schaffner and colleagues (8) look extremely promising; using fast IR detectors and a "point-by-point" flash photolysis technique they have been able to probe the short-lived intermediates in the solution chemistry of $Cr(CO)_6$. There is also the possibility of exploiting pulsed Resonance Raman spectroscopy, but there will be problems with photolysis and fluorescence being more significant than Raman scattering (9). Stopped flow and related techniques may be considered as a "slow" form of *instantaneous techniques*.

In both *continuous generation* and *flash photolysis*, the intermediate starts to react as soon as it is formed and if it is very reactive (e.g. "naked" $Cr(CO)_5$, $Ni(CO)_3$) it may not be possible to detect.

If there is a significant activation energy for these reactions, reduction in temperature may slow the reaction rate to such an extent that the intermediate can be examined by more "leisurely" spectroscopic methods (e.g. IR), but simply lowering the temperatures will be ineffective for very reactive species which have activation energies close to zero. However, if the species can be isolated in a rigid matrix, then reaction may be prevented altogether by introducing a high activation energy for diffusion. The technique of *matrix isolation* - either in frozen glasses at 77K or in frozen noble gases at ∿4-20K - has provided structural information about many organometallic intermediates (10-13), but, of course, without any kinetic or energetic data.

In what follows, we select a number of systems of current interest and try to illustrate the relationship between the results obtained by the various techniques mentioned above.

$Cr(CO)_5$...X and Related Species

It is generally believed (1) that the photochemical substitution reactions of $Cr(CO)_6$ can be summarised:

$$Cr(CO)_6 \xrightarrow{h\nu} Cr(CO)_5 + CO \xrightarrow{L} Cr(CO)_5L + CO$$

However, it will become apparent that the process is more complex than this simple scheme suggests. The objective is to determine the structure of the intermediate, examine its behavior and to unravel the photochemical processes. This can only be achieved by a combination of techniques.

Matrix Studies. In the early 60's Sheline and co-workers (14,15) used IR spectroscopy to follow the photochemical behavior of frozen solutions of $M(CO)_6$ in glasses at 77K. Clear evidence for the generation of carbonyl fragments was obtained but the conclusions were necessarily somewhat speculative. Perutz and Turner (16) definitively established the structure of the primary photoproduct, $Cr(CO)_5$, by UV photolysis of $Cr(CO)_6$ in noble gas and methane matrices at 20K. Figure 1 illustrates both IR and visible spectra. Certain features of this work are worth noting. The stoichiometry, symmetry and even bond angles of the carbonyl fragments could be determined reliably, using isotopic ^{13}CO and $C^{18}O$ enrichment combined with the simple frequency factored force field (17) and simple intensity arguments (18). Thus, the primary photoproduct is $Cr(CO)_5$ with C_{4v} symmetry, in agreement with theoretical predictions. (19)

[Preparing isotopically enriched carbonyls and related species is not always a trivial problem. We have recently developed (20) a method which looks particularly promising in some cases. CW CO_2 laser irradiation of a gas phase mixture containing SF_6 as an energy transfer agent can promote thermal chemistry without complications due to wall reactions, e.g.

$$Fe(CO)_5 + {}^{13}CO + SF_6 \xrightarrow{CO_2\ \text{laser}} Fe({}^{12}CO)_{5-x}({}^{13}CO)_x$$

The isotopically enriched carbonyl can be separated from the CO and SF_6 in a circulating vacuum system.]

It was observed that photolysis could be reversed by irradiation of the matrix with light at a wavelength corresponding to the visible absorption band of $Cr(CO)_5$.

$$\underset{[O_h]}{Cr(CO)_6} \underset{\text{Visible}}{\overset{\text{UV}}{\rightleftharpoons}} \underset{[C_{4v}]}{Cr(CO)_5} + CO$$

Although the IR spectrum of $Cr(CO)_5$ was only slightly dependent upon the matrix, the position of the visible band of $Cr(CO)_5$ was extremely sensitive to matrix (21,22). Figure 2 illustrates this effect. Mixed matrix experiments were used to prove that this effect arises from interaction between the matrix species and the

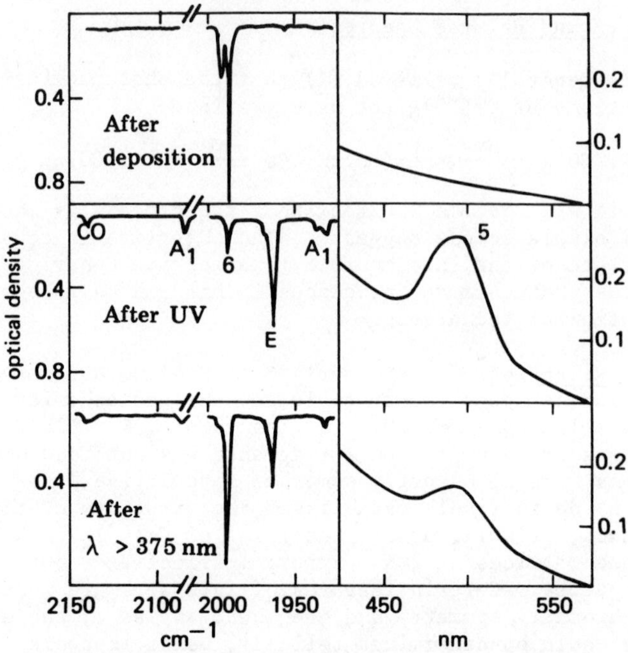

Figure 1. Photolysis of $Cr(CO)_6$ in CH_4 at 20 K (IR and visible spectra).

Key: top, deposition of $Cr(CO)_6$ (T_{1u} mode marked 6); middle, 15 s photolysis with unfiltered Hg arc showing production of $Cr(CO)_5$ and molecular CO [$Cr(CO)_5$ has three IR bands (marked A_1, A_1, and E) and a visible band (marked 5), and the UV band of $Cr(CO)_5$ is not shown]; and bottom, 2 min photolysis with Hg arc + λ >375-nm filter, showing regeneration of $Cr(CO)_6$. The spectra above 2050 cm^{-1} and the visible spectra are taken with about five times as much material as the spectra below 2050 cm^{-1}. (Reproduced from Ref. 21. Copyright 1975, American Chemical Society.)

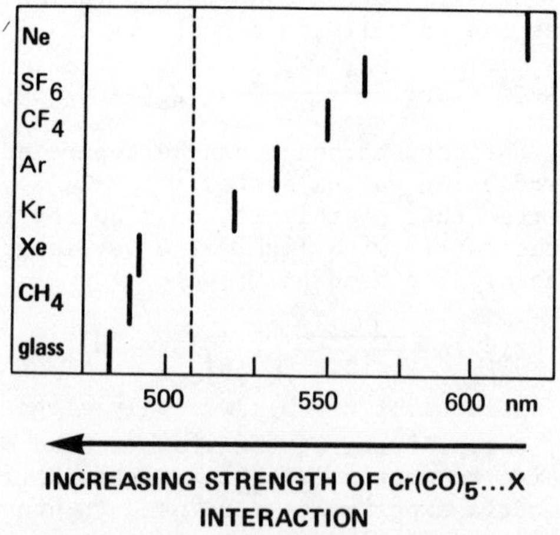

Figure 2. Diagrammatic representation of the position of the visible band of $Cr(CO)_5$ in different matrices (plot is linear in cm^{-1}). (Reproduced from Ref. 21. Copyright 1975, American Chemical Society.)

empty 6th coordination site of $Cr(CO)_5$. Of these species the weakest interaction will clearly involve $Cr(CO)_5$...Ne and the strongest $Cr(CO)_5$...CH_4, so although the shift in λ_{max} is due mostly to perturbation of orbitals only populated in the *excited state* (12,13), the trend in λ_{max} also reflects *ground state* interaction (23). This interaction is of very great importance in understanding the solution behavior of "$Cr(CO)_5$" - in particular, it suggests that no solvent will be completely innocuous towards "naked" $Cr(CO)_5$ and that trace impurities may be extremely important in the photochemistry. An important technique for probing the photochemical mechanism involves the combination of matrix isolation and the use of plane polarized light for both photolysis and IR/UV-vis spectroscopy (24-26). The polarized photolysis can generate preferentially oriented molecules in the matrix which display dichroic absorption of light. In the absence of photochemical stimulation, the molecules are held rigidly in the matrix cage and so the photochemically induced dichroism can subsequently be probed by spectroscopic techniques. [This method is somewhat akin to the use of nematic solvents by Gray and colleagues (27) for the assignment of UV spectra, e.g. $Mn_2(CO)_{10}$.] In this way it was shown that the visible absorption band in $Cr(CO)_5$...X has a transition moment of E symmetry and the detailed photochemical behavior could be explained by the scheme shown in Figure 3. It will be noted that this mechanism involves the initial generation from $Cr(CO)_6$ of C_{4v} $Cr(CO)_5$ in an excited singlet state (^1E); this matches Hay's calculations, (28) although it has caused some surprise (29). The excited C_{4v} molecule decays via a D_{3h} structure to ground state C_{4v} ($^1A_1$) and subsequent photochemically induced interconversion between $Cr(CO)_5$ in different orientations proceeds via this same D_{3h} intermediate. Thus the regeneration of $Cr(CO)_6$ from $Cr(CO)_5$ arises because the $Cr(CO)_5$ fragment is "stirred" by the irradiation until the empty coordination site encounters the originally ejected CO. This mechanism is relevant to solution chemistry and perhaps fortuitously predicts a quantum yield for $Cr(CO)_6 \xrightarrow{UV} Cr(CO)_5$...X of $^2/_3$ which is in exact agreement with a recent determination (30) of the quantum yield for the solution reaction of $Cr(CO)_6$ with pyridine.

The matrix experiments thus reveal some complex photochemistry of relevance to solution chemistry but the experiments do not provide information about kinetics. For this we need a fluid medium e.g. gas or liquid, and we consider such experiments in the next two sections. Flash photolysis suggests itself as the technique for detecting a species as reactive as $Cr(CO)_5$ but before describing these experiments we show what can be achieved from low-temperature solutions.

Low-temperature Solutions. The matrix spectroscopic data for $Cr(CO)_5$...X (Figure 2) suggest that the interaction between $Cr(CO)_5$ and saturated hydrocarbons may be quite substantial,

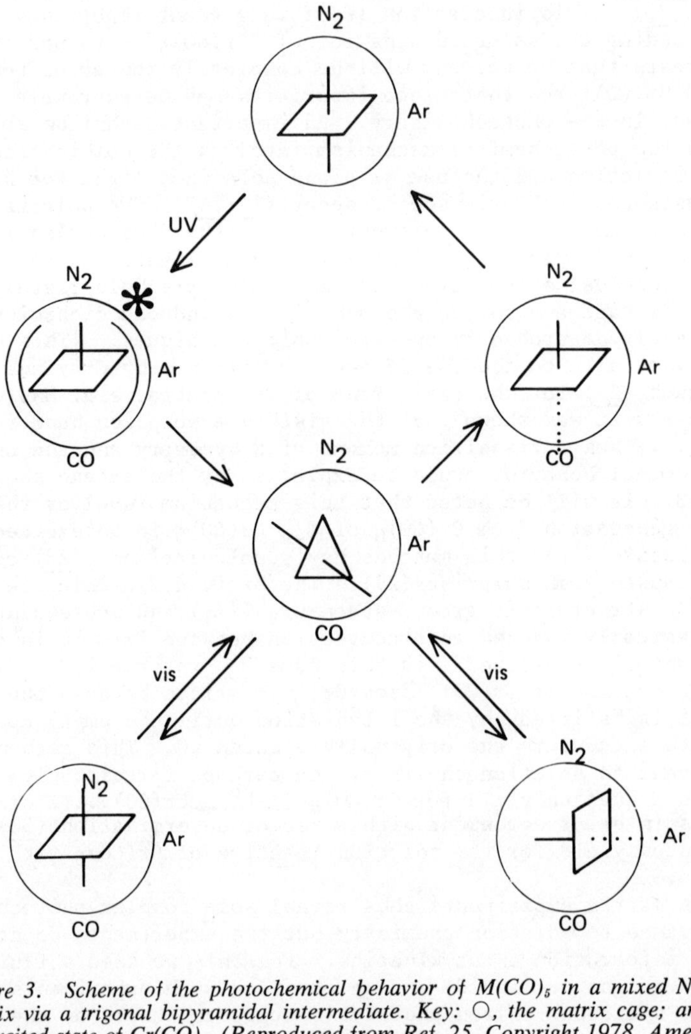

*Figure 3. Scheme of the photochemical behavior of $M(CO)_5$ in a mixed N_2/Ar matrix via a trigonal bipyramidal intermediate. Key: ○, the matrix cage; and *, ^1E excited state of $Cr(CO)_5$. (Reproduced from Ref. 25. Copyright 1978, American Chemical Society.)*

perhaps large enough to stabilize $Cr(CO)_5$...(hydrocarbon) species at low temperatures in fluid solutions. Tyler and Petrylak (31) have indeed shown, using IR spectroscopy, that in extraordinarily well purified methylcyclohexane the species $M(CO)_5$...MCH (M = Cr, Mo, W; MCH = methylcyclohexane) can be photochemically generated with a lifetime of about an hour at -78°C. The degree of purity of MCH required is extremely high since even trace concentrations of unsaturated hydrocarbons promote the production of π-olefin complexes. Although there is little doubt that Tyler and Petrylak produced $Cr(CO)_5$...MCH, a more sensitive test of these weak interactions is UV-visible spectroscopy (21) and it is hoped such measurements will be made.

We have recently been investigating an alternative approach, *liquid* noble gases. These seem to be ideal solvents for investigating organometallic photochemistry; the noble gases are completely transparent over a very wide spectral range (hence long pathlengths are possible). By working in a specially designed pressure cell (32) it is possible to cover the temperature range ∿80K to 240K, using Ar, Kr or Xenon, and these noble gas solvents, particularly Ar (Figure 2), are likely to be weakly interacting.

Surprisingly, metal carbonyls are soluble in these liquids e.g. $Cr(CO)_6$ in liquid Xe, and $Fe(CO)_5$ even in liquid Ar. Our initial experiments have involved N_2 complexes, since these are expected to be moderately stable, and in the case of $Cr(CO)_5N_2$, allow comparison with both matrix isolation and room temperature flash photolysis experiments. We have generated (33,34) the species $Cr(CO)_{6-x}(N_2)_x$ (x = 1 to 5) by photolysis of $Cr(CO)_6$ in *liquid* xenon at -80°C doped with N_2. Figure 4 shows both the photolytic production of $Cr(CO)_5N_2$ and also photochemical regeneration of $Cr(CO)_6$, i.e. exactly mimicking the matrix behavior (22). Prolonged photolysis of $Cr(CO)_6$ in liquid xenon with higher concentration of added N_2 generates more highly N_2-substituted species. Figure 5 shows how the stabilities of the different $Cr(CO)_{6-x}(N_2)_x$ species depend on the degree of N_2 substitution and suggests that kinetic parameters might be obtainable. For thermally less stable compounds, sequential photolysis and spectral analysis are inadequate because the compounds decompose before they are detected. However, using a special 4-way cell we have been able to generate the very unstable species $Ni(CO)_3N_2$, by photolysis of $Ni(CO)_4$ in liquid Kr doped with N_2 at -170°C, and record the IR spectrum during UV irradiation. On switching off the UV lamp the IR spectrum of $Ni(CO)_3N_2$ decays and by monitoring the rate of this decay we have measured kinetic and activation energy parameters (35) for $Ni(CO)_3N_2 + CO \longrightarrow Ni(CO)_4 + N_2$. UV photolysis of $Cr(CO)_6$ in pure liquid xenon (i.e. in the absence of N_2) produced IR bands of a transient species ($t_{\frac{1}{2}}$ ∿2 sec at -78°C) which may well be unstable $Cr(CO)_5$...Xe. We believe that such experiments will provide an important extra window on solution kinetics monitored by flash techniques to which we now turn.

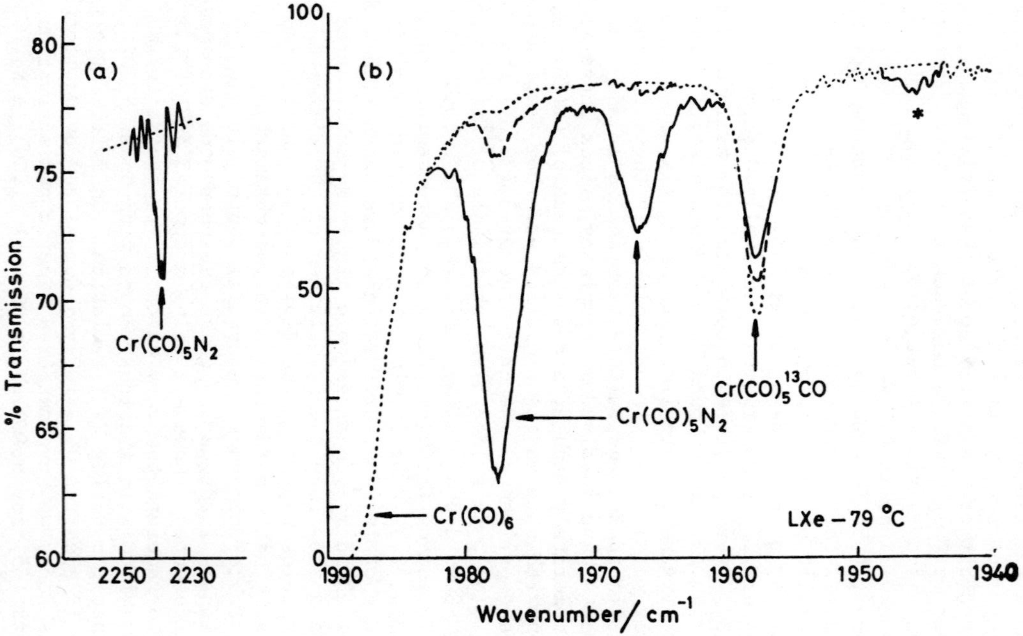

Figure 4. Absorption spectra of $Cr(CO)_5N_2$ and $Cr(CO)_6$ dissolved in liquid Xe at −79 °C. Key: • • •, before photolysis; ——, after 3.3 min irradiation with an unfiltered Hg arc lamp placed 28 cm from the center of the cryostat; and – – –, after irradiation for about 12 min with the Hg arc lamp and Balzers 367-nm filter about 14 cm from the center of the cryostat. (Where the curves are indistinguishable, only the dotted curve is shown.) For clarity, the noise has been omitted from the left curve. The feature marked with an asterisk (*) is assigned to $Cr(CO)_4(N_2)_2$. (Reproduced with permission from Ref. 33. Copyright 1980, Royal Society of Chemistry.)

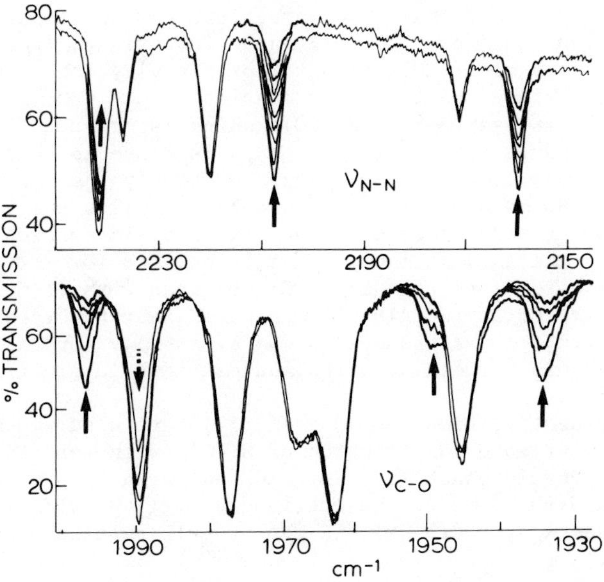

Figure 5. Thermal decomposition of $Cr(CO)_3(N_2)_3$ in liquid Xe at −79 °C. Time dependence of the intensity of IR absorptions in both the N–N and C–O stretching regions on warming the solution to −79 °C is shown.

Spectra are taken repetitively over 3-min intervals. Bands marked ↑ all decay simultaneously, and the band at ~2240 cm^{-1} demonstrates an underlying feature coincident with the decaying absorption. The band marked ↓ is due to $Cr(CO)_6$ and it increases in intensity with time, probably due to undissolved $Cr(CO)_6$ dissolving slowly at the higher temperature. Note that the other bands due to $Cr(CO)_5N_2$ and $Cr(CO)_4(N_2)_2$ remain unchanged in intensity throughout, indicating that these compounds are thermally stable at −79 °C. (Reproduced from Ref. 34. Copyright by American Chemical Society.)

Flash Photolysis in Solution. Studies on $Cr(CO)_6$ by flash photolysis in solution have been plagued with impurities and some of the earlier work has been shown to be wrong for this reason. It is worth noting that one of the great advantages of the matrix technique is that trace impurities are rarely a problem because there is little or no diffusion in the matrix. Recently Bonneau and Kelly (36) have obtained definitive results on this solution system, partly by reference to data obtained in solid matrices (21), which suggest that saturated perfluorocarbons should interact with $Cr(CO)_5$ less than other practicable room temperature solvents. Thus, Bonneau and Kelly have investigated the laser flash photolysis of $Cr(CO)_6$ in perfluoromethylcyclohexane (C_7F_{14}) at room-temperature. A transient species (λ_{max} ~620 nm) formed by the flash, is assigned to $Cr(CO)_5$ which is presumably only very weakly coordinated to C_7F_{14}, since λ_{max} is close to that of matrix isolated $Cr(CO)_5...Ne$ (624 nm) where presumably the interaction is very small. Bonneau and Kelly were able to follow the reaction of $Cr(CO)_5...C_7F_{14}$ with other dissolved species, in particular with cyclohexane (the 'normal' innocuous solvent!) to form $Cr(CO)_5$... cyclohexane with CO to regenerate $Cr(CO)_6$ and with $Cr(CO)_6$ to form $Cr(CO)_5...Cr(CO)_6$. Also it was noted that the rate of decay of $Cr(CO)_5...C_7F_{14}$ was enhanced by N_2 presumably forming $Cr(CO)_5N_2$. The relationship between solution and matrix data are illustrated in Figure 6.

More recently, Lees and Adamson (37) have used flash photolysis to investigate the reaction of $W(CO)_6$ with 4-acetylpyridine (L) in methylcyclohexane (S). A transient with λ_{max} = 425 nm is assigned to $W(CO)_5S$ (c.f. the matrix data (21) $W(CO)_5...CH_4$ in solid CH_4 has λ_{max} = 415 nm) and the overall reaction can be summarised:

$$W(CO)_6 \xrightarrow{h\nu} W(CO)_5S + CO$$

$$W(CO)_5S \underset{}{\overset{S}{\rightleftharpoons}} W(CO)_5 + S$$

$$W(CO)_5 + L \longrightarrow W(CO)_5L$$

The detailed analysis of this scheme leads to a number of surprises; for instance, the rate constant for reaction of $W(CO)_5$ with L is 270 times larger than with S. Their analysis did, however, ignore possible complications due to impurities, particularly H_2O.

The limitation in all of these flash experiments is that only broad featureless UV/vis bands are observed and hence assignment has to rely on comparison with matrix data and/or kinetic consistency. How much more informative vibrational spectroscopy would be! There is good reason to be optimistic as in the recent work of Schaffner (8), where, incidentally, it is shown how important a role is played by traces of H_2O in the detailed mechanism of the photochemistry of $Cr(CO)_6$.

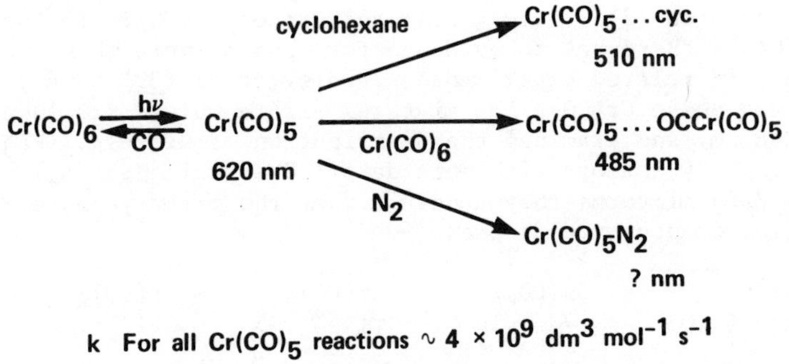

Figure 6. Summary of the results of flash photolysis of $Cr(CO)_6$ in C_7F_{14} in solution. (Reproduced from Ref. 36. Copyright 1980, American Chemical Society.)

One might conclude this first section by commenting that the combination of room-temperature and cryogenic techniques now available promises to unravel some photochemical problems of considerable complexity. Before turning to some further examples, it is worth noting some very interesting recent experiments involving the gas phase photochemistry of $Cr(CO)_6$.

Flash Photolysis in the Gas Phase. The only way of obtaining *completely* "naked" $Cr(CO)_5$ must be in a low pressure gas, and Breckenridge (38) has described preliminary gas phase experiments with the ultimate objective of following the behavior of naked $Cr(CO)_5$. Flash photolysis of $Cr(CO)_6$ vapour with the frequency tripled output of a pulsed Nd YAG laser at 355 nm shows a transient species (λ_{max} 485 nm) with a lifetime of several hundred µsec - presumably $Cr(CO)_5...Cr(CO)_6$ (c.f. Bonneau and Kelly (36)). [With highly focussed lasers, multiphoton effects are observed and species such as Cr, Cr^+ detected.] In the presence of Ar or CH_4, gas phase transients were observed at wavelengths corresponding to the matrix species (21) $Cr(CO)_5...Ar$ and $Cr(CO)_5...CH_4$. In these experiments there was no evidence for loss of more than one CO. However, in related experiments, Yardley et al (39) have photolysed gas phase $Cr(CO)_6/PF_3$ mixtures with a pulsed KrF laser (λ = 248 nm) and examined the distribution of $Cr(CO)_{6-x}(PF_3)_x$ products. By analogy with more detailed experiments (40) on $Fe(CO)_5/PF_3$ mixtures they conclude that the primary photofragmentation quantum yields are:

$Cr(CO)_5$	$Cr(CO)_4$	$Cr(CO)_6$	$Cr(CO)_2$
.03 ± .03	.73 ± .16	.14 ± .05	0.1 ± .06

Yardley and co-workers conclude that a 248 nm proton has sufficient energy (115 kcal mol^{-1}/481 kJ mol^{-1}) to break four Cr-CO bonds of average strength 25.0 kcal mol^{-1}. It is suggested that in the gas phase the production of lower fragments is possible by sequential CO loss, following absorption of one photon because there is a high degree of internal energy retention in the remaining carbonyl fragment, whereas in condensed phases (liquid or matrix) rapid dissipation of excess energy due to intermolecular interaction prevents additional fragmentation. This is why KrF photolysis of $Cr(CO)_6$/1-butene mixtures in the gas phase can result in isomerization to 2-butene (39). Complexation of a low carbonyl fragment to the olefin produces a π-complex which is still coordinatively unsaturated and which is thus capable of additional unimolecular or bimolecular reaction.

The difference between these two sets of gas phase experiments is very interesting and a full explanation is awaited. Meanwhile it is worth commenting that matrix isolation work has already provided information about these lower carbonyl fragments. Prolonged photolysis of $M(CO)_6$ (M = Cr, Mo, W) produces sequential CO loss to give $M(CO)_5$, $M(CO)_4$, $M(CO)_3$ and $M(CO)_2$ (41), with the loss of each CO group presumably requiring the absorption of one UV photon. Isotopic substitution and IR band intensity measurements suggest the following structures for these species: $M(CO)_5$-C_{4v}; $M(CO)_4$-(cis di-vacant) C_{2v}; $M(CO)_3$-C_{3v}; $M(CO)_2$ - uncertain. These structures are in agreement with the theoretical work of both Burdett and Hoffman (19). It should be noted that carbonyl fragments in matrices can be generated by either "molecular striptease" involving prolonged photolysis of parent - e.g. $Fe(CO)_5$ in argon eventually produces Fe atoms (42) - or by cocondensation of metal atoms with CO/Ar mixtures, a technique exploited by several workers but particularly in the laboratories of Ozin and Moskovits (10).

Note added in typing. In a very recent paper (81) Vaida and co-workers have used picosecond laser photolysis to show that, in cyclohexane solution, $Cr(CO)_5$...cyclohexane (λ_{max} 497 nm) is formed within 25 ps of the photolysis of $Cr(CO)_6$. This suggests that, in solution, the primary photoproduct is $Cr(CO)_5$ and that there is essentially no activation energy for the reaction of $Cr(CO)_5$ with the solvent. Clearly, experiments with pulsed KrF lasers on carbonyls in solution and matrix may be very revealing.

Some Other Carbonyls and Related Species

In this section, we outline a number of studies by various workers in which matrix isolation experiments have contributed to the understanding of some photochemical pathways and where the objective was to relate the experiments to room-temperature photochemistry. We then consider briefly some current problems involving dinuclear and polynuclear carbonyls.

Organometallic Intermediates by CO Loss. There are many studies where the primary photoproduct generated by CO loss has been characterized in a matrix. One example will illustrate. Rest (43) and Wrighton (44) have irradiated $(\eta^5\text{-}C_5H_5)W(CO)_3R$ in noble gas and paraffin matrices respectively; the observed product, identified by IR spectroscopy, is $(\eta^5\text{-}C_5H_5)W(CO)_2R$ and there is no evidence for a radical pathway probably because of its low quantum yield and the ready recombination of radical pairs in the frozen environment. By warming the paraffin matrix, Wrighton (44)was able to observe for R = n-pentyl, in addition to reformation of the starting species, a significant conversion of the coordinatively unsaturated 16e species to *trans* $(\eta^5\text{-}C_5H_5)W(CO)_2$(1-pentene)H i.e. the β-H transfer was observed at temperatures as low as -100°C, a result of potentially great significance for catalytic cycles (vide infra).

Selective Photochemistry. It is well known that for many substituted metal carbonyl compounds in solution, photochemical loss of either CO or the substituent L can be promoted depending on the wavelength of irradiation used (1). It has often been supposed that in a frozen matrix the "cage effect" would preclude the observation of any intermediates produced by loss of a bulky ligand L because "in cage" recombination would be very facile. In an elegant series of experiments involving both independent and joint work Rest and Oskam and their colleagues have shown how solution chemistry involving even large ligands can be mimicked in a matrix but with some differences not yet resolved. A nice example (45) involves $Cr(CO)_5$(pyrazine) isolated in Ar. Irradiation at 229, 254, 280 or 313 nm produces cis-$Cr(CO)_4$(pyrazine) (i.e. CO loss) whereas irradiation with λ = 405 or 436 nm gives $Cr(CO)_5$ (i.e. pyrazine loss). Similarly (46) short-wavelength photolysis of matrix isolated $M(CO)_5$pip complexes (M = Cr, Mo, W; pip = piperidine) gives $M(CO)_4$pip species of C_s symmetry and long wavelength photolysis results in $M(CO)_5$.

<u>H-Containing Compounds</u> (47). The key intermediate in the photochemistry of $(\eta^5\text{-}C_5H_5)_2WH_2$ in solution is believed to be tungstenocene, $(\eta^5\text{-}C_5H_5)_2W$ (48). Photolysis of the dihydride species in a CO matrix shows (49) the formation of $(\eta^5\text{-}C_5H_5)_2W$, as well as some $(\eta^5\text{-}C_5H_5)_2WCO$ but no HCO. This is surprising since it is known that H atoms react readily with CO to form HCO. Thus, it is concluded that the loss of H_2 from $(\eta^5\text{-}C_5H_5)_2WH_2$ is close to concerted and no H atoms are produced. In contrast (50) photolysis of $(\eta^5\text{-}C_5H_5)_2ReH$ in a CO matrix does generate HCO as expected as well as rhenocene, $(\eta^5\text{-}C_5H_5)_2Re$.

Photolysis (51) of $H_2Fe(CO)_4$ in matrices at 10K with 254 nm light produces first $Fe(CO)_4$ (52) and subsequently lower carbonyl species $Fe(CO)_3$ (53) etc. The process is readily reversed, particularly in H_2-doped matrices, e.g. by visible light and even on warming the matrix to 20K; this latter observation suggests that the activation energy for the oxidative addition of H_2 to $Fe(CO)_4$ must be extremely small. At first sight estimating the relative quantum yields for H loss and CO loss from $HM(CO)_x$ species is impossible since, although the H atoms readily escape, CO reversal is so facile (see $Cr(CO)_5$ section). However, by carrying out the photolysis in a solid ^{13}CO matrix, the approximate relative quantum yields can be obtained by comparing the rate of production of the H-loss fragment with the rate of incorporation of ^{13}CO into the parent. For both $HMn(CO)_5$ (54) and $HCo(CO)_4$ (55), CO loss is about an order of magnitude more important than H loss, even though the apparent primary product in each case is $Mn(CO)_5$ and $Co(CO)_4$. It is for this reason that the definitive identification of the spectrum and structure of the important hydroformylation intermediate, $HCo(CO)_3$, has been so difficult (56,57).

<u>Interaction with Hydrocarbons</u>. Wrighton (1), in particular, has shown how photolysis of suitable precursors in solution can generate higher concentrations of catalytic intermediates than

obtained in conventional thermal reactions. One of the aims of homogeneous catalysis is the easy activation of hydrocarbons (58), and one of the most interesting recent relevant experiments is photochemically induced C-H activation in a completely saturated hydrocarbon (59).

There are some interesting matrix experiments of relevance to this problem. Although the interaction of CH_4 with $Cr(CO)_5$ is only slight (vide supra) the interaction between CH_4 and $Fe(CO)_4$ is sufficiently great to alter the geometry of the $Fe(CO)_4$ fragment considerably (52,60). We are planning with new FTIR techniques, to probe this interaction in more detail. In a recent extension of earlier experiments (vide supra) Kazlauskas and Wrighton (61) have detected *two* types of intermediate on photolysis of $(\eta^5-C_5H_5)W(CO)_3R$ in paraffin matrices. With R = CH_3 the intermediate is blue, but when R contains β hydrogens the intermediate is yellow; by analogy with the $Cr(CO)_5$ interaction data (21) it is concluded that this yellow intermediate adopts a cyclic structure in which a β-H is coordinated back to the metal centre.

The most potent species for activating hydrocarbons are likely to be naked transition metal atoms. Margrave (62) and Ozin (63) have shown that cocondensation of Fe atoms or Cu atoms with methane at ∿10K followed by UV irradiation produces definite spectroscopic evidence for the insertion of the metal atoms into the C-H bond.

It is likely that further experiments along these lines will open up new catalytic possibilities.

Rearrangements. It will have been obvious from the $Cr(CO)_5$... X discussion above that the matrix technique is particularly well suited to the examination of photochemically induced rearrangements and isomerization since the matrix holds the participating species in fixed positions and hence allows spectroscopic study of each, yet allows ready photochemical interconversion.

One of the more detailed studies of this type (64) has involved the unravelling of the rearrangement modes of the coordinatively unsaturated species $W(CO)_4CS$, which was generated by UV photolysis of $W(CO)_5CS$; the experiments involved IR/UV spectroscopy, wave-selective photolysis and analysis based on ^{13}CO substitution. The results are summarised in Figure 7. These experiments also confirm that $W(CO)_4CS$ is initially formed in an excited state, as predicted by the scheme outlined in Figure 3.

Even more complex experiments have been performed on matrix isolated $Fe(CO)_4$, generated by UV photolysis of $Fe(CO)_5$. Isotopic labelling coupled with CW-CO laser pumping (65) of the CO stretching vibrations (∿1900 cm^{-1}) showed that the rearrangement mode of $Fe(CO)_4$ follows an inverse Berry pseudo-rotation as shown in Figure 8.

Similarly, in experiments with $Fe(CO)_3L$ (generated in a matrix from $Fe(CO)_4L$; L = NMe_3), the C_{3v} and C_s forms of the species can be interconverted with light of appropriate wavelength (66).

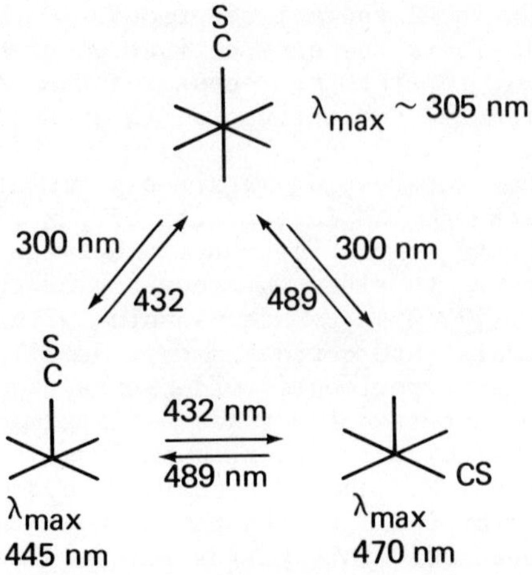

Figure 7. Summary of the photochemistry of $W(CO)_5CS$ in an Ar matrix. (Reproduced from Ref. 64. Copyright 1976, American Chemical Society.)

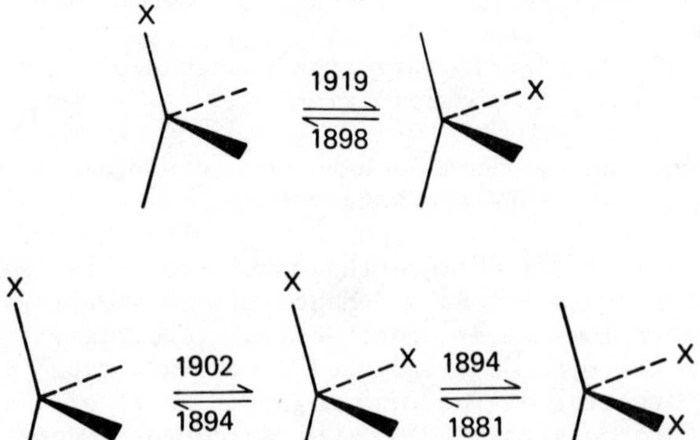

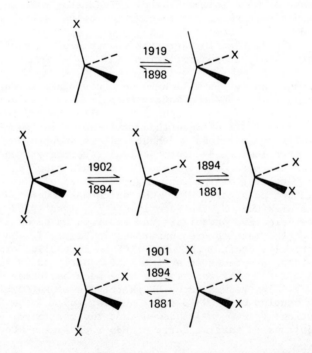

Figure 8. The observed IR-laser induced isomerizations of $Fe(^{12}C^{16}O)_{4-x}(^{13}C^{18}O)_x$ species in an Ar matrix. X represents the $^{13}C^{18}O$ group, and the numbers represent the wavenumbers of the CO laser lines that induce the particular isomerizations.

Photochemical Routes. Finally, two sets of experiments from Rest's laboratory which demonstrate the subtle implications of quite complex matrix photochemistry experiments. The first (67) involves $(\eta^5\text{-}C_5H_5)Co(CO)_2$ and can be summarised

$$(\eta^5 C_5H_5)Co(CO)_2 \underset{\lambda\, >\, 290\ \text{nm}}{\overset{\lambda\, <\, 280\ \text{nm/CO matrix}}{\rightleftarrows}} (\eta^3\text{-}C_5H_5)Co(CO)_3$$

[There was no evidence for production of $(\eta^5\text{-}C_5H_5)Co(CO)$ in Ar matrices, presumably because of very ready recombination with CO in the matrix cage. Whether a matrix isolated carbonyl molecule can eject CO *without* automatic recombination probably requires the photochemical path to lead to a ground state fragment in which the empty coordination site is oriented away from the photo-ejected CO - see the $Cr(CO)_5$ photochemical scheme (Figure 3).] In a CO matrix, therefore, the primary product involves an expanded coordination number ("ring-slippage") and it is argued that such a species is consistent with the associative mechanism proposed for room temperature substitution reactions of $(\eta^5\text{-}C_5H_5)Co\text{-}(CO)_2$.

By contrast, when $(\eta^5\text{-}C_5H_5)Fe(CO)_2(COR)$ is photolysed (68), even in CO matrices, only CO loss is observed with $(\eta^5\text{-}C_5H_5)Fe\text{-}(CO)(COR)$ being formed en route to $(\eta^5\text{-}C_5H_5)Fe(CO)_2R$.

Dinuclear Carbonyls. One of the characteristic features in the UV spectra of metal-metal bonded dinuclear carbonyls and substituted species is an intense band in the 350 nm region (1). For example, for $Mn_2(CO)_{10}$ this band, at 336 nm, is assigned to a transition from the filled σ M-M bonding molecular orbital to the corresponding antibonding orbital. Confirmation of this assignment comes from an elegant experiment (27) based on the argument that such a transition must be polarized along the Mn-Mn axis (i.e. of transition moment B_2) confirmed by examining the dichroic properties of IR and UV bands of $Mn_2(CO)_{10}$ in a nematic solvent.

A great deal of photochemistry of dinuclear carbonyls is consistent with the generation of radicals following M-M bond schism on irradiation into the $\sigma \longrightarrow \sigma^*$ band. However, even for the much studied $Mn_2(CO)_{10}$ the picture is still obscure. For instance, pulse radiolysis (69) and flash photolysis (70-72) combine to suggest that in addition to generation of two $Mn(CO)_5$ radicals other photochemical routes may involve bridged-$Mn_2(CO)_{10}$, $Mn_2(CO)_9$, $Mn_2(CO)_8$; however, in view of the enormous importance of solvent interactions and impurities (see $Cr(CO)_5$ above) one wonders what the correct explanation is going to be.

Matrix isolation has already provided a valuable insight into the behavior of dinuclear carbonyls. In the first experiments of this kind we were able (73) to show that, on photolysis in solid Ar, $Fe_2(CO)_9$ loses CO to form $Fe_2(CO)_8$ in both bridged and unbridged forms and the behavior of these fragments was studied;

interestingly Hoffmann ($\underline{74}$) has recently drawn attention to this $Fe_2(CO)_8$ species and to Stone's experiments involving polynuclear complexes of $Fe_2(CO)_8$. More recently, Sweany and Brown ($\underline{75}$) have shown that UV photolysis of $Co_2(CO)_8$ in argon generates unbridged $Co_2(CO)_7$, while in CO matrices there is evidence for $Co(CO)_4$. Thus, there is evidence that matrix studies could help with $Mn_2(CO)_{10}$; unfortunately, such experiments have so far been rather unrevealing.   Presumably the Cage Effect encourages the recombination of any $Mn(CO)_5$ radicals generated by photolysis. Indeed the best way of generating $Mn(CO)_5$ photochemically is by UV photolysis of $HMn(CO)_5$ in a CO matrix (54). The structure of $Mn(CO)_5$ as a C_{4v} fragment has now been established ($\underline{54}$) by isotopic IR and, more recently, confirmed by ESR ($\underline{76}$). It is, however, our belief that subtle wavelength-dependent photolysis experiments, particularly using polarized light, will eventually unravel the problem of $Mn_2(CO)_{10}$.

Another dinuclear carbonyl which presents interesting problems is $[(\eta^5-C_5H_5)Fe(CO)_2]_2$. Does the photochemistry proceed exclusively through homolytic fission to produce two $(\eta^5-C_5H_5)Fe-(CO)_2$ radicals or by other possible routes? The discussion of this reaction has involved mechanistic and synthetic studies ($\underline{77}$), flash photolysis ($\underline{78}$) and low-temperature photolysis ($\underline{79}$) - the latter work, in THF or ethyl chloride at $-78°C$, invokes an intermediate in which the Fe-Fe direct bond is broken but the two halves of the molecule are held together by a CO bridge. Clearly such an intriguing problem merits more detailed investigations.

Polynuclear Clusters. The photochemistry of polynuclear clusters is a very active field, partly because of the potential catalytic importance of these compounds. It is still not clear how to predict the conditions which will lead to declusterification as opposed to internal rearrangement or substitution (1). There is an absence of good definitive evidence for specific intermediates and we close with a hint. Some years ago in rather crude unpublished work ($\underline{80}$) we showed that the photolysis of $Fe_3(CO)_{12}$ isolated in a matrix led to production of some CO, but more particularly, complete disappearance of IR bands due to the bridging CO groups and the appearance of new terminal CO IR bands. We believe that careful studies of this kind, taking advantage of sophisticated FTIR methods, will provide valuable insight into even complex photochemical intermediates.

Acknowledgments

We thank all those who have had discussions with us and are grateful to the S.E.R.C. for generous support of our work. We wish to acknowledge the help of all our colleagues in Nottingham, particularly Mr P.W. Lemeunier and Dr M.A. Healy.

Literature Cited

1. For an excellent recent review see Geoffroy, G.L.; Wrighton, M.S. "Organometallic Photochemistry", Academic Press: New York, 1979.
2. Kidd, D.R.; Cheng, C.P.; Brown, T.L. J. Am. Chem. Soc. 1978, 100, 4103.
3. McCullen, S.B.; Brown, T.L. J. Am. Chem. Soc. in press
4. Whyman, R. J. Organomet. Chem. 1975, 94, 303, and earlier references quoted therein.
5. Koerner von Gustorf, E.A.; Leenders, L.H.G.; Fischler, I; Perutz, R.N. Adv. Inorg. Rad. Chem. 1976, 19, 65.
6. Chase, D.B.; Weigert, F.J. J. Am. Chem. Soc. 1981, 103, 977.
7. Kazlauskas, R.J.; Wrighton, M.S. J. Am. Chem. Soc. in press.
8. Hermann, H.; Grevels, F.W.; Henne, A.; Schaffner, K., to be published.
9. Brus, L.E.; personal communication.
10. Moskovits, M.; Ozin, G.A. Eds.; "Cryochemistry"; Wiley: New York, 1976
11. Barnes, A.J.; Müller, A.; Orville-Thomas, W.J. Eds. "Matrix Isolation Spectroscopy"; Deidel, 1981.
12. Turner, J.J.; Burdett, J.K.; Perutz, R.N.; Poliakoff, M. Pure and Appl. Chem. 1977, 49, 271.
13. Burdett, J.K. Coord. Chem. Rev. 1978, 27, 1.
14. Stolz, I.W.; Dobson, G.R.; Sheline, R.K. J Am. Chem. Soc. 1962, 84, 3589.
15. Stolz, I.W.; Dobson, G.R.; Sheline, R.K. J Am. Chem. Soc. 1963, 85, 1013.
16. Perutz, R.N.; Turner, J.J. Inorg. Chem. 1975, 14, 262.
17. Burdett, J.K.; Poliakoff, M.; Timney, J.A.; Turner, J.J. Inorg. Chem. 1978, 17, 948.
18. Burdett, J.K. Inorg. Chem. 1981, 20, 2607.
19. Burdett, J.K. "Molecular Shapes"; Wiley-Interscience: New York, 1980.
20. Ryott, G.J. Ph.D., Thesis, Nottingham University, Nottingham, 1982.
21. Perutz, R.N.; Turner, J.J. J. Am. Chem. Soc. 1975, 97, 4791.
22. Burdett, J.K.; Downs, A.J.; Gaskill, G.P.; Graham, M.A.; Turner, J.J.; Turner, R.F. Inorg. Chem. 1978, 17, 523.
23. Demuynck, J.; Kochanski, E.; Veillard, A. J. Am. Chem. Soc. 1979, 101, 3467.
24. Burdett, J.K.; Perutz, R.N.; Poliakoff, M.; Turner, J.J. J.C.S. Chem. Commun. 1975, 157.
25. Burdett, J.K.; Grzybowski, J.M.; Perutz, R.N.; Poliakoff, M; Turner, J.J.; Turner R.F. Inorg. Chem. 1978, 17, 147.
26. Baird, M.S.; Dunkin, I.R.; Hacker, N.; Poliakoff, M.; Turner, J.J. J. Am. Chem. Soc. 1981, 103, 5190.
27. Levenson, R.A.; Gray, H.B.; Ceasar, G.P. J. Am. Chem. Soc. 1970, 92, 3653.
28. Hay, P.J. J. Am. Chem. Soc. 1978, 100, 2411.

29. Adamson, A.W.; personal communication.
30. Nasielski, J.; Colas, A. J. Organomet. Chem. 1975, 101, 215.
31. Tyler, D.R.; Petrylak, D.P. J. Organomet. Chem. 1981, 212, 289.
32. Beattie, W.H.; Maier, W.B.; Holland, R.F.; Freund, S.M.; Stewart, B. Proc. SPIE (Laser Spectroscopy) 1978, 158, 113.
33. Maier, W.B.; Poliakoff, M.; Simpson, M.B.; Turner, J.J. J.C.S. Chem. Commun. 1980, 587.
34. Maier, W.B.; Poliakoff, M.; Simpson, M.B.; Turner, J.J. Inorg. Chem., in press.
35. Maier, W.B.; Poliakoff, M.; Simpson, M.B.; Turner, J.J. to be published.
36. Bonneau, R.; Kelly, J.M. J. Am. Chem. Soc. 1980, 102, 1220.
37. Lees, A.J.; Adamson, A.W. Inorg. Chem. 1981, 20, 4381.
38. Breckenridge, W.H.; Sinal, N. J. Phys. Chem. 1981, 85, 3557.
39. Tumes, W.; Gitlin, B.; Rosan, A.M.; Yardley, J.T. J. Am. Chem. Soc. 1982, 104, 55.
40. Nathanson, G.; Gitlin, B.; Rosan, A.; Yardley, J.T. J. Chem. Phys. 1981, 74, 361, 370.
41. Perutz, R.N.; Turner, J.J. J. Am Chem. Soc. 1975, 97, 4800.
42. Poliakoff, M.; Turner, J.J. J.C.S. Faraday Trans. II, 1974, 70, 93.
43. Mahmoud, K.A.; Narayanaswamy, R.; Rest, A.J. J. Chem. Soc., Dalton, 1981, 2199.
44. Kazlauskas, R.J.; Wrighton, M.S. J. Am. Chem. Soc. 1980, 102, 1727.
45. Boxhoorn, G.; Stufkens, D.J.; Oskam, A. Inorg. Chim. Acta, 1979, 33, 215.
46. Boxhoorn, G.; Schoemaker, G.C.; Stufkens, D.J.; Oskam, A.; Rest, A.J.; Darensbourg, D.J. Inorg. Chem. 1980, 19, 3455.
47. Geoffroy, G.L. Prog. Inorg. Chem. 1980, 27, 123.
48. Green, M.L.H. Pure Appl. Chem. 1978, 50, 27.
49. Grebenik, P.; Downs, A.J.; Green, M.L.H.; Perutz, R.N. J.C.S. Chem. Commun. 1979, 742.
50. Chetwynd-Talbot, J.; Grebenik, P.; Perutz, R.N. J.C.S. Chem. Commun. 1981, 452.
51. Sweany, R.L. X Cong. Organomet. Chem., Toronto, 1981, Abstract IC05, 44.
52. Poliakoff, M; Turner, J.J. J.C.S. Dalton, 1974, 2276.
53. Poliakoff, M. J.C.S. Dalton, 1974, 210.
54. Church, S.P.; Poliakoff, M.; Timney, J.A.; Turner, J.J. J. Am. Chem. Soc. 1981, 103, 7515.
55. Sweany, R.L. Inorg. Chem. 1982, 21, 752.
56. Wermer, P.; Ault, B.S.; Orchin, M. J. Organomet. Chem. 1978, 162, 189.
57. Sweany, R.L. Inorg. Chem. 1980, 19, 3512.
58. Parshall, G.W. "Homogeneous Catalysis"; Wiley Interscience: New York, 1980.
59. Janowicz, A.H.; Bergman, R.G. J. Am. Chem. Soc. 1982, 104, 352.

60. Poliakoff, M. Chem. Soc. Rev. 1978, 7, 528.
61. Kazlauskas, R.J.; Wrighton, M.S. J. Am. Chem. Soc. in press.
62. Billups, W.E.; Konarski, M.M.; Hauge, R.H.; Margrave, J.L. J. Am. Chem. Soc. 1980, 102, 7394.
63. Ozin, G.A.; McIntosh, D.F.; Mitchell, S.A.; Garcia-Prieto, J. J. Am. Chem. Soc. 1981, 103, 1574.
64. Poliakoff, M. Inorg. Chem. 1976, 15, 2022, 2892.
65. Poliakoff, M.; Turner, J.J. "Chemical and Biochemical Applications of Lasers", Moore, C.B. Ed; Academic Press; New York, 1980; Vol. 5, p.175.
66. Boxhoorn, G.; Cerfoutain, M.B.; Stufkens, D.J.; Oskam, A. J.C.S. Dalton, 1980, 1336.
67. Crichton, O.; Rest, A.J.; Taylor, D.J. J.C.S. Dalton, 1980, 167.
68. Fettes, D.J.; Narayanaswamy, R.; Rest, A.J. J.C.S. Dalton, 1981, 2311.
69. Waltz, W.L.; Hackelberg, O.; Dorfman, L.M.; Wojcicki, A. J. Am. Chem. Soc. 1978, 100, 7259.
70. Hughey, J.L.; Anderson, C.P.; Meyer, T.J. J. Organomet. Chem. 1977, 125, C49.
71. Wegman, R.W.; Olsen, R.J.; Gard, D.R.; Faulkner, L.R.; Brown, T.L. J. Am. Chem. Soc. 1981, 103, 6089.
72. Yasufuku, K.; Yesaka, H.; Kobayashi, T.; Yamazaki, H.; Nagakura, S. Proc. 10th Conf. Organomet. Chem., Toronto, 1981, Abstract IC02, p41.
73. Poliakoff, M.; Turner, J.J. J. Chem. Soc. A 1971, 2403.
74. Hoffmann, R. Nobel Lecture, 1981.
75. Sweany, R.; Brown, T.L. Inorg. Chem. 1977, 16, 421.
76. Symons, M.C.R.; Sweany, R.L. Organometallics, in press.
77. Abrahamson, H.B.; Palazzotto, M.C.; Reichel, C.L.; Wrighton, M.S. J. Am. Chem. Soc. 1979, 101, 4123.
78. Caspar, J.V.; Meyer, T.J. J. Am. Chem. Soc., 1980, 102, 7794.
79. Tyler, D.R.; Schmidt, M.A.; Gray, H.B. Inorg. Chem. 1979, 14, 2753.
80. Poliakoff, M. Ph.D., Thesis, Cambridge,University, Cambridge, England, 1972.
81. Welch, J.A.; Peters, K.S.; Vaida, V. J. Phys. Chem., 1982, 86, 1941.

RECEIVED August 3, 1982

Discussion

A.W. Adamson, University of Southern California: You spoke of an excited state C_{4v} product. We have never been able to observe solution photochemistry of a coordination compound in which ligand dissociation produces an excited state product. Further, while chemiluminescent reactions are known, the ones involving coordination compounds produce light only in quite low yield, indicating that the path through an excited state is not a favored one. Do you know of any direct experimental evidence for the $[C_{4v}]^*$ product?

J.J. Turner, University of Nottingham: The short answer is No(!); but the first point to make is that we don't claim that C_{4v} $Cr(CO)_5$ in an *excited* state is a 'product' - undoubtedly the 'product' is C_{4v} $Cr(CO)_5$ in the *ground* state. What the experiments tell us is this. The polarized photolysis/spectroscopic experiments (ref. 24,25 of above) prove that the $Cr(CO)_5X$ species are 'stirred' by visible light, first via an excited state of the C_{4v} $Cr(CO)_5X$ and then, almost certainly, via the D_{3h} intermediate state (see figure 9 in ref. 25), i.e. all the pathways in figure 3 (above) are established except that from $Cr(CO)_6$ to D_{3h} $Cr(CO)_5$ via the excited C_{4v} fragment. However, in matrix experiments, my colleague, Martyn Poliakoff (ref. 64 above), showed that photolysis of stereospecifically ^{13}CO-labelled *trans*-$(^{13}CO)W(CO)_4CS$ yields, as the major product, square-pyramidal *cis*-$(^{13}CO)W(CO)_3CS$ with the CS group in the axial position, i.e. the principal path involves loss of equatorial CO *and* rearrangement of the remaining CS and four CO groups. Whether this is imagined as CO loss followed by rearrangement (i.e. analogous to the excited state path in figure 3) *or* as a concerted process does not affect the argument - we certainly don't consider the excited C_{4v} fragment as a 'thexi' state, we just find it easier to picture the 6 ⟶ 5 process via the excited fragment.

G.A. Ozin, University of Toronto: In our Cr/CO matrix cocondensation experiments (Angew. Chem., Int. Ed. Eng. 1975, 14, 292), we reported evidence for the facile formation of a binuclear chromium carbonyl complex $Cr_2(CO)_{10}$ or $Cr_2(CO)_{11}$ which could be described as square pyramidal $Cr(CO)_5$ weakly interacting with either a $Cr(CO)_5$ or $Cr(CO)_6$ moiety in the vacant (sixth) site. As a result, the infrared spectrum of this "weakly-coupled" binuclear species closely resembled that of the mononuclear fragment $Cr(CO)_5$. I would like to ask you, whether or not you have any evidence for the existence of such a binuclear species in your $Cr(CO)_6$/Xe cryogenic solutions following various photolysis treatments.

J.J. Turner, University of Nottingham: You are quite correct that species such as $Cr(CO)_5...X$, where X is a weakly coordinating 'ligand' such as Ar, Xe, CH_4, $Cr(CO)_6$, show barely distinguishable IR spectra in the carbonyl region. However, the visible band is extremely sensitive to X whether measured in matrices (Ar, 533 nm; Xe, 492 nm; CH_4, 489 nm (see ref. 21 above); $Cr(CO)_6$, $\sim$460 nm (J. Amer. Chem. Soc., 1975, 97, 4805)) or in solution (cyclohexane, 510 nm; $Cr(CO)_6$, 485 nm (see ref. 36 above)). The position of this band reflects the strength of $Cr(CO)_5/X$ interaction which suggests that $Cr(CO)_6$ and Xe will interact similarly with $Cr(CO)_5$; given that in liquid Xe, $Cr(CO)_6$ is extremely dilute ($\sim$1 ppm) it seems unlikely that one would observe $Cr(CO)_5...Cr(CO)_6$ in preference to $Cr(CO)_5...Xe$. Nevertheless, proof of this must await UV data which we are in the process of obtaining.

Two further related comments. Firstly, since N_2 is a "good" ligand (λ_{max}, 364 nm), with N_2-doped Xe there is no trace even of $Cr(CO)_5...Xe$ since $Cr/CO/N_2$ species predominate. Secondly, during experiments with $Ni(CO)_4/N_2$/liquid Kr (see above), photolysis in the complete *absence* of dissolved N_2 led to the appearance of a transient carbonyl species with IR bands similar to those assigned to $Ni_2(CO)_7$ (J.E. Hulse and M. Moskovits; unpublished data) - presumably the interaction of $Ni(CO)_3$ with $Ni(CO)_4$ is considerably stronger than with Kr.

4

Fuel and Electricity Generation from Illumination of Inorganic Interfaces

MARK S. WRIGHTON

Massachusetts Institute of Technology, Department of Chemistry, Cambridge, MA 02139

Semiconductor-based photoelectrochemical devices represent good systems for the sustained, direct conversion of light to chemical or electrical energy. The interfacial structure, energetics, and redox kinetics control the overall performance of such systems. Examples of improvements in efficiency and durability of photoelectrochemical cells stemming from chemical manipulations at semiconductor/liquid electrolyte interfaces illustrate the critical importance of understanding interface properties.

Inorganic chemistry at interfaces is crucial to a large number of processes, systems, and devices that have practical consequence now and into the 21st century. Heterogeneous catalyst systems, batteries, fuel cells, field effect transistors, optical and acoustical recorders, sensors, solar cells, and even natural plant photosynthesis are all dependent on interfaces. It is becoming evident that characterization, synthesis, manipulation, and understanding of interfacial properties will comprise a significant fraction of the fundamental effort undergirding many practical applications of inorganic chemistry. With the advent of an arsenal of new spectroscopic probes it is apparent that chemically complex interfaces can be characterized with good resolution. The impact of the structural characterization of interfaces is likely to be as great as from the structural characterization that is commonplace for relatively small molecular entities. Elucidation of molecular structure plays a central role in the understanding of reactions and leads to insight into details of mechanism and electronic structure. Exploitation of interfaces in inorganic-based systems can continue with no new insights. However, as the 21st century approaches inorganic chemists have the opportunity to contribute heavily to the understanding of

0097-6156/83/0211-0059-$09.25/0

interfaces and rational application of basic knowledge to better use interfaces.

One current basic research endeavor involves the use of illuminated interfacial chemical systems to bring about the sustained conversion of optical energy to chemical or electrical energy. In fact, semiconductor-based photoelectrochemical cells like that in Scheme I represent the best chemical systems for the direct conversion of optical energy to electrical energy or to chemical energy in the form of high energy redox products.(1) In such devices the light-absorbing semiconductor electrode immersed in an electrolyte solution comprises a photosensitive interface where thermodynamically uphill redox processes can be driven with optical energy. Depending on the nature of the photoelectrode, either a reduction or an oxidation half-reaction can be light-driven with the counterelectrode being the site of the accompanying half-reaction. N-type semiconductors are photoanodes, p-type semiconductors are photocathodes, (2,3,4) and intrinsic materials (5) can be either photoanodes or photocathodes depending on the nature of the contact by both the liquid and the wire of the external circuit. Within the past decade remarkable progress has emerged from conscious efforts to understand and improve semiconductor-based photoelectrochemical cells. Systems based on n-type GaAs,(6,7) n-type WSe_2, (8,9) and p-type InP (10,11) have been shown to be able to convert sunlight with greater efficiency than the widely regarded minimum useful efficiency of 10% for large scale energy generation from sunlight in the United States.(12)

The aim of this presentation is both to outline recent advances and to identify problems associated with illuminated semiconductor electrodes for optical energy conversion. Results from this laboratory will be highlighted. Practical applications, if they come at all, will be important in the 21st century and beyond. However, though the prospects for large scale energy generation from sunlight may appear dim, it is clear that existing fossil fuel reserves are finite, fission nuclear power has an uncertain future, and fusion does not work now and may not work in the future. Solar chemical conversion schemes do work on a large scale as evidenced by the natural photosynthetic apparatus. Existing solar insolation is far greater than man's needs. The research effort required to fully investigate interfacial solar conversion schemes is worth expending when viewed against the possible return and the prospect that other alternatives may prove unacceptable from a technical, social, political, economic, or safety standpoint.

Semiconductor/Liquid Electrolyte Interface Energetics

Schemes II and III represent the equilibrium interface energetics for ideal n- and p-type semiconductors, respectively, for

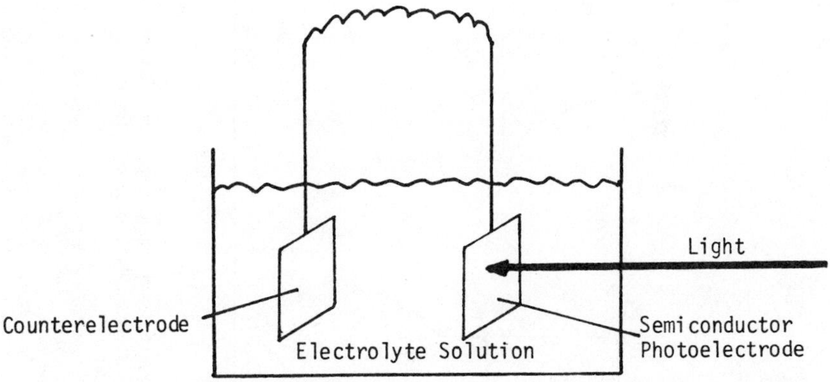

Scheme I. Semiconductor-based photoelectrochemical cell. Energy output may be in the form of electricity by putting a load in series in the external circuit, or the output can be in the form of chemical energy as redox products formed at the electrodes. N-type semiconductors effect uphill oxidations upon illumination, and p-type semiconductors effect uphill reductions under illumination. Either or both electrodes in the cell can be a photoelectrode.

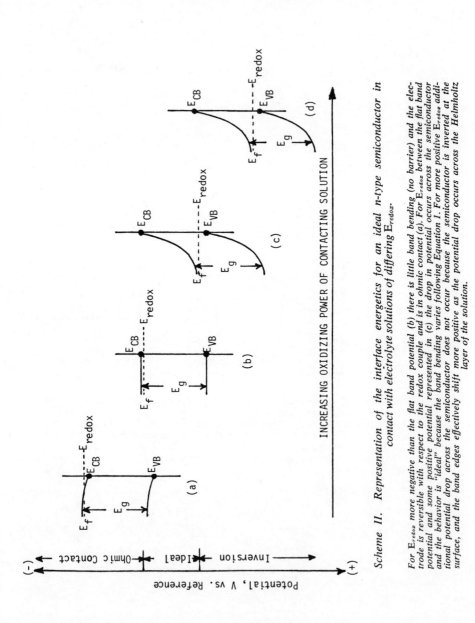

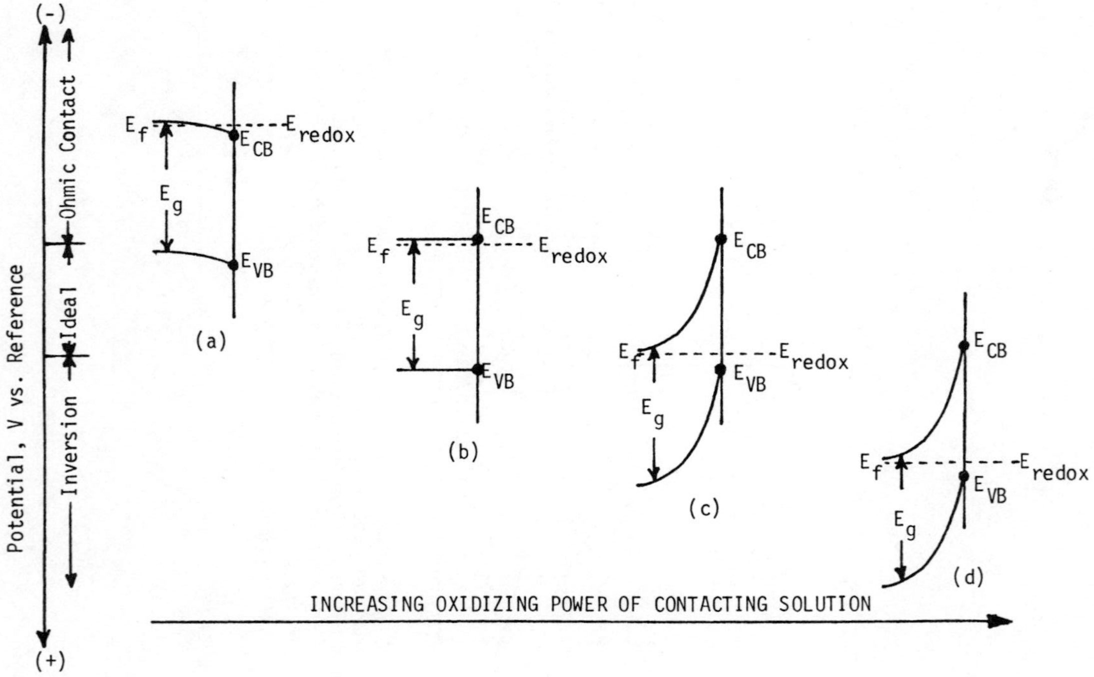

Scheme II. Representation of the interface energetics for an ideal n-type semiconductor in contact with electrolyte solutions of differing E_{redox}.

For E_{redox} *more negative than the flat band potential (b) there is little band bending (no barrier) and the electrode is reversible with respect to the redox couple and is in ohmic contact (a). For* E_{redox} *between the flat band potential and some positive potential represented in (c) the drop in potential occurs across the semiconductor and the behavior is "ideal" because the band bending varies following Equation 1. For more positive* E_{redox} *additional potential drop across the semiconductor does not occur because the semiconductor is inverted at the surface, and the band edges effectively shift more positive as the potential drop occurs across the Helmholtz layer of the solution.*

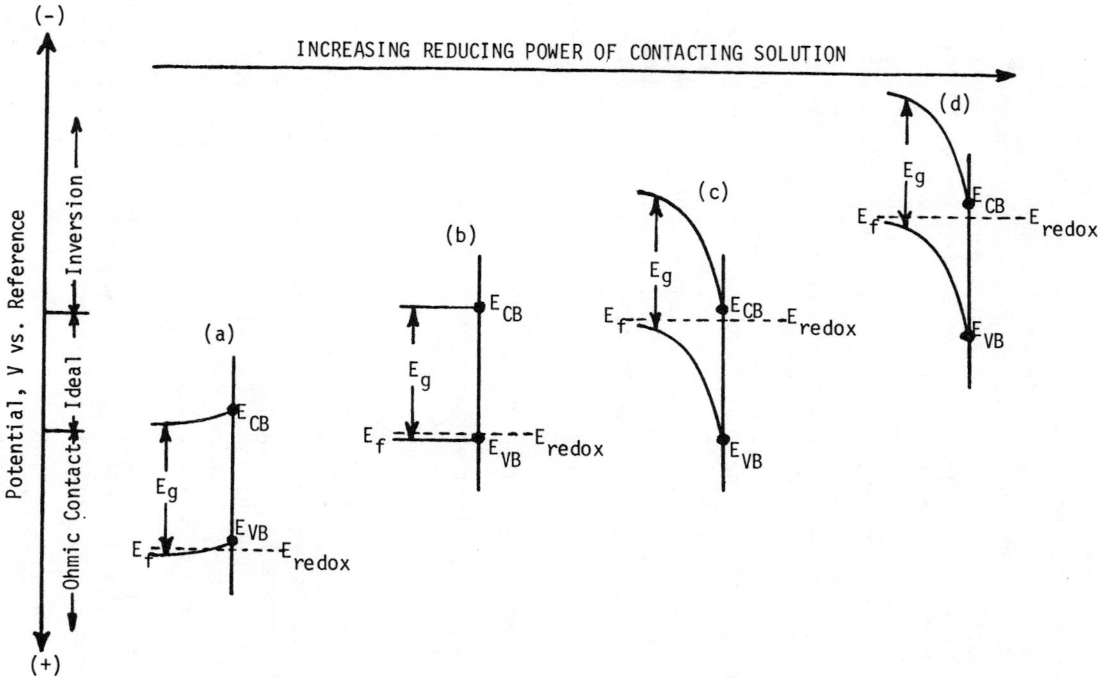

Scheme III. Representation of the interface energetics for an ideal p-type semiconductor in contact with electrolyte solutions of differing E_{redox}.

The situation is analogous to the n-type semiconductor (Scheme II) except that the more positive E_{redox} *gives an ohmic contact and the most negative* E_{redox} *gives inversion. For* E_{redox} *between the flat band potential (b) and some negative potential, the behavior is "ideal" because the band bending varies following Equation 1.*

contact by electrolyte solutions containing different redox reagents that vary in electrochemical potential, E_{redox}.(2,3,4,13) For E_{redox} sufficiently negative for n-, or sufficiently positive for p-type, semiconductors the electrode behaves as a metallic electrode, not blocking the flow of electrons in either direction. This situation is analogous to the criteria for forming an ohmic contact to an n- or p-type semiconductor.(14) When E_{redox} is between the top of the valence band, E_{VB}, and the bottom of the conduction band, E_{CB}, the p-type semiconductor is blocking to reductions and the n-type semiconductor is blocking to oxidations in the dark. The minority carrier (e^- or h^+ for p- or n-type semiconductors, respectively) is only available upon photoexcitation with $>E_g$ light and is driven to the interface, owing to the field in the semiconductor near the surface. The availability of the minority carrier at the interface upon photoexcitation allows oxidation with h^+ or reduction with e^-, and importantly the oxidizing power of the h^+ for n- and the reducing power for p-type semiconductors can be greater than that associated with the electrochemical potential (Fermi level), E_f, in the bulk of the semiconductor. This means that light can be used to drive thermodynamically non-spontaneous redox processes. The extent to which a process can be driven in an uphill sense is the photovoltage, E_V, that is given by equation (1) for ideal

$$E_V = |E_{FB} - E_{redox}| \quad (1)$$

(For E_{redox} between E_{CB} and E_{VB})

semiconductors where E_{FB} is the flat-band potential, i.e. the value of E_f where the bands are not bent. For commonly used carrier concentrations E_{FB} is within 0.1 V of E_{VB} for p-type semiconductors, and for n-type semiconductors E_{FB} is within 0.1 V of E_{CB}. Thus, $>E_g$ illumination of a semiconductor electrode at open-circuit tends to drive E_f to E_{FB}, more positive for p-type electrodes and more negative for n-type electrodes compared to dark equilibrium where $E_f = E_{redox}$.

Ideally, the maximum value of E_V approaches the band gap, E_g, of the semiconductor. However, ~0.3 V of band bending is required to efficiently separate the e^- - h^+ pairs created by light. The separation of e^- - h^+ pairs is essential to obtain a high quantum yield for net electron flow, Φ_e. In any energy conversion application the efficiency, η, for the photoelectrode is given by equation (2), and since i is proportional to Φ_e it is

$$\eta = \frac{E_V \times i}{\text{Input Optical Power}} \quad (2)$$

i = photocurrent

desirable to obtain a large value of Φ_e. Good E_V at i = 0 (illuminated, open-circuit) or good i at E_V = 0 (illuminated, short-circuit) are both situations that give η = 0. The objective is to optimize the product of E_V and i in order to achieve the maximum efficiency. Figure 1 shows a steady-state photocurrent-voltage curve (15) for an n-type WS_2 ($E_g \approx 1.4$ eV) photoanode based cell for conversion of light to electricity employing the Br_2/Br^- redox couple and the inset shows the full cell energetics for operation at the maximum power point, the value of E_f when E_V x i is maximum. Figure 2 shows a similar curve for a cell based on p-type WS_2 employing $Fe(\eta^5\text{-}C_5Me_5)_2^{+/0}$ as the redox couple.(16) In solar energy applications an E_g in the range 1.1-1.7 eV is desirable, since a single photoelectrode based device would have optimum solar efficiency, exceeding 20, for such band gaps.(12)

The WS_2 electrodes represent n- and p-type semiconductors that behave relatively ideally (15,16) with respect to interface energetics in that the value of E_V does vary with E_{redox} according to equation (1) for E_{redox} within ~0.8 V of E_{FB}. However, for many electrode materials the variation in E_V does not follow equation (1).(17) For example, n-type CdTe (E_g = 1.4 eV) can give either a constant value of E_V (independent of E_{redox}) or a nearly ideal variation in E_V, Figure 3, depending on the pretreatment of the surface prior to use.(18) The ability to improve the value of E_V from the constant value of ~0.5 V to ~0.9 V using a reducing surface pretreatment is clearly desirable, but at the same time certain oxidation reactions can be driven in an uphill sense using the oxidizing pretreatment that could not be done with the CdTe pretreated with the reducing reagent.

For CdTe it is evident that the surface pretreatment chemistry is related to the value of E_V. In fact, Auger and X-ray photoelectron spectroscopy reveal a correlation of the E_V vs. E_{redox} behavior related to the composition of the surface of CdTe.(18) The oxidizing surface etch leaves a Te-rich overlayer that causes the CdTe to behave as if it is coated with a metal having a work function that gives a Schottky barrier height of ~0.6 V. In the liquid junction system the analogue of Schottky barrier height is the $E_{redox} - E_{CB}$ or $E_{redox} - E_{VB}$ separation for p- or n-type semiconductors. If a Schottky barrier is immersed into a liquid electrolyte solution the value of E_V should be independent of E_{redox}, as found from n-type CdTe after an oxidizing pretreatment.

Generally, when E_V is fixed for a wide range of E_{redox} the semiconductor is said to be "Fermi level pinned".(17) Fermi level pinning simply means that the value of E_f at the surface of the semiconductor is pinned to some value relative to the band edge positions, independent of the electrochemical potential of a contacting solution or work function of a contacting metal. The pinning of the E_f is due to surface states of sufficient density

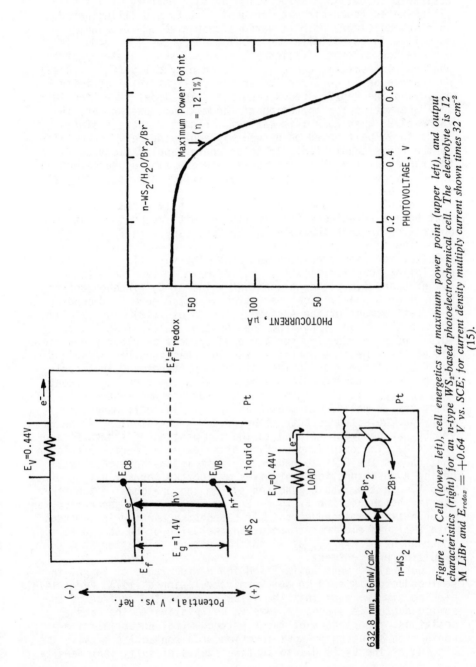

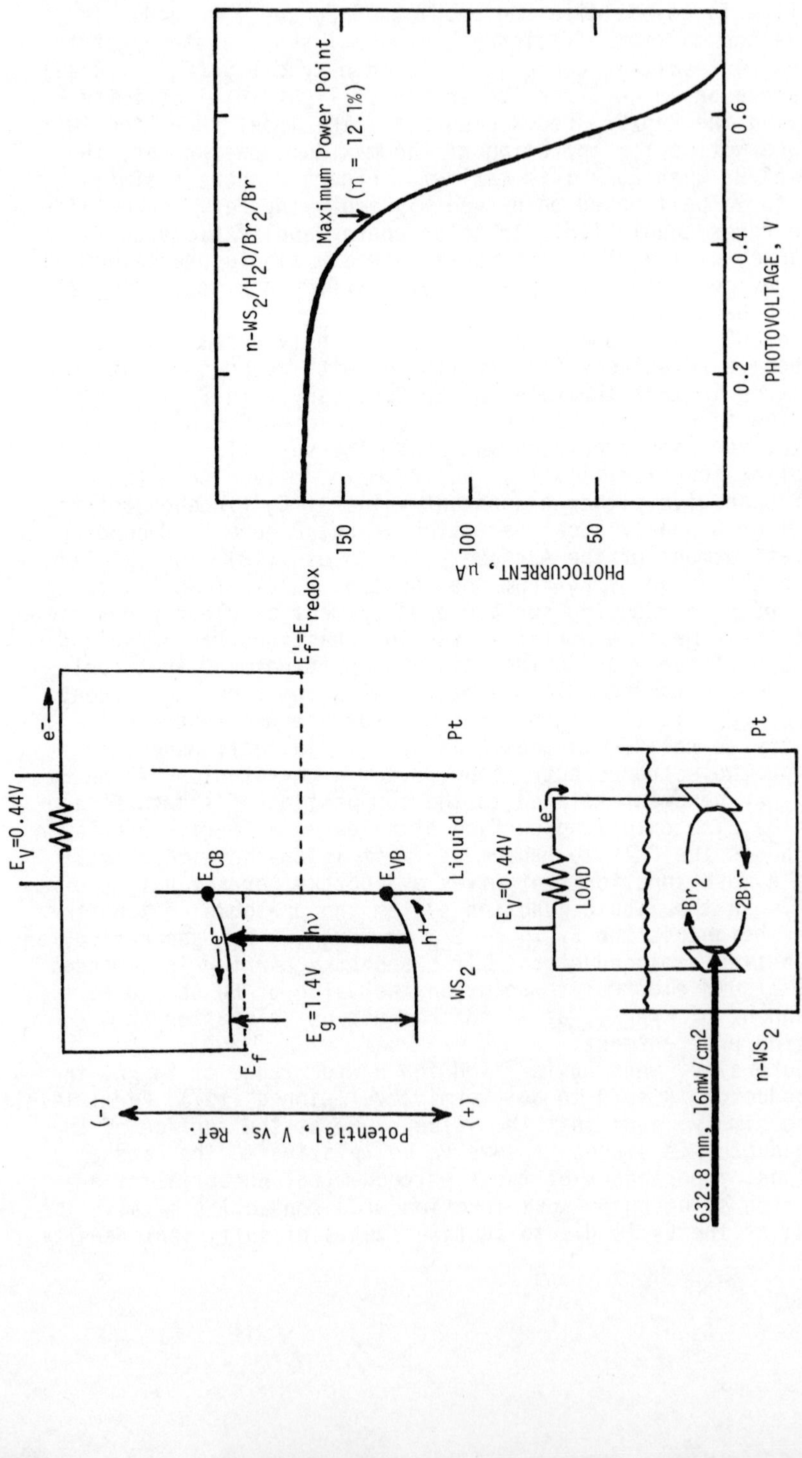

Figure 1. Cell (lower left), cell energetics at maximum power point (upper left), and output characteristics (right) for an n-type WS_2-based photoelectrochemical cell. The electrolyte is 12 M LiBr and $E_{redox} = +0.64$ V vs. SCE; for current density multiply current shown times 32 cm^{-2} (15).

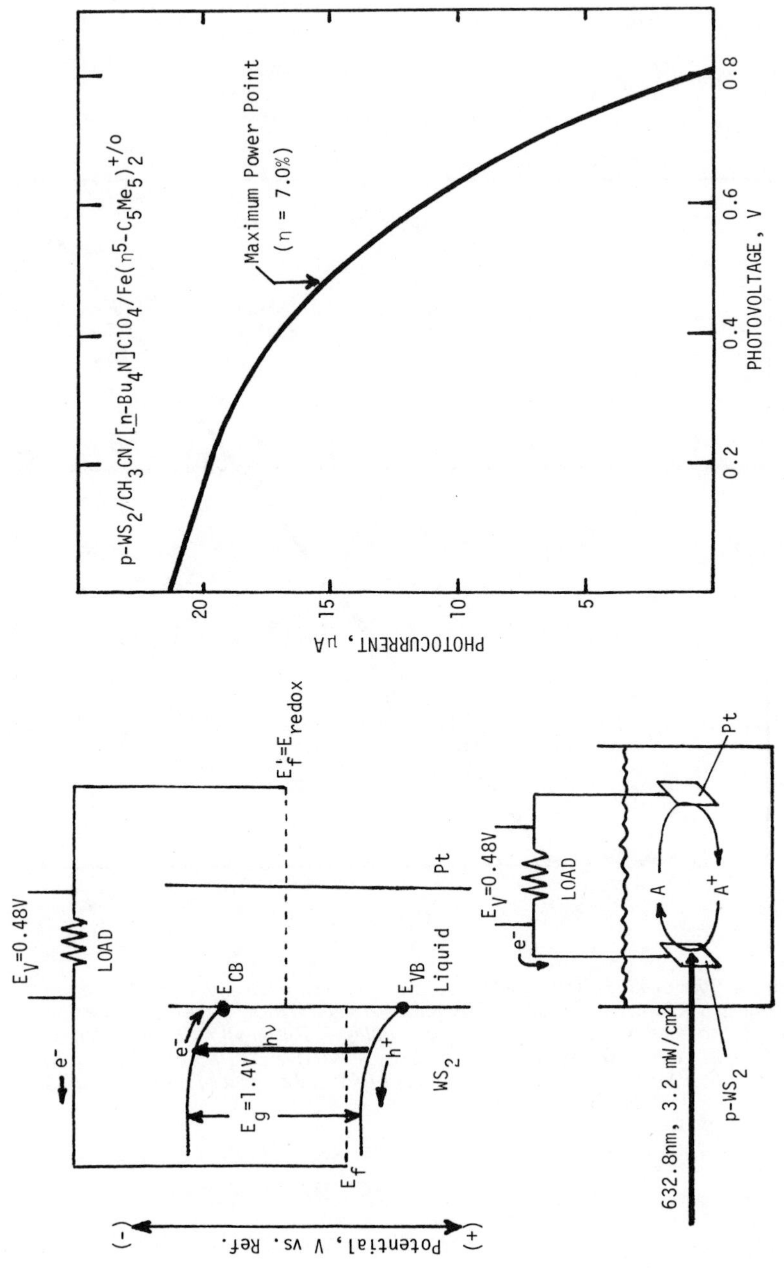

Figure 2. Cell (lower left), cell energetics at maximum power point (upper left), and output characteristics (right) for a p-type WS_2-based photoelectrochemical cell. The E_{redox} = −0.38 V vs. Ag^+/Ag; for current density multiply current shown by 32 cm^{-2} (16).

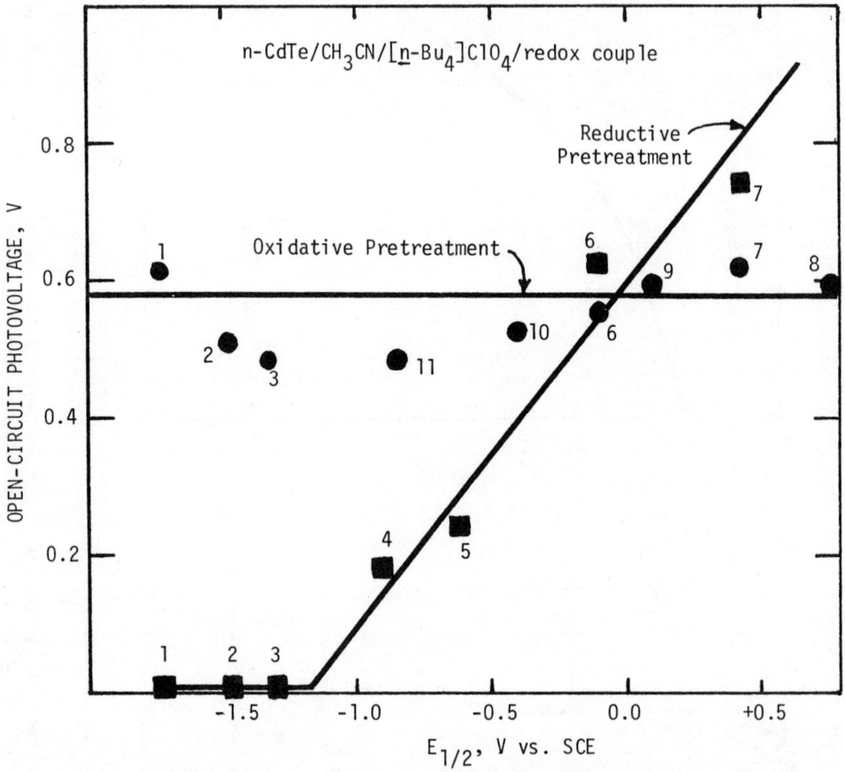

Figure 3. Photovoltage from n-CdTe etched with an oxidizing etch (●) or an oxidizing etch followed by a reducing treatment ($NaOH/S_2O_4^{2-}$) (■) as a function of $E_{1/2}$ of a contacting couple. Key to redox couples: 1, $Ru(bipy)_3^{0/-}$; 2, $Ru(bipy)_3^{+/0}$; 3, $Ru(bipy)_3^{2+/+}$; 4, $TQ^{+/0}$; 5, $TQ^{2+/+}$; 6, $Fe(\eta^5\text{-}C_5Me_5)_2^{+/0}$; 7, $Fe(\eta^5\text{-}C_5H_5)_2^{+/0}$; 8, $TMPD^{2+/+}$; 9, $TMPD^{+/0}$; 10, $MV^{2+/+}$; and 11, $MV^{+/0}$. (Reproduced from Ref. 18.)

and distribution. Thus, the presence of surface states can alter the behavior of a semiconductor photoelectrode with respect to output efficiency vs. E_{redox} owing to a limitation on E_V. The degree of alteration in behavior depends on the surface state distribution. For n-type CdTe the value of E_V is fixed to ~0.5 V, a relatively small fraction of the band gap. For p-type InP (E_g = 1.3 eV) it would appear that the surface state distribution is such that the value of E_V is fixed to ~0.8 V, (11) a much larger fraction of the band gap. Understanding and controlling surface states is thus crucial to development of efficient semiconductor-based devices.

Even for ideal (surface-state free) semiconductors the behavior with respect to E_V vs. E_{redox} can be confusing. For the ideal p- or n-type semiconductor sufficiently negative or positive E_{redox}, respectively, will result in carrier inversion at the surface of the semiconductor, Schemes II and III.(14,19) In the region of E_{redox} where there is inversion the semiconductor photoelectrode can still effect uphill redox processes upon illumination and E_V can be independent of E_{redox}. The Fermi level, relative to the band edge positions, simply cannot be driven significantly more negative than E_{CB} or more positive than E_{VB}. Thus, for a range of E_{redox} the band edge positions vary with E_{redox} in a manner similar to when surface states between E_{CB} and E_{VB} pin the Fermi level. In the ideal case it is the high density of states from the valence oθ conduction band that eventually gives an E_V that is independent of E_{redox}.

When the observed value of E_V is a large fraction of E_g it is not easy to determine whether surface states pin E_f or whether a tailing density of valence or conduction band states effect pinning. Whenever the E_V exceeds $1/2E_g$ there is a measure of carrier inversion at the surface at dark equilibrium, but this does not mean that strong inversion is possible. Strong inversion can only occur when the region between E_{CB} and E_{VB} is sufficiently free of states that E_V can approach E_g. Generally, it is difficult to effect strong inversion when the semiconductor is in contact with any conductor, including liquid electrolyte solutions. The metal chalcogenides, MoS_2, $MoSe_2$, WS_2, and WSe_2, seem to be closest to ideal of the semiconductors studied, though there still seems to be a role for surface states.(15,16,20) A material such as n-type CdTe that gives a fixed E_V at $<1/2E_g$ must be one that is Fermi level pinned by states between E_{CB} and E_{VB}. Fermi level pinning would also appear to apply to photoelectrode materials such as $SrTiO_3$, (21) TiO_2, (21) InP,(11) GaAs, (22) and Si.(23)

Manipulating surface states of semiconductors for energy conversion applications is one problem area common to electronic devices as well. The problem of Fermi level pinning by surface states with GaAs, for example, raises difficulties in the development of field effect transistors that depend on the ability to move the Fermi level at the surface. Interface electronic states need to be understood in all semiconductor-based devices and will continue to be the object of study for the foreseeable future.

The conclusions from these considerations are that semiconductor photoelectrodes can be used to effect either reductions (p-type semiconductors) or oxidations (n-type semiconductors) in an uphill fashion. The extent to which reaction can be driven uphill, E_V, is no greater than E_g, but may be lower than E_g owing to surface states between E_{CB} and E_{VB} or to an inappropriate value of E_{redox}. Both E_g and E_{FB} are properties that depend on the semiconductor bulk and surface properties. Interestingly, E_V can be independent of E_{redox} meaning that the choice of E_{redox} and the associated redox reagents can be made on the basis of factors other than theoretical efficiency, for a given semiconductor. Thus, the important reduction processes represented by the half-reactions (3)-(5) could, in principle, be effected with the same efficiency at a Fermi level pinned (or

$$N_2 + 6H^+ + 6e^- \rightarrow 2NH_3 \quad (3)$$

$$CO_2 + 2H^+ + 2e^- \rightarrow HCOOH \quad (4)$$

$$2H^+ + 2e^- \rightarrow H_2 \quad (5)$$

carrier inverted) p-type semiconductor photocathode. However, as is usual in chemical systems, thermodynamics relates what is possible, but kinetics rule whether a thermodynamically spontaneous process will occur at a practically useful rate.

Each of the reduction processes represented in (3)-(5) has great potential practical significance if it could be done efficiently using solar energy. However, each is a multi-electron process having poor heterogeneous kinetics at illuminated p-type semiconductors. Thus, the ability to exploit the available driving force from illumination of a semiconductor will depend on improvements in heterogeneous kinetics for these and other multi-electron redox processes. Success in this particular area will have practical consequence in the future even if semiconductor-based photoelectrochemical devices fail to prove useful. Development of better fuel cells depends on the improvement of heterogeneous redox kinetics of multi-electron processes. If fusion works and provides inexpensive electricity, chemical fuel formation via electrolytic processes may be useful. Again, heterogeneous kientics must be improved. And clearly, heterogeneous catalytic chemistry will continue to be the key to efficient chemical production. The problem of poor heterogeneous kinetics for most fuel-forming reactions is thus one facet of a generic problem pervading much of chemistry.

One additional problem at semiconductor/liquid electrolyte interfaces is the redox decomposition of the semiconductor itself.(24) Upon illumination to create e^- - h^+ pairs, for example, all n-type semiconductor photoanodes are thermodynamically unstable with respect to anodic decomposition when immersed in the liquid electrolyte. This means that the oxidizing power of the photogenerated oxidizing equivalents (h^+'s) is sufficiently great that the semiconductor can be destroyed. This thermodynamic instability is obviously a practical concern for photoanodes, since the kinetics for the anodic decomposition are often quite good. Indeed, no non-oxide n-type semiconductor has been demonstrated to be capable of evolving O_2 from H_2O (without surface modification), the anodic decomposition always dominates as in equations (6) and (7) for

$$CdS + 2h^+ \longrightarrow Cd^{2+}_{(aq)} + S \quad (6)$$

$$Si + 4h^+ + 2H_2O \longrightarrow SiO_2 + 4H^+_{(aq)} \quad (7)$$

n-type CdS(25) and Si, (26) respectively. Protecting visible light responsive n-type semiconductors from photoanodic decomposition has been a major activity in the past half dozen years and will continue to be an issue of concern. The p-type semiconductors have not been plagued by gross durability problems, but thermodynamic instability can be a problem in some cases.(24,27) Interestingly, the redox decomposition processes of semiconductors are multi-electron processes that can be sufficiently slow that kinetic competition with desired redox processes can be successful to bring about sustained generation of energy-rich products or electricity from photoexcitation of a photoanode.

Suppression of Photocorrosion of Photoanodes and Manipulation of Kinetics for Anodic Processes

In 1976 the first sustained conversion of visible light to electricity using an n-type semiconductor-based cell (CdS (E_g = 2.4 eV) or CdSe (E_g = 1.7 eV)) was reported.(25,27) In the ensuing six years remarkable progress has been realized in this area. The key has been to find reducing reagents, A, that can capture photogenerated oxidizing equivalents, h^+, at a rate that precludes decomposition of the semiconductor, equation (8). The reverse process, equation (9), can then be effected at the

$$A + h^+ \longrightarrow A^+ \quad (8)$$

$$A^+ + e^- \longrightarrow A \quad (9)$$

counterelectrode to complete a chemical cycle involving no net chemical change, but yield significant efficiency for electricity generation. Now, a variety of n-type semiconductor/solvent/electrolyte/A^+/A/counterelectrode systems are known to comprise efficient, durable visible light energy to electrical energy.(6-9,15,25,27,28)

Often, there is a potential regime where the process represented by (8) is completely dominant compared to the anodic decomposition of the semiconductor. In some cases, e.g. CdS/S_n^{2-}, (29) $CdTe/Te_n^{2-}$, (29) and $MoSe_2/I_3^-$, (30) the redox species interact strongly with the electrode material resulting in changes in E_{FB}. Such strong interactions can be useful in protecting the semiconductor. In the examples cited above, E_{FB} shifts more negative reducing the tendency for anodic decomposition and opening a wider potential regime where the desired oxidation process can be effected without competition from anodic decomposition.

The unique interactions of a semiconductor with solution species, such as $MoSe_2$ with I_3^-, (30) are very likely the sorts of situations that will lead to the first applications of semiconductor photoelectrochemical devices for photochemical synthesis of redox products. Generally, the practical competition will come from conventional, including electrochemical, methods for producing redox reagents. When the semiconductor electrode has unique surface chemistry this can change the product distribution and in some cases it may be that the semiconductor may be the only surface at which a desired reaction will occur efficiently. However, even when a semiconductor is the electrode material of choice it is not clear that light would be used. The desirable interactions that exist for a p-type photocathode, for example, would likely exist as well for the oppositely doped, n-type, material. The reductions that require light at the p-type electrode can be effected in the dark at the n-type material. Also, degenerately doped semiconductors often behave well in the dark and are not blocking to any redox processes. However, the unique chemistry of semiconductor surfaces needs to be elucidated before a verdict can be reached regarding the practical consequences.

The strong interaction of the I_3^-/I^- redox system with the metal dichalcogenide materials was recently exploited (31) to bring about the visible light-driven process represented by equation (10). In 50% by weight H_2SO_4 the reaction as written

$$SO_2 + H_2O \xrightarrow[\sim 50\%\ H_2SO_4,\ \sim 10^{-3}\ \underline{M}\ I^-]{h\nu} H_2SO_4 + H_2 \qquad (10)$$

requires ~0.3 V or driving force, a good match to the photovoltage at the maximum power point for n-type WS_2-based cells for the oxidation of I^-.(15) But the oxidation of SO_2 to SO_4^{2-} in H_2SO_4 has poor kinetics (32) and does not compete with photoanodic decomposition of the electrode. Further, $E_{redox}(SO_4^{2-}/SO_2)$ is such that equation (1) predicts a very small photovoltage.(32) The I_3^-/I^-, though, interacts to shift E_{FB} of WS_2 favorably to give a good E_V with respect to $E_{redox}(SO^{4-}/SO_2)$ and simultaneously provides a mechanism for the oxidation of SO_2, since the I_3^- rapidly oxidizes SO_2 to SO_4^{2-}, Scheme IV. Thus, the I_3^-/I^- serves as a redox mediator and favorably alters the interface energetics to give a good E_V and a potential window of durability. The visible light-driven reaction represented by equation (10) is one of the most efficient optical to chemical energy conversions (up to ~14% from 632.8 nm light) known. The system illustrates the complex relationships that must be understood in order to efficiently drive multi-electron redox processes.

The suppression of photoanodic corrosion is not always difficult. For example, metal dichalcogenides are not durable in aqueous 0.1 M KCl; equation (11) represents the photoanodic

$$MoS_2 + 18h^+ + 8H_2O \longrightarrow 16H^+ + 2SO_4^{2-} + Mo^{6+} \quad (11)$$

decomposition process for MoS_2.(33) However, the oxidation of low concentrations of Cl^- occurs in CH_3CN solution with 100% current efficiency.(34) Even in aqueous solution the oxidation of Cl^- can be effected at sufficiently high Cl^- activity.(15,35) In aqueous 15 M LiCl the oxidation of Cl^- has 100% current efficiency. The high LiCl concentration yields very high Cl^- activity and lower activity of H_2O, both contributing to the improved durability of illuminated n-type MoS_2.(35) Thus, these experiments show, not surprisingly, (24) that the medium in contact with the semiconductor can alter the overall interfacial chemistry, even though Cl^- is available as a reductant in each case. The ability to effect the sustained generation of Cl_2 at an illuminated interface shows that potent oxidants can be made photochemically; Cl_2 is thermodynamically more potent, and kinetically more aggressive, than O_2.

The ability to manipulate the anodic corrosion problem using high concentrations of redox active electrolyte also makes possible the sustained oxidation of Br^- at illuminated metal dichalcogenide-based cells, Figure 1.(15) The use of high concentrations of electrolyte has proven valuable in situations involving other photoanode materials, notably n-type Si.(36,37)

Reducing photoanodic corrosion with high concentrations of redox active materials led to the conclusion that redox reagents covalently anchored to the photoelectrode might prove useful.(38) For example, research showed that n-type Si could be a durable photoanode in EtOH/0.1 M $[n\text{-}Bu_4N]ClO_4/Fe(\eta^5\text{-}C_5H_5)_2^{+/0}$ for the generation of electricity.(26) The $Fe(\eta^5\text{-}C_5H_5)_2$ is a fast,

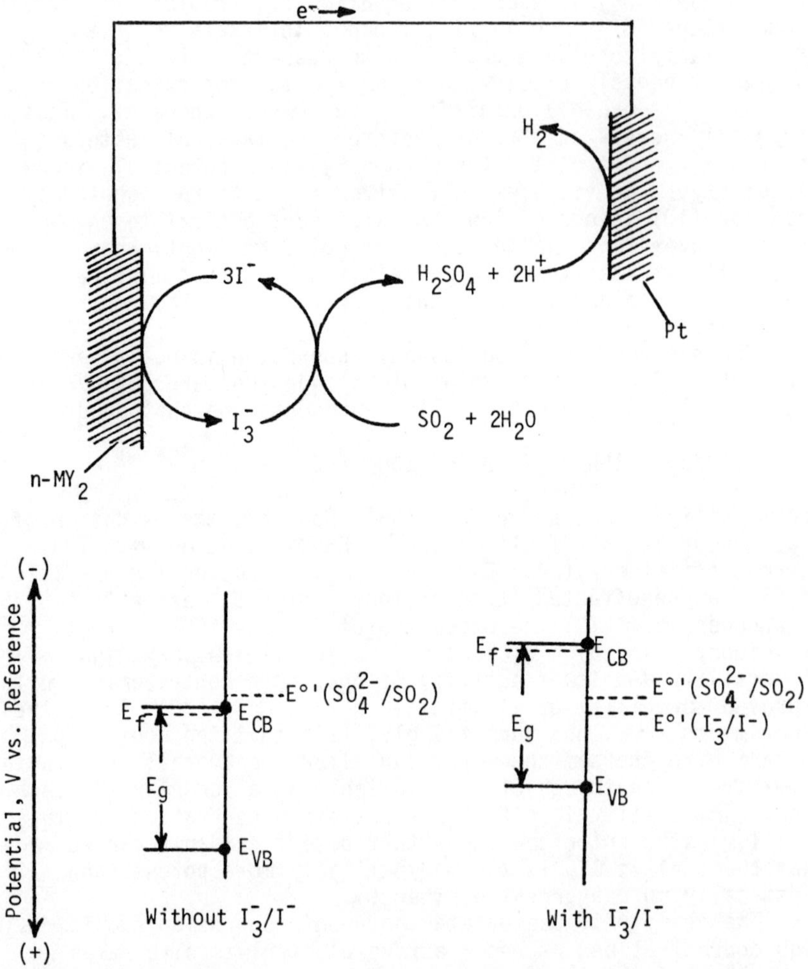

Scheme IV. Representation of the I_3^-/I^- mediated oxidation of SO_2 at illuminated metal dichalcogenide photoanodes (top) and interface energetics with and without I_3^-/I^- in 6 M H_2SO_4/1 M SO_2 for MoS_2 (bottom) (31). In the absence of the mediator system the photovoltage for the SO_2 oxidation is expected to be negligible. The adsorption of the I_3^-/I^- is unaffected by the SO_2 so the negative shift of the flat band potential can be exploited to give a photovoltage for the desired process with the I_3^-/I^- system simultaneously providing an acceleration of the SO_2 oxidation.

one-electron reductant that is durable and gives a durable oxidation product. Subsequent study showed that the reagent represented by I can be polymerized and attached to the surface

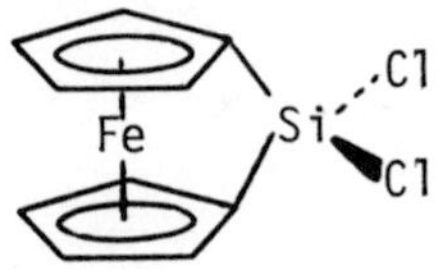

I

of n-type Si to protect it from photoanodic corrosion.(38,39,40) The important result is that once the electrode is made durable using the surface-confined redox system, $[A^+/A]_{surf.}$, then the photoelectrode can be used to sustain many oxidation processes, say $B \rightarrow B^+$, where B itself is not successful in competing with the anodic corrosion of the electrode. The ability to photooxidize B would then depend on the thermodynamics and kinetics for the interfacial process represented by equation (12)

$$[A^+]_{surf.} + B \rightarrow B^+ + [A]_{surf.} \qquad (12)$$

and not the kinetics for h^+ capture by B. It has been shown that n-type Si derivatized with I is capable of effecting the photoassisted oxidation of a variety of reagents in H_2O solvent, where the decomposition (7) is most severe.(40) In several cases it has been established that the process represented by (12) represents the dominant path for production of B^+. Thus, in principle, the molecular properties of the surface-confined reagent could be exploited to effect specific reactions. For the surface species derived from I the oxidizing power is fixed to ~+0.45 V vs. SCE, the formal potential, E°', of the surface reagent.(41) However, the $Fe(\eta^5\text{-}C_5H_5)_2^{+/0}$ systems are outer-sphere reagents and do not offer any basis for selectivity other than the potential. However, such species may prove useful in facilitating the redox reactions of biological reagents, *vide infra*.

Several groups have recently shown (36,42,43,44) that photoanode materials can be protected from photoanodic corrosion by an anodically formed film of "polypyrrole".(45) The work has been extended (46) to photoanode surfaces first treated with reagent II that covalently anchors initiation sites for the formation of polypyrrole. The result is a more adherent polypyrrole film that better protects n-type Si from photocorrosion. Unlike the material derived from polymerization of I, the anodically formed polypyrrole is an electronic conductor.(45) This may prove ultimately important in that the rate of ion transport of redox polymers may prove to be too slow

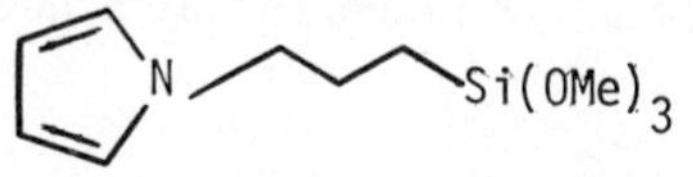

II

to be useful in attaining useful photocurrent densities. For the electronically conducting polymer the rate would not be limited by ion transport.

At this point it is evident that there are many approaches to the sustained conversion of visible light using photoanodes. The approaches based on strong interaction, e.g. $WS_2/I_3^-/I^-$,(15) or modified surfaces, e.g. derivatized n-type Si, (38,39,40) seem most interesting since the unique properties of the interface can be exploited at the molecular level. Significantly, the generation of potent oxidants such as Br_2 or Cl_2 can be effected using visible light and with reasonably good efficiency but without electrode deterioration.

Improving the Kinetics for Hydrogen Generation from P-Type Semiconductors

No naked semiconductor photocathode has been demonstrated to have good kinetics for the evolution of H_2, despite the fact that the position of E_{CB} in many cases has been demonstrated to be more negative than $E°'(H_2O/H_2)$. This means that electrons excited to the conduction band have the reducing power to effect H_2 evolution, but the kinetics are too poor to compete with $e^- - h^+$ recombination. The demonstration that N,N'-dimethyl-4,4'-bipyridinium, MV^{2+}, could be efficiently photoreduced at illuminated p-type Si to form MV^+ in aqueous solution under conditions where $E°'(MV^{2+/+}) = E°'(H_2O/H_2)$ when no H_2 evolution occurs establishes directly that the thermodynamics are good, but the kinetics are poor, for H_2 evolution.(23,47) The ability to efficiently reduce MV^{2+} to MV^+ at illuminated p-type Si led to studies of the surface derivatizing reagent III

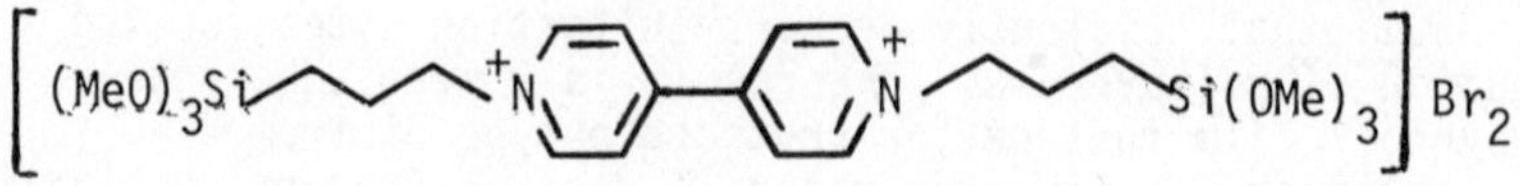

III

for use as an electron acceptor on photocathode surfaces.(48) Subsequent deposition of Pd(0) by electrochemical reduction of low concentrations of aqueous $PdCl_4^{2-}$ leads to the interface represented in Scheme V.(49) The Pd(0) is crucial to bring about equilibration of the surface-confined viologen reagent, $[(PQ^{2+/+})_n]_{surf.}$, with the H_2O/H_2 couple. The $[(PQ^{2+/+})_n]_{surf.}$/Pd(0) catalyst system can, in principle, be used on any photocathode surface to improve H_2 evolution kinetics.

The improvement in photoelectrochemical H_2 evolution efficiency using the $[(PQ^{2+/+})_n]_{surf.}$/Pd(0) system is reflected by the data in Figure 4. For the naked surface, H_2 evolution barely onsets at $E°'(H_2O/H_2)$. For the photoelectrode bearing $[(PQ^{2+/+})_n]_{surf.}$/Pd(0) the onset for H_2 evolution is up to ~500 mV more positive than $E°'(H_2O/H_2)$. The extent to which the onset is more positive than $E°'(H_2O/H_2)$ is E_V. It is obvious that the modified photoelectrode gives superior performance.

The mechanism of the catalysis for the p-Si/$[(PQ^{2+/+})_n]_{surf.}$/Pd(0) system is represented by equations (13) and (14). The $E°'[(PQ^{2+/+})_n]_{surf.} = -0.55 \pm 0.5$ V vs. SCE,

$$[(PQ^{2+})_n]_{surf.} \xrightarrow{ne^-} [(PQ^+)_n]_{surf.} \quad (13)$$

$$nH^+ + [(PQ^+)_n]_{surf.} \xrightarrow{Pd(0)} [(PQ^{2+})_n]_{surf.} + 1/2nH_2 \quad (14)$$

(50) independent of pH, whereas $E°'(H_2O/H_2)$ varies with pH. Thus, the catalyzed process represented by (14) is only downhill for sufficiently low pH; at the lower pH's the driving force is greater and the rate is faster. However, as the pH is lowered E_V becomes smaller. This leads to an optimum in overall energy conversion efficiency at pH ≃ 4. For monochromatic 632.8 nm light the efficiency for photoassisted H_2 evolution is up to ~5% whereas naked electrodes have negligible efficiency.(49,50)

The $[(PQ^{2+/+})_n]_{surf.}$ system can also be employed with Pt(0) (50) as the catalyst instead of Pd(0). Both Pt(0) and Pd(0) have excellent H_2 evolution kinetics. One virtue of Pd(0) is that it is much more easily detected by Auger electron spectroscopy than is Pt(0).(49) Auger electron spectra taken while sputtering the surface with Ar^+ ions have led to the establishment of interface structures like that represented in Scheme V. Typically, 10^{-8} mol/cm^2 of PQ^{2+} centers and ~10^{-8} mol/cm^2 of Pd(0) are used to give an overlayer of ~2000 Å in dimension. A key feature of the interface represented by Scheme V is that there is no Pd(0) at the p-type Si/SiO_x surface. This means that the only mechanism for equilibrating H_2O/H_2 with the photogenerated reducing equivalents is via the $[(PQ^{2+/+})_n]_{surf.}$ system. Studies have also been done with Pt(0) or Pd(0) dispersed throughout the polymer from III via the sequence represented by equations (15)

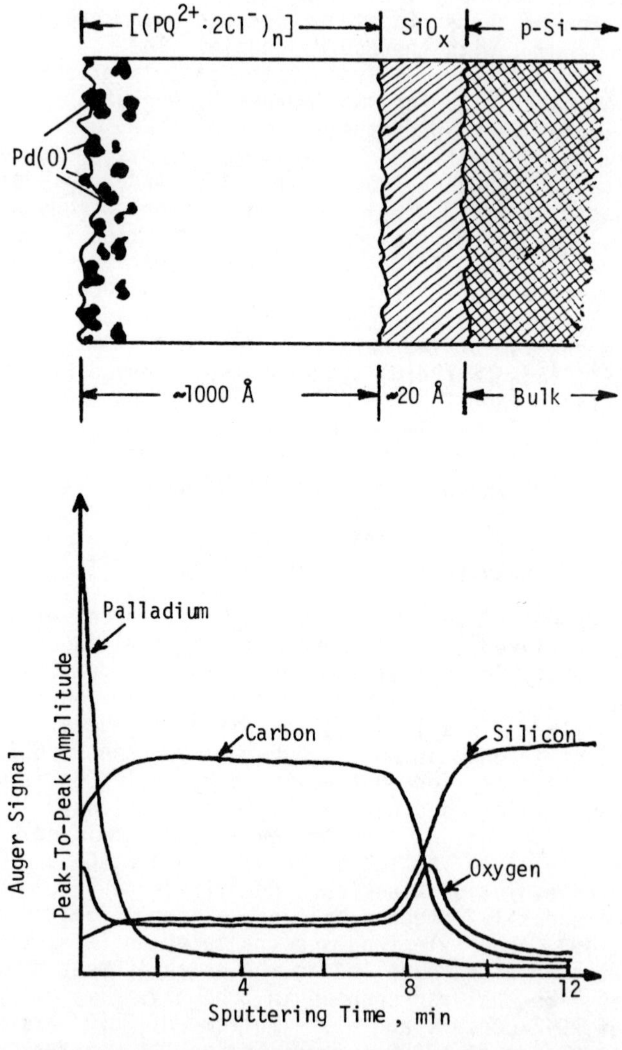

Scheme V. Representation of the catalytic p-type Si photocathode for H_2 evolution prepared by derivatizing the surface first with Reagent III followed by deposition of approximately an equimolar amount of Pd(0) by electrochemical deposition. The Auger/depth profile analysis for Pd, Si, C, and O is typical of such interfaces (49) for coverages of approximately 10^{-8} mol PQ^{2+}/cm^2.

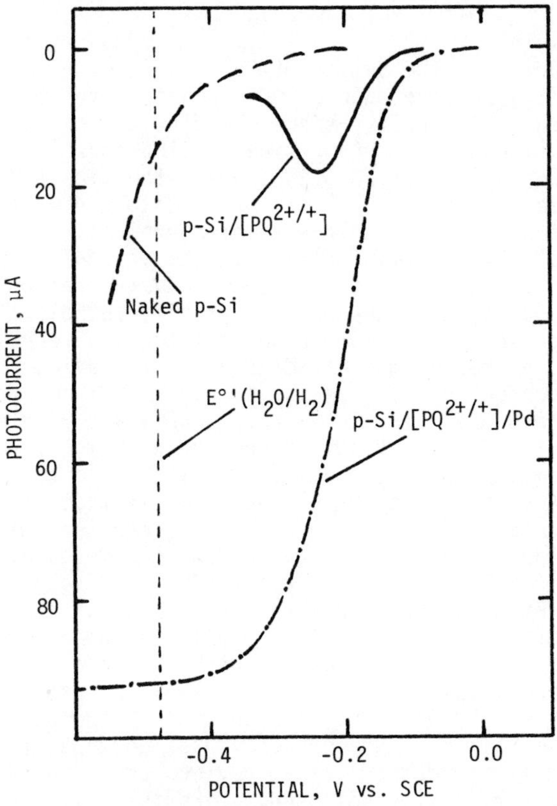

Figure 4. Comparison of photocathodic current (632.8 nm; ~6 mW/cm²) for naked p-type Si (– – –), p-type Si bearing $[(PQ^{2+})_n]_{surf}$ (——), and p-type Si bearing $[(PQ^{2+})_n]_{surf}/Pd(0)$ (– • – •) at pH = 4. The photocathodic current in the last case is associated with H_2 evolution that occurs more positive than $E^{0\prime}$ (H_2O/H_2). For current density multiply values shown by 10 cm^{-2}. The current peak for the smooth curve is associated with the uphill reduction $[(PQ^{2+})_n]_{surf} \rightarrow [(PQ^{+})_n]_{surf}$. (Reproduced from Ref. 49. Copyright 1982, American Chemical Society.)

$$[(PQ^{2+}\cdot 2Br^-)_n]_{surf.} + nPdCl_4^{2-} \longrightarrow [(PQ^{2+}\cdot PdCl_4^{2-})_n] + 2nBr^- \quad (15)$$

and (16) for Pd(0).(49) In such cases there is at least a small

$$[(PQ^{2+}\cdot PdCl_4^{2-})_n]_{surf.} \xrightarrow{\text{reduce}} \xrightarrow{\text{oxidize}} [(PQ^{2+}\cdot 2Cl^- . Pd(0)_n]_{surf.} \quad (16)$$

amount of Pd(0) at the p-type Si/SiO_x surface. The Pd(0) at the p-type Si/SiO_x surface should be responsible for some H_2 evolution current density, since the direct deposition of Pd(0) or Pt(0) (without III) onto p-type Si/SiO_x does improve H_2 evolution dramatically.(50,51) Indeed, the direct deposition of catalytic metals onto p-type InP has led to a demonstration of ~12% efficiency for the solar-assisted production of H_2.(52)

The ability to efficiently catalyze the H_2 evolution with a metal deposited onto the p-type semiconductor raises the legitimate question of why use III at all, let alone attempt to prepare and exploit the more ordered interface represented by Scheme V. In theory, the efficiency of all devices based on a given semiconductor would be the same. The direct deposition of Pt(0) or the use of $[(PQ^{2+/+})_n]_{surf.}$/Pd(0) are both ways of catalyzing the H_2 evolution. A problem with the redox polymers, as already mentioned, is that rate is likely to be ion transport limited. A problem with Pt(0) and Pd(0) is that they often give an ohmic contact, rather than a Schottky barrier, with p-type semiconductors. For example, in attempting to catalyze H_2 evolution from illuminated p-type WS_2 by electrochemically depositing Pt(0) or Pd(0) a large percentage of the electrodes give an ohmic contact.(16) This results in no photoeffects from the electrode. A uniform coating with the redox polymer from III gives a reproducible, photosensitive surface that can be used to generate H_2 via Pd(0) or Pt(0) deposited on the outermost surface. Neither of the approaches has led to sufficiently durable catalysts that practical devices are at hand; the Pt(0) or Pd(0) is very easily poisoned. Surprisingly, the $[(PQ^{2+/+})_n]_{surf.}$ polymer does not suffer deterioration on the timescale of loss of activity of the Pd(0) or Pt(0) catalyst.

The synthesis of catalytic photocathodes for H_2 evolution provides evidence that deliberate surface modification can significantly improve the overall efficiency. However, the synthesis of rugged, very active catalytic surfaces remains a challenge. The results so far establish that it is possible, by rational means, to synthesize a desired photosensitive interface and to prove the gross structure. Continued improvements in photoelectrochemical H_2 evolution efficiently can be expected, while new surface catalysts are needed for N_2 and CO_2 reduction processes.

A Role for Biological Redox Catalysts?

The enzymes hydrogenase, nitrogenase, and formate dehydrogenase can be used to equilibrate reducing reagents with H_2O/H_2, N_2NH_3, and $CO_2/HCOOH$, respectively.(53) In no case do the enzymes involve expensive noble metals as catalysts. Practical considerations aside, the multi-electron transfer catalysis effected by enzymes provides an existence proof for desired photoelectrode catalysts. One of the major difficulties is that large biological redox reagents are often unresponsive at electrode surfaces. For a variety of reasons the heterogeneous electron transfer kinetics for large biological reagents are poor. However, small redox reagents dissolved in solution do equilibrate rapidly with the large biological reagents.(54) Interestingly, MV^+, for example, will effect reduction of H_2O, N_2, or CO_2 when the proper enzyme is present as a catalyst.(53) The use of surface-confined, fast, one-electron, outer-sphere redox reagents like those derived from I or III as redox mediators for biological reagents would seem to represent an excellent approach to the equilibration of the electrode with the biological reagents.

Experiments relating to the oxidation and reduction of ferro- and ferricytochrome c, cyt c(red) and cyt $c_{(ox)}$ from horse heart, establish that photoelectrodes derivatized with molecular reagents can give significantly improved response to biological redox reagents.(55,56) The cyt c provides an example of a readily accessible biomolecule that generally has poor kinetics at electrode surfaces.(57) The first experiments concerned electrodes derivatized with III.(55) The fact that MV^+ rapidly reduces cyt $c_{(ox)}$ to cyt $c_{(red)}$ led to the use of the $[(PQ^{2+/+})_n]_{surf.}$ for this purpose.(58) The E°'(cyt $c_{(ox)}$/cyt $c_{(red)}$) is ~+0.02 V vs. SCE (59) and the reaction represented by equation (17) is thus downhill by ~0.5 V. Interestingly, the

$$[(PQ^+)_n]_{surf.} + n \text{ cyt } c_{(ox)} \longrightarrow [(PQ^{2+})_n]_{surf.} + n \text{ cyt } c_{(red)} \quad (17)$$

process represented by (17) was shown to account for the reduction of cyt c at illuminated p-type Si functionalized with III.(55) The reduction of cyt $c_{(ox)}$ at electrodes derivatized with III is mass transport limited and independent of coverage of PQ^{2+} centers on the electrode from ~10^{-10} to 10^{-8} mol/cm^2. Naked electrodes do not respond to the cyt $c_{(ox)}$ in the same potential range. Inasmuch as adsorption of cyt c, or impurities contained in it, onto most electrodes leads to overall poor kinetics, it is particularly noteworthy that high concentrations of cyt $c_{(ox)}$ can be reduced with good kinetics via the $[(PQ^+)_n]_{surf.}$. Thus, the modification of electrode surfaces with III brings about improvement in response with respect to reduction of cyt $c_{(ox)}$.

While experiments with the cyt $\underline{c}_{(ox)}$ at $[(PQ^{2+/+})_n]_{surf.}$ do establish a point, the disparity E°' of the reagent and the mediator precludes the claims that electrodes can in fact equilibrate with the biological reagent. The synthesis of IV has

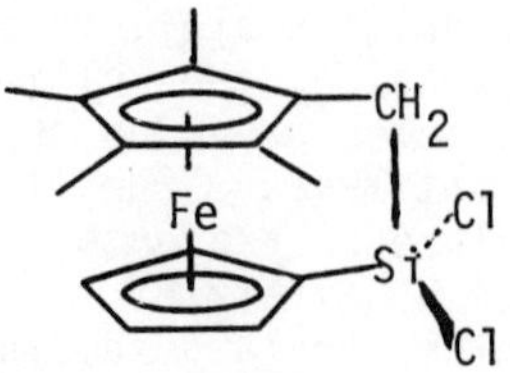

IV

led to the demonstration that derivatized electrodes can be equilibrated with biological redox reagents.(56) Representing the surface species from IV by $[PMFc^{+/0}]_{surf.}$, the E°'$[PMFc^{+/0}]_{surf.}$ = +0.04 V vs. SCE, that is very close to the E°' for cyt c. Conventional Pt electrodes derivatized with IV can be used to oxidize cyt $\underline{c}_{(red)}$, or reduce cyt $\underline{c}_{(ox)}$ near the E°' of cyt c, via the equilibrium process represented by equation (18). Importantly, n-type Si electrodes derivatized with IV can

$$[PMFc^{+}]_{surf.} + \text{cyt } \underline{c}_{(red)} \rightleftarrows [PMFc^{0}]_{surf.} + \text{cyt } \underline{c}_{(ox)} \quad (18)$$

be used in aqueous electrolyte solution at pH = 7 and the process represented by equation (19) can be effected in an uphill sense,

$$[PMFc^{0}]_{surf.} + h^{+} \longrightarrow [PMFc^{+}]_{surf.} \quad (19)$$

$E_V \approx 300$ mV. As on the Pt surface, the $[PMFc^{+}]_{surf.}$ on n-type Si is capable of effecting the oxidation of cyt $\underline{c}_{(red)}$. Thus, the n-type Si/$[PMFc^{+/0}]_{surf.}$ electrode can be used to effect the uphill oxidation of cyt $\underline{c}_{(red)}$. The rate constant for reaction represented by equation (20) is $>7 \times 10^3$ $\underline{M}^{-1}s^{-1}$. The observed

$$[PMFc^{+}]_{surf.} + \text{cyt } \underline{c}_{(red)} \longrightarrow [PMFc^{0}]_{surf.} + \text{cyt } \underline{c}_{(ox)} \quad (20)$$

heterogeneous electron rate constant is $>1 \times 10^{-4}$ cm/s for a variety of electrodes independently prepared, representing substantial improvement compared to naked electrodes that give negligible rates under the same conditions.(56) Significantly, as for $[(PQ^{2+/+})_n]_{surf.}$, the $[PMFc^{+/0}]_{surf.}$ is useful at high concentrations of cyt c. Very pure, low concentration cyt c apparently responds well at conventional electrodes, but small amounts of decomposition or impurities cause severe problems from adsorption.(60) The surface reagents from III or IV apparently minimize the adsorption problems, while providing a mechanism for exchanging electrons with the electrode.(55,56) Preliminary results have established that hydrogenase can equilibrate with the polymer derived from III, suggesting that the one-electron polymers from III and IV can in fact come into redox equilibrium with multi-electron transfer catalysts for reactions of possible importance in energy conversion.

Large Area Photosensitive Materials

The studies described so far have concerned relatively small, ~0.1-1 cm^2, single-crystal photoelectrode materials. Promising results have been obtained in that there are a variety of durable, efficient paths to generation of high energy chemicals or electricity. However, single-crystal photoelectrode materials are likely to remain too expensive for significant practical development. The question is whether the basic results from single-crystal systems can be applied to large area photosensitive materials not necessarily fabricated from single crystals. In this area as well some promising results have been obtained. Thin film and/or polycrystalline GaAS (61) and CdX (62) photoanodes have been shown to have relatively good efficiency compared to their single-crystal analogues.

Recently, results from amorphous hydrogenated silicon, a-Si:H, E_g = 1.7 eV, obtained in a glow discharge of SiH_4 show significant promise. Solid state devices for solar to electrical energy conversion based on absorption of light by a-Si:H have been shown to have almost 10% efficiency (63) and it is believed that a-Si:H can be produced inexpensively and uniformly in large areas.

In principle, intrinsic photoconductors such as a-Si:H can be good photoelectrodes.(5,64) Scheme VI shows the approximate interface situation for a recently reported a-Si:H-based cell for the generation of electricity.(64) Critically, the intrinsic thin film (1-4 μ) of a-Si:H was deposited onto a very thin (~200 Å) heavily n-doped a-Si:H layer on stainless steel in order to assure that the Fermi level contacts the bottom of the conduction band. This means that E_{redox} positions of E_{CB} will result in a field across the photoconductor such that photogenerated h^+'s will be driven toward the electrolyte/redox couple solution. For E_{redox} close to E_{VB} the E_V would be expected to approach E_g as for an n-type semiconductor photoanode. As for any other photoanode the a-Si:H is susceptible to photoanodic decomposition, but the corrosion can be completely suppressed by using the 0.1 M $[n\text{-}Bu_4N]ClO_4$/EtOH/$Fe(\eta^5\text{-}C_5H_5)_2{}^{+/0}$ electrolyte/redox couple solution, Figure 5. Interestingly, the sustained conversion of 632.8 nm light is just as efficient for the intrinsic a-Si:H photoanode as for single-crystal n-type Si electrodes under the same conditions.(5)

The surface of a-Si:H can also be derivatized with reagent I and the E_V is ~750 mV compared to $E_V \approx$ 500 mV on single crystal

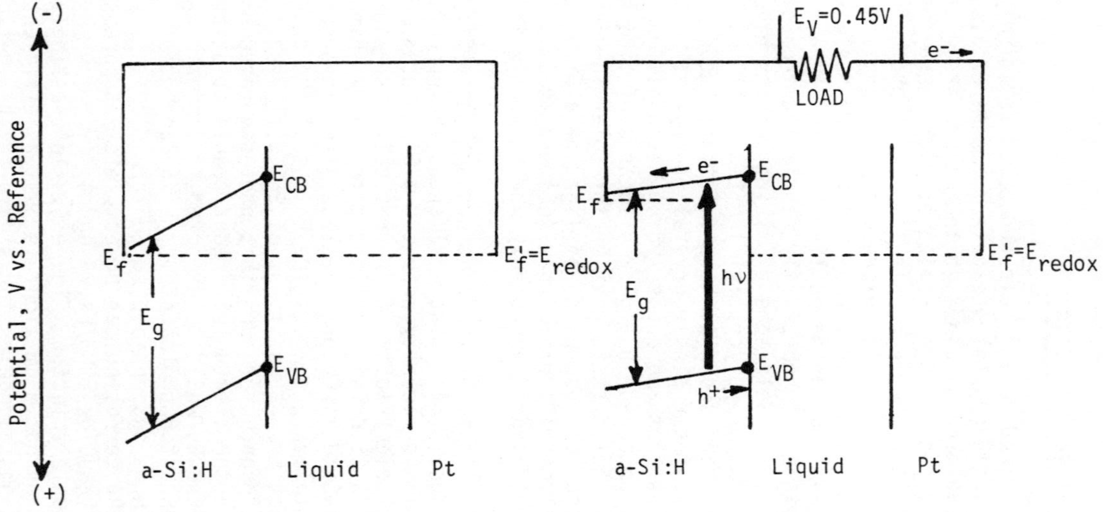

Scheme VI. Representation of the interface energetics for intrinsic a–Si:H at short circuit, dark equilibrium with ferricenium/ferrocene in EtOH, electrolyte solution (left) and under illumination with 632.8-nm light with a load in series in the external circuit (right). The diagrams are adapted from data in Reference 65 for intrinsic a–Si:H (1–4-μ thick) on stainless steel first coated with heavily n-doped a–Si:H (200-Å thick) to ensure an ohmic contact near the bottom of the conduction band. In typical experiments E_{redox} *= 0.4 V vs. SCE.*

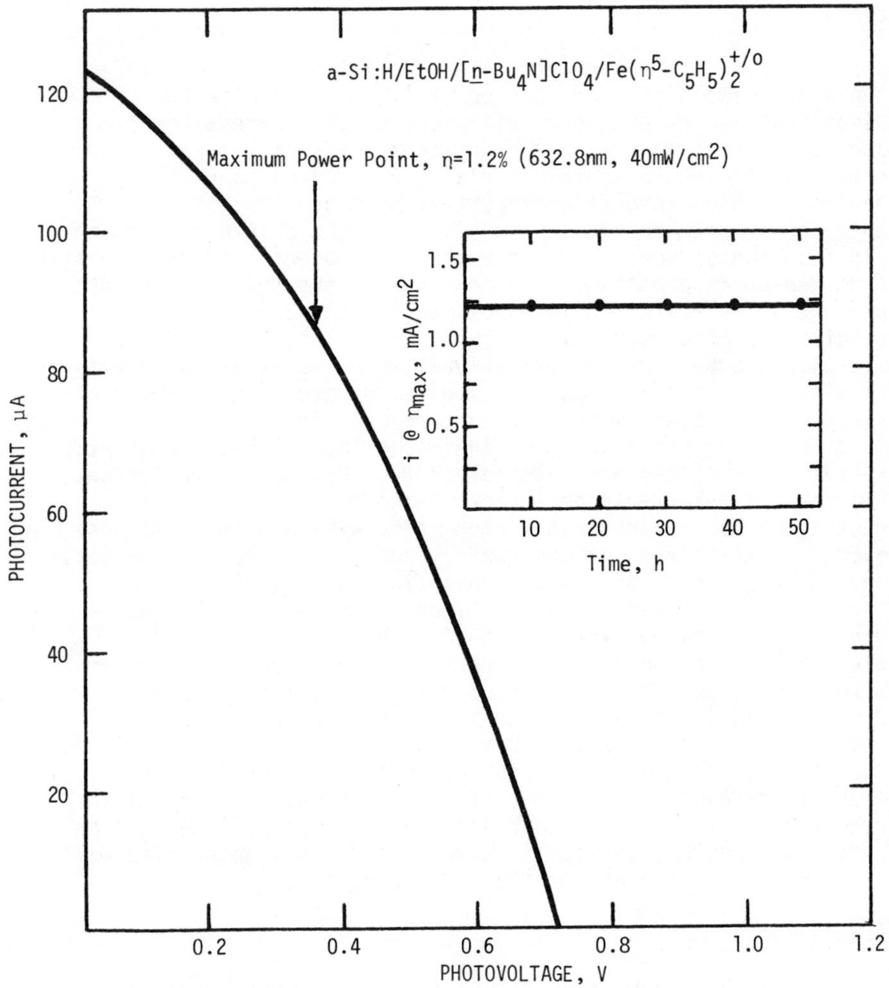

Figure 5. Output characteristics and photocurrent density at maximum power point against time (inset) for an intrinsic a-Si:H photoanode-based cell. (Reproduced from Ref. 5.)

n-type Si.(5) This result suggests that a-Si:H could be durable in aqueous solutions via protection by the surface reagent. As for n-type Si this would allow the use of large area, efficient a-Si:H to effect a variety of light-driven oxidation processes.

Summary

Semiconductor-based photoelectrochemical cells can effect the sustained, direct conversion of light to chemical or electrical energy with good efficiency. There are several approaches to suppressing the photocorrosion of n-type semiconductor photoanodic materials, all depending on the manipulation of interface properties such as structure, energetics, and kinetics. The visible light-driven generation of Cl_2 from photoanodes represents the most potent oxidant generated from non-oxide electrodes. Output parameters depend on surface properties as reflected in experiments with reducing vs. oxidizing pretreatments for n-type CdTe. Interfacial redox kinetics can be modified by rational means as illustrated with results for photocathodes modified to improve H_2 evolution kinetics. However, much more work remains to be done on multi-electron processes to bring about improvements in kinetics. Certain enzymes may prove useful in N_2, CO_2, or H_2O reduction. Progress in relatively efficient, large area, inexpensive photoelectrode materials has been made, with a-Si:H being one example. At this point the performance of interfacial inorganic chemistry systems for energy conversion is sufficiently good that they cannot be ruled out as contenders for large scale energy generation. The near-term, pre-2000 charter is to fully elaborate the basic science underlying the interfacial systems with a conscious effort directed toward efficient (>10%), durable, and inexpensive systems for the direct production of energy-rich redox products from abundant, inexpensive resources such as H_2O and CO_2. It is too early to focus on fuel vs. electricity generation as the ultimate objective. Many of the requirements for both outputs are the same but fuel generation poses the greatest challenge, since useful fuel generation will require new multi-electron transfer catalysts.

Acknowledgments

Research support from the United States Department of Energy, Office of Basic Energy Sciences, Division of Chemical Sciences is gratefully acknowledged. Support from GTE Laboratories, Inc. and Dow Chemical Company, U.S.A. for aspects of this work is also acknowledged. Partial support from the Office of Naval Research for work on the surface chemistry of CdTe is appreciated.

Literature Cited

1. Wrighton, M.S. Chem. Eng. News, 1979, 57, Sept. 3, p. 29.
2. (a) Wrighton, M.S. Acc. Chem. Rev., 1979, 12, 303; (b) Nozik, A.J. Ann. Rev. Phys. Chem., 1978, 29, 189.
3. Gerischer, H. J. Electroanal. Chem., 1975, 68, 263 and in "Physical Chemistry: An Advanced Treatise", Eyring, H., Henderson, D., and Jost, W., eds., Academic Press: New York, 1970, Vol. 9A, Chapter 5.
4. Bard, A.J. Science, 1980, 207, 139 and J. Phys. Chem., 1982, 86, 172.
5. (a) Rose, A. Phys. Stat. Sol., 1979, 56, 11, (b) Calabrese, G.S.; Lin, M.S.; Dresner, J.; Wrighton, M.S. J. Am. Chem. Soc., 1982, 104, 2412.
6. (a) Heller, A.; Parkinson, B.A.; Miller, B. Appl. Phys. Lett., 1978, 33, 521; (b) Heller, A.; Lewerenz, H.J.; Miller, B. Ber. Bunsenges. Phys. Chem., 1980, 84, 592.
7. Ellis, A.B.; Bolts, J.M.; Kaiser, S.W.; Wrighton, M.S. J. Am. Chem. Soc., 1977, 99, 2848.
8. Kline, G.; Kam, K.; Canfield, D.; Parkinson, B.A. Solar Energy Mtls., 1981, 4, 301.
9. Fan, F.R.F.; White, H.S.; Wheeler, H.S.; Bard, A.J. J. Electrochem. Soc., 1980, 127, 518.
10. Heller, A.; Miller, B.; Lewerenz, H.J.; Bachman, K.J. J. Am. Chem. Soc., 1980, 102, 6555.
11. Dominey, R.N.; Lewis, N.S.; Wrighton, M.S. J. Am. Chem. Soc., 1981, 103, 1261.
12. "Solar Photovoltaic Energy Conversion", Ehrenreich, H., ed., The American Physical Society, 335 East 45th Street, New York, New York 10017, 1979.
13. Jaeger, C.D.; Gerischer, H.; Kautek, W. Ber. Bunsenges. Phys. Chem., 1982, 86, 20.
14. Sze, S.N., "Physics of Semiconductor Devices", Wiley: New York, 1969.
15. Baglio, J.A.; Calabrese, G.S.; Kamieniecki, E.; Kershaw, R.; Kubiak, C.P.; Ricco, A.J.; Wold, A.; Wrighton, M.S.; Zoski, G.D. J. Electrochem. Soc., 1982, 129, 0000.
16. Baglio, J.A.; Calabrese, G.S.; Harrison, D.J.; Kamieniecki, E.; Ricco, A.J.; Wrighton, M.S.; Zoski, G.D., submitted for publication.
17. Bard, A.J.; Bocarsly, A.B.; Fan, F.R.F.; Walton, E.G.; Wrighton, M.S. J. Am. Chem. Soc., 1980, 102, 3671.
18. (a) Tanaka, S.; Bruce, J.A.; Wrighton, M.S. J. Phys. Chem., 1981, 85, 3778; (b) Aruchamy, A.; Wrighton, M.S. J. Phys. Chem., 1980, 84, 2848.

19. (a) Turner, J.A.; Manassen, J.; Nozik, A.J. Appl. Phys. Lett., 1980, 37, 488 and ACS Symposium Ser., 1981, 146, 253;
(b) Kautek, W.; Gerischer, H. Ber. Bunsenges. Phys. Chem., 1980, 84, 645;
(c) Nagasubramanian, G.; Bard, A.J. J. Electrochem. Soc., 1981, 128, 1055.
20. (a) Schneemeyer, L.F.; Wrighton, M.S. J. Am. Chem. Soc., 1980, 102, 6964;
(b) White, H.S.; Fan, F.R.F.; Bard, A.J. J. Electrochem. Soc., 1981, 128, 1045.
21. Lin, M.S.; Hung, N.; Wrighton, M.S. J. Electroanal. Chem., 1982, 135, 122.
22. Fan, F.R.F.; Bard, A.J. J. Am. Chem. Soc., 1980, 102, 3677.
23. Bocarsly, A.B.; Bookbinder, D.C.; Dominey, R.N.; Lewis, N.S.; Wrighton, M.S. J. Am. Chem. Soc., 1980, 102, 3683.
24. Bard, A.J.; Wrighton, M.S. J. Electrochem. Soc., 1977, 124, 1706.
25. Ellis, A.B.; Kaiser, S.W.; Wrighton, M.S. J. Am. Chem. Soc., 1976, 98, 1635.
26. Legg, K.D.; Ellis, A.B.; Bolts, J.M.; Wrighton, M.S. Proc. Natl. Acad. Sci. USA, 1977, 74, 4116.
27. (a) Gerischer, H. J. Electroanal. Chem., 1977, 82, 133;
(b) Park, S.M.; Barber, M.E. J. Electroanal. Chem., 1979, 99, 67.
28. Wrighton, M.S. in "Chemistry in Energy Production", Wymer, R.G. and Keller, O.L., eds., National Technical Information Service, U.S. Depart-ment of Commerce, 5285 Port Royal Road, Springfield, Virginia 22161.
29. Ellis, A.B.; Kaiser, S.W.; Bolts, J.M.; Wrighton, M.S. J. Am. Chem. Soc., 1977, 99, 2839.
30. (a) Tributsch, H. J. Electrochem. Soc., 1978, 125, 1086;
(b) Gobrecht, J.; Tributsch, H.; Gerischer, H. J. Electrochem. Soc., 1978, 125, 2085.
31. Calabrese, G.S.; Wrighton, M.S. J. Am. Chem. Soc., 1981, 103, 6273.
32. Lu, P.W.T.; Ammon, R.L. J. Electrochem. Soc., 1980, 127, 2610.
33. Tributsch, H.; Bennet, J.C. J. Electroanal. Chem., 1977, 81, 97.
34. Schneemeyer, L.F.; Wrighton, M.S. J. Am. Chem. Soc., 1979, 101, 6496.
35. Kubiak, C.P.; Schneemeyer, L.F.; Wrighton, M.S. J. Am. Chem. Soc., 1980, 102, 6898.
36. Fan, F.R.F.; Wheeler, B.L.; Bard, A.J. J. Electrochem. Soc., 1981, 128, 2042.

37. Results for platinum silicide/Si photoanodes appear very promising in combination with high concentration electrolytes; Bard, A.J., private communication.
38. Wrighton, M.S.; Austin, R.G.; Bocarsly, A.B.; Bolts, J.M.; Haas, O.; Legg, K.D.; Nadjo, L.; Palazzotto, M.C. J. Am. Chem. Soc., 1978, 100, 1602.
39. Bolts, J.M.; Bocarsly, A.B.; Palazzotto, M.C.; Walton, E.G.; Lewis, N.S.; Wrighton, M.S. J. Am. Chem. Soc., 1977, 101, 1378.
40. Bocarsly, A.B.; Walton, E.G.; Wrighton, M.S. J. Am. Chem. Soc., 1980, 102, 3390.
41. Wrighton, M.S.; Palazzotto, M.C.; Bocarsly, A.B.; Bolts, J.M.; Fischer, A.B.; Nadjo, L. J. Am. Chem. Soc., 1978, 100, 7264.
42. Noufi, R.; Tench, D.; Warren, L.F. J. Electrochem. Soc., 1980, 127, 2310 and 1981, 128, 2596.
43. Skotheim, T.; Lundstrom, I.; Prejza, J. J. Electrochem. Soc., 1981, 128, 1625.
44. Noufi, R.; Frank, A.J.; Nozik, A.J. J. Am. Chem. Soc., 1981, 103, 1849.
45. (a) Kanazawa, K.K.; Diaz, A.F.; Geiss, R.H.; Gill, W.D.; Kwak, J.F.; Logan, J.A.; Rabolt, J.; Street, G.B. J. Chem. Soc., Chem. Commun., 1979, 854;
(b) Diaz, A.F.; Castillo, J.J. J. Chem. Soc., Chem. Commun., 1980, 397;
(c) Kanazawa, K.K.; Diaz, A.F.; Gill, W.D.; Grant, P.M.; Street, G.B.; Gardini, G.P.L. Kwak, J.F. Synth. Met., 1979/1980, 2, 329;
(d) Diaz, A.F.; Vasquez Vallejo, J.M.; Martinez Duran, A. IBM J. Res. Dev., 1981, 25, 42.
46. Simon, R.A.; Ricco, A.J.; Wrighton, M.S.; J. Am. Chem. Soc., 1982, 104, 2031.
47. Bookbinder, D.C.; Lewis, N.S.; Bradley, M.G.; Bocarsly, A.B.; Wrighton, M.S. J. Am. Chem. Soc., 1979, 101, 7721.
48. Bookbinder, D.C.; Wrighton, M.S., J. Am. Chem. Soc., 1980, 103, 5123.
49. Bruce, J.S.; Murahashi, T.; Wrighton, M.S. J. Phys. Chem., 1982, 86, 1552.
50. Dominey, R.N.; Lewis, N.S.; Bruce, J.A.; Bookbinder, D.C.; Wrighton, M.S. J. Am. Chem. Soc., 1982, 104, 467.
51. Nakato, Y.; Abe, K.; Tsubomura, H. Ber. Bunsenges. Phys. Chem., 1976, 80, 1002.
52. Heller, A.; Vadimsky, R.G. Phys. Rev. Lett., 1981, 46, 1153.
53. Summers, L.S., "The Bipyridinium Herbicides", Academic Press: London, 1980, pp. 122-124.
54. (a) Szentrimay, R.; Yeh, P.; Kuwana, T., ACS Symposium Series, 1977, 38, 143;

(b) Heineman, W.R.; Meckstroth, M.L.; Norris, B.J.; Su. C.-H. Bioelectrochem. Bioenerg., 1979, 6, 577;
(c) Kuwana, T.; Heiniman, W.R. Acc. Chem. Res., 1976, 9, 241.
55. Lewis, N.S.; Wrighton, M.S. Science, 1981, 241, 944.
56. Chao, S.; Robbins, J.L.; Wrighton, M.S. J. Am. Chem. Soc., 1982, 104, 0000.
57. Margoliash, E.; Schejtor, A., in "Advances in Protein Chemistry", Anfinsen, C.B.; Anson, M.L.; Edsall, J.T.; Richards, F.M., eds, Academic Press: New York, 1969, vol. 21, chapter 2.
58. Land, E.J.; Swallow, A.J. Ber. Bunsenges. Phys. Chem., 1975, 79, 436.
59. Margalit, R.; Schejter, A. Eur. J. Biochem., 1973, 32, 492.
60. Bowden, E.F.; Hawkridge, F.M.; Chlebowski, J.F.; Bancroft, E.E.; Thorpe, C.; Blount, H.N., submitted for publication and private communication.
61. Heller, A.; Miller, B.I.; Chu, S.S.; Lee, Y.T. J. Am. Chem. Soc., 1979, 101. 7632.
62. Hodes, G. Nature, 1980, 285, 29.
63. Dresner, J., private communication.
64. Cahen, D.; Hodes, G.; Manassen, J.; Vainas, B.; Gibson, R.A.G., J. Electrochem. Soc., 1980, 127, 1209.

RECEIVED August 9, 1982

Discussion

C.H. Langford, Concordia University: Dr. Wrighton's paper maintains appropriate discretion with respect to the twenty-first century. As Neils Bohr said, "Prediction is difficult, especially about the future". However, he does call attention to a number of problems in solid state and interfacial chemistry which are certainly important now and would not have been likely to be picked out for emphasis in a meeting of inorganic chemists held ten years ago and oriented to the theme of inorganic chemistry in 1990. This makes me reflect on the one aspect of the activities of the majority of the inorganic chemists present where we are required, like it or not, to predict the future. I refer, of course, to the teaching function. In teaching, we make choices designed to prepare the next generation of chemists. Typically, the impact of these choices can be expected to endure over 40 years. This means that consciously or unconsciously we engage in predictive activity.

After reading the preprint of Wrighton's stimulating paper, I did a quick and incomplete survey of the inorganic chemistry textbooks on my shelf. Of five that I examined, only one had a serious treatment of semiconductors and one seemed to me to develop a "Materials" perspective as a subtheme or sidelight. I am troubled by this. Surely, the materials that do and will support the development of electronic materials and novel catalysts offer a fruitful field for research by inorganic chemists and represent a field which can benefit from the participation of inorganic chemists. We should tell our students.

On another point, Wrighton raises the question of electronic versus ionic conductance in films and offers the thought that only the first will be sufficiently fast for effective device development. I think that there is a particular subspecies of electronic conductance that is likely to dominate the behavior of chromophore films involving molecular units. It is a "hopping" or polaron mechanism which is closely related to the homogeneous electron transfer process which has been extensively studied by inorganic chemists. I'd like to raise the question of the limit on the rate of this mechanism where diffusion together of the partners is not required. Here, we may find the test of the existance and significance of the elusive "inverted region".

M.S. Wrighton: I concur with your observation that students in inorganic chemistry receive relatively little formal instruction in materials science. This will likely change as the inorganic chemists begin to exert their influence on the practical aspects of materials-based devices.

Regarding the question of the rate of electron transport through polymer films, it is not yet clear what ultimate rate can be achieved. In solar energy applications the important issue is whether the rate can be high enough so that the net electron transfer rate is light intensity limited.

5

Integrated Chemical Systems: *n*-Silicon/Silicide/Catalyst Systems

ALLEN J. BARD, FU-REN F. FAN, G. A. HOPE, and R. G. KEIL

The University of Texas, Department of Chemistry, Austin, TX 78712

The paper by Wrighton describes semiconductor systems which incorporate other components, such as polymer layers, to produce useful electrode structures. The development of such multicomponent, multiphase systems, which we call "integrated chemical systems" by analogy to the integrated circuits used in semiconductor devices, clearly represents an important new trend in chemistry. The design of useful semiconductor electrodes and powders will require surface modification to passivate surface states to improve efficiency, to protect the surface from photodecomposition, to catalyze desired reactions and to provide sensitizers. For example the system p-GaAs/viologen polymer/Pt can be used for the photodriven evolution of hydrogen. The development of such systems will probably require the application of techniques very different from those normally used in chemical synthesis, e.g., molecular beam epitaxy, sputtering, ion implantation, spin coating, and other methods borrowed from solid state physics and semiconductor technology. These integrated chemical systems will have properties different and, we hope, more useful, than that of the individual components. Such synergistic effects are well-known in biological systems where the overall behavior of the complex structure is usually more than the simple sum of the parts.

I would like to describe briefly an integrated chemical system recently under investigation in our laboratory based on n-type silicon which illustrates some of the above features (1,2). As Wrighton points out in his paper, n-Si electrodes are usually unstable in aqueous solutions, because they tend to form a passivating oxide film under irradiation. We have found that by forming a platinum silicide layer on the surface of the Si electrode (by flash evaporation of Pt on the pretreated Si surface followed by annealing) the electrode will show stable operation in aqueous photoelectrochemical (PEC) cells. The current-voltage curves for an n-Si electrode (coated with 40 angstroms of Pt and annealed at 400° C in vacuum for 10 minutes) in a solution of 1 M $FeCl_2$, 0.1 M

0097-6156/83/0211-0093$06.00/0

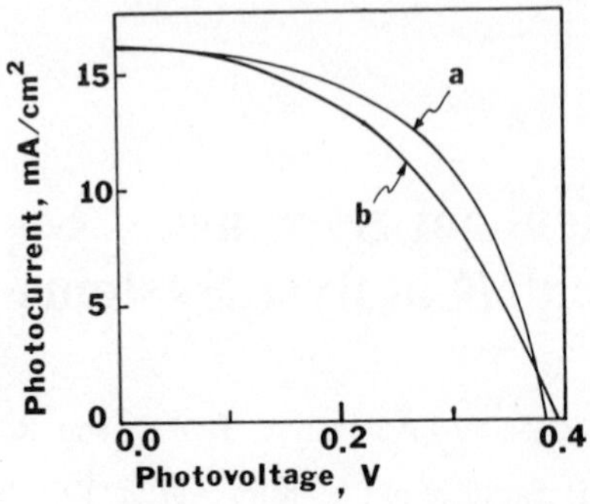

Figure 1. Photocurrent–photovoltage characteristics of the cell n-Si (Pt silicide coated)/1.0 M *$FeCl_2$, 0.1* M *$FeCl_3$, 1* M *HCl/Pt at 65 nW/cm² illumination. Pt thickness deposited ~ 40 Å and annealing temperature 400 °C at ~ 10^{-6} torr for 10 min. Key: a, before long-term stability test; and b, after long-term stability test.*

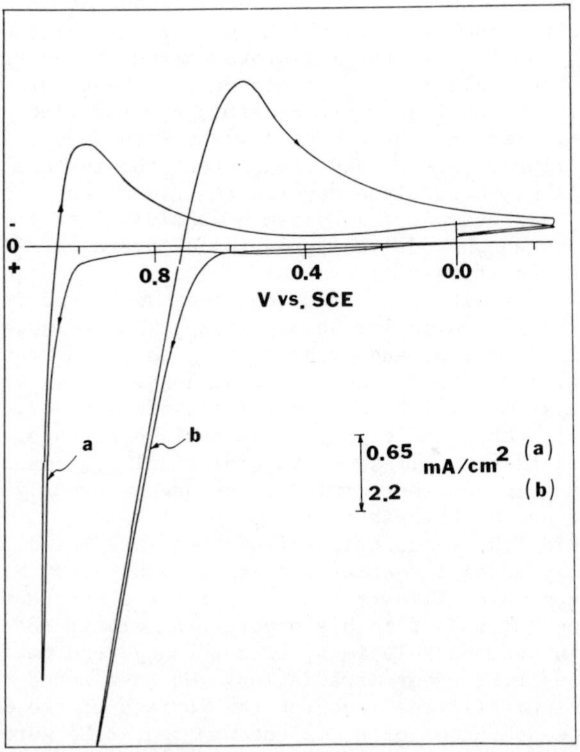

Figure 2. Voltammetric curves of Pt (curve a) and n-Si(Ir)/RuO_2 (curve b) electrode in 11 M *LiCl at pH = 7. Light intensity 65 nW/cm², and scan rate 100 mV/s. (Reproduced with permission from Ref. 1. Copyright 1982, The Electrochemical Society, Inc.)*

$FeCl_3$, and 1 M HCl under illumination of 65 mW/cm^2 (tungsten-halogen lamp) are shown in Figure 1. This cell has operated for more than 20 days with no appreciable change in performance or decomposition at a maximum power conversion efficiency of about 6%. To stabilize such an electrode to oxidants stronger than iron(III) a catalyst such as RuO_2 must be added to the surface to allow the rapid transfer of photogenerated holes to a solution species before Si reactions occur. Under these conditions even chlorine and bromine evolution are possible on Si. For example an iridium silicide coated n-Si electrode with RuO_2 catalyst in 11 M LiCl solution will evolve chlorine at potentials about 0.4V less positive than the reversible chloride/chlorine potential (Figure 2) and show no appreciable deterioration after seven days of continuous irradiation.

In looking towards the 21st Century, I predict that interfacial photochemical and electrochemical processes at designed and integrated chemical systems will play an important role in the development of energy conversion and other devices.

LITERATURE CITED

1. Fan, F. R., Hope, G. A., and Bard, A. J. J. Electrochem. Soc., in press.
2. Fan, F. R., Keil, R. G., and Bard, A. J., J. Am. Chem. Soc., submitted.

RECEIVED December 30, 1982

THERMAL AND PHOTOCHEMICAL ELECTRON TRANSFER REACTIONS

6

Metalloporphyrins

Catalysts for Dioxygen Reduction and P-450-Type Hydroxylations

DAVID DOLPHIN and BRIAN R. JAMES

University of British Columbia, Department of Chemistry, Vancouver, British Columbia V6T 1Y6 Canada

The functions of the heme proteins, cytochrome oxidase, cytochrome P-450, catalase, and peroxidase, are discussed with special reference to the mechanisms of enzymatic action and the development of protein-free in vitro catalysts. The use of dimeric face-to-face cobalt porphyrins for the electrochemical four-electron reduction of dioxygen to water is limited, probably as a result of destruction of the porphyrin macrocycle by intermediate hydrogen peroxide that is released. The cytochrome oxidase system prevents hydrogen peroxide release, possibly by utilizing it in much the same manner as catalase and peroxidase, as well as P-450 that more commonly functions as a hydroxylating reagent using dioxygen and a two-electron reducing source. The mechanisms of each of the enzyme systems can be pictured as involving formation of a ferric peroxide that breaks down, via cleavage of the oxygen-oxygen bond, into an oxo iron(IV) porphyrin cation radical species with release of water. The oxo intermediate, that in the case of P-450 can be generated also via the iron(III) state and iodosylbenzene or peracids, is considered to be the key species governing the reactivity pattern of the enzyme.

Figure 1 shows in bold type the trace elements in nature essential for the proper functioning of biological processes. The importance of metals has been long known but it is only in the past three decades that some of their specific roles have begun to be elucidated. It is perhaps not surprising that iron, the most naturally abundant of all metals, should play many important roles in nature. We shall present here one small aspect of this rapidly expanding area of inorganic chemistry, namely that of the functioning of iron when coordinated to porphyrins (1, 2, 3). Figure 2 shows the major heme proteins

0097-6156/83/0211-0099$06.00/0

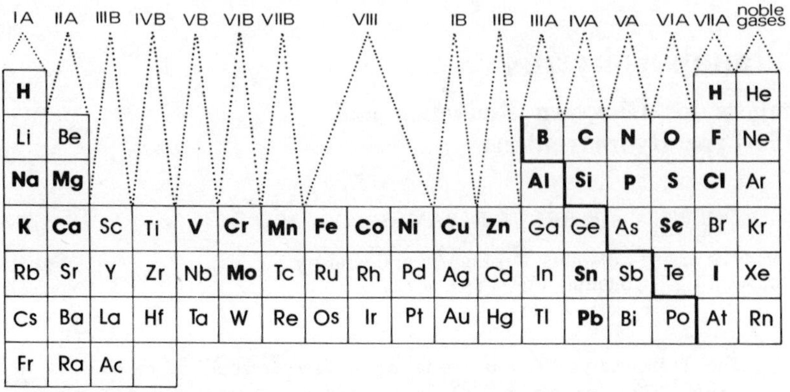
IA IIA IIIB IVB VB VIB VIIB VIII IB IIB IIIA IVA VA VIA VIIA noble gases
H H He
Li Be B C N O F Ne
Na Mg Al Si P S Cl Ar
K Ca Sc Ti V Cr Mn Fe Co Ni Cu Zn Ga Ge As Se Br Kr
Rb Sr Y Zr Nb Mo Tc Ru Rh Pd Ag Cd In Sn Sb Te I Xe
Cs Ba La Hf Ta W Re Os Ir Pt Au Hg Tl Pb Bi Po At Rn
Fr Ra Ac

Figure 1. The periodic chart: essential trace elements in nature are shown in bold.

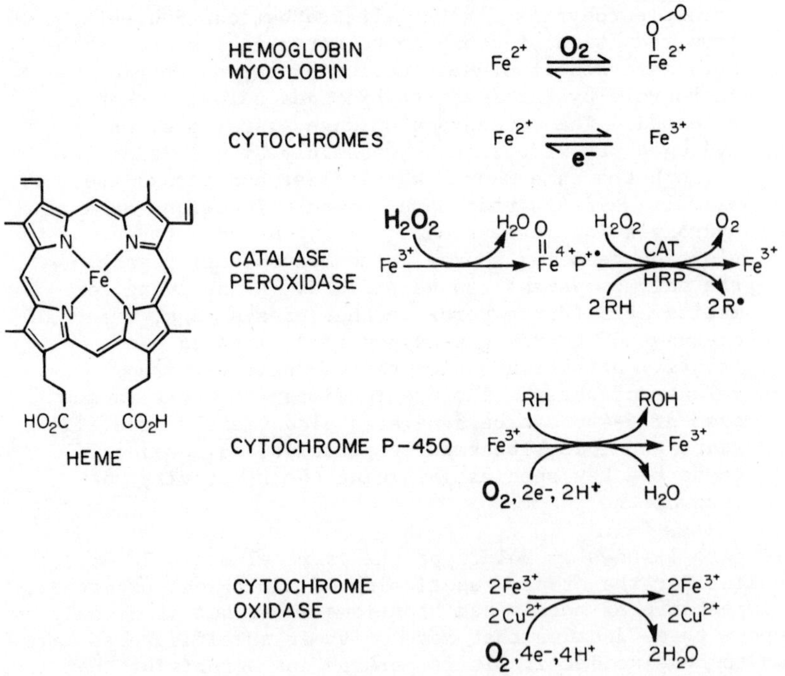

Figure 2. The major heme proteins and their biochemical functions.

and their biochemical functions. In addition to functioning as transport and storage proteins for electrons and oxygen, hemoproteins play other important roles. A major theme played by heme enzymes is the control and utilization of oxygen and reduced oxygen derivatives (4). Figure 3 shows the complex chemistry of oxygen in terms of the reduction potentials governing its oxidation states at both neutral (pH 7.0) and acid (pH 0.0) conditions; shown also are the areas where heme proteins exert their influence.

Nature has had a long time to perfect her inorganic chemistry, and understanding the mechanisms of enzymatic action can suggest ways for the inorganic chemist to develop corresponding *in vitro* catalysts. We shall show how knowledge of cytochrome oxidase and cytochrome P-450 has led to such possibilities.

Cytochrome oxidase is the terminal electron acceptor in the respiratory chain of all oxygen-breathing organisms (5). Nature uses cytochrome oxidase to bring about the "concerted" four-electron reduction of dioxygen to water without the release of the toxic one- and two-electron reduction products (superoxide and hydrogen peroxide, respectively). The study of cytochrome oxidase is complicated by the fact that it is a multi-component enzymatic system which functions within and across the biological membrane. Until Sanger's group (6) recently sequenced part of human mitochondrial genome, even the size of each component of the enzyme, let alone the individual protein sequences, was unknown. The whole enzyme complex, however, is known to contain two heme porphyrins and two copper centers. While these four-electron redox centers allow for speculation on the storage of the four electrons provided via cytochrome *c* (another heme protein), the mechanism of their delivery to dioxygen to form water is far from clear.

The practical need to reduce dioxygen to water without the release of peroxide is important in fuel-cell technology, where the fabrication of a cheap and robust oxygen electrode is still a major goal. The use of metalloporphyrins for this purpose has been widely studied and some considerable success has been achieved recently. By using cofacially linked dimeric cobalt porphyrins (**1**), Anson, Collman et al. (7, 8) have brought about the "direct" four-electron reduction of dioxygen to water. However, the catalytic lifetime of the dimeric porphyrins on the carbon electrodes is short (9). We supposed that hydrolysis of the amide bridging groups, under the strongly acidic conditions employed for oxygen reduction, accounted for the loss of catalytic activity. We have prepared an analogous series of dimeric cofacial porphyrins linked by polymethylene chains (**2**) (10) which cannot, of course, undergo hydrolysis. These complexes parallel those of Anson's and Collman's with respect to their ability to promote the four-electron reduction of water. Unfortunately, their catalytic lifetime is also short, suggesting that the loss of activity results not from the breaking of the linking groups but rather from destruction of the porphyrin macrocycle by hydrogen peroxide (11).

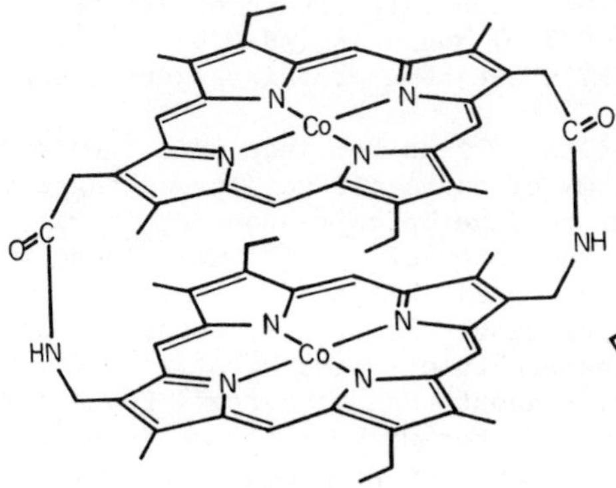

1

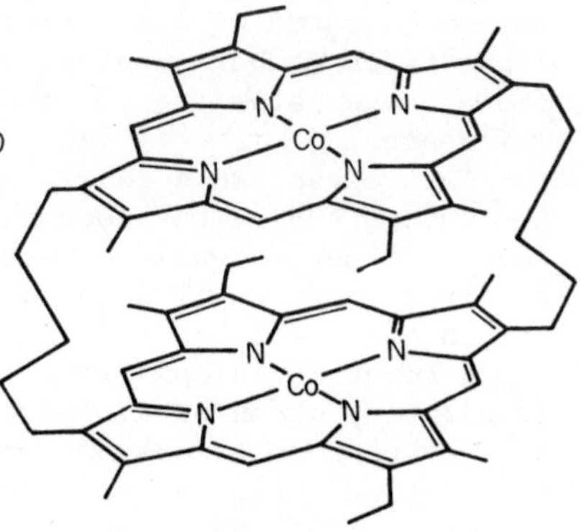

2

How does nature prevent the release of hydrogen peroxide during the cytochrome oxidase-mediated four-electron reduction of dioxygen? It would appear that cytochrome oxidase behaves in the same manner as other heme proteins which utilize hydrogen peroxide, such as catalase and peroxidase (vide infra), in that once a ferric peroxide complex is formed the oxygen-oxygen bond is broken with the release of water and the formation of an oxo iron(IV) complex which is subsequently reduced to the ferrous aquo state (12). Indeed, this same sequence of events accounts for the means by which oxygen is activated by cytochromes P-450.

Cytochromes P-450 occur in both procaryotes and eucaryotes (13) and, while many different specific enzymatic reactions are known to be controlled by these enzymes, they can all be generalized into two main categories; namely, the epoxidation of olefins (including aromatic systems) which gives rise to the carcinogenicity of fused polycyclic aromatic hydrocarbons, and the oxidation of unactivated carbon-hydrogen bonds to the corresponding alcohol, such as the conversion of cyclohexane to cyclohexanol (14).

The enzymatic cycle for cytochrome P-450 is shown in Figure 4. Binding of the substrate (RH) to the ferric hemoprotein changes the iron from low to high spin; a subsequent one-electron reduction gives the ferrous heme protein. Like myoglobin (Mb), ferrous P-450 binds both oxygen and carbon monoxide, but the analogy ends there. The majority of heme proteins exhibit a strong Soret absorption band at approximately 420 nm. However, the CO complex of P-450 is atypical in that it exhibits a split Soret band with one transition at 450 nm (from which the enzyme

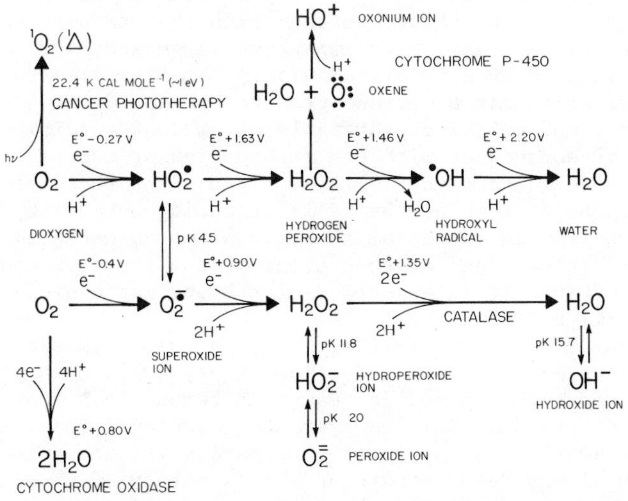

Figure 3. Standard reduction potentials associated with the chemistry of oxygen; values in upper and lower halves of diagram refer to pH 0.0 and pH 7.0 conditions, respectively.

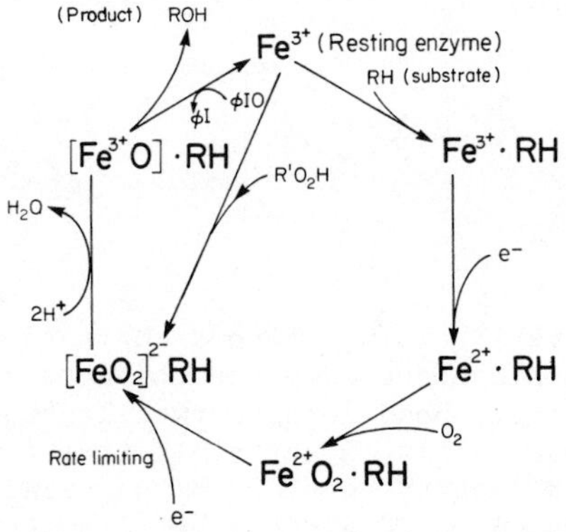

Figure 4. The enzymatic cycle for cytochrome P-450.

was named (15)) and a second higher energy transition at 350 nm. This unusual optical spectrum has been shown to result from the unique axial ligation of a thiolate anion provided by a cysteine residue of the protein (16-19). Oxy P-450 also exhibits such a split Soret band and this, together with pH-titration data (19), show that this enzymatically important oxygenated derivative is also coordinated by a thiolate anion. The high electron density on sulfur, which can be transmitted to the coordinated oxygen, makes oxy P-450 relatively unstable at ambient conditions, and the loss of superoxide with the reformation of the ferric complex is a facile process (19). At first sight this instability appears to be a waste of the cells' reducing capability, but we suspect that it is a compromise between the unwanted side-effect of the thiolate ligand at this stage of the reaction and its positive effects at later stages of the enzymatic cycle, that we shall detail below.

The next, and rate limiting step, in the enzymatic reaction is a further one-electron reduction of oxy P-450. As yet, little is known about the enzymatic reactions beyond this stage. We (20) and others (21, 22), however, have explored this further chemistry using model systems. Enzymatically, the one-electron reduction of oxy P-450 occurs at approximately -0.20 V (23); this value is vs. Ag/AgCl, which has a potential of about +0.23 V w.r.t. the S.H.E. We have found that a one-electron reduction of oxy ferrous octaethylporphyrin in DMSO/CH_3CN at -20° can be brought about electrochemically at a platinum electrode at a very similar potential, -0.24 V (vs. Ag/AgCl), to give the η^2-peroxy complex (**3**) (20, 24). However, in our hands, and those of others (25), this model monoanionic complex exhibits none of the oxidizing powers of P-450. This may, of course, mean that a

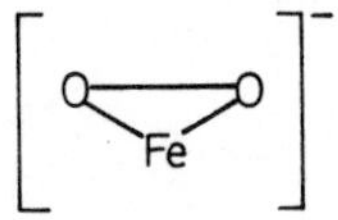

3

complex of the type (**3**) is not formed enzymatically. More likely though is that within the enzyme the thiolate now plays a critical role. Any single bond between two electronegative elements is necessarily weak. This is certainly true for peroxides and the added electron density supplied by an axially coordinate thiolate anion ligand would aid in the cleavage of the peroxy bond in an enzymatic species corresponding to complex **3**. In addition all of the substrates for cytochromes P-450 are hydrophobic, and the binding site for substrate, which is necessarily close to the iron porphyrin, must also be hydrophobic. Yet the addition of the last electron in the enzymatic cycle could generate a species such as **3**, that in fact would have a net double negative charge. In such a non-polar environment the peroxo ligand will be strongly nucleophilic (basic) and this could ease O-O bond cleavage either by direct protonation or acylation (26). By whatever mechanism the bond cleavage occurs, one oxygen atom must leave at the oxidation level of water. Scheme 1 shows that the iron oxygen complex (**4**) remaining, which is presumed to be the oxidizing agent, is formally an Fe^{3+} oxene (oxygen atom) complex. The chemistry of oxenes is hardly explored, although oxygen atoms can bring about P-450-like hydrocarbon oxidation (27). By analogy, however, the chemistry of oxenes can be expected to parallel that of nitrenes and carbenes. Here direct insertion into C-H bonds and the addition across olefins without loss of stereochemistry is well defined (28). Cytochromes P-450 do not appear to function via direct insertion, and the oxene analogy is probably not a reasonable one (29).

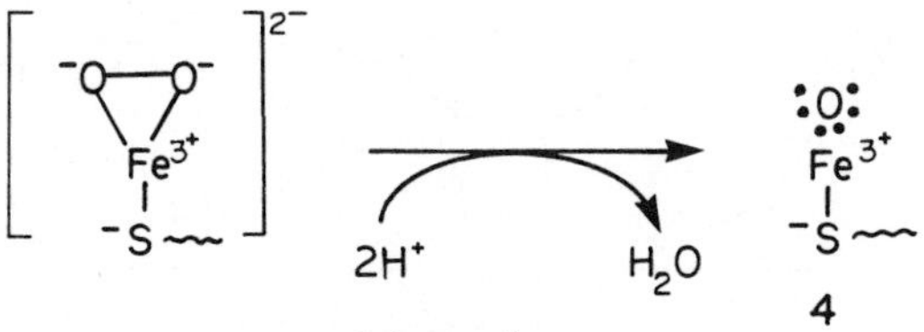

Scheme 1

If the active oxidizing agent is not formulated as a ferric oxene complex, the questions then remain as to what is its electronic ground state and how does hydrocarbon oxidation occur?

Answers to these questions were initiated over a decade ago during our studies on catalase (CAT) and horseradish peroxidase (HRP) (30). Both native enzymes are ferric hemoproteins and both are oxidized by hydrogen peroxide. These oxidations cause the loss of two electrons and generate active enzymatic intermediates that can be formally considered as Fe^{5+} complexes. Until very recently no complexes of Fe^{5+} were known, and so we examined the electronic ground state of these highly oxidized hemoproteins, which are known as the compound I derivatives, CAT I and HRP I.

An examination of their optical spectra (Figure 5) suggested that both CAT I and HRP I contained porphyrin π-cation radicals in which the porphyrin ring had undergone a one-electron oxidation (31). Many porphyrins and metalloporphyrins can be oxidized to π-cation radicals and these species can occupy one of two ground states ($^2A_{1u}$ or $^2A_{2u}$). These two ground states have different but characteristic optical spectra and can be interchanged as a function of axial ligation of the metalloporphyrin (Figure 5) (30). Catalase is axially coordinated by a phenoxide of tyrosine (32), and HRP by an imidazole of histidine (33), and this may well explain their different electronic ground states

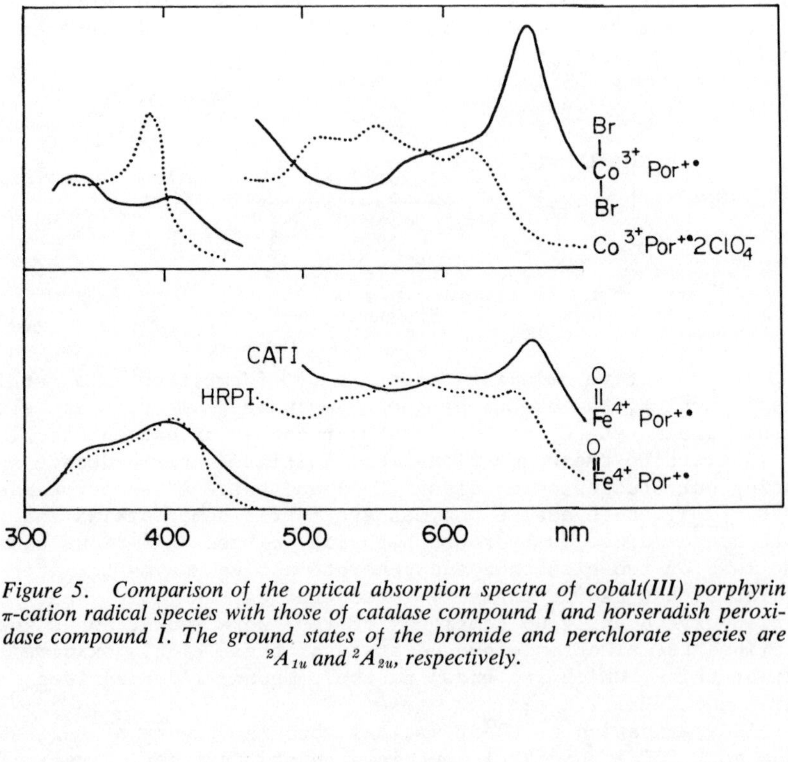

Figure 5. Comparison of the optical absorption spectra of cobalt(III) porphyrin π-cation radical species with those of catalase compound I and horseradish peroxidase compound I. The ground states of the bromide and perchlorate species are $^2A_{1u}$ and $^2A_{2u}$, respectively.

even though both enzymes contain iron protoporphyrin. Oxidation to the π-cation radical accounts for the loss of one of the two electrons involved in the peroxide-mediated oxidation. The other electron is lost from the iron, as exemplified by optical and Mossbauer spectroscopy (34). Hence the electronic configuration of CAT I and HRP I is believed to be that of an Fe^{4+} porphyrin π-cation radical; moreover, labelled oxygen (35) and ENDOR studies (36) have shown that the complex contains an oxo ligand (**5**).

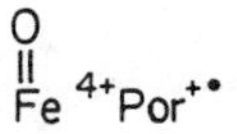

5

The ferric oxene complex (**6**) is, of course, just one resonance formulation of **5** (see scheme 2). Moreover, the addition of dioxygen and two electrons to P-450 is equivalent to using

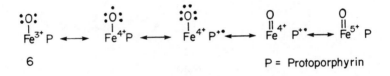

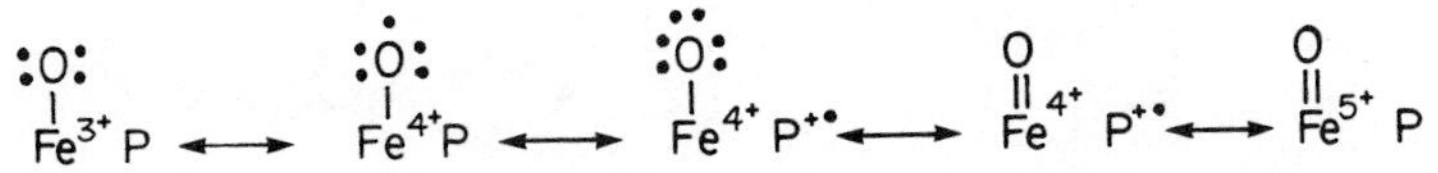

6

P = Protoporphyrin

Scheme 2

peroxide, and indeed the cytochrome P-450 catalysis can be shunted by using alkyl hydroperoxides (28), Figure 4. It is now apparent that the hemoprotein families of the catalases, peroxidases, and cytochromes P-450 are closely related. Scheme 3 illustrates the suggested similarities for each of these separate enzymes. In addition to activating P-450 by the use of alkyl hydroperoxides, both iodosylbenzene (**7**) and peracids (28) can also be used.

Groves et al. have elegantly followed up such findings, and have found that ferric porphyrins and iodosylbenzene provide a system, catalytic in porphyrin, for the room temperature oxidation of hydrocarbons (37), see Figure 4. More recently this group has shown that ferric tetra-mesitylporphyrin (**8**), when treated with a peracid, yields an isolable oxo intermediate which is the active hydrocarbon oxidizing agent (38). Depending upon the axial ligation, this intermediate exists as either the Fe^{4+} π-cation radical or the Fe^{5+} complex (39). These observations add further support to the hypothesis concerning the electronic structure of the active enzymatic oxidizing agent,

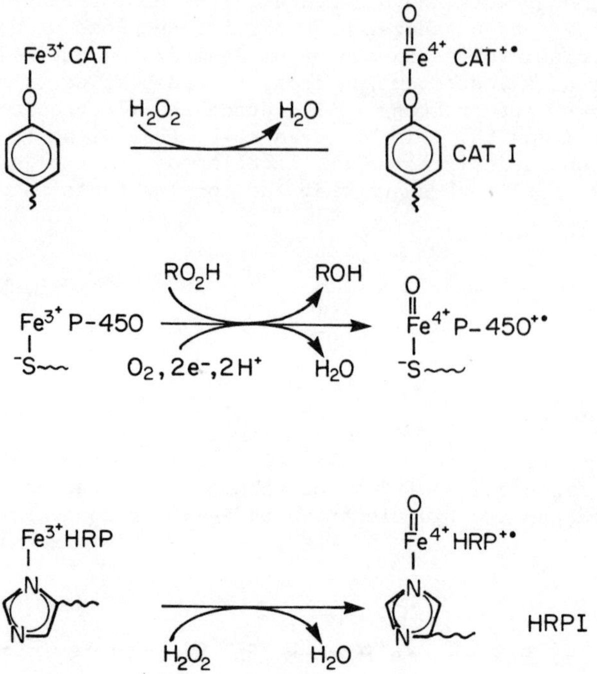

Scheme 3

as outlined in schemes 2 and 3. We can now attempt to describe in more detail the iron porphyrin complexes that are involved in the enzymatic cycle of cytochrome P-450 (Figure 6, cf. Figure 4).

7

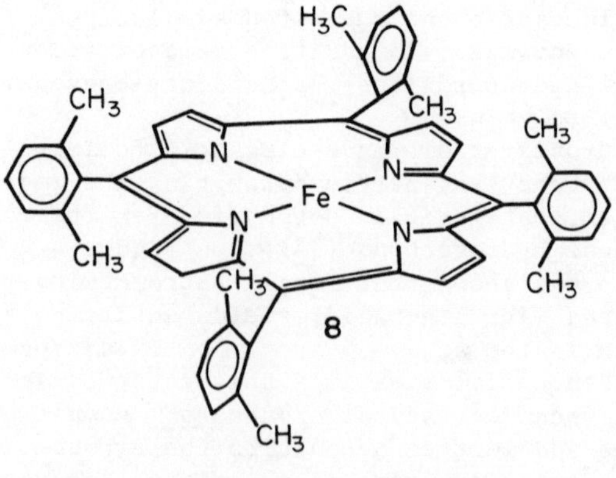

8

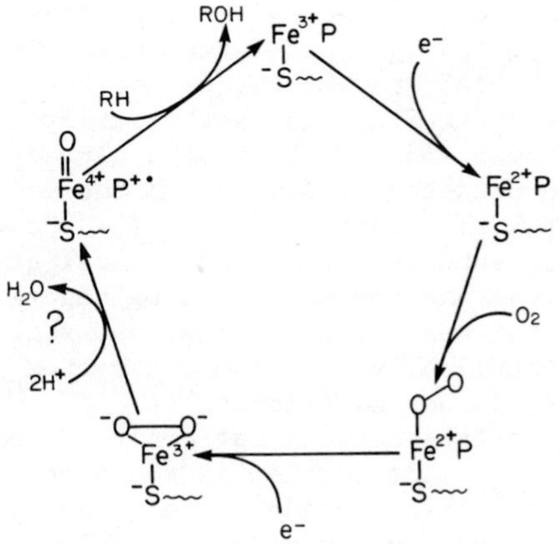

Figure 6. The enzymatic cycle for cytochrome P-450, detailing possible electronic structure of intermediates. (P represents protoporphyrin.)

Studies with both cytochromes P-450 and model systems suggest that hydrocarbon oxidation proceeds via the intermediacy of a substrate radical (28, 29); the hydrophobic nature of the binding site would also argue for an uncharged (radical) substrate intermediate. The storage of unpaired spins at iron, oxygen, and the porphyrin macrocycle, would allow for the stabilization of radical intermediates during the enzymatic oxidation. The abstraction of a substrate hydrogen atom, and the subsequent transfer of a hydroxyl radical to substrate from the iron porphyrin complex, account for the observations concerning hydrocarbon oxidation.

It is interesting to note that in so many areas of oxygen chemistry, nature uses iron porphyrins for the transport, storage, and utilization of dioxygen and its reduced derivatives. To this end, cytochrome P-450 may be considered as nature's equivalent of Fenton's reagent; in the enzyme system, highly reactive intermediates such as hydroxyl radicals and oxenes appear to be controlled by their interaction with hemoprotein.

What can all these studies suggest to the inorganic chemist interested in the controlled and facile catalytic oxidation of hydrocarbons? Groves and coworkers have already shown that iron porphyrins in the presence of iodosylbenzene and peracids can be used for such catalytic reactions (37, 38). However, the cost of the oxidants makes such reaction uneconomical at this time. Nature uses dioxygen and two electrons to generate the active oxidizing agent, but we have shown that one such intermediate generated electrochemically with a model system (lacking the thiolate, however) is not an oxidant.

The only difference between the active oxidizing and a ferric porphyrin hydroxide complex is two electrons (scheme 4). Indeed, the electrochemical oxidation of hydroxy ferric tetramesitylporphyrin shows two reversible one-electron oxidations (40), and, in principle, use of water and an electrode should allow development of a system capable of catalytically oxidizing hydrocarbons.

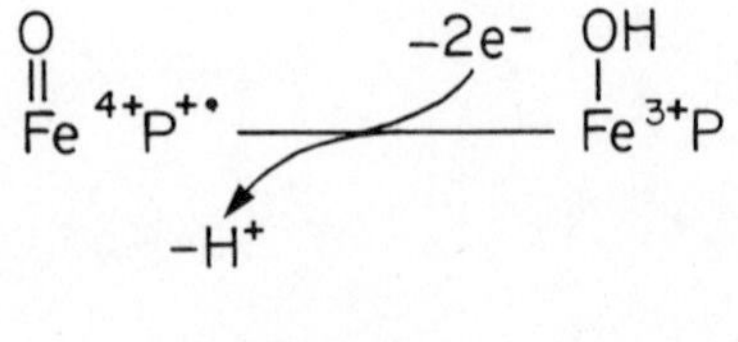

Scheme 4

Acknowledgments

This is a contribution from the Bioinorganic Chemistry Group, which was supported by grants from the Canadian Natural Sciences and Engineering Research Council and the United States National Institutes of Health (AM 17989).

Literature Cited

1. Sigel, H., Ed.; "Metal Ions in Biological Systems"; Vol. 7, Marcel Dekker, New York, 1978.
2. Hughes, M.N. "The Inorganic Chemistry of Biological Processes"; 2nd Ed., Wiley, New York, 1981, Chapters 5, 7.
3. Ochiai, E. "Bioinorganic Chemistry. An Introduction"; Allyn and Bacon, Boston, 1977, Chapters 5-7, 10.
4. Spiro, T.G., Ed.; "Metal Ion Activation of Dioxygen"; Wiley-Interscience, New York, 1980.
5. Chance, B. Current Topics in Cellular Regulation, 1981, 18, 343.
6. Anderson, S.; Bankier, A.T.; Barrell, B.G.; de Bruijn, M.H.L.; Coulson, A.R.; Drouin, J.; Eperon, I.C.; Nierlich, D.P.; Roe, B.A.; Sanger, F.; Schreier, P.H.; Smith, A.J.H.; Staden, R.; Young, I.G. Nature, 1981, 290, 457.
7. Collman, J.P.; Marrocco, M.; Denisevich, P.; Koval, C.; Anson, F.C. J. Electroanal. Chem., 1979, 101, 117.
8. Collman, J.P.; Denisevich, P.; Konai, Y.; Marrocco, M.; Koval, C.; Anson, F.C. J. Am. Chem. Soc., 1980, 102, 6027.
9. Anson, F.C.; Durand, R.R., Jr.; personal communication.
10. Dolphin, D.; Hiom, J.; Paine, J.B., III; unpublished results.
11. Durand, R.R., Jr.; Anson, F.C. J. Electroanal. Chem., 1982, 134, 273.
12. Chan, S.I.; personal communication.
13. Griffin, B.W.; Peterson, J.A.; Esterbrook, R.W. in "The Porphyrins"; Vol. VII, Dolphin, D., Ed.; Academic Press, New York, 1979, p. 333.
14. Chang, C.K.; Dolphin, D. in "Bioorganic Chemsitry"; Vol. IV, van Tamelen, E.E., Ed.; Academic Press, New York, 1978, p. 37.
15. Omura, T.; Sato, R. J. Biol. Chem., 1962, 237, 1375.
16. Collman, J.P.; Sorrell, T.N. J. Am. Chem. Soc., 1975, 97, 4133.
17. Chang, C.K.; Dolphin, D. J. Am. Chem. Soc., 1975, 97, 5948.
18. Hanson, L.K.; Eaton, W.A.; Sligar, S.G.; Gunsalus, I.C.; Gouterman, M.; Connell, C.R. J. Am. Chem. Soc., 1976, 98, 2672.
19. Dolphin, D.; James, B.R.; Welborn, H.C. J. Mol. Catal., 1980, 7, 201.
20. Welborn, H.C.; Dolphin, D.; James, B.R. J. Am. Chem. Soc., 1981, 103, 2869.
21. Reed, C.A. in Adv. Chem. Series; Vol. 201, Kadish, K.M., Ed.; Am. Chem. Soc., Washington, D.C., 1982, Chapter 15.
22. McCandlish, E.; Miksztal, A.R.; Nappa, M.; Sprenger, A.Q.; Valentine, J.S.; Stong, J.D.; Spiro, T.G. J. Am. Chem. Soc., 1980, 102, 4268.
23. Gunsalus, I.C.; personal communication.
24. Dolphin, D.; James, B.R.; Welborn, H.C. in Adv. Chem. Series; Vol. 201, Kadish, K.M., Ed.; Am. Chem. Soc., Washington, D.C., 1982, Chapter 23.
25. Groves, J.T.; Valentine, J.S.; personal communications.

26. Sligar, S.G.; Kennedy, K.A.; Pearson, D.C. Proc. Natl. Acad. Sci. U.S.A., 1980, 77, 1240.
27. Hori, A.; Takamuku, S.; Sakurai, H. J. Org. Chem., 1977, 42, 2318.
28. Coon, M.J.; White, R.E. in "Metal Ion Activation of Dioxygen"; Spiro, T.G., Ed.; Wiley-Interscience, New York, 1980, p. 73.
29. Groves, J.T.; ibid., p. 125.
30. Dolphin, D.; Forman, A.; Borg, D.C.; Fajer, J.; Felton, R.H. Proc. Natl. Acad. Sci. U.S.A., 1971, 68, 614.
31. Dolphin, D.; Felton, R.H. Acc. Chem. Res., 1974, 7, 26.
32. Reid, T.J., III; Murphy, M.R.N.; Sicignano, A.; Tanaka, N.; Musick, W.D.L.; Rossmann, M.G. Proc. Natl. Acad. Sci. U.S.A., 1981, 78, 4767.
33. Yonetani, T.; Yamamoto, H.; Erman, J.E.; Leigh, J.S., Jr.; Reed, G.H. J. Biol. Chem., 1972, 247, 2447.
34. Dolphin, D. Isr. J. Chem., 1981, 21, 67.
35. Hager, L.P.; Doubek, D.L.; Silverstein, R.M.; Hargis, J.H.; Martin, J.C. J. Am. Chem. Soc., 1972, 94, 4364.
36. Roberts, J.E.; Hoffman, B.M.; Rutter, R.; Hager, L.P. J. Am. Chem. Soc., 1981, 103, 7654.
37. Groves, J.T.; Nemo, T.E.; Myers, R.S. J. Am. Chem. Soc., 1979, 101, 1032.
38. Groves, J.T.; Haushalter, R.C.; Nakamura, M.; Nemo, T.E.; Evans, B.J. J. Am. Chem. Soc., 1981, 103, 2884.
39. Groves, J.T.; personal communication.
40. Dolphin, D.; unpublished observations.

RECEIVED August 10, 1982

Discussion

N. Sutin, Brookhaven National Laboratory: In the reduction of oxygen by cytochrome c oxidase, is there any evidence for cooperativity between heme a_3 and Cu_B? In particular, is there evidence for the formation of a peroxo-bridged a_3-Cu_B intermediate?

D. Dolphin: Clearly the fact that the a_3-Cu_B pair can be antiferromagnetically coupled shows that these two centres can communicate with each other. It is not clear, however, that this particular "intermediate" has any enzymatic significance. I know of no definitive evidence concerning the formation of a μ-peroxo complex between a_3 and Cu_B, and if the analogy between cytochrome oxidase and the other heme proteins, I have referred to, holds then I see no need to suggest such peroxo-bridging. I suspect that Cu_B is there as an electron carrier and not as a ligand for dioxygen in any of its oxidation states during its reduction to water.

D.R. McMillin, Purdue University: Like cytochrome oxidase, the blue copper protein laccase catalyzes the reduction of dioxygen to water, albeit at a somewhat slower rate. Given the structural analogies that exist between the proteins -- each contains four metal ions, a site involving a coupled pair of metal ions, etc. -- one might expect there to be analogies in their mechanisms of action as well. On the other hand, it seems unlikely that a copper ion could achieve the valency changes you ascribe to the a_3 heme. Do you have any comment regarding the laccase system?

D. Dolphin: I agree with you that I find it unlikely either the copper ion, or the copper ion and its attendant ligands, could achieve the high oxidation states analogous to those postulated for the a_3 heme.

T.J. Kemp, Warwick: Noting the critical role of higher oxidation states of Fe in the functioning of cytochrome P-450, it is interesting to note the recent u.v.-spectral characterisation of transient intermediate in the oxidation of $[Fe^{II}(CN)_6]^{4-}$ by both HOCl and XeF_2 in aqueous solution.[1,2] This species is assigned to an Fe^{IV} species by Purmal et al.,[1,2] and in association with his laboratory, Dr. Peter Moore and I have obtained fully time-resolved u.v.-visible spectra of these systems in a stopped-flow apparatus coupled to a rapid scanning spectrometer, as exemplified in the figure.[3]

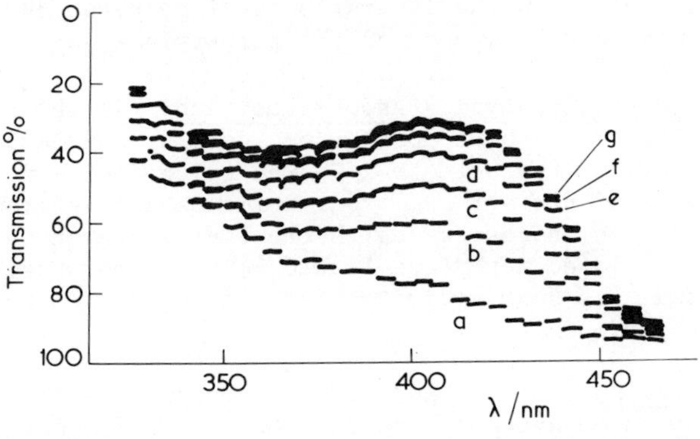

Spectral profile of the initial part of the reaction between $Fe(CN)_6^{4-}$ (2.5×10^{-4} mol dm^{-3}) and NaOCl (2.5×10^{-3} mol dm^{-3}) at pH 2.25; trace *a* represents spectrum on mixing (i.e. of reactants), while *b*,*c*, ... are spectra at successive intervals of 150 ms. *g* is the spectrum of the intermediate, attributed to $Fe^{IV}(CN)_6^{2-}$.

1. Kozlov, Yu.N.; Moravskii, A.P.; Purmal', A.P.; Shuvalov, V.F.; Zh. Fiz. Khim. 1981, 55, 764.
2. Kozlov, Yu.N.; Vorob'eva, T.P.; Purmal'; A.P. Zh. Fiz. Khim. 1981, 55, 2279.
3. Kemp, T.J.; Kozlov, Yu.N.; Moore, P.; Purmal', A.P.; Silver, G.R., to be published.

D. Dolphin: A very interesting observation. There are, of course, many such examples of the Fe^{IV} oxidation state when the iron is coordinated to a porphyrin.

T.J. Kemp, Warwick: Is it not the case that the product distribution in hydroxylation of organic molecules by Fenton's reagent differs in considerable detail from that realised by cytochrome P-450?

D. Dolphin: Groves and his co-workers[1,2] have shown that ferrous perchlorate and hydrogen peroxide can carry out hydroxylations which bear close parallels to the P-450 mediated systems.

1. Groves, J.T.; McClusky, G.A. J. Am. Chem. Soc. 1976, 98, 859.
2. Groves, J.T.; Van Der Puy, M. J. Am. Chem. Soc. 1976, 98, 5290.

M.S. Wrighton, M.I.T.: Why are the 4e-O_2 reduction catalysts decomposed by H_2O_2 when it appears that these catalysts operate at potentials where it is thermodynamically impossible to make H_2O_2?

D. Dolphin: Such potentials refer to free hydrogen peroxide (H_2O_2). The potentials at which coordinated peroxide can be formed are lower, but subsequent protonation of such coordinated peroxide can result in the liberation of free peroxide which can then react with the periphery of the porphyrin.

E.R. Evitt, Catalytica Associates, Inc: In their work with amide-linked cofacial porphyrins, Collman and Anson have only observed 100% four-electron O_2 reduction in the four-atom-bridged member of the series. Your comments implied four-electron O_2 reduction is observed with the five and eight as well as four-atom bridged methylene-linked cofacial porphyrins you have prepared. Is such activity observed for the compounds with longer linkages?

D. Dolphin: Collman and Anson have found with their five-atom bridged amide systems that the reduction of dioxygen is partitioned roughly one third undergoing a four-electron reduction to water and the rest a two-electron reduction to peroxide. We see the same result; indeed the 5-5 systems of ours and that of Collman-Anson are indistinguishable. Neither type of cofacial dimers with chains greater than five show any four-electron reduction.

E.R. Evitt, Catalytica Associates, Inc.: Have you prepared polymethylene-linked cofacial porphyrins that contain two different metals in the cofacial unit? Is it possible to prepare polymethylene-linked cofacial porphyrins that are unsymmetrical by virtue of two different length bridges in the same molecule?

D. Dolphin: For reasons that are at present unclear both the rates of insertion, or removal, of metals within the cofacial systems are different so that mixed metal complexes can be conviently prepared. We have, as yet, not looked at the unsymmetrically linked systems.

7

Electron Transfer Mechanisms

R. J. KLINGLER, S. FUKUZUMI, and J. K. KOCHI

Indiana University, Department of Chemistry, Bloomington, IN 47405

The finely tunable steric and polar properties inherent to alkyl ligands can be exploited in both the homogeneous and the heterogeneous electron transfers from several classes of organometals. Outer-sphere mechanisms pertain to electron transfer with iron(III) oxidants such as $Fe(phen)_3^{3+}$, since the oxidations of various organometals are singularly unaffected by steric effects and follow the free energy correlation established by Marcus Theory. Likewise the rates of heterogeneous electron transfer at a platinum electrode are shown to be directly related to the homogeneous chemical oxidation with $Fe(phen)_3^{3+}$ provided they are evaluated at the same potential, i.e., driving force. The free energy relationship for the outer-sphere rates of electron transfer, which can be measured by the electrochemical method over an extended region far from the equilibrium potential, shows the asymptotic behavior at both the endergonic and exergonic limits describable by the empirical Rehm-Weller equation. The contrasting inner-sphere mechanism for electron transfer applies to the oxidation of the same series of organometal donors RM by either hexachloroiridate(IV) or tetracyanoethylene (TCNE), in which steric effects play an important kinetic role. The observation of transient, charge-transfer absorption bands from metastable [RM TCNE] complexes can be used to evaluate the work term for ion-pair formation with the aid of the Mulliken formulation. A single, unified free energy relationship applicable to inner-sphere processes relates the activation free energy to the standard free energy change for electron transfer and the work term. The large variations in the apparent Brønsted slopes (from unity to even negative values) can be attributed to changes in the

0097-6156/83/0211-0117$10.75/0

work term arising from steric effects in ion-pair formation.

Traditionally, electron transfer processes in solution and at surfaces have been classified into outer-sphere and inner-sphere mechanisms (1). However, the experimental basis for the quantitative distinction between these mechanisms is not completely clear, especially when electron transfer is not accompanied by either atom or ligand transfer (i.e., the bridged activated complex). We wish to describe how the advantage of using organometals and alkyl radicals as electron donors accrues from the wide structural variations in their donor abilities and steric properties which can be achieved as a result of branching the alkyl moiety at either the α- or β-carbon centers.

Structural Variations in Electron Donors

We consider the four structurally diverse classes of organometals I-IV, in which the configuration and coordination about the metal centers vary systematically from octahedral, square planar, tetrahedral to linear, respectively.

ORGANOMETALS:

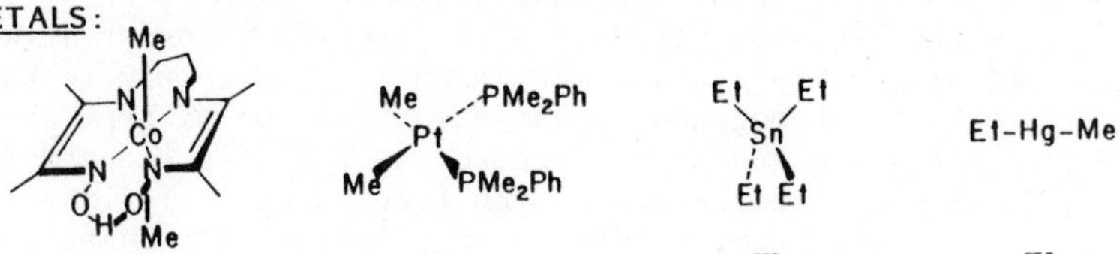

I II III IV

These organometals are especially desirable for kinetic studies since they are all sufficiently substitution stable in solution to allow meaningful measurements to be made. Moreover, for these neutral organometal donors, the work term of the reactants w_r is considered to be unimportant. An additional benefit derived from the use of organometal donors lies in the transient character of many of the oxidized organometal cations. As a result, the back electron transfer is generally minimal, and the electron transfer process is overall irreversible in these systems. As relatively volatile and electron-rich compounds, the photoelectron spectra of the alkylmetals III and IV are readily accessible, and the vertical ionization potentials I_D can be accurately measured. For example, the photoelectron spectra in Figure 1 illustrates how the lowest energy band of a homologous series of dialkylmercurials undergoes large, systematic variations merely by branching at the α-carbon of the alkyl ligand (2).

In these alkylmetals of the main group elements, ionization occurs from the highest occupied molecular orbital (HOMO) which has σ-bonding character, i.e., they are σ-donors. Consequently

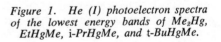

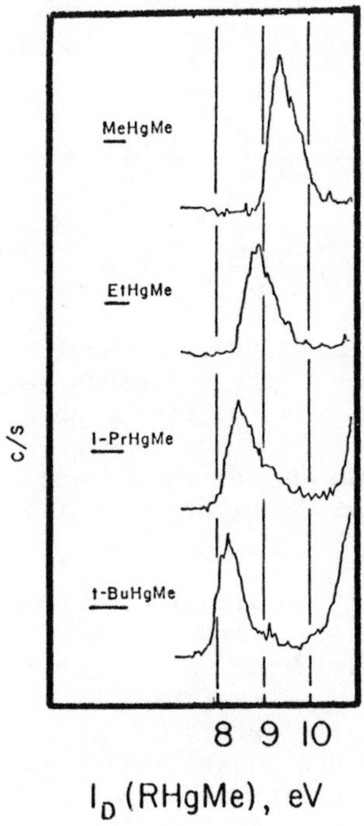

Figure 1. He (I) photoelectron spectra of the lowest energy bands of Me_2Hg, EtHgMe, i-PrHgMe, and t-BuHgMe.

alkyl ligands exert a large dominating influence on the ionization potentials and the steric properties of alkylmetals. Both trends are illustrated in Figure 2 for the α-branched ligands: methyl, ethyl, isopropyl, and tert-butyl on the left, as well as the β-branched ligands: ethyl, n-propyl, isobutyl, and neopentyl on the right (3). Note that the steric and polar effects generally increase together in the α-branched alkyl ligands, whereas variations in the steric effect dominate in the β-branched alkyl ligands.

   Alkyl radicals share many of the desirable properties of organometals described above, insofar as electron transfer reactions are concerned. Thus the steric properties of alkyl radicals with α- and β-branches follow the trends in Figure 2. Moreover, the direct parallel in their donor properties is shown in Figure 3 by

ALKYL RADICALS:

α-Branched	β-Branched

$$\cdot CH_3 \quad \cdot CH_2 \quad \cdot \underset{\underset{Me}{|}}{\overset{\overset{Me}{|}}{C}}H \quad \cdot \underset{\underset{Me}{|}}{\overset{\overset{Me}{|}}{C}}\text{-Me} \qquad \cdot CH_2CH_3 \quad \cdot CH_2CH_2 \quad \cdot CH_2\underset{\underset{Me}{|}}{\overset{\overset{Me}{|}}{C}}H \quad \cdot CH_2\underset{\underset{Me}{|}}{\overset{\overset{Me}{|}}{C}}\text{-Me}$$

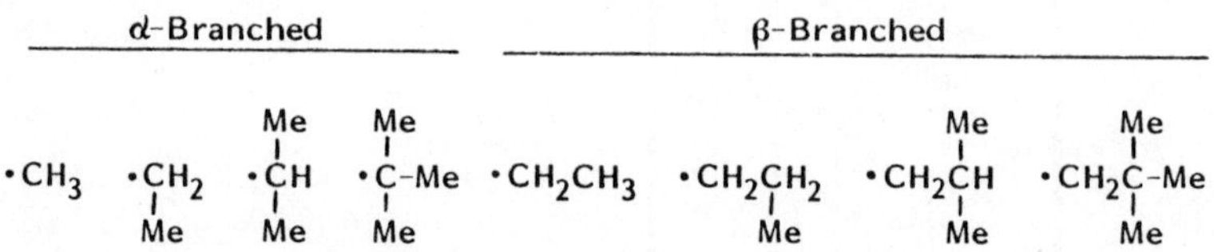

a comparison of the ionization potentials of the series of α-branched alkyl radicals (R·) with the I_D of the corresponding alkylmethylmercurials (RHgMe) and dialkyldimethyltin compounds (R$_2$-SnMe$_2$) (4). The same parallel behavior is also observed with the β-branched alkyl series, although the variations in the absolute values of I_D are not as large (see Figure 2, right).

Outer-Sphere Electron Transfer

   The minimal interpenetration of the coordination spheres of the reactants is inherent in any mechanistic formulation of the outer-sphere process for electron transfer. As such, steric effects provide a basic experimental criterion to establish this mechanism. Therefore we wish to employ the series of structurally related donors possessing the finely graded steric and polar properties described in the foregoing section for the study of both homogeneous and heterogeneous processes for electron transfer.

   Homogeneous Processes with Tris-phenanthroline Metal(III) Oxidants. The rates of electron transfer for the oxidation of these organometal and alkyl radical donors (hereafter designated generically as RM and R·, respectively, for convenience) by a series of tris-phenanthroline complexes ML$_3{}^{3+}$ of iron(III), ruthenium(III), and osmium(III) will be considered initially, since they have been previously established by Sutin and others as outer-sphere oxidants (5).

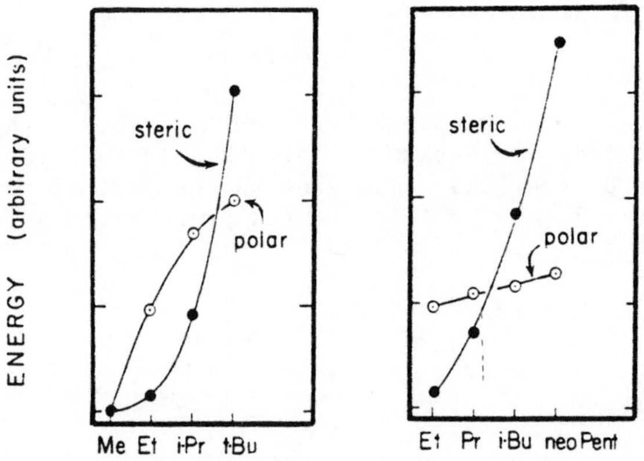

Figure 2. Steric and polar effects of alkyl ligands with α-branching (left) and β-branching (right).

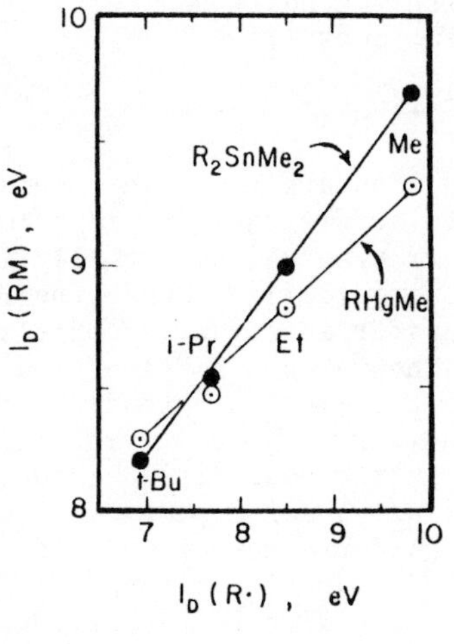

Figure 3. Direct relationship between the ionization potentials of alkyl radicals (R•) with I_D of the alkylmetals: RHgMe (⊙) and R_2SnMe_2 (●).

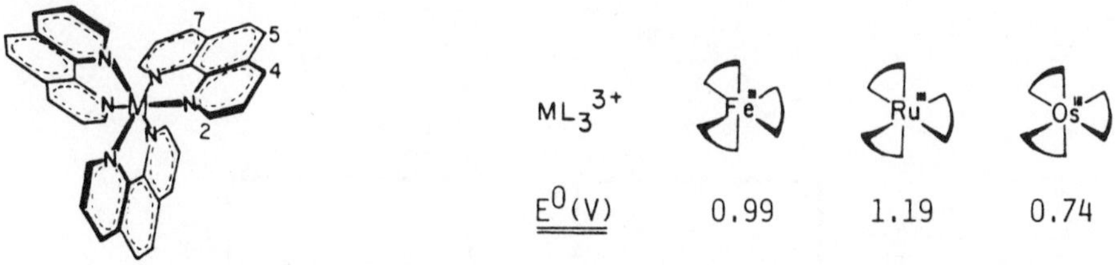

The iron(III) complexes FeL_3^{3+}, where L = 2,2'-bipyridine and various substituted 1,10-phenanthrolines, cleave a variety of organometals in acetonitrile according to the general reaction mechanism in Scheme I (6). The activation process for oxidative cleavage is represented by the electron transfer step in eq 1. The organometal cation RM^+ is a transitory intermediate which subsequently undergoes a rapid fragmentation in eq 2, rendering the electron transfer irreversible.

$$RM + FeL_3^{3+} \xrightarrow{k_{Fe}} RM^+ + FeL_3^{2+} \qquad (1)$$

$$RM^+ \xrightarrow{\text{fast}} R\cdot + M^+ \text{, etc.} \qquad (2)$$

Scheme I

The same iron(III) complexes also oxidize alkyl radicals, particularly those with secondary and tertiary centers, to the corresponding carbonium ions (7).

$$R\cdot + FeL_3^{3+} \longrightarrow R^+ + FeL_3^{2+} \qquad (3)$$

Since these carbonium ions are reactive, the subsequent followup steps with solvent are sufficiently rapid to make electron transfer in eq 3 rate determining and essentially irreversible.

For a particular iron(III) oxidant, the rate constant (log k_{Fe}) for electron transfer is strongly correlated with the ionization potential I_D of the various alkylmetal donors in Figure 4 (left) (6). The same correlation extends to the oxidation of alkyl radicals, as shown in Figure 4 (right) (7). [The cause of the bend (curvature) in the correlation is described in a subsequent section.] Similarly, for a particular alkylmetal donor, the rate constant (log k_{Fe}) for electron transfer in eq 1 varies linearly with the standard reduction potentials E^0 of the series of iron(III) complexes FeL_3^{3+}, with L = substituted phenanthroline ligands (6).

In order to account for the foregoing kinetic behavior, we rely on the Marcus theory for outer-sphere electron transfer to provide the quantitative basis for establishing the free energy relationship (8), i.e.,

$$\Delta G^{\ddagger} = w_r + \Delta G_0^{\ddagger}\left[1 + \frac{\Delta G}{4\Delta G_0^{\ddagger}}\right]^2 \qquad (4)$$

where $\Delta G = \Delta G^0 + w_p - w_r$ and ΔG^0 is the standard free energy change accompanying electron transfer. The intrinsic barrier $\Delta G_0^{\ddagger}$ represents the activation free energy for electron transfer at $\Delta G = 0$, and w_p is the work required to bring the products together. For neutral reactants such as the organometal examined in this study, the work term of the reactants w_r is considered to be nil, and the rate constants are converted to the activation free energies by,

$$\Delta G^{\ddagger} = -RT \ln (k/Z) \tag{5}$$

where the collision frequency Z is taken to be $3 \times 10^{10}\ M^{-1}\ s^{-1}$. In order to apply the Marcus equation to electron transfer from neutral organometals RM to the series of FeL_3^{3+} oxidants, eq 4 can be rewritten in an alternative form as (6),

$$\sqrt{\Delta G^{\ddagger}} = [\sqrt{\Delta G_0^{\ddagger}} + \frac{\mathcal{F}}{4\sqrt{\Delta G_0^{\ddagger}}}(E^0_{RM} + \frac{w_p}{\mathcal{F}})] - \frac{\mathcal{F}}{4\sqrt{\Delta G_0^{\ddagger}}} E^0_{Fe} \tag{6}$$

since $\Delta G^0 = (E^0_{RM} - E^0_{Fe})$ and $w_r = 0$. E^0_{RM} and E^0_{Fe} are the standard electrode potentials of the alkylmetal and FeL_3^{3+}, respectively, and $\mathcal{F}$ is the Faraday constant. The experimental correlation between $\Delta G^{\ddagger}$ and E^0_{Fe} is linear for the various alkylmetals in Figure 5 (left), and the slopes of the correlations represent $\mathcal{F}/4\sqrt{\Delta G_0^{\ddagger}}$. It is important to note that the most sterically hindered alkylmetal examined, viz., tetra-neopentyltin, is precisely included with the other alkylmetals in the correlations shown in both Figures 4 and 5. In other words, alkylmetals are not distinguished from each other on the basis of their steric properties when they undergo oxidation with FeL_3^{3+} -- the electron donor property of the alkylmetal being the dominant factor. Such a conclusion accords with an outer-sphere mechanism for electron transfer in eq 1. The activation free energies $\Delta G^{\ddagger}$ obtained from the kinetic data are plotted against the driving force in Figure 5 (right). The dashed line in the figure represents $\Delta G^{\ddagger}$ calculated from the Marcus eq 4 using the single value of $\Delta G_0^{\ddagger} = 10$ kcal mol^{-1}. Thus electron transfer from alkylmetals to FeL_3^{3+} accord well with the Marcus theory. The absence of notable steric effects in this outer-sphere process must be emphasized.

Heterogeneous Processes at a Platinum Electrode. The series of organometals I-IV are also readily oxidized electrochemically (9). Thus we can apply the same steric probe to the corresponding anodic process for which the analogous mechanism for heterogeneous electron transfer at an electrode [E] is represented by an electrochemical EC sequence shown in Scheme II.

$$RM \xrightarrow{[E]} RM^{+} + e \tag{7}$$

$$RM^{+} \xrightarrow{fast} R\cdot + M^{+}, \text{ etc.} \tag{2}$$

Scheme II

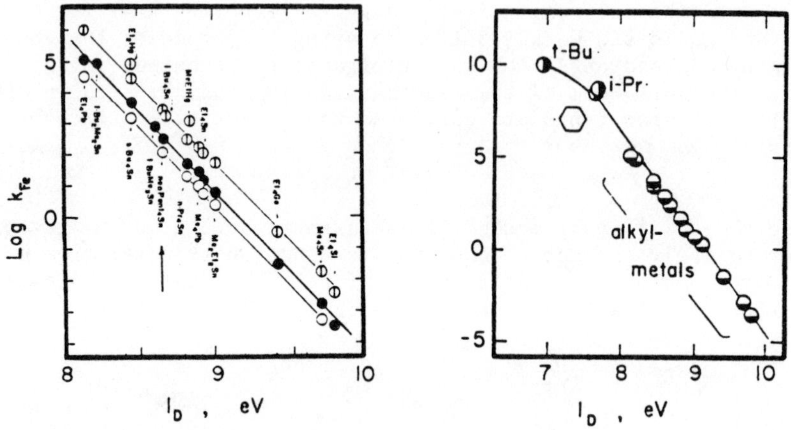

Figure 4. Correlation of the ionization potentials of alkylmetal donors with the electron-transfer rate constant (log k_{Fe}) for $Fe(phen)_3^{3+}$ (●), $Fe(bpy)_3^{3+}$ (○), and $Fe(Cl\text{-}phen)_3^{3+}$ (⦶), (left). The figure on the right is the same as the left figure for $Fe(phen)_3^{3+}$ except for the inclusion of electron-transfer rates for some alkyl radicals as identified. (Note the expanded scale.)

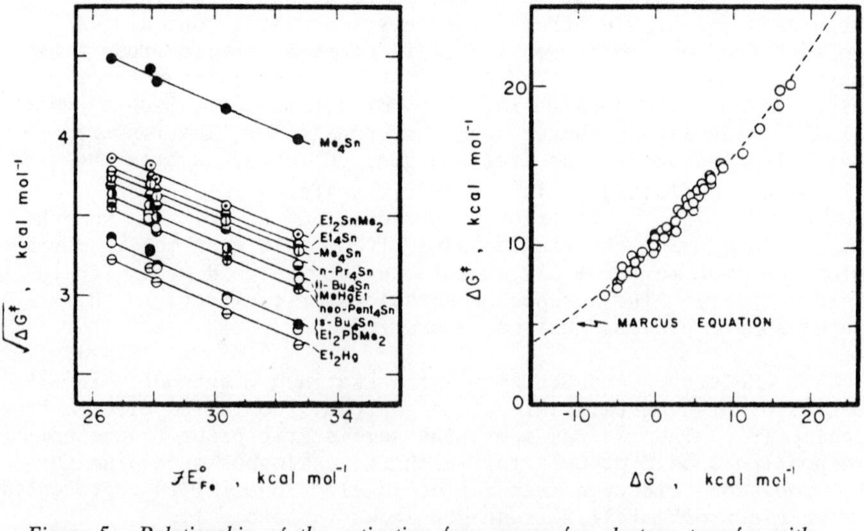

Figure 5. Relationship of the activation free energy for electron transfer with the electrode potentials of various FeL_3^{3+} according to Equation 6 (left), and the driving force according to the Marcus Equation 4 (right).

The heterogeneous rates of electron transfer in eq 7 were measured by two independent electrochemical methods: cyclic voltammetry (CV) and convolutive potential sweep voltammetry (CPSV). The utility of the cyclic voltammetric method stems from its simplicity, while that of the CPSV method derives from its rigor.

For a totally irreversible electrochemical process, the heterogeneous rate constant k_e for electron transfer at the CV peak potential E_p is given by

$$k_e(E_p) = 2.18[D\beta n\mathcal{F}v/RT]^{1/2} \qquad (8)$$

where n is the number of electrons transferred in the rate-limiting step, and v is the scan rate. The diffusion coefficient D and the transfer coefficient β are measured by independent techniques. Equation 8 is derived from Fick's laws of diffusion together with the usual expression for the linear potential dependence of the heterogeneous rate constant k_e for electron transfer, viz.,

$$k_e(E) = k_s \exp[(\beta n\mathcal{F}/RT)(E - E^0)] \qquad (9)$$

where E^0 is the standard potential for the organometal which fixes the value of k_s. [Alternatively stated, k_s is the intrinsic rate constant for electron transfer at the equilibrium potential, i.e., $E = E^0$. It is thus related to the intrinsic barrier $\Delta G_0^{\ddagger}$ in Marcus theory.] The shift in the current maximum with the sweep rate accords with standard electrochemical theory (10), and may be used with the aid of eq 8 to determine the heterogeneous rate constant k_e for electron transfer at different values of the applied potential (driving force). The results of this analysis are shown in Figure 6 for the representative organometals I-IV (11). The linear plots in Figure 6 correspond to the free energy relationship in eq 9, with the slope representing the electrochemical transfer coefficient β, i.e.,

$$\beta = 2.3RT/n\mathcal{F}[\partial \log k_e/\partial E] \qquad (10)$$

A closer scrutiny of Figure 6 reveals the persistence of small, but consistent curvature in all of the plots. In order to verify the curvature, the transfer coefficient β was also determined independently from the width of the CV wave, as described by Nicholson and Shain (10). The potential dependence of β obtained in this manner is shown in Figure 7. The slopes ∂β/∂E represent the unmistakable presence of curvature in Figure 6.

We deemed it necessary to confirm the CV results by the alternate method using convolutive potential sweep voltammetry, which requires no assumptions as to the form of the free energy relationship and is ideally suited for an independent analysis of curvature revealed in Figure 7. In convolutive linear sweep voltammetry, the heterogeneous rate constant k_e is obtained from the cur-

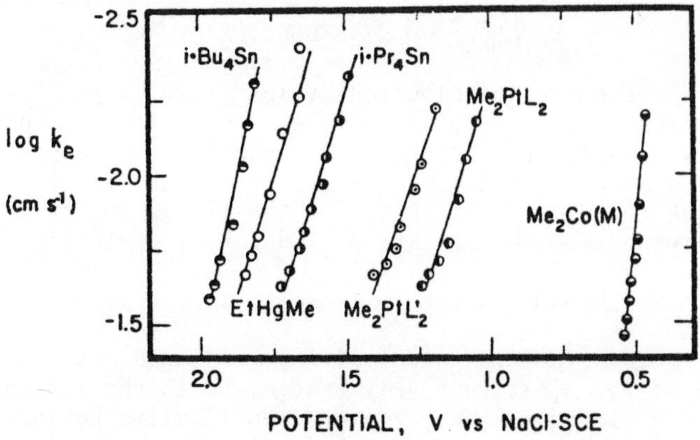

Figure 6. Variation of the heterogeneous rate constant (log k_e) with the applied potential for some representative organometals.

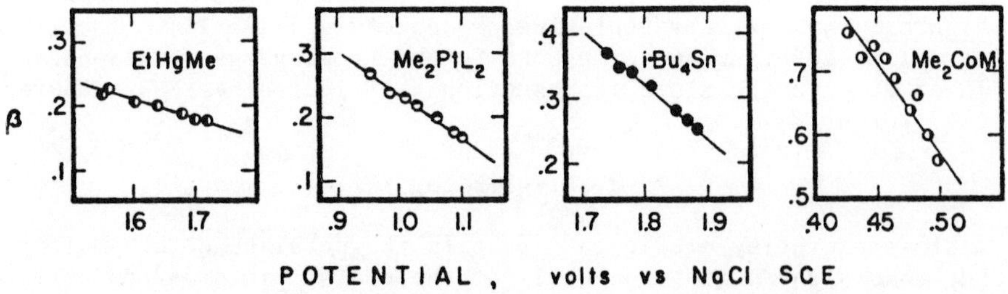

Figure 7. Dependence of the heterogeneous transfer coefficient β with the applied electrode potential for the representative class of organometals I–IV.

rent i(E) at a given applied potential by

$$k_e(E) = (n\mathcal{F}A)^{-1} i(E)/C(t) \qquad (11)$$

where A is the area of the electrode and C(t) is the time-dependent concentration of the organometal at the electrode surface (12). The values of C(t) can be evaluated from the current-time history using the convolution integral

$$C(t) = C^0[1 - \frac{10^3}{n\mathcal{F}AC(\pi D)^{1/2}} \int_0^t \frac{i(\eta)}{(t - \eta)^{1/2}} d\eta] \qquad (12)$$

where C^0 is the initial concentration of the organometal in the bulk solution. Indeed the CPSV method, which is kinetically rigorous, affords values of the rate constant k_e, the transfer coefficient β, and the curvature $\partial\beta/\partial E$, which are all in excellent agreement with those obtained by the more direct CV method (11).

Comparison of Homogeneous and Heterogeneous Electron Transfers. Let us now compare the values of the heterogeneous rate constants k_e for the anodic oxidation of organometals with the homogeneous rate constants k_{Fe} for the chemical oxidation of the same organometals with FeL_3^{3+}. The direct comparison of the electrochemical and chemical oxidations of organometals requires that they be evaluated at the same thermodynamic driving force, i.e., the applied electrode potential must be reasonably similar to the standard reduction potential E^0 of FeL_3^{3+}. This situation is unfortunately difficult to obtain experimentally since the anodic currents for the electrochemical oxidation of alkylmetals at the standard reduction potentials of FeL_3^{3+} (ranging from 0.92 to 1.18 V) are too small to be reliably measured. Conversely, at the applied electrode potentials convenient for the CV measurements of alkylmetals (ranging from 1.5 to 1.9 V), the homogeneous chemical rates of oxidation would be too fast to measure (even if such powerful oxidants were available). Thus, basically the problem is to find the free energy relationship which will allow the electron transfer rate constant, measureable at one potential, to be evaluated at another potential. We consider two free energy relationships: the well-accepted linear form in eq 9 and the quadratic Marcus eq 4.

The linear free energy relationship is obtained from the CV data by the combination of eq 8 and 9, i.e.,

$$k_e(E) = 2.18[\frac{D\beta n\mathcal{F}v}{RT}]^{1/2} \exp[\frac{\beta n\mathcal{F}}{RT}(E - E_p)] \qquad (13)$$

The striking correlation of the homogeneous and heterogeneous rate constants, when both processes are measured under conditions of equivalent thermodynamic driving force, is presented in Figure 8 (left) (9). Note the slope of 0.76. We now turn to the quadratic

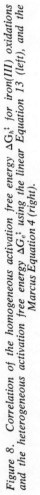

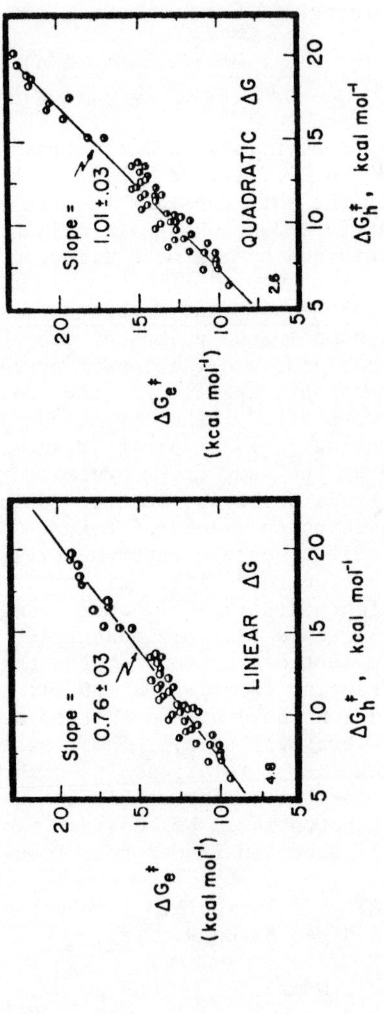

Figure 8. Correlation of the homogeneous activation free energy $\Delta G_h^{\ddagger}$ for iron(III) oxidations and the heterogeneous activation free energy $\Delta G_e^{\ddagger}$ using the linear Equation 13 (left), and the Marcus Equation 4 (right).

form of the Marcus theory to provide an alternative free energy relationship for the heterogeneous rate constants. The intrinsic barrier $\Delta G_0^{\ddagger}$ required in eq 4 can be evaluated from the measured values of the transfer coefficient β by the relationship

$$\Delta G_0^{\ddagger} = \Delta G^{\ddagger}/4\beta^2 \qquad (14)$$

which allows $\Delta G^{\ddagger}$ to be determined at any other potential from a knowledge of $\Delta G^{\ddagger}$ at the CV peak potential. The heterogeneous values of $\Delta G^{\ddagger}$ evaluated at the standard reduction potential of the iron(III) oxidants are plotted against the homogeneous values in Figure 8 (right) (11). The linear correlation with the slope of 1.01 (in contrast to 0.76 obtained with the aid of eq 13) represents strong support for a common outer-sphere mechanism for homogeneous and heterogeneous electron transfer from these organometals. An important feature in Figure 8 is the striking absence of steric effects arising from the organometal. Thus Figure 8 establishes the essential unity of outer-sphere mechanisms in homogeneous and heterogeneous electron transfer from organometals.

Historically, the slopes of the free energy relationships have been referred to separately as the Brønsted coefficient α for homogeneous processes and as the transfer coefficient β for heterogeneous processes. For example, to illustrate their distinction we noted that the homogeneous Brønsted slopes α lie in the range of 0.71 to 0.53 for the oxidation of sec-Bu_4Sn by various FeL_3^{3+}, whereas the electrochemical transfer coefficients fall in the significantly lower range of 0.2 and 0.35. However, when the differences in the driving forces are also taken into account, the values of α and β follow the same relationship, as shown in Figure 9. Three important conclusions can be drawn from the analysis in Figure 9. First, the difference between the electrochemical β and the chemical α slopes arises from the fact that the two are measured in different potential regions, necessitated solely by experimental limitations. Otherwise there is no inherent difference between α and β in the outer-sphere oxidation of these organometals. Second, the importance of the second-order term in the potential, i.e., the curvature in Figure 6, is confirmed for both the heterogeneous and the homogeneous rate processes. Thus a single value of either α or β cannot be used to describe both the homogeneous and heterogeneous rate data. Third, the accordance of β with α demonstrates that factors contributing to the potential dependence of the heterogeneous process must fundamentally be the same as that in the homogeneous electron transfer proceeding via an outer-sphere mechanism.

Electron Transfer Far From Equilibrium. We have shown how the Marcus Theory of electron transfer provides a quantitative means of analysis of outer-sphere mechanisms in both homogeneous and heterogeneous systems. It is particularly useful for predicting electron transfer rates near the equilibrium potential, but its applicability unfortunately decreases as the free energy change is extended to the exergonic and endergonic limits. The desirability of establishing an extended free energy relationship stems from its importance to systems which undergo interesting chemistry attendant upon electron transfer. These include an ever increasing variety of organic and organometallic systems in which questions relating to the involvement of (single) electron transfer or "SET" mechanisms have arisen. For these chemically irreversible systems, electron transfer rates are often measureable only at potentials (driving forces) far removed from equilibrium.

Most electron transfer studies have heretofore encompassed only a limited range in ΔG since they are largely circumscribed in homogeneous systems by the availability of chemical oxidants and reductants with standard reduction potentials of sufficient span. On the other hand, the electrochemical methods described in the foregoing section provide an excellent opportunity to examine such a range of electron transfer rates, simply by continuously tuning the applied electrode potential E in eq 9 over an extended region. Among the various organometals, the organocobalt macrocycle typified by $Me_2Co(M)$ shown in Figure 10 is unusually well suited in that the essential kinetic features of Scheme II have all been determined (13). Thus the direct comparison with electron transfer effected by various chemical oxidants in solution has established the fundamental unity of outer-sphere electron transfer between heterogeneous electrochemical and homogeneous chemical processes in this system. Furthermore, the first-order decay of the cation ($RM^+ = Me_2Co(M)^+$ in eq 2) has been independently verified to proceed with the first-order rate constant of 10^5 s^{-1}. The potential dependence of the electron transfer rate constant k_e determined by the CPSV method is noteworthy for two reasons. First, at the low potential end in Figure 10, i.e., the endergonic region, the slope of the relationship between the activation free energy ($\log k_e$) and the driving force (E) is close to unity, as shown by the line of unit slope (14). Second, at the other limiting extreme, of high potential, the net forward rate constant for electron transfer is observed to be clearly leveling off and becoming independent of the applied potential. The electron transfer rate behavior so graphically illustrated in Figure 10 is unique, and we know of no other kinetic measurement in which both limiting kinetic extremes have been covered, particularly in a single system. Consequently, we now wish to exploit the experimental information contained in Figure 10 to develop the free energy relationship for electron transfer over an extended range. In order to determine the functional dependence of the rate constant for electron transfer with the free energy change, we must first clearly delineate the distinction between the experimental rate constant k_e and the intrinsic rate constant k_1 for forward electron transfer in Scheme III. Indeed the relationship between the two rate constants is critically dependent on the lifetime of $Me_2Co(M)^+$, which is determined from the magnitudes of the rate con-

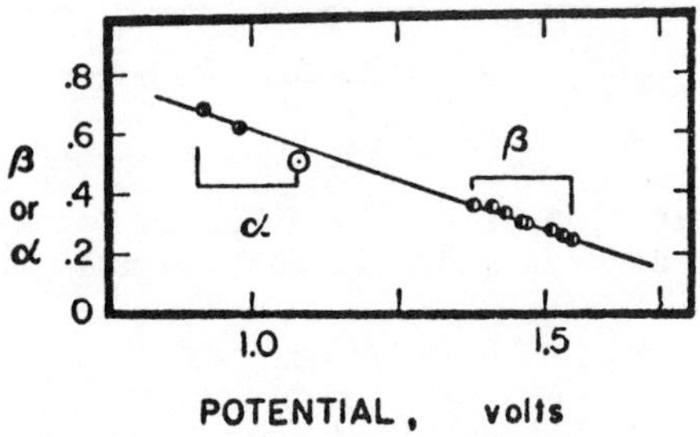

Figure 9. The direct relationship between the homogeneous Brønsted coefficient α (⊙) and the heterogeneous transfer coefficient β (◐) with the electrode potentials, as measured for sec-Bu_4Sn.

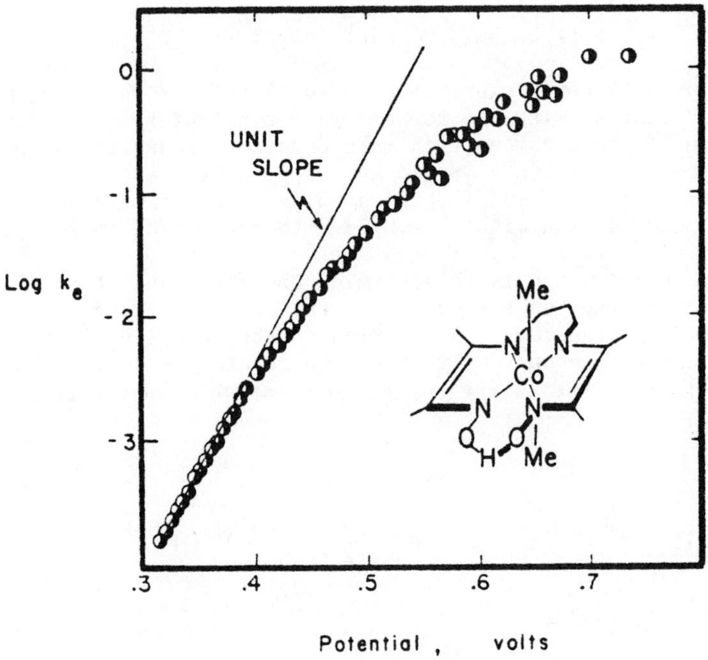

Figure 10. Variation in the experimental rate constant (log k_e*) for heterogeneous electron transfer as a function of the applied potential.*

stant k_{-1} for back electron transfer and the rate constant k_2 for decomposition, as graphically illustrated below.

$$Me_2Co(M) \underset{k_{-1}}{\overset{k_1}{\rightleftharpoons}} e^- + Me_2Co(M)^+ \xrightarrow{k_2} \text{products} \quad (\text{overall: } k_e)$$

Scheme III

The rigorous combination of the kinetics in Scheme III with Fick's laws of diffusion affords the relationship among these rate constants as:

$$k_e = k_1[k_2'/(k_{-1} + k_2')] \qquad (15)$$

which is the same as that obtained from a straightforward steady state treatment. [Note that the homogeneous rate constant k_2 in Scheme III has been converted to its heterogeneous equivalent k_2' for use in eq 15.] The evaluation of the intrinsic rate constant k_1 proceeds from the rearrangement of eq 15, followed by the combination with the Nernst expression to yield

$$k_1 = k_e k_1'\{k_2' - k_e \exp[(-n\mathcal{F}/RT)(E - E^0)]\}^{-1} \qquad (16)$$

The complete kinetic expression in eq 16 relates the experimental rate constant k_e with the forward rate constant k_1, as a direct function of the decomposition rate constant k_2' and the standard reduction potential E^0. Since an independent measurement of $k_2' = 1.2\ cm\ s^{-1}$ (or $k_2 = 10^5\ s^{-1}$) is available for $Me_2Co(M)^+$, it can be used in conjunction with $E^0 = 0.53$ V to convert k_e to k_1, shown in Figure 11 (14).

Our problem now is to determine the functional form of this experimental free energy curve for the intrinsic rate constant k_1 for electron transfer. In addition to the Marcus eq 4, two other relationships are currently in use to relate the activation free energy to the free energy change in electron transfer reactions (15,16).

Marcus:
$$\Delta G^{\ddagger} = \Delta G_0^{\ddagger}[1 + \frac{\Delta G}{4\Delta G_0^{\ddagger}}]^2 \qquad (4)$$

Rehm-Weller:
$$\Delta G^{\ddagger} = \frac{\Delta G}{2} + [(\frac{\Delta G}{2})^2 + (\Delta G_0^{\ddagger})^2]^{1/2} \qquad (17)$$

Marcus-Levine-Agmon:
$$\Delta G^{\ddagger} = \Delta G + \frac{\Delta G_0^{\ddagger}}{\ln 2}\ln\{1 + \exp[-\frac{\Delta G \ln 2}{\Delta G_0^{\ddagger}}]\} \qquad (18)$$

The functional form of the free energy relationships among these three formulations is typically illustrated in Figure 12 (17).

Each of these free energy relationships employs the intrinsic barrier $\Delta G_0^{\ddagger}$ as the disposable parameter. [The intrinsic barrier represents the activation energy for electron transfer when the driving force is zero, i.e., $\Delta G^{\ddagger} = \Delta G_0^{\ddagger}$ at $\Delta G = 0$ or the equili-

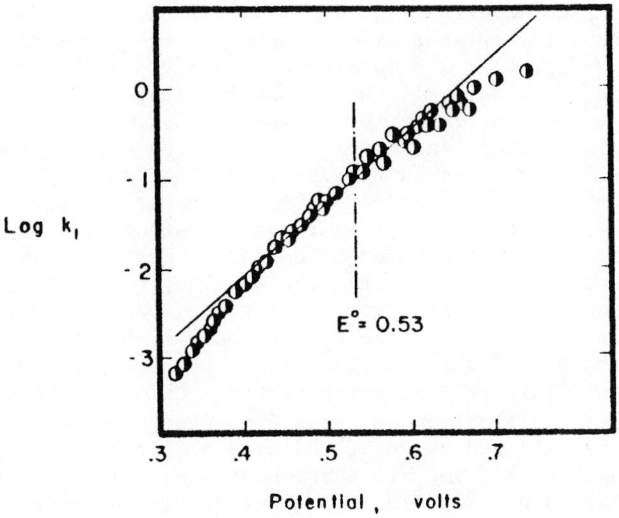

Figure 11. Free energy relationship for the intrinsic rate constant (log k_l) for heterogeneous electron transfer.

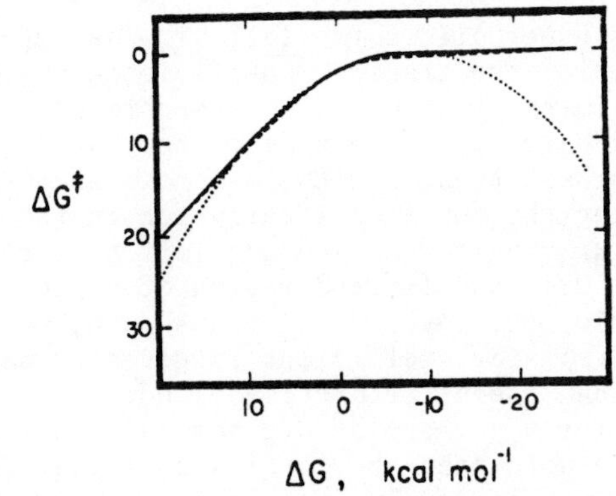

Figure 12. The shapes of the free energy relationships for electron transfer according to Marcus (Equation 4) (• • •), Rehm–Weller (Equation 17) (– – –), and Marcus–Levine–Agmon (Equation 18) (——).

brium potential E^0.] Since the intrinsic barrier is the single parameter in eqs 4, 17, and 18, the applicability of these relationships to the experimental free energy curve in Figure 11 is best carried out by testing each for the consistency of $\Delta G_0^{\ddagger}$, which was calculated from the experimental value of $\Delta G^{\ddagger}$ at each ΔG. The result is graphically illustrated in Figure 13. Several features in Figure 13 are noteworthy. First, in the equilibrium region of $E^0 = 0.53$ V, all three relationships yield a consistent value of the intrinsic barrier as $\Delta G_0^{\ddagger} = 6.3$ kcal mol^{-1}. [Note that the scatter of points in the exergonic region arises from the experimental difficulties in the measurement of k_e.] Second, the applicability of the Marcus eq 4 is limited to the region of the free energy change about the equilibrium potential. Significant deviations occur in the endergonic region, particularly at potentials less than ~0.4 V. Third, the Rehm-Weller and the Marcus-Levine-Agmon relationships are equally applicable over the entire span of the experimental free energy change. Both relationships yield values of $\Delta G_0^{\ddagger}$ which deviate less than 0.3 kcal mol^{-1} only at the extrema. This test thus represents a unique example of the experimental verification of these free energy relationships over an unusually extended range of the driving force. Owing to its simpler functional form, all subsequent discussions of the free energy relationship for electron transfer will be based on the Rehm-Weller relationship.

The Rehm-Weller formulation of the free energy relationship in eq 17 can be rewritten as

$$\Delta G^{\ddagger}(\Delta G^{\ddagger} - \Delta G) = \Delta G_0^{\ddagger 2}$$

to emphasize its hyperbolic form (18). The asymptotes of $\Delta G^{\ddagger} = 0$ and $\Delta G^{\ddagger} = \Delta G$ predict the transfer coefficients in the exergonic and endergonic limits to be 0 and 1, respectively. The magnitude of the intrinsic barrier $\Delta G_0^{\ddagger}$ determines how rapidly the transfer coefficients approach these limits, as shown in Figure 14 -- the right side showing the functional variation in the activation free energy with the driving force, and the left side the actual variation of β (19). Over the limited region of ΔG in which the Marcus eq 4 is applicable, the inverse relationship between the intrinsic barrier and the transfer coefficient is given by eq 14. Indeed, this remarkable and useful prediction is borne out by the constancy of $\beta^2 \Delta G_0^{\ddagger}$ for a variety of organometals in Table I (11). It is interesting to note that the intrinsic barrier for the octahedral $Me_2Co(M)$ is significantly less than those for either the tetrahedral tetralkyltin and lead compounds or the square planar dimethylplatinum(II) complex.

Inner-Sphere Electron Transfer

The series of organometal donors I-IV are also readily oxidized by the one-equivalent oxidant hexachloroiridate(IV) as well as by the organic acceptor tetracyanoethylene (TCNE) (3). In both

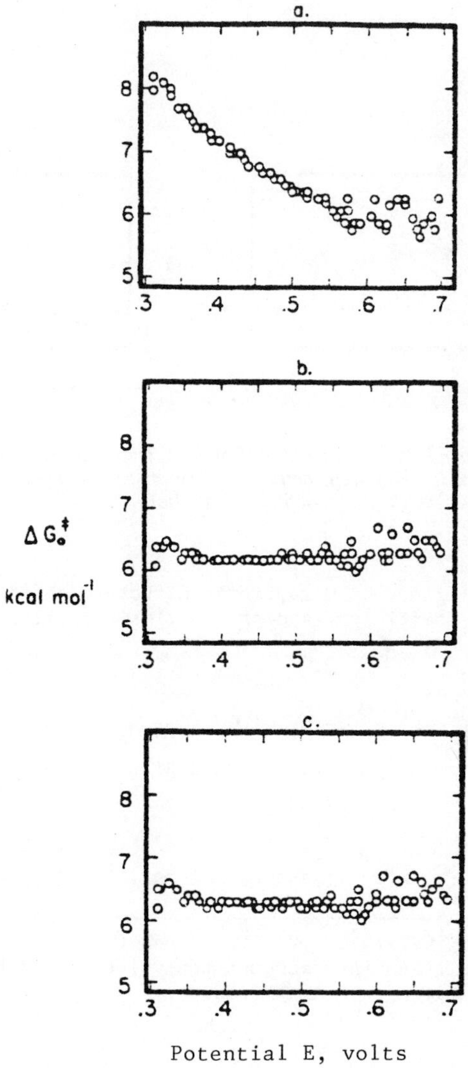

Figure 13. Test for the consistency of $\Delta G_0^{\ddagger}$ evaluated from the intrinsic rate constant (log k_I) using Marcus Equation 4 (a), Rehm–Weller Equation 17 (b), and Marcus–Levine–Agmon Equation 18 (c) at various potentials.

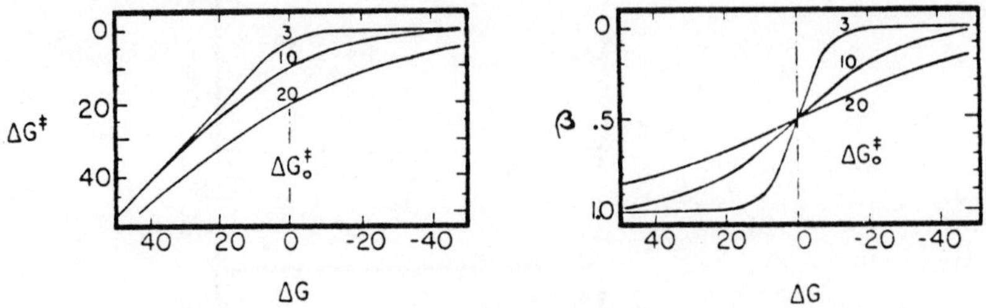

Figure 14. Dependence of the Rehm–Weller free energy relationship on the intrinsic barrier $\Delta G_0^{\ddagger}$. Key: left, activation free energy change; and right, transfer coefficient variation.

Table I. Effect of Organometal Structure on the Transfer Coefficient and the Intrinsic Barrier.

Organometal	$\Delta G^{\ddagger}$	β[a]	$\Delta G_0^{\ddagger}$	$\beta^2 \Delta G_0^{\ddagger}$
$Me_2Co(DpnH)$	7.46	0.73	2.1[b]	1.1
$(s\text{-}Bu)_4Sn$	7.65	0.31	10.0	1.0
Et_4Pb	7.68	0.28	10.0	0.8
$Me_2Pt(PMe_2Ph)_2$	7.69	0.27	13.0	0.9
EtHgMe	7.70	0.20	22.0	0.9

[a]Measured by CV at 100 mV s^{-1}. [b]When the effect of reversibility is taken into account the value is 6.3 kcal mol^{-1} for rate-limiting electron transfer.

cases, the rate-limiting step involves electron transfer, i.e.,

$$IrCl_6^{2-} + RM \longrightarrow IrCl_6^{3-} + RM^+ \quad (19)$$

$$TCNE + RM \longrightarrow TCNE^- + RM^+ \quad (20)$$

which is essentially the same stoichiometry described for the tris-phenanthrolinemetal(III) oxidants in eq 1.

Reactivity Patterns for $IrCl_6^{2-}$ and TCNE as Oxidants. Although the reactivity pattern for $IrCl_6^{2-}$ and TCNE are similar, they are quite distinct from the outer-sphere pattern established for FeL_3^{3+}. The contrast is graphically illustrated in Figure 15, in which the second-order rate constants (log k) for each of the three oxidants are plotted against the ionization potentials of the same series of alkylmetal donors (3). The uniquely linear correlation for $Fe(phen)_3^{3+}$ was discussed in the foregoing section for outer-sphere oxidation. The marked deviations of all the points for $IrCl_6^{2-}$ and TCNE from any analogous relationship is unmistakable. In particular, the most sterically hindered organometal tetra-neopentyltin (entered as entry 18 in Figure 15) shows the most pronounced deviant behavior. Furthermore in Figure 16, the experimental values of $\Delta G^{\ddagger}$ for $IrCl_6^{2-}$ (right) and TCNE (left) consistently fall below the solid lines representing the expected outer-sphere rates which were calculated from the Marcus eq 4 by taking into account the differences in the reorganization energies of $IrCl_6^{2-}$ and TCNE (3). With the exception of t-Bu_2SnMe_2 (entry 17) all the experimental rates are faster than the calculated values. The magnitudes of the deviations vary from 12.5 kcal mol^{-1} for the least sterically hindered Me_4Sn to ~0 for the hindered t-Bu_2SnMe_2. Indeed the departure from the outer-sphere correlation is the most pronounced with the last hindered alkylmetals (see Figure 2 for a qualitative expectation of steric effects). Thus among the symmetrical tetraalkyltin compounds R_4Sn, the least hindered methyl and n-alkyl derivatives all lie farthest from the outer-sphere correlation. Conversely those tetraalkyltin compounds with α- and β-branched alkyl ligands consistently lie closest to the lines in Figure 16.

The same steric dichotomy persists in the oxidation of alkyl radicals by tris-phenanthrolineiron(III), in which two competing processes occur -- viz., the oxidation to alkyl cations, previously identified as an outer-sphere process in eq 3,

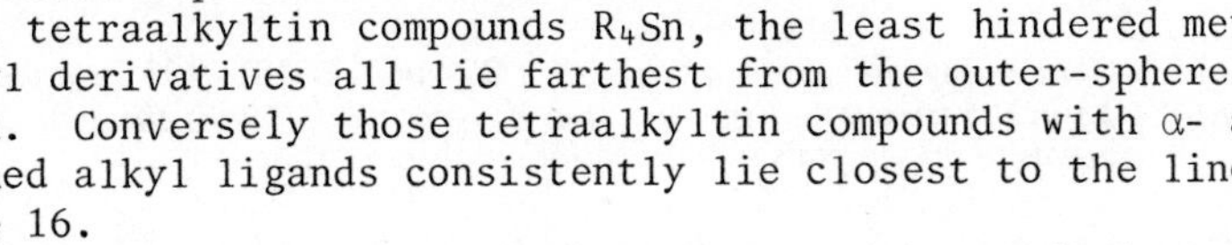

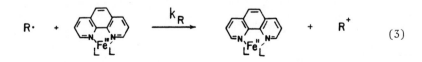

(3)

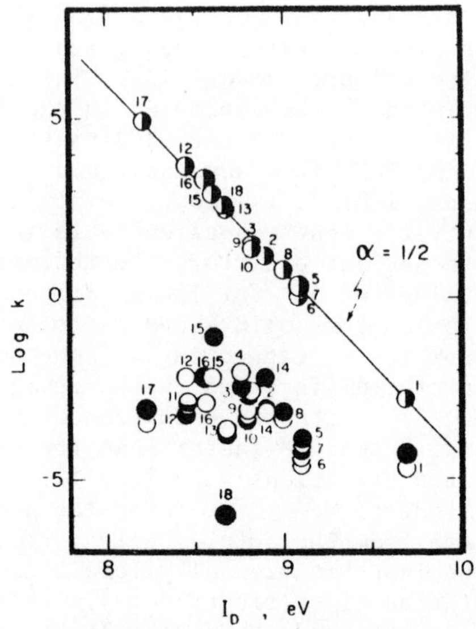

Figure 15. Contrasting behavior of $IrCl_6^{2-}$ (●) and TCNE (○) relative to $Fe(phen)_3^{3+}$ (◑) in the correlation of rates (log k) of oxidation with the ionization potentials of alkylmetals (identified by numbers in Reference 3).

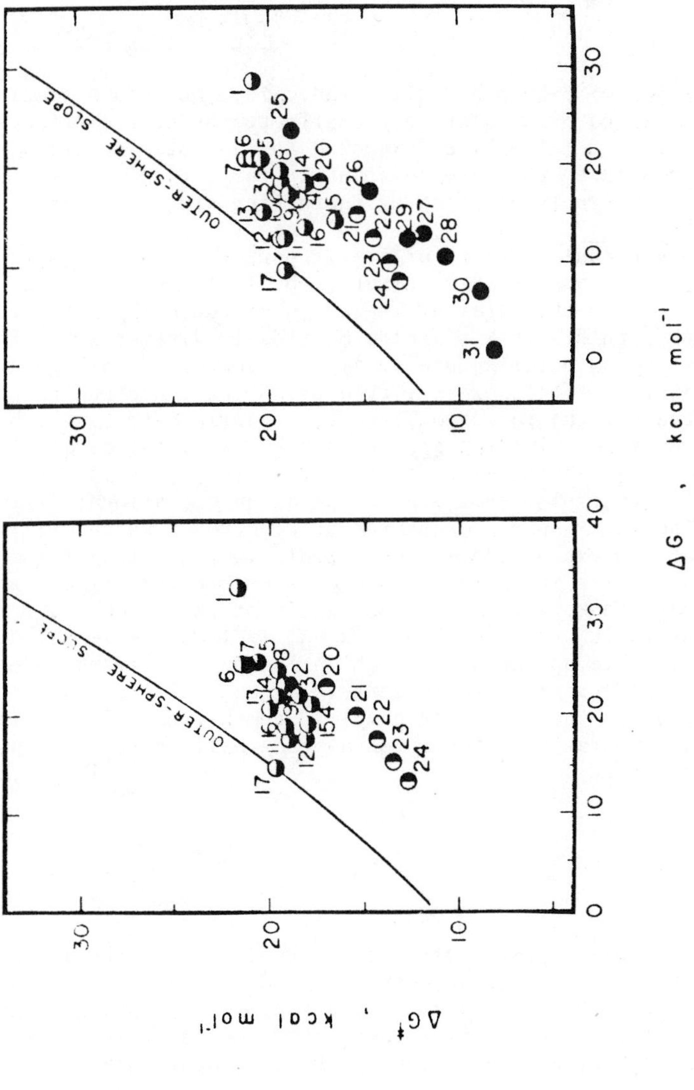

Figure 16. Relationship between the activation free energy and the driving force for electron transfer for alkylmetals to TCNE (left) and $IrCl_6^{2-}$ (right) according to Marcus Equation 4.

and aromatic substitution on the phenanthroline ligand as in eq 21 (7).

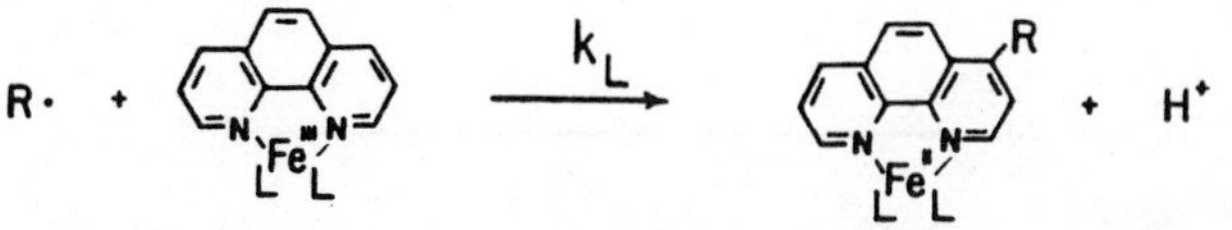

(21)

For the series of β-branched alkyl radicals, the second-order rate constant k_R in eq 3 is relatively unaffected by steric effects [compare Figure 2 (right)] as expected for an outer-sphere process. In strong contrast, the rate constant k_L for ligand substitution in eq 21 is adversely affected by increasing steric effects, as shown in Figure 17.

The measureable effects of steric perturbation on the rates of oxidation of encumbered alkylmetals by $IrCl_6^{2-}$ and TCNE in Figure 16, and on the rates of ligand substitution by alkyl radicals on FeL_3^{3+} in Figure 17, are to be clearly distinguished from the corresponding outer-sphere processes identified in Figure 4, in which steric effects exert little or no influence on the rates. Such a trend for the former must reflect steric effects which perturb the inner coordination sphere of the alkylmetal or alkyl radical donor in the transition state for electron transfer. Indeed we wish to employ this conclusion as an operational definition for the inner-sphere mechanism of electron transfer (3).

In the context of the Marcus formulation, the lowering of the activation barrier in an inner-sphere process could arise from the reduction of the work term w_p as a result of the strong interaction in the ionic products, e.g., $[R_4Sn^+\ IrCl_6^{3-}]$ and $[R_4Sn^+\ TCNE^-]$. The electrostatic potential in such an ion pair is attractive and may cause the tetraalkyltin to achieve a quasi five-coordinate configuration in the precursor complex, reminiscent of a variety of trigonal bipyramidal structures already well-known for tin(IV) derivatives, i.e.,

The extent to which steric effects adversely affect the attainment of such intimate ion-pair structures would be reflected in an increase in the work term and concomitant diminution of the inner-sphere rate. This qualitative conclusion accords with the reactivity trend in Figure 16. However, Marcus theory does not provide a quantitative basis for evaluating the variation in the work term of such ion pairs. To obtain the latter we now turn to the Mulliken theory of charge transfer in which the energetics of ion-pair formation evolve directly, and provide quantitative informa-

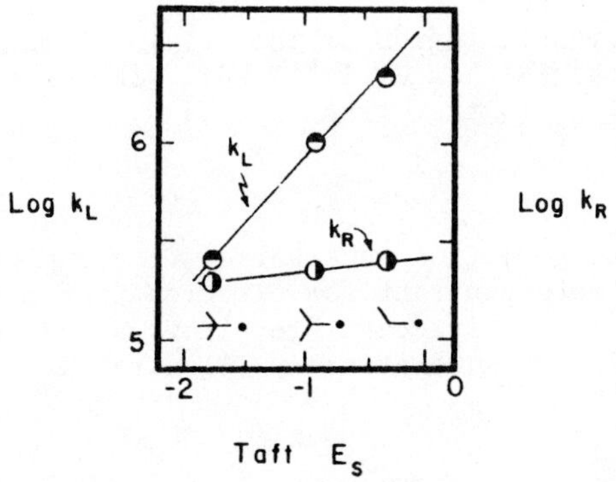

Figure 17. Steric effects (E_s) on the rates (log k_R) for the outersphere oxidation (◑) and the rates (log k_L) for the ligand substitution (◓) of $Fe(phen)_3^{3+}$ by β-branched alkyl radicals.

tion on the steric effects. Before doing so, however, let us formulate the problem in a more general context in which we consider organometals as electron donors D and the oxidants $IrCl_6^{2-}$ and TCNE as electron acceptors A.

Mechanistic Formulation of Electron Transfer. The Importance of the Work Term. Accordingly, the electron transfer mechanism can be considered in the light of the standard potentials E^0 for each redox couple, i.e., E^0_{ox} for the oxidation of the donor ($D \rightarrow D^+ + e^-$) and E^0_{red} for the reduction of the acceptor ($A + e^- \rightarrow A^-$). Thus the general reaction scheme for an irreversible process is represented by (20):

$$D + A \underset{}{\overset{K_{DA}}{\rightleftharpoons}} [D\ A] \xrightarrow{k_{et}} [D^+A^-] \xrightarrow{fast} \text{Products}$$

in which electron transfer in the encounter complex [D A] is rate-limiting. This scheme is particularly applicable to systems which lie in the endergonic region of the driving force [i.e., where the standard free energy change, $\Delta G^0 \gg 0$]. The observed second-order rate constant k_{obs} is then given by:

$$k_{obs} = k_{et}\, K_{DA} \qquad (22)$$

where K_{DA} is the formation constant of the complex and k_{et} is the intramolecular rate constant for electron transfer. The thermochemical cycle for an irreversible electron transfer in the highly endergonic region is schematically illustrated below.

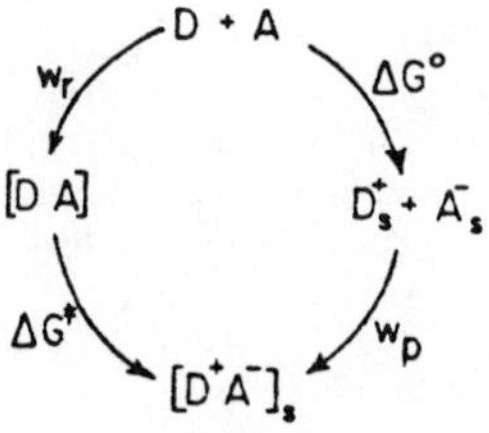

Scheme IV

According to the thermochemical cycle in Scheme IV, the activation free energy for electron transfer $\Delta G^\ddagger$ in the encounter complex is given by (16):

$$\Delta G^\ddagger = \Delta G^0 + w_p - w_r \qquad (23)$$

ΔG^0 corresponds to the standard free energy change of the redox process: $D + A \rightarrow D^+ + A^-$, and the work terms w_p and w_r represent the energy required to bring together the products and reactants, respectively, to within the mean separation r_{DA} in the ion pair $[D^+ A^-]$. The observed second-order rate constant in eq 22 can be expressed in terms of the activation free energy $\Delta G^{\ddagger}$ in eq 23 which leads to the free energy relationship for electron transfer as:

$$\log k_{obs} = -\frac{1}{2.3RT}(\Delta G^0 + w_p) + C_1 \tag{24}$$

where $C_1 = \log (\kappa T/h)$. Since the free energy change for electron transfer ΔG^0 is obtained from the standard potentials of the donor and acceptor [i.e., $\Delta G^0 = \mathcal{F}(E^0_{ox} - E^0_{red})$], the free energy relationship in eq 24 may be represented by an equivalent form in eq 25,

$$\log k_{obs} = -\frac{1}{2.3RT}(\mathcal{F}E^0_{ox} + w_p) + \text{constant} \tag{25}$$

when one is dealing with a series of electron transfer reactions with a fixed acceptor. Thus eq 24 and 25 state that the observed rate constant ($\log k_{obs}$) for electron transfer can be expressed as the sum of the standard free energy change (ΔG^0) and the work term of the products (w_p). Previous considerations of electron transfer have tended to focus only on the magnitude of ΔG^0, without explicitly taking into account the energetics of ion pair $[D^+ A^-]$ formation, i.e., the work term w_p. Such an oversimplication is perhaps understandable if one considers the experimental difficulty of directly evaluating the work terms of ion pairs by the usual procedures. The problem is severely compounded by the exceedingly short lifetimes of the transient ion pairs which must be involved in irreversible electron transfer. We now wish to discuss how the observation of charge transfer spectra in these systems can be used to evaluate such ion-pairing energies.

Evaluation of the Work Term from Charge Transfer Spectral Data. The intermolecular interaction leading to the precursor complex in Scheme IV is reminiscent of the electron donor-acceptor or EDA complexes formed between electron donors and acceptors (21). The latter is characterized by the presence of a new absorption band in the electronic spectrum. According to the Mulliken charge transfer (CT) theory for weak EDA complexes, the absorption maximum $h\nu_{CT}$ corresponds to the vertical (Franck-Condon) transition from the neutral ground state to the polar excited state (22).

$$[D\ A] \xrightarrow{h\nu_{CT}} [D^+A^-]^* \tag{26}$$

The asterisk identifies an excited ion pair with the same mean separation r_{DA} as that in the precursor or EDA complex. The thermochemical cycle for such a charge transfer transition is schematically illustrated below.

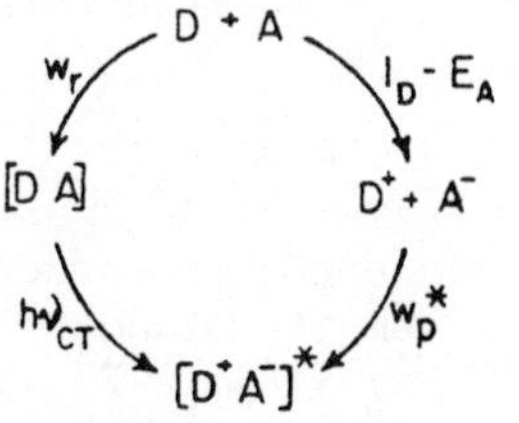

Scheme V

The vertical ionization potential of the electron donor is represented by I_D, and the electron affinity of the acceptor by E_A.

The charge transfer Scheme V is akin to the adiabatic electron transfer cycle in Scheme IV. In this case the work term w_p^* required to bring the products D^+ and A^- together to the mean separation r_{DA} in the CT excited state is given by:

$$w_p^* = h\nu_{CT} - I_D + C_2 \quad (27)$$

where $C_2 = E_A + w_r$ and considered to be constant for a fixed acceptor (23). In the limit of simple ion-pair interactions, the work term is largely coulombic, i.e., $w_p^* \cong -e^2/r_{DA}$. We relate the work term w_p^* to that of a reference donor w_p^{0*} since the series of organometals in Figure 15 involves either TCNE or $IrCl_6^{2-}$ as the common accpetor. This comparative procedure allows the acceptor component in a given series to drop out by cancellation, and it follows from eq 27 that:

$$\Delta w_p^* = w_p^* - w_p^{0*} = \Delta h\nu_{CT} - \Delta I_D \quad (28)$$

In eq 28, $\Delta h\nu_{CT}$ refers to the relative CT transition energy with a common acceptor, and ΔI_D is the difference in the ionization potentials of the donor and the chosen reference. Thus by selecting a reference donor, we can express all changes in the ion-pairing energies in the CT excited state (including steric, distortional, and other effects in the D/A pair) by a single, composite energy term Δw_p^*.

We now return to the thermal electron transfer reaction in eq 20, in which the rate-limiting activation process has been shown to proceed from the electron donor acceptor complex (23), i.e.,

$$[RM\ TCNE] \longrightarrow [RM^+\ TCNE^-] \quad (29)$$

The variation in the driving force in eq 29 derives from two factors: (1) the differences in the oxidation potentials of the alkylmetals and (2) the changes in the work terms in the ion pairs resulting from differences in the steric effects (i.e., mean separations). The first is obtained from the free energy change for electron transfer (i.e., $RM \rightarrow RM^{+} + e$) in the outer-sphere process (see eq 6). The change in the work term is equated to Δw_p^*, determined from the CT interaction in eq 28 (24). Thus the linear correlation in Figure 18 (left) between the driving force and the activation free energy represents a linear free energy relationship for electron transfer expressed as:

$$\Delta G^{\ddagger} = \Delta G^0 + \Delta w_p^* + C \tag{30}$$

where the constant C represents the reference terms (3). Furthermore, the same relationship applies to $IrCl_6^{2-}$, as shown by the correlation in Figure 18 (right). [The applicability of the values of Δw_p^* obtained from TCNE to the inner-sphere oxidation with $IrCl_6^{2-}$ is probably fortuitous. An indication of their similarity is, however, apparent from a comparison of Figure 15 or 16.]

Applications of the Ion-Pair Formulation. Electron transfer mechanisms are receiving increased attention in a wide variety of organic and organometallic reactions such as (1) the addition of halogens to alkenes, (2) the electrophilic aromatic substitution, (3) the Diels-Alder reaction, and (4) the halogenolysis of alkylmetals (24). Such mechanisms have been referred to as single electron transfer or SET since they derive their driving force from one component ($A = Br_2$, $HgCl_2$, $(NC)_2C{=}C(CN)_2$, I_2) acting as an one-electron acceptor relative to the other component ($D = C_6H_6$, $CH_2{=}CH_2$, C_6H_8, Me_2Hg) which is then considered to be the electron donor. In each case, a charge transfer interaction can be identified in the precursor complex (24), and typical variations in the relative work term Δw_p^* are shown in Figure 19 as a function of the ionization potential of the donor (25).

Several features in Figure 19 are noteworthy. First, the magnitude and the variation of the work term depends strongly on the electron donor as well as the acceptor. Thus w_p^* is rather constant for TCNE when it interacts with various substituted-anthracene donors (Figure 19a), but it decreases significantly as the ionization potential of the organometal donors increases (Figure 19b). Furthermore, the variation in w_p^* is the largest with organometals, when considered among other donors such as arenes and alkenes. With a particular series of donors, the variation of w_p^* is the largest with the mercury(II) complexes $Hg(OAc)_2$ and $HgCl_2$, among other acceptors such as TCNE and Br_2.

We associate such variations in the work term w_p^* with changes in the mean separation r_{DA} in the EDA complexes. Qualitatively, such changes may be viewed as steric effects which hinder the close approach of the acceptor and the donor. For example, the constancy of w_p^* for the substituted-anthracene donors accords with the minor steric perturbation a substituent is expected to exert

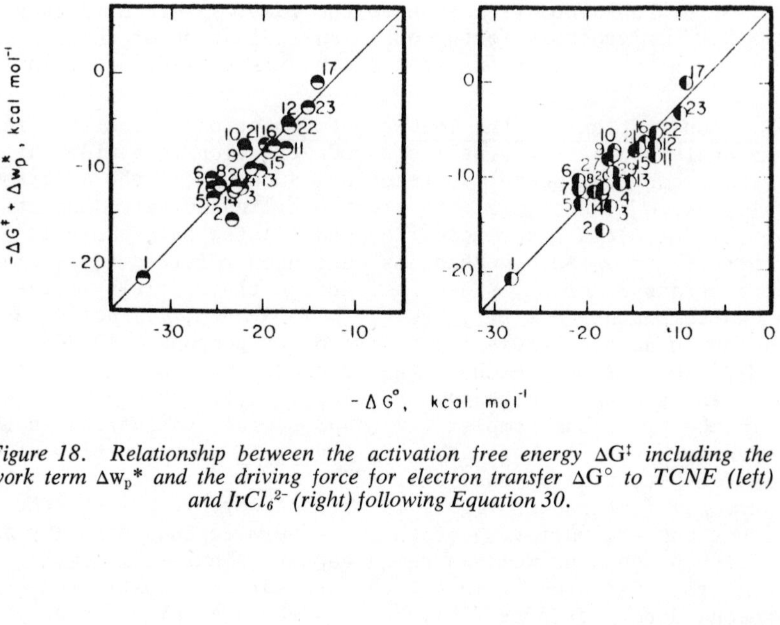

Figure 18. Relationship between the activation free energy $\Delta G^{\ddagger}$ including the work term $\Delta w_p{}^*$ and the driving force for electron transfer ΔG° to TCNE (left) and $IrCl_6{}^{2-}$ (right) following Equation 30.

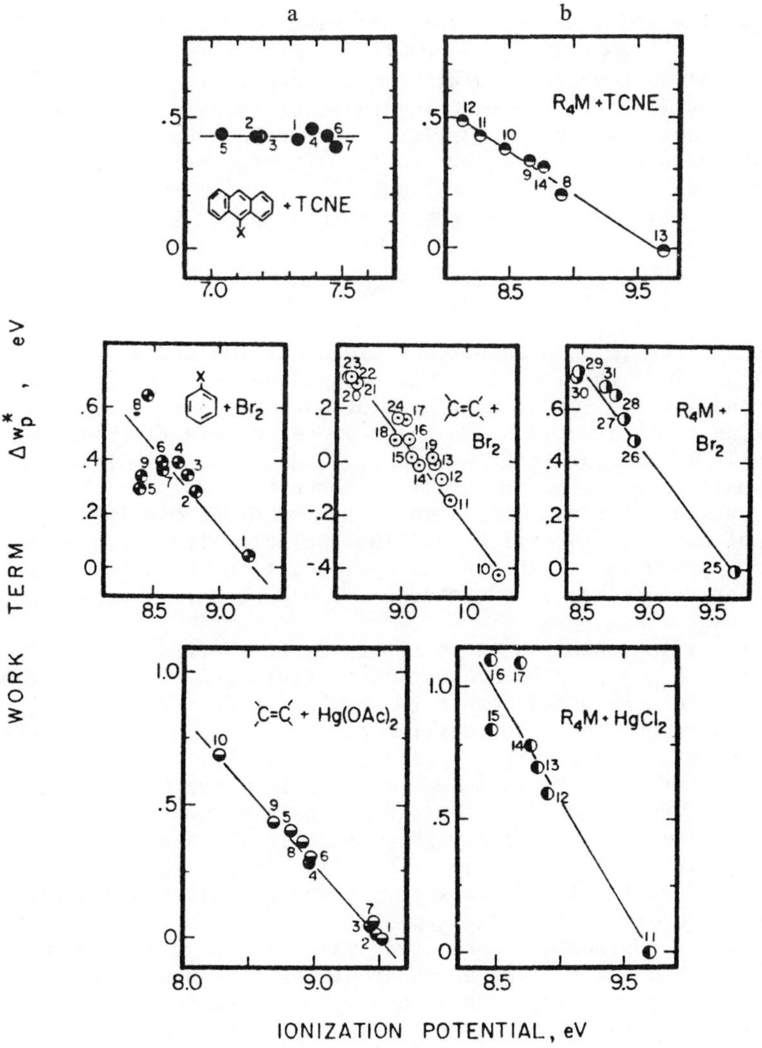

Figure 19. Variation of the work term Δw_p^* evaluated for various charge-transfer complexes.

on a large diffuse π-donor such as a polycyclic aromatic network. By contrast, bulky substituents on alkylmetals can effect large changes in w_p^* owing to the localized nature of the interaction in these σ-donors.

These work terms can be employed in the free energy relationship in the same manner as that described for eq 20 above. For example, in Figure 20 (top), we find a direct relationship between the E^0_{ox} of anthracenes and alkylmetals and their reactivity ($\log k_{obs}$) toward TCNE. The experimental plot shows two linear, but separate correlations (25,26). It is noteworthy that the two different correlations in Figure 20 (top) emerge as a single correlation when the work term Δw_p^* is included. The result in right figure is a linear free energy relationship,

$$\log k_{obs} = -\frac{1}{2.3RT}\left[\mathcal{F}E^0_{ox} + \Delta w_p^*\right] + \text{constant} \qquad (31)$$

which applies to both anthracene cycloaddition and alkylmetal insertion. Note the line is arbitrarily drawn with a slope of unity to emphasize the fit of both sets of data to eq 31. Such an unification is remarkable since these processes clearly represent inherently dissimilar reaction types. Thus the Diels-Alder reaction is usually considered to involve a concerted [2π + 4π] cycloaddition, whereas the alkylmetal insertion requires a multistep scission of a σ carbon-metal bond. The applicability of eq 31 to such diverse processes as anthracene cycloaddition and alkylmetal insertion derives from a correspondence of the work term variation, i.e., $\Delta w_p^* = \Delta w_p$.

The relationship between the rates of electrophilic bromination ($\log k_{obs}$) and the standard potentials (E^0_{ox}) of the arene, alkene, and alkylmetal donors is shown in Figure 20 (middle). Although a rough linear correlation exists for each series, there is no clearcut interrelationship among any of them. However, if the variation of the work term Δw_p^* derived from the CT data is included, the three separate relationships become a single correlation. The result in the right figure is the same linear free energy relationship expressed in eq 31. Indeed the line in the figure is arbitrarily drawn with a slope of unity to emphasize the correlation of the data with eq 31.

Figure 19 (bottom) shows that the cleavage of alkylmetals by mercuric chloride follows an apparent negative Brønsted slope α (left figure). However, when the work term w_p^* is included with E^0_{ox}, the curved free energy relationship with negative slopes is transformed into the linear correlation shown on the right.

Significance of the Brønsted Slope in Electron Transfer. Linear free energy relationships have been extensively studied for electron transfer and related reactions in both inorganic and organic systems. For highly endergonic reactions, the Brønsted slope α is close to unity. In many cases, however, the more or

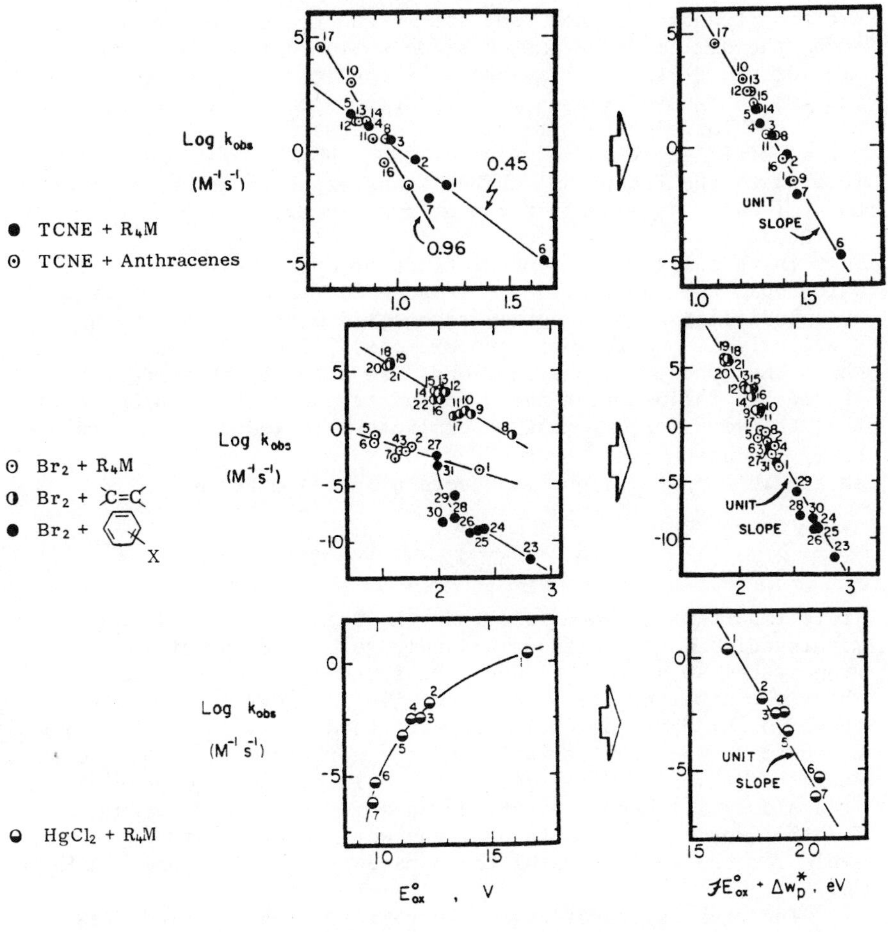

Figure 20. Unified free energy relationship for ion-pair formation. Key: left, free energy relationships between the rates of reaction (log k_{obs}) and the oxidation potential $E_{ox}°$ of the donor; and right, after inclusion of the work term following Equation 31. (Keys to symbols are located at the far left.)

less linear plots of $\Delta G^{\ddagger}$ (or a related quantity such as the rate constant log k_{obs}) and ΔG^0 (or some related quantity such as the polarographic $E_{1/2}$ potentials) have been observed with slopes significantly less than unity (e.g., typically $0.1 < \alpha < 0.5$) in systems showing irreversible electron transfer in the endergonic region. Such contradictions have led to the explanation that electron transfer is not complete in the transition state. As such, the diminished Brønsted slope would correspond to the fraction of charge transfer in the activated complex. Alternatively, the small Brønsted slopes may be ascribed to large intrinsic barriers $\Delta G_0^{\ddagger}$ for electron transfer, since the limit of α will be 0.5 for a sufficiently large value of $\Delta G_0^{\ddagger}$. For the latter, α is not related to the fractional charge transfer which is taken to be unity in all cases, but it cannot readily explain those cases with $\alpha < 0.5$.

In this light, our observations of apparent Brønsted slopes which span the limit of $\alpha = 1$ [Figure 20 (top)] to $\alpha \sim 0$ [Figure 20 (middle)] cannot be easily reconciled with either explanation especially if one includes the examples [Figure 20 (bottom)] in which the apparent Brønsted slopes are actually negative. However it clearly follows from the electron transfer cycle in Scheme IV that the work term w_p must be explicitly included in the free energy relationship. Such a FER is expressed by eq 23, and the intrinsic Brønsted slope is accordingly evaluated as:

$$\alpha = \frac{\partial \Delta G^{\ddagger}}{\partial \Delta G^0} = 1 + \frac{\partial w_p}{\partial \Delta G^0} \qquad (32)$$

It is important to emphasize that the Brønsted slope is not a meaningful quantity without a knowledge of the variation of w_p from system to system. Indeed the systems shown in Figure 20, for which the apparent Brønsted slopes lie in the range $0 < \alpha < 1$, are those in which variation in the work term is $-1 < (\partial w_p/\partial \Delta G^0) < 0$. Furthermore when the variation of the work term is large enough such that $(\partial w_p/\partial \Delta G^0)$ is less than -1 in eq 32, the apparent Brønsted slope will become negative, as observed in Figure 20 (bottom). In other words the variation of the work term is largely responsible for the magnitudes of the apparent Brønsted slopes in Figure 20.

The right sides of Figure 20 represent the unified free energy relationship in eq 31 and uniformly support the use of the CT work term w_p^* in the experimental verification of the thermochemical cycle in Scheme IV (eq 23). Thus the work term w_p^* is sufficient to unify all those systems with small and even negative values of the apparent Brønsted slope into a single general relationship in eq 31. At this juncture, we might enquire as to why the work term evaluated from the CT interaction in Scheme V serves as such an adequate representation of the adiabatic work term in Scheme IV.

By employing the variation of the work term Δw_p^* in eq 31, we are actually focussing in large part on the electrostatic interaction in the ion pair, since the other structural factors drop out by cancellation in the comparative procedure used in eq 28. Thus the contribution to Δw_p^* consists largely of a coulombic potential (i.e., $-e^2/r_{DA}$) in which the mean separation r_{DA} in the ion pair is the determinant factor. It is then likely that steric effects which mediate the magnitude of r_{DA} in the EDA complex, and thus the CT excited ion-pair state, parallel those in the adiabatic ion pair in Schemes V and IV, respectively.

Conclusions and Outlook

The activation processes for outer-sphere electron transfer in both homogeneous and heterogeneous systems are well provided by Marcus theory, particularly when they are examined near the equilibrium potential. However, the applicability of the Marcus equation decreases as the free energy change is removed from this benchmark into the exergonic and endergonic limits. Although the Rehm-Weller formulation provides a closer functional relationship between $\Delta G^{\ddagger}$ and ΔG over an extended range of the driving force, as an empirical equation it does not lend any additional insight as to the causes of the failure of eq 4 to predict these rates.

The susceptibility to steric effects can serve to differentiate the outer-sphere mechanism for electron transfer from the more ubiquitous inner-sphere processes for organometals. If the Marcus theory is applied directly to the inner-sphere mechanism, a large reduction in the work term is necessitated in order to accommodate rates which can be many orders of magnitude faster than the predictions of eq 4. This change in the adiabatic work term w_p as a result of steric effects is functionally equivalent to the change in the charge transfer work term w_p^* evaluated from Mulliken theory. Operationally, the unification of Marcus electron transfer and Mulliken charge transfer theories can provide a quantitative basis for predicting the rates of inner-sphere electron transfer with organometals (3). An alternative approach to the free energy relationship for inner-sphere processes derives from the standard free energy change and the inclusion of the work term for ion-pair formation evaluated from the CT data. Owing to its strong dependence on the experimental determination of standard free energy changes from irreversible systems, the limitations of this approach must be scrutinized by further studies.

The prospects for electron transfer mechanisms clearly extend beyond inorganic chemistry into the broad regions of organometallic and organic systems. Pushed to these limits, adequate quantitative criteria will be needed to delineate outer-sphere from inner-sphere mechanisms. However, the extent to which theoretical studies will provide more concrete guidelines of predictive value will determine whether electron transfer processes will form the basis of reaction mechanisms into the next century.

Acknowledgment

We wish to thank our coworkers, especially C.L. Wong and K.L. Rollick, for their invaluable contributions to the work we have presented, and the National Science Foundation for financial support of this research and funds to purchase the computer.

Literature Cited

1. For recent reviews, see Cannon, R.D. "Electron Transfer Reactions"; Butterworths: London, 1980; and Sutin, N. In "Inorganic Biochemistry"; Eichhorn, G.L., Ed.; Elsevier: Amsterdam, 1973; Vol. 2.
2. Fehlner, T.P.; Ulman, J.; Nugent, W.A.; Kochi, J.K. Inorg. Chem. 1976, 15, 2544.
3. Fukuzumi, S.; Wong, C.L.; Kochi, J.K. J.Am.Chem.Soc. 1980, 102, 2928.
4. Houle, F.A.; Beauchamp, J.L. J.Am.Chem.Soc. 1979, 101, 4067.
5. Dulz, G.; Sutin, N. Inorg.Chem. 1963, 2, 917. Diebler, H.; Sutin, N. J.Phys.Chem. 1964, 68, 174, and related papers.
6. Wong, C.L.; Kochi, J.K. J.Am.Chem.Soc. 1979, 101, 5593.
7. Rollick, K.L.; Kochi, J.K. J.Am.Chem.Soc. 1982, 104, 1319.
8. Marcus, R.J.; Zwolinski, B.J.; Eyring, H. J.Phys.Chem. 1954, 58, 432. Marcus, R.A. J.Chem.Phys. 1956, 24, 966, and related papers.
9. Klingler, R.J.; Kochi, J.K. J.Am.Chem.Soc. 1980, 102, 4790.
10. Nicholson, R.S.; Shain, I. Anal.Chem. 1964, 36, 706.
11. Klingler, R.J.; Kochi, J.K. J.Am.Chem.Soc. 1981, 103, 5839.
12. Saveant, J.M.; Tessier, D. J.Phys.Chem. 1978, 82, 1723. Imbeaux, J.C.; Saveant, J.M. Electroanal.Chem. 1973, 44, 169.
13. Tamblyn, W.H.; Klingler, R.J.; Hwang, W.S.; Kochi, J.K. J.Am.Chem.Soc. 1981, 103, 3161.
14. Klingler, R.J.; Kochi, J.K. J.Am.Chem.Soc. 1982, 104, 0000.
15. Rehm, D.; Weller, A. Ber.Bunsenges.Phys.Chem. 1969, 73, 834.
16. Marcus, R.A. J.Phys.Chem. 1968, 72, 891. Agmon, N.; Levine, R.D. Chem.Phys.Lett. 1977, 52, 197.
17. Scandola, F.; Balzani, V. J.Am.Chem.Soc. 1979, 101, 6140.
18. Lewis, E.S.; Shen, C.C.; More O'Ferrall, R.A. J.Chem.Soc., Perkin Trans.2 1981, 1084.
19. Agmon, N. Int.J.Chem.Kinet. 1981, 13, 333.
20. See Scandola, F.; Balzani, V.; Schuster, G.B. J.Am.Chem.Soc. 1981, 103, 2519.
21. Compare Foster, R. "Organic Charge Transfer Complexes"; Academic Press: New York, 1969.
22. Mulliken, R.S. J.Am.Chem.Soc. 1952, 74, 811. Mulliken, R.S.; Person, W.B. "Molecular Complexes"; Wiley: New York, 1969.
23. (a) Fukuzumi, S.; Mochida, K.; Kochi, J.K. J.Am.Chem.Soc. 1979, 101, 5961. (b) Gardner, H.C.; Kochi, J.K. J.Am.Chem. Soc. 1976, 98, 2460. (c) Chen, J.Y.; Gardner, H.C.; Kochi, J.K. J.Am.Chem.Soc. 1976, 98, 6150.

24. Fukuzumi, S.; Kochi, J.K. J.Phys.Chem. 1980, 84, 608, 617; 1981, 85, 648; 1980, 84, 2246, 2254; J.Am.Chem.Soc. 1981, 103, 2783; J.Org.Chem. 1981, 46, 4116; Tetrahedron 1982, in press.
25. The various donors included in the figures are identified in references 23, 24, and 26.
26. Fukuzumi, S.; Kochi, J.K., to be submitted.

RECEIVED August 10, 1982

Discussion

N. Sutin, Brookhaven National Laboratory: The linear free energy plots that you obtain are of considerable interest. It seems to me that implicit in these plots is the assumption that the reorganization parameter λ is constant for the particular series of organometallics. This might not be a good assumption if there is a large variation in the size of the organometallics. The latter would give rise to changes in the outer-sphere (solvent) barrier.

J.K. Kochi: Since the HOMOs of these organometals are centered on the carbon-metal bonds, the configuration about the metal is the most important factor in determining the magnitude of the reorganization energies. Thus, for this homologous set of organometals, it is not unreasonable for the reorganization energies to be rather invariant.

N. Sutin, Brookhaven National Laboratory: Strictly speaking, the outer-sphere and inner-sphere designations refer to limiting cases. In practice, reactions can have intermediate outer-sphere or inner-sphere character: this occurs, for example, when there is extensive interpenetration of the inner-coordination shells of the two reactants. Treating this intermediate situation requires modification of the usual expressions for outer-sphere reactions -- particularly those expressions that are based upon a hard-sphere model for the reactants.

J.K. Kochi: I agree. The quantitative treatment of inner-sphere mechanisms is difficult from a purely theoretical point of view. The phenomemological approach describes the activation barrier for inner-sphere process quantitatively, but provides no theoretical basis, unfortunately.

D.R. McMillin, Purdue University: You have suggested that certain oxidants are more sensitive to steric effects when reacting with the tin alkyls. Is there evidence for preferential attack of the less bulky alkyls when unsymmetrical reagents such as SnR_2R_2' or SnR_3R' are used?

J.K. Kochi: Yes, the problem has been treated somewhat in J. Phys. Chem. 84, 608-16 (1980). For example, at the same values of I_D, the interaction energy of the unsymmetrical $RSnMe_3$ will be larger than that of the symmetrical $R_4'Sn$ with a given acceptor [when R is α-branched].

Dr. R.H. Morris, University of Toronto

My question to Dr. J. Kochi is whether it is possible to correlate the steric factor in his equations describing the oxidation of alkylmetal compounds to some measure of the bulkiness of the alkyl groups such as cone angles similar to ones suggested by Dr. C.A. Tolman for tertiary phosphine ligands.

Jay Kochi: In principle, yes. However, the problem is quite a bit more complicated than that encountered with phosphines, in which steric effects derive from a coordination to metal with minimal changes in configuration, e.g.,

In contrast, with organometals there may be substantial reorganization changes in inner-sphere processes. With the tetraalkyltins, for example, it could involve a conversion of a tetrahedral structure to a trigonal bipyramidal structure, i.e.,

Thus in addition to the sheer bulk of the alkyl group in the tetrahedral structure, one must include the compressional change in going to the trigonal bipyramidal structure. We tried to calculate the latter with Jeremy Burdett (Chicago), but with no notable success, as yet.

T.J. Kemp, Warwick: Have you determined the slopes in the plots of log k_{ET} versus E^o? The point is that strict adherence to modified Marcus theory[1] should produce a slope of $- 16.9\ V^{-1}$. Your comprehensive study of thermal oxidation of metal alkyls by tris-phenanthroline complexes is complemented by a briefer one where electron-transfer rates from two series of metallocenes[2] and metal carbonyls[3] to excited UO_2^{2+} ion were determined by laser flash photolysis. Data are combined in the plot of log k_{ET} versus E^o, which is linear, but with a slope of only $- 2.07 \pm 0.04\ V^{-1}$.

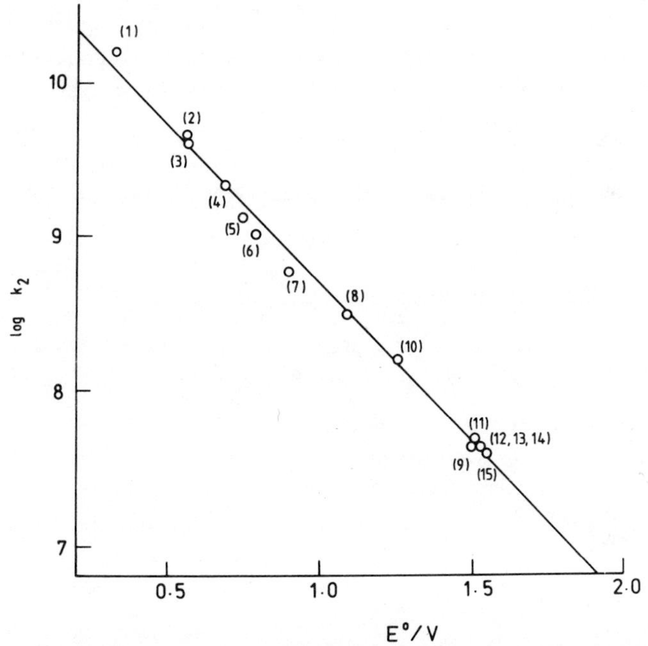

Key: 1, [Fe(η-C₅H₅)₂]; 2, [Fe(η-C₅H₅)(η-C₅H₄COPh)]; 3, [Fe(η-C₅H₅)(η-C₅H₄-COMe)]; 4, [Ru(η-C₅H₂)₂]; 5, [Ru(η-C₅H₅)(η-C₅H₄CHOHPh)]; 6, [Fe(η-C₅H₄-COMe)₂]; 7, [Ru(η-C₅H₅)(η-C₅H₄COPh]; 8, [Ru(η-C₅H₄COPh)₂]; 9, [Os(η-C₅H₅)₂]; 10, Ni(CO)₄; 11, Fe(CO)₅; 12, W(CO)₆; 13, Mo(CO)₆; 14, Cr(CO)₆; and 15, Mn₂(CO)₁₀.

Correlation of log k_2 ($/dm^3\ mol^{-1}\ s^{-1}$) versus E^o for the quenching of the excited UO_2^{2+} ion by organometallic molecules in acetone solution.

1. R.A. Marcus, *Ann. Rev. Phys. Chem.*, 15, 155 (1964).
2. D. Rehm and A. Weller, *Isr. J. Chem.*, 8, 259 (1970).
3. O. Traverso, R. Rossi, L. Magon, A. Cinquantini, and T.J. Kemp, *J. Chem. Soc. Dalton Trans.*, 569 (1978).
4. S. Sostero, O. Traverso, P.D. Bernardo, and T.J. Kemp, *J. Chem. Soc. Dalton Trans.*, 658 (1979).

J.K. Kochi: For outer-sphere electron transfer, the slope of the plot of the rate constant ($\log k_{ET}$) versus E° varies with the magnitude of the driving force for electron transfer. For example, the slope is expected to be -16.9 V^{-1} (which corresponds to a Brønsted coefficient of $\alpha = 1$) in the endergonic region of the driving force.[1] For the oxidations of organometals by FeL_3^{3+} examined in our studies,[2] the slope is -8.5 V^{-1} or $\alpha = 0.5$, as predicted by Marcus theory for electron transfer in the region about $\Delta G = 0$.[3] A slope of only -2.07 V^{-1} corresponds to $\alpha = 0.12$, which lies in the exergonic region according to the Rehm-Weller formulation, graphically underscored in Figure 14 with $\alpha = \beta$. [Note that the span in ΔG over which the slope is linear depends on the magnitude of the intrinsic barrier $\Delta G_0^{\ddagger}$.] Are the relevant standard potentials and intrinsic barriers for the metallocenes, carbonylmetals, and excited uranyl ion known?

(1) Marcus, R.A. J.Phys.Chem. **1968**, 72, 91.
(2) Wong, C.L.; Kochi, J.K. J.Am.Chem.Soc. **1979**, 101, 5593.
(3) Klingler, R.J.; Kochi, J.K. J.Am.Chem.Soc. **1982**, 104, 4186.

T.J. Meyer, University of North Carolina: One of the most remarkable things to come out of this work is the fact that as time passes organic chemists have been forced to consider and then have discovered many classical reactions which appear to involve electron transfer. Do you have any comments on the extent of one-electron transfer processes in some of these classical organic transformations?

J.K. Kochi: I do believe that the applicability of the concepts is pervasive to organic chemistry. However, there are many skeptics, and we will have to provide ironclad experiments for each system.

8

Excited-State Electron Transfer

THOMAS J. MEYER

The University of North Carolina, Department of Chemistry,
Chapel Hill, NC 27514

Many of the features of electron transfer reactions involving excited states can be understood based on electron transfer theory.

In other contributions in this symposium, a clear case has been made for the possible value of molecular excited states as a basis for solar energy conversion processes. Amongst possible approaches in using molecular excited states is the following sequence of events: 1) Optical excitation to give an excited state. 2) Electron transfer quenching of the excited state to give separated redox products. 3) Utilization of the separated redox products as a basis either for a photovoltaic application where the stored chemical redox energy appears as a photopotential, or a photochemical application where the stored energy appears as high energy redox products.

In order to illustrate the approach suggested above, it is of value to consider a specific case. Visible or near-UV excitation of the complex $Ru(bpy)_3^{2+}$ results in excitation and formation of the well-characterized metal to ligand charge transfer (MLCT) excited state $Ru(bpy)_3^{2+*}$. The consequences of optical excitation in the Ru-bpy system in terms of energetics are well established. and are summarized in eq. 1 in a Latimer type diagram where the potentials are versus the normal hydrogen electrode (NHE) and are

$$(1)\quad \begin{array}{ccccc} Ru(bpy)_3^{3+} & \xrightarrow{1.26} & Ru(bpy)_3^{2+} & \xrightarrow{-1.26} & Ru(bpy)_3^{+} \\ & & \Big\downarrow{+2.1V} & & \\ Ru(bpy)_3^{3+} & \xrightarrow{-0.84} & Ru(bpy)_3^{2+*} & \xrightarrow{+0.84} & Ru(bpy)_3^{+} \end{array}$$

written as reduction potentials (1). In the reduced form of the couple, $Ru(bpy)_3^{+}$, the added electron is in a π*(bpy) level and in the oxidized form of the chromophore, $Ru(bpy)_3^{3+}$, the electron has been removed from a $d\pi$ level.

0097-6156/83/0211-0157$06.00/0

The redox potential diagram in eq. 1 illustrates that the effect of optical excitation is to create an excited state which has enhanced properties both as an oxidant and reductant, compared to the ground state. The results of a number of experiments have illustrated that it is possible for the excited state to undergo either oxidative or reductive electron transfer quenching (2). An example of oxidative electron transfer quenching is shown in eq. 2 where the oxidant is the alkyl pyridinium ion, paraquat (3).

(2) $Ru(bpy)_3^{2+} \xrightarrow{+h\nu} Ru(bpy)_3^{2+*}$

$Ru(bpy)_3^{2+*} + PQ^{2+} \rightarrow Ru(bpy)_3^{3+} + PQ^{+}$

$(PQ^{2+} = CH_3-N\bigcirc\!\!-\!\!\bigcirc N-CH_3^{2+})$

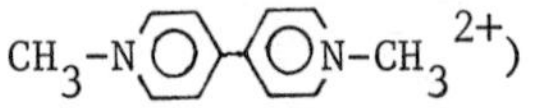

Eq. 2 is notable in the context of the energy conversion scheme proposed above because the separated redox products, quite remarkably, have the thermodynamic ability over a broad pH range to either oxidize water (using $Ru(bpy)_3^{3+}$) or to reduce water (using PQ^{+}).

Reductive electron transfer quenching of $Ru(bpy)_3^{2+*}$ has also been observed, e.g., eq. 3 (4).

(3) $Ru(bpy)_3^{2+*} + DMA \longrightarrow Ru(bpy)_3^{+} + DMA^{+}$

(DMA = Me_2NPh)

In the case of either eq. 2 or eq. 3, even though excited state energy has been converted into stored chemical redox energy, the storage is temporary because of back electron transfer between the separated redox products, e.g., eq. 4.

(4) $Ru(bpy)_3^{3+} + PQ^{+} \longrightarrow Ru(bpy)_3^{2+} + PQ^{2+}$

A major dilemma in any approach to energy conversion processes based on electron transfer reactions of molecular excited states is utilization of the stored redox products before back electron transfer can occur.

The net quenching reaction in eq. 2, which leads to separated redox products capable of oxidizing and reducing water, relies on a series of electron transfer steps. The basic theme of this account is excited state and related electron transfer events which occur in such systems and the basis that we have for understanding them both experimentally and theoretically.

In order to begin it is useful to consider Scheme 1, in which the quenching reaction in eq. 2 is considered in kinetic detail. In the scheme, the assumption is made that the only important quenching event is electron transfer and that energy transfer quenching is negligible. The series of electron transfer events in the scheme are initiated by optical excitation to give the excited state and the electron transfer reactions which occur following excitation are: 1) Excited state decay to the ground state occuring by a combination of radiative and nonradiative processes ($1/\tau_o$ in Scheme 1). Nonradiative decay is by a ligand to metal

Scheme 1

$$\mathrm{RuB_3}^{2+*} + \mathrm{PQ}^{2+} \underset{k_{-D}}{\overset{k_D}{\rightleftharpoons}} \mathrm{RuB_3}^{2+*},\mathrm{PQ}^{2+} \underset{k_{-1}}{\overset{k_1}{\rightleftharpoons}} \mathrm{RuB_3}^{3+},\mathrm{PQ}^{+} \underset{k'_D}{\overset{k'_{-D}}{\rightleftharpoons}} \mathrm{RuB_3}^{3+} + \mathrm{PQ}^{+}$$

$$\mathrm{RuB_3}^{2+} + \mathrm{PQ}^{2+} \underset{1/\tau_o}{\overset{+h\nu}{\rightleftharpoons}} \mathrm{RuB_3}^{2+*} + \mathrm{PQ}^{2+}$$

$$\mathrm{RuB_3}^{3+},\mathrm{PQ}^{+} \xrightarrow{k_2} \mathrm{RuB_3}^{2+},\mathrm{PQ}^{2+}$$

$$\mathrm{RuB_3}^{2+} + \mathrm{PQ}^{2+} \xleftarrow{k''_{-D}} \mathrm{RuB_3}^{2+},\mathrm{PQ}^{2+}$$

$$(\mathrm{RuB_3}^{2+} = \mathrm{Ru(bpy)_3}^{2+})$$

electron transfer reaction, $(\mathrm{b\bar{p}y})\mathrm{Ru^{III}(bpy)_2}^{2+*}(d\pi^5\pi^*) \rightarrow (\mathrm{bpy})\mathrm{Ru^{II}}$-$(\mathrm{bpy})_2^{2+}(d\pi^6)$. 2) Electron transfer quenching (k_1). 3) Back electron transfer to repopulate the excited state (k_{-1}). 4) Back electron transfer to give the ground state rather than the excited state (k_2).

In any photoredox application based on electron transfer quenching, like the one in Scheme 1, the critical factors determining device performance are the per photon efficiency with which the separated redox products appear, and the utilization of the separated redox products before back electron transfer between them can occur. The separation efficiency is determined in large part by the series of electron transfer steps described above: 1) Quenching must occur before excited state decay. 2) Following quenching, the efficiency of separation of the redox products shown by k'_{-D} in Scheme 1 must be rapid compared to back electron transfer to give the ground state, k_2. Clearly it is essential to understand the factors which control the rate constants for such electron transfer events.

Electron Transfer Theory

The factors that determine the rate of electron transfer between chemical sites are the extent of electronic coupling between electron donor and acceptor sites and the extent of vibrational trapping of the exchanging electron by both intramolecular and medium vibrations. Vibrational trapping is a natural consequence of the effects of changes in electron content on molecular structure. The point is illustrated in Figure 1 where a plot is shown of potential energy versus a normal coordinate for the electron transfer system, $D,A \rightarrow D^+,A^-$. The plot is for a trapping vibration assuming the harmonic oscillator approximation. In order for a vibration to be a trapping vibration, the displacement between the bottoms of the potential wells in Figure 1, ΔQ_{eq}, must be nonzero. In the classical limit, electron transfer can only occur at the intersection between the two potential curves because it is only at that point that energy is conserved before and after the electron trans-

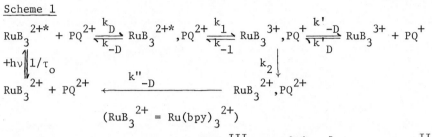

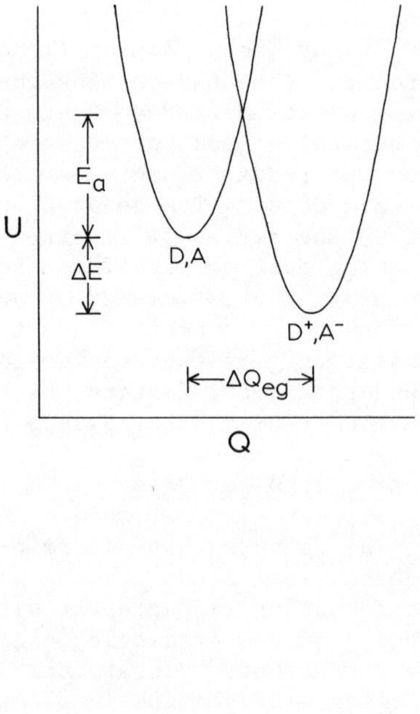

Figure 1. Plot of potential energy vs. normal coordinate for a trapping vibration for the reaction: D,A → D^+, A^-.

fer event. The classical energy of activation will be determined by the thermal energy needed to reach the intersection region, for all of the normal vibrations which respond to the change in electron distribution before and after electron transfer occurs. As noted above, there are two contributions to vibrational trapping, one from the intramolecular vibrations of the molecule, which are normally in the frequency range 200-4000 cm^{-1} and the collective vibrations of the surrounding medium, which for a solvent are of low frequency and in the range 1-10 cm^{-1}.

In fact, electron transfer occurs at the microscopic level where quantum mechanics provides the necessary description of the phenomenon (5-13). In the quantum mechanical solution, associated with the potential curves in Figure 1 are quantized energy levels, $E_j = (v_j + 1/2)\hbar\omega_j$, where v_j and $\omega_j = 2\pi\nu_j$ are the vibrational quantum number and angular frequency for vibration j. The corresponding vibrational wavefunctions are χ_j. In the quantum mechanical picture the advantage of thermal activation to the intersection region is enhanced vibrational overlap between the vibrational wavefunctions for the initial (D,A) and final (D^+,A^-) states. The importance of vibrational overlap is that where it occurs the vibrational configurations of both states are allowed and electron transfer can occur.

The presence of the electron acceptor site adjacent to the donor site creates an electronic perturbation. Application of time dependent perturbation theory to the system in Figure 1 gives a general result for the transition rate between the states D,A and D^+,A^-. The rate constant is the product of three terms: 1) $2\pi V^2/\hbar$ where V is the electronic resonance energy arising from the perturbation. 2) The vibrational overlap term. 3) The density of states in the product vibrational energy manifold.

In the classical limit where the condition $\hbar\omega_M << k_BT$ is met for the trapping vibrations, the rate constant for electron transfer is given by eq. 6. In eq. 6, $\chi/4$ is the classical vibrational trapping energy which includes contributions from both intramolecular (χ_i) and solvent (χ_o) vibrations (eq. 5). In eq. 6 ΔE is the internal energy difference in the reaction, ν_n is the frequen-

$$\chi = \chi_i + \chi_o \tag{5}$$

cy or an averaged frequency of the trapping vibration or vibrations, and κ is the single passage Landau-Zener transmission coefficient. In the limit that electron transfer is slow on the

$$k_{et} = \nu_n \kappa \exp{-E_a/RT} \tag{6}$$

$$E_a = \frac{(\chi+\Delta E)^2}{4\chi}$$

vibrational timescale, eq. 6 becomes eq. 7.

$$k_{et} = \nu_{et} \exp -E_a/RT \tag{7}$$

$$\nu_{et} = \frac{2\pi V^2}{\hbar}\left(\frac{1}{4\pi k_B T \chi}\right)^{1/2}$$

Although eq. 5-7 are derived in the classical limit, related equations are available which include in a specific way the contribution of high frequency vibrations (6,8-11).

Excited State Quenching. The Normal Region. In order to maximize the energy stored in a device based on Scheme 1, it is important that the energetics of the quenching step be such that the overall quenching reaction occur with ΔG near zero. An electron transfer reaction in this domain is said to occur in the "normal" region in that the minima of the potential curves before and after electron transfer are on different sides of the intersection region, as is the case in Figure 1. In terms of the quantities which appear in the definition of E_a in eq. 6, the normal region is defined by the condition that $-\Delta E < \chi$. The classical result in eq. 6 is valid in the normal region, at least for low frequency trapping vibrations and the equation has a direct application to some of the electron transfer events in Scheme 1. In particular, the quenching step (k_1) and the step involving repopulation of the excited state (k_{-1}) are usually carried out in the normal region.

In eq. 8 are shown the results of a kinetic analysis of the series of reactions in Scheme 1. The analysis is based on the quenching rate constant k'_q, corrected for diffusional effects, which would be measured for the quenching of $Ru(bpy)_3^{2+*}$ by PQ^{2+}.

$$k'_q = k_1 K_A \left(\frac{k'_{-D} + k_2}{k'_{-D} + k_2 + k_{-1}}\right) \qquad \left(K_A = \frac{k_D}{k_{-D}}\right) \tag{8}$$

Assuming that $\Delta E \sim \Delta G$, the rate constants for the quenching reaction and its reverse (k_1 and k_{-1} in eq. 8) can be written using the theoretical results of eq. 5-7. In order to make the situation clear it is useful to consider two limiting cases. In the first limit, $k_{-1} << k_2, k'_{-D}$, back electron transfer to give the excited state, is slow relative to separation or back electron transfer to give the ground state, and eq. 9 applies.

$$k'_q = k_1 K_A = \nu_n \kappa K_A \exp -\left[\frac{(\chi + \Delta G_1)^2}{4\chi RT}\right] \tag{9}$$

$$RT \ln k'_q = RT\left(\ln \nu_n \kappa K_A - \frac{\chi}{4}\right) - \frac{\Delta G_1}{2}\left(1 + \frac{\Delta G_1}{2\chi}\right)$$

$$= RT \ln k_q(0) - \frac{\Delta G_1}{2}\left(1 + \frac{\Delta G_1}{2\chi}\right)$$

In the second limit, $k_{-1} >> k_2, k'_{-D}$, back electron transfer to give the excited state is rapid, and eq. 10 applies.

$$RT\ln k'_q = RT\ln(k'_{-D} + k_2)K_A - \Delta G_1 \quad (10)$$

In the equations, $k_q(0)$ is the hypothetical quenching rate constant when ΔE ($=\Delta G$) $= 0$ and ΔG_1 is the free energy change on quenching.

Eqs. 9 and 10 make clear predictions about the dependence of quenching rate constants on the free energy change in the quenching step. One way of testing the theory is to observe the quenching of the excited state by a series of related quenchers where the parameters $k_q(0)$, K_A, and k'_{-D} should remain sensibly constant and yet where the potentials of the quenchers as oxidants or reductants can be varied systematically. Such experiments have been carried out, most notably with the MLCT excited state, $Ru(bpy)_3^{2+*}$ (1). The experiments have utilized both a series of oxidative nitroaromatic and alkyl pyridinium quenchers, and a series of reductive quenchers based on aniline derivatives. From the data and known redox potentials for the quenchers, plots of $RT\ln k'_q$ vs. $E°'$ for the quencher couple show regions of slope 1/2 and slope 1 as predicted by eq. 9 and 10. In fact, the theoretical equations appear to account for the observed variation of $\ln k'_q$ with $E°'$ satisfactorily. Given the agreement with theory, it follows that if an excited state is thermally equilibrated, it can be viewed as a typical chemical reagent with its own characteristic properties, and that those properties can be accounted for by using equations and theoretical developments normally used for ground state reactions.

Transition Between the Normal and Inverted Regions. In the previous section, electron transfer events were considered in the framework of available theory for the normal region where $-\Delta E < \chi$. According to eq. 6, as $-\Delta E$ approaches χ in magnitude, $E_a \rightarrow 0$, and there is no longer a vibrational trapping barrier for the exchanging electron. With a further increase in $-\Delta E$, $-\Delta E > \chi$ and as shown in Figure 2, one potential curve is "embedded" within the other. In this, the inverted region, the situation is quite different from that for an electron transfer reaction in the normal region. In the normal region, the intersection region occurs outside the potential curves. In the quantum mechanical view, thermal activation is desired in order to reach regions near the intersection region where vibrational overlap is maximized. In the inverted region, the two potential curves are within each other and vibrational overlap plays a more important role. In addition, because of the embedded nature of the potential curves, emission can occur in the inverted region.

The relationship between electron transfer in the normal and inverted regions is illustrated in Figure 3 for the case of quenching of $Ru(bpy)_3^{2+*}$ by a nitroaromatic quencher. Excitation

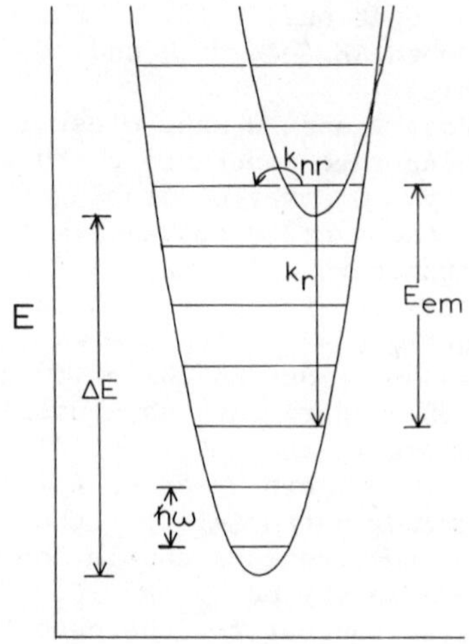

Figure 2. Energy-coordinate diagram for the "inverted region" where $-\Delta E > \chi$.

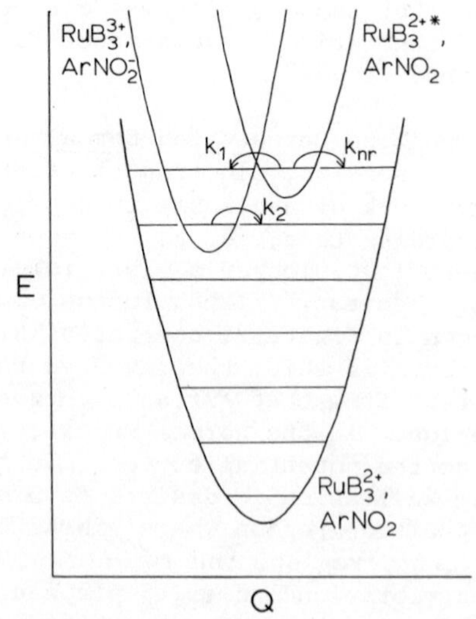

Figure 3. Energy-coordinate diagram for quenching of $Ru(bpy)_3^{2+*}$ (RuB_3^{2+*}) by a nitroaromatic ($ArNO_2$).

of an association complex between the ruthenium complex and the quencher, $Ru(bpy)_3^{2+}$,$ArNO_2$, would lead to $Ru(bpy)_3^{2+*}$ in close contact with the quencher. Excited state decay of this state by nonradiative decay (k_{nr} in Figure 3) can occur as an electron transfer process in the inverted region. Electron transfer quenching and excited state repopulation (k_1 and k_{-1}) are reactions in the normal region. The quenching step can be viewed as an interconversion between *intramolecular* and *intermolecular* charge transfer excited states--$(bpy)_2Ru^{III}(b\bar{p}y)^{2+*}$,$ArNO_2$ $\rightarrow$ $(bpy)_2Ru^{III}$-$(bpy)^{2+}$,$ArNO_2^-$. The final process to consider in Figure 3 is back electron transfer to give the excited state (k_2)--$Ru(bpy)_3^{3+}$, $ArNO_2^-$ $\rightarrow$ $Ru(bpy)_3^{2+}$,$ArNO_2$. This is also a reaction in the inverted region and can be viewed as nonradiative decay of an "outer-sphere", charge transfer excited state.

Electron Transfer in the Inverted Region. Decay of an excited state and electron transfer in the inverted region are conceptually the same process (14). As suggested by the potential curves in Figure 2, there are three limiting pathways available for excited state decay, but all three are subject to the limitations imposed by energy conservation. Decay can occur by emission, in which case energy conservation is met by the loss of an emitted photon. Decay could occur nonradiatively by thermal activation to the intersection region between potential curves, which is the pathway in the classical limit. However, decay can also occur from low-lying vibrational levels of the excited state to high energy levels of the ground state (Figure 2). Because of energy conservation, the electronic energy lost in the transition between states, e.g., $(b\bar{p}y)Ru^{III}(bpy)_2^{2+*}$ $\rightarrow$ $(bpy)Ru^{II}(bpy)_2^{2+}$, must appear as vibrational energy in the product. A critical feature in determining the rate of the transition is the extent of vibrational overlap between the vibrational wavefunctions for the states before and after electron transfer. Vibrational overlap is obviously of importance in the inverted region because of the nesting of the potential curves as shown in Figure 2, and its magnitude is influenced by: 1) The vibrational quantum spacing for the acceptor vibration or vibrations, $\hbar\omega_M$. As the vibrational quantum number v increases, the vibrational amplitude increases near the potential curve. For a high frequency vibration, the energy conservation condition is met for vibrations with lower v values. 2) The extent of distortion of the acceptor vibration in the excited state compared to the ground state, ΔQ_{eq}. As ΔQ_{eq} increases, vibrational overlap increases. 3) The energy gap between the ground and excited state, ΔE. As ΔE decreases, energy conservation is met for vibrational levels of lower v number where the amplitudes are greater toward the center of the potential curve for the ground state.

In the limit that $\hbar\omega_M >> k_BT$, the rate constant for nonradiative decay is simply the product of the square of the vibrational overlap integral $<\chi_{initial}|\chi_{final}>$ and an electronic term for the transition between states. In the limit that $\hbar\omega_M >> k_BT$ and assuming a relatively small excited state distortion, the rate constant for nonradiative excited state decay is given by eq. 11 (11).

$$k_{nr} = C^2\omega_k\left(\frac{\pi}{2\hbar\omega_M|\Delta E|}\right)^{1/2}\exp{-S_M}\exp-\left(\frac{\gamma|\Delta E|}{\hbar\omega_M}\right) \qquad (11)$$

ΔE; the internal energy change accompanying excited state decay.

$\omega_M = 2\pi\nu_M$; the angular frequency of the acceptor vibration or vibrations.

$S_M = \frac{1}{2}\Delta_M{}^2$; a measure of the distortion of the acceptor vibration in the excited state. Δ_M is the dimensionless, fractional displacement in normal vibration M between the thermally equilibrated excited and ground states. It is related to ΔQ_{eq} by $\Delta_M = \Delta Q_{eq}(\frac{M\omega}{\hbar})^{1/2}$, where M is the reduced mass for the vibration.

$\gamma = \ln(|\Delta E|/S_M\hbar\omega_M)-1$

$C^2\omega_k$; note below.

To first order, the transition between excited and ground state is forbidden because they are both solutions of the same molecular Hamiltonian and their wavefunctions are orthogonal. However, the transition can be induced by exciting a promoting vibration or vibrations (ω_k) which when activated result in a change in electronic overlap between the electron donor and acceptor sites. The magnitude of the vibrationally induced mixing of the states is given by C; the angular frequency of the promoting vibration is ω_k.

When contributions from the low frequency, collective vibrations of the solvent are included, eq. 11 becomes eq. 12. In eq.

$$k_{nr} = C^2\omega_k\left(\frac{\pi}{2\hbar\omega_M E_{em}}\right)^{1/2}\exp{-S_M}\exp-\frac{\gamma_o E_{em}}{\hbar\omega_M}\exp+\frac{\chi_o}{\hbar\omega_M}\left[\frac{k_BT}{\hbar\omega_M}(\gamma_o+1)^2\right] \qquad (12)$$

12 the assumption is made that the experimentally observed emission energy at λ_{max}, E_{em}, is related to ΔE and χ_o as in eq. 13.

$$E_{em} \sim |\Delta E| - \chi_o \qquad (13)$$

The assumption is reasonable in the limit that the distortion in ω_M is not too great in the excited state.

When written in logarithmic form and noting that the pre-exponential and γ_o terms in eq. 12 are only weakly dependent on E_{em}, eq. 12 can be written in the form of the "energy gap law" (12a, 12b) as shown in eq. 14.

$$\ln k_{nr} = \ln k_{et} = \ln\beta_o - \frac{\gamma_o E_{em}}{\hbar\omega_M} + \frac{b\chi_o}{\hbar\omega_M} \qquad (14)$$

$$\beta_o = C^2\omega_k\left(\frac{\pi}{2\hbar\omega_M E_{em}}\right)^{1/2}$$

$$b = \frac{k_B T}{\hbar\omega_M}(\gamma_o+1)^2$$

One way to test the "energy gap law" and therefore the application of electron transfer to the inverted region is by measuring radiationless decay rate constants for a series of closely related excited states. An especially useful series has turned out to be the complexes $(bpy)OsL_4{}^{2+}$ or $(phen)OsL_4{}^{2+}$ (L = py, RCN, 1/2 bpy, 1/2 phen, PR_3...), all of which contain an MLCT (Os(dπ)-bpy(π*)) based chromophore. The range of ligands that can serve the role of L is large, and systematic changes in L can cause relatively large variations in the emission energy. The primary origin of the variations in E_{em} appears to be in the backbonding abilities of L in stabilizing the ground state by metal to ligand backbonding. In any case, for a series of related chromophores in a constant solvent, the terms β_o, $\hbar\omega_M$, and χ_o in eq. 14 are expected to remain sensibly constant and plots of $\ln k_{nr}$ vs. E_{em} are expected to be linear. In addition, in low temperature (77°K) emission measurements vibrational structure can be observed. From the energy spacings between the vibrational components, $\hbar\omega_M \sim$ 1300-1400 cm^{-1} for the acceptor vibration(s). It is also possible to calculate S_M ($= \frac{1}{2}\Delta_M{}^2$) from the relative intensities of the vibrational components (15).

Experimentally, E_{em} is available from the emission spectrum and k_{nr} from a combination of excited state lifetime (τ_o) and emission quantum yield (ϕ_r) measurements as shown in eq. 15. An

$$\frac{1}{\tau_o} = k_{nr} + k_r \tag{15}$$

$$\phi_r = \frac{k_r}{k_r + k_{nr}}$$

extended series of osmium complexes is available, and the results of experiments where both emission energies and k_{nr} allow plots of $\ln k_{nr}$ to be made (16). As the data show, the linear relationship predicted by eq. 14 is observed. It is even more striking that the slopes of the lines of the plots are also in agreement with eq. 14 although a detailed analysis of the data is required. The similarity in slopes and vibrational spacings in low temperature emission spectra for the two series show that the acceptor vibrations are ring-stretching in nature. The different in intercepts for the two series is not surprising. The intercept includes the term C^2 which is dependent on the electronic structure of the chromophoric ligand.

Given the change in electronic distribution between MLCT excited and ground states, e.g., $(b\bar{p}y)Os^{III}L_4 \rightarrow (bpy)Os^{II}L_4$, a second approach to testing the energy gap law should be through solvent variations. In the classical dielectric continuum limit, ΔE is predicted to vary with the static dielectric constant of the medium (D_S), as in equation 16, and χ_o with D_S and the optical di-

$$(16) \qquad \Delta E = \left(\frac{\mu_f^2-\mu_i^2}{a^3}\right)\left(\frac{1-D_S}{2D_S+1}\right)$$

electric constant D_{op} as in eq. 17 (17). In eq. 16 and 17 μ_i and μ_f are the dipole moments of the metal-ligand electronic distributions before and after electron transfer has occurred. In the

$$(17) \qquad \chi_o = \left[\left(\frac{1-D_{op}}{2D_{op}+1}\right)-\left(\frac{1-D_S}{2D_S+1}\right)\right]\left(\frac{\mu_f^2-\mu_i^2}{a^3}\right)$$

equations a is the radius of a sphere enclosing the metal-ligand dipole.

According to eq. 14, if variations in χ_o are relatively small through a series of solvents, plots of $\ln k_{nr}$ vs. E_{em} should be linear for a single excited state. It has been shown that for a series of polar organic solvents, the prediction is borne out for several of the $(phen)OsL_4^{2+}$-type MLCT excited states, including $Os(phen)_3^{2+*}$ (18). It is especially striking that the slopes of the plots are the same within experimental error as in the experiments described above where L was varied.

However, for hydroxylic solvents such as methanol or water, specific solvent effects exist, the dielectric continuum result in eq. 17 is no longer applicable, and variations in χ_o are appreciable. Even so, eq. 14 still applies in that for a series of excited states like $(bpy)OsL_4^{2+*}$, plots of $\ln k_{nr}$ vs. E_{em} remain linear and have the same slope as the plots for polar organic solvents. The difference is that the lines are parallel but offset, because the χ_o term appears in the intercept and χ_o is non-negligible for the hydroxylic solvents.

Studies like those mentioned here on the osmium complexes are more difficult for related complexes of ruthenium because of the intervention of a lowlying, thermally populable d-d excited state. However, it is possible to separate the two contributions to excited state decay by temperature dependent measurements. In the case of $Ru(bpy)_3^{2+*}$, temperature dependent lifetime studies have been carried out in a series of solvent, and the results obtained for the variation of k_{nr} with E_{em} are in agreement with those obtained for the Os complexes (19).

A closely related test of the energy gap law for Ru complexes has come from temperature dependent lifetime and emission measurements for a series of complexes of the type $Ru(bpy)_2L_2^{2+}$ (L = py, substituted pyridines, pyrazine...). From the data, the variation in $\ln k_{nr}$ with E_{em} predicted by the energy gap law has been observed and it has been possible to observe the effect of changing the ligands L on the transition between the MLCT and dd states (20).

The discussion in this section has been oriented toward the use of intramolecular reactions and excited state decay to test electron transfer theory in the inverted region. However, the same ideas should apply to intermolecular electron transfer where electron hopping occurs between different chemical sites, as long as the conditions which define the inverted region are appropriate.

One striking prediction of the energy gap law and eq. 11 and 14 is that in the inverted region, the electron transfer rate constant ($k_{nr} = k_{et}$) should *decrease* as the reaction becomes more favorable ($\ln k_{nr} \propto -\Delta E$). Some evidence has been obtained for a fall-off in rate constants with increasing $-\Delta E$ (or $-\Delta G$) for intermolecular reactions (21). Perhaps most notable is the pulse radiolysis data of Beitz and Miller (22). Nonetheless, the applicability of the energy gap law to intermolecular electron transfer in a detailed way has yet to be proven.

Application of the energy gap law to the energy conversion mechanism in Scheme 1 leads to a notable conclusion with regard to the efficiency for the appearance of separated redox products following electron transfer quenching. From the scheme, the separation efficiency, ϕ_{sep}, is given by eq. 18. Diffusion apart of the

$$\phi_{sep} = \frac{k'_{-D}}{k_2 + k'_{-D}} \qquad (18)$$

quenching products once the quenching step has occurred (k'_{-D}) is dictated by such features as the charge types on the reactants, and the solvent polarity and viscosity. However, back electron transfer to give the ground state (k_2) is an electron transfer reaction. According to the energy gap law, as the quenching step produces redox products where the energy stored becomes more and more favorable, $-\Delta E$ increasing, k_2 should decrease. One therefore reaches the striking conclusion that in a properly designed system both the efficiency of separation of the redox products and the amount of energy stored should increase as the reduction potential of the quencher approaches that of the excited state (2b).

Directed Electron Transfer. This account was begun by considering the possibility of creating energy conversion schemes based on simple electron transfer processes involving excited states. One obvious extension is to more complex systems where some of the demands of an overall conversion mechanism could be spread amongst different chemical sites. In order to build more complex, coupled systems where light absorption and chemical redox events are separated, it is necessary to understand and control the molecular characteristics which can lead to directed, intramolecular electron or energy transfer. What is meant by directed electron transfer is that following optical excitation of a chromophore, the excited electron or electron hole, or perhaps both, drift away from the initial chromophoric site by intramolecular electron transfer. If the rate of recombination of the excited electron-electron hole pair is sufficiently slow, it is possible to extract them from different parts of the molecular at a later time, before recombination can occur (23). As an example of such a system, consider the complex $[Ru(bpy)_2(N\bigcirc\bigcirc N^+-Me)_2]^{4+}$ in which there are both Ru(bpy) chromophoric sites and attached, intramolecular electron acceptor pyridinium sites. There is a clear similarity between this intramolecular redox complex and the reaction in Scheme 1 based on paraquat. In the complex, the bimolecular, intermolecular quenching reaction of Scheme 1 has been converted into an intramolecular quenching reaction. Experimentally, optical excitation of the Ru–bpy chromophore in a frozen solution at 77°K leads to excited state decay by an emission which in terms of emission energy, vibrational structure and lifetime closely resembles the emission from $Ru(bpy)_3^{2+*}$. However, at room temperature a strongly red-shifted, weaker, short-lived emission is observed from the molecule (24). The most straightforward interpretation of the sequence of events that occurs following excitation at room temperature is summarized in Scheme 2.

Scheme 2

$$CT_2 \quad\text{———}\quad (bpy)(b\bar{p}y)(L)Ru^{III}(N\bigcirc\bigcirc N^+-Me]^{4+*}$$

$$CT_1 \quad\text{———}\quad (bpy)_2(L)Ru^{III}(N\bigcirc\bigcirc N^{\cdot}-Me]^{4+*}$$

$$+h\nu \quad\big|\quad -h\nu'$$

$$GS \quad\text{———}\quad (bpy)_2(L)Ru^{II}(N\bigcirc\bigcirc N^+-Me)]^{4+}$$

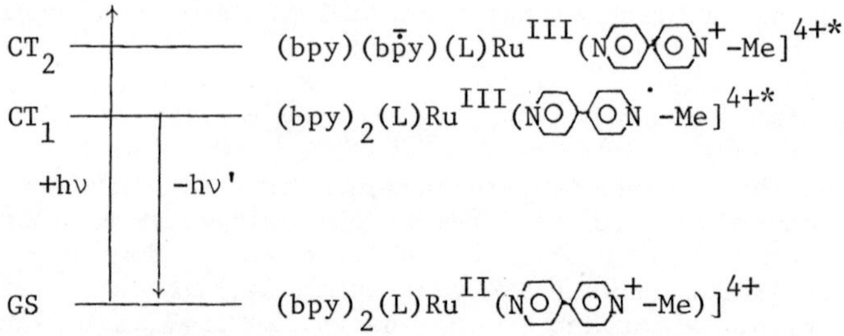

At room temperature thermally activated electron transfer occurs from the bpy ligand to the remote pyridinium site followed by decay of the lower, pyridinium-based CT state. The electron transfer step is the intramolecular analog of the paraquat quenching of $Ru(bpy)_3^{2+*}$.

A feature of interest in the sequence of events outlined in Scheme 2 is that following optical excitation, directed electron transfer does occur away from the chromophoric site to a remote ligand. In that sense, the excited state acts as a molecular photodiode (23). With the proper molecular design it may be possible to build systems where the excited electron and/or electron hole cascade through a series of redox levels leading to the spatial separation of the electron–electron hole pair and slow back electron transfer. It is important to realize that the observation of directed electron transfer is subject to the "rules" of electron transfer as described above. For the case of the pyridinium complex, nonradiative decay from the initial CT state, CT_2, $[(bpy)(b\bar{p}y)(L)Ru^{III}(N\bigcirc\bigcirc N^+-Me)]^{4+*} \longrightarrow [(bpy)_2(L)Ru^{II}(N\bigcirc\bigcirc N^{+\underline{\cdot}}$ Me)]$^{4+}$, is an electron transfer process in the inverted region subject to the energy gap law. On the other hand, the bpy $\longrightarrow$ pyridinium electron transfer corresponding to the transition between excited states, $CT_2 \rightarrow CT_1$, appears to be a thermally activated electron transfer which occurs in the normal region.

A second system which is closely related to the pyridinium

complex mentioned above is based on the series of dimeric complexes, $(PhSCH_2CH_2SPh)_2ClRu^{II}(L)Ru^{III}Cl(bpy)_2^{3+}$ (L = N⟨O⟩⟨O⟩N, N⟨O⟩–C(H)=C(H)–⟨O⟩N, N⟨O⟩–CH_2CH_2–⟨O⟩N). In the series of mixed-valence dimers the use of the dithioether ligand has the effect of localizing the site of oxidation on the bpy site and, in addition, it introduces no lowlying, visible chromophoric sites. In the Ru(II)-Ru(III) mixed-valence dimers the only significant chromophore in the visible region arises from optical transitions from the Ru(II) site to the bridging pyridine-type ligands. In the dimers, optical excitation into the $d\pi(Ru^{II}) \rightarrow \pi^*(L)$ chromophore leads to emission spectra and excited state lifetimes which are characteristic of those for the related monomeric complexes, $(L)Ru^{II}Cl(bpy)_2^{+}$ (25). The only reasonable explanation is that following optical excitation into the Ru → L chromophore, the resulting MLCT excited state undergoes intramolecular electron transfer quenching from $\pi^*(L)$ to $\pi^*(bpy)$ on the second site, $(PhSCH_2CH_2SPh)Ru^{II}(L)Ru^{III}(bpy)_2^{3+}$ $\xrightarrow{+h\nu}$ $(PhSCH_2CH_2SPh)_2Ru^{III}(\bar{L})Ru^{III}Cl(bpy)_2^{3+*} \rightarrow (PhSCH_2CH_2SPh)_2$-$Ru^{III}(L)Ru^{III}Cl(b\bar{p}y)(bpy)^{3+*}$. The excited state resulting from intramolecular electron transfer is a typical MLCT excited state based on the Ru-bpy chromophore, and it subsequently decays by a combination of radiative and nonradiative decay processes. The net effect is clearly closely related to that shown in Scheme 2. However, in this case optical excitation at one chromophore leads to emission and subsequent decay from a second chromophore. The two observations are connected by the fact that initial excitation is followed by one-electron transfer events of an intramolecular nature. Both suggest the possibility of building such features into more complex systems in a systematic way.

Acknowledgements are made to the National Science Foundation under grant no. CHE-8008922 for support of this research.

Literature Cited

1. a) Bock, C. R.; Connor, J. A.; Gutierrez, A. R.; Meyer, T. J.; Whitten, D. G.; Sullivan, B. P.; Nagle, J. K. J. Am. Chem. Soc. 1979, 101, 4815, Chem. Phys. Lett. 1979, 61, 522.
   b) Ballardini, R.; Varani, G.; Indelli, M. T.; Scandola, F.; Balzani, V. J. Am. Chem. Soc. 1978, 100, 7219.
2. a) Balzani, V.; Bolleta, F.; Gandolfi, M. T.; Maestri, M. Top. Curr. Chem. 1978, 75, 1.
   b) Sutin, N.; Creutz, C. Pure Appl. Chem. 1980, 52, 2717.
   c) Sutin, N. J. Photochem. 1979, 10, 19.
   d) Meyer, T. J. Accts. Chem. Res. 1978, 11, 94.
   e) Meyer, T. J. Isr. J. Chem. 1977, 15, 200.
3. Bock, C. R.; Meyer, T. J.; Whitten, D. G. J. Am. Chem. Soc. 1974, 96, 4710.
4. a) Anderson, C. P.; Salmon, D. J.; Meyer, T. J.; Young, R. C. J. Am. Chem. Soc. 1977, 99, 1980. b) Maestri, M.; Gratzel, M. Ber. Bunsenges Phys. Chem. 1977, 81, 504.

5. a) Kubo, R.; Toyozawa, Y. Prog. Theor. Phys. (Osaka), 1955, 13, 161. b) Lax, B. J. Chem. Phys. 1952, 20, 1752. c) Huang, K.; Rhys, A. Proc. Phys. Soc., Ser. A, 1951, 204, 413.
6. Ulstrup, J.; Jortner, J. J. Chem. Phys. 1975, 63, 4358. b) Efrima, S.; Bixon, M. Chem. Phys. 1976, 13, 447. c) Van Duyne, R. P.; Fischer, S. F. Chem. Phys. 1974, 5, 183.
7. Holstein, T. Philos. Mag. 1978, 37, 49, Ann. Phys. (Leipzig), 1959, 8, 343.
8. Kestner, N. R.; Logan, J.; Jortner, J. J. Phys. Chem. 1974, 78, 2148.
9. Brunschwig, B. S.; Logan, J.; Newton, M. D.; Sutin, N. J. Am. Chem. Soc. 1980, 102, 5798.
10. Ulstrup, J. "Charge-Transfer Processes in Condensed Media," Springer-Verlag: West Berlin, 1979.
11. a) Englman, R.; Jortner, J. Mol. Phys. 1970, 18, 145, b) Freed, K. F.; Jortner, J. J. Chem. Phys. 1970, 52, 6272.
12. a) Robinson, G. W.; Frosch, R. P. J. Chem. Phys. 1963, 38, 1187, b) Henry, B. R.; Siebrand, W. in "Organic Molecular Photophysics," J. B. Birks, ed., Wiley: New York, 1973, Vol. 1, c) Fong, F. K., "Theory of Molecular Relaxation," Wiley: New York, 1975.
13. Meyer, T. J. in "Mixed-Valence Compounds," D. B. Brown, ed., D. Reidel Co.: Dordrecht, 1980, pp. 75-113.
14. Meyer, T. J., Prog. Inorg. Chem., S. Lippard, ed., Wiley: New York, Vol. 30.
15. Caspar, J. V.; Kober, E. M.; Sullivan, B. P.; Meyer, T. J., manuscript in preparation.
16. a) Kober, E. M.; Sullivan, B. P.; Dressick, W. J.; Caspar, J. V.; Meyer, T. J. J. Am. Chem. Soc. 1980, 102, 7383, a) Caspar, J. V.; Kober, E. M.; Sullivan, B. P.; Meyer, T. J. J. Am. Chem. Soc. 1982, 104, 630.
17. a) Marcus, R. A. Discuss. Faraday Soc. 1960, 29, 21, Ann. Rev. Phys. Chem. 1964, 15, 155, J. Chem. Phys. 1965, 43, 679. b) Hush, N. S. Trans. Faraday Soc. 1961, 57, 155; Electrochim. Acta 1968, 13, 995; Kirkwood, J. G. J. Chem. Phys. 1934, 2, 351.
18. Caspar, J. V.; Meyer, T. J. Chem. Phys. Lett., in press.
19. Caspar, J. V.; Meyer, T. J., submitted.
20. Caspar, J. V.; Meyer, T. J., submitted.
21. Creutz, C.; Sutin, N. J. Am. Chem. Soc. 1977, 99, 241.
22. Beitz, J. V.; Miller, J. R., "Tunneling in Biological Systems," Academic Press: New York, 1974, p. 269.
23. Nagle, J. K.; Bernstein, J. S.; Young, R. C.; Meyer, T. J. Inorg. Chem. 1981, 20, 1760.
24. Sullivan, B. P.; Abruna, H. D.; Finklea, H. O.; Salmon, D. J.; Nagle, J. K.; Meyer, T. J.; Sprintschnik, H. Chem. Phys. Lett. 1978, 58, 389.
25. Curtis, J. C.; Bernstein, J. S.; Schmehl, R. H.; Meyer, T. J. Chem. Phys. Lett. 1981, 81, 48.

RECEIVED October 21, 1982

Discussion

N. Sutin, Brookhaven National Laboratory: The correlations of the lifetimes of the osmium(II) excited states with the energy gap law are very impressive. In this regard, it is noteworthy that the energy gap law works well in rationalizing the lifetimes of charge-transfer states, but seems less successful in predicting the rates of very exothermic bimolecular electron transfer processes. The formation of excited state products could be responsible for the apparent breakdown of the energy gap law in some of these cases. A change in reaction mechanism might be responsible in others.

D.R. McMillin, Purdue University: You indicated that emission can sometimes be detected from luminophors fixed near a platinum electrode. Isn't this surprising given the fact that metal electrodes are generally excellent quenchers of excited states?

T.J. Meyer: The observation of emission from the films is not surprising. The films described are rather thick and consist of the equivalent of many molecular monolayers. No doubt we are observing emission from layers well away from the electrode surface. Emission is not observed from thin films, suggesting that efficient quenching of excited states in the layers near the electrode does occur.

M.S. Wrighton, M.I.T.: Your H_2O_2, Br_2 generation is a good example of how one can couple one-electron reagents with interfacial systems to do the desired reaction. Most photochemical systems will be one-electron initially, since one photon excites one electron. Also, under what conditions does your system work?

T.J. Meyer: At this point the H_2O_2/Br_2 cell is restricted to acetonitrile solution, the chromophore used is not an intense visible absorber, and, in terms of sustained operation, the cell could be limited by oxidation of H_2O_2 by $N(\underline{p}\text{-}C_6H_4Br)_3^+$ as H_2O_2 builds up in solution. Nonetheless, there are some viable approaches for developing aqueous solution cells based on a series of possible chromophores.

C.H. Brubaker, Michigan State University: How did you deposit the polypyridine ligand on the electrode surface?

T.J. Meyer: The procedure used involved evaporation of the preformed metallopolymer from a solution in a polar organic solvent like methanol.

A.J. Bard, University of Texas: The fact that one can generate chemiluminescence in polymer films containing Ru-(bpy)$_3{}^{2+}$ implies that the excited state may not be quenched completely by electron transfer reactions. Are the photoreactions you describe thermodynamically uphill (i.e., with chemical storage or radiant energy) or are they photocatalytic?

T.J. Meyer: In the production of H_2O_2 and Br_2, energy is stored; $\sim 0.4V$ in aqueous solution. In the experiments with polymer films, the excited state quenching is irreversible since Co(III) is reduced to Co(II) with the loss of ligands.

N. Sutin, Brookhaven National Laboratory: In some instances it may be desirable to change or control the direction of electron flow between a pair of reactants. Some of the factors determining the direction of electron transfer can be illustrated by considering the quenching of the metal-to-ligand charge-transfer excited state of $RuL_3{}^{2+}$ (L = 4,7-(CH$_3$)$_2$phen) by Co(bpy)$_3{}^{2+}$. Since the excited state may act as an electron donor or as an electron acceptor, two electron-transfer quenching pathways, reductive and oxidative, are possible.

$$[Ru^{II}(L)_3]^{2+} \xrightarrow{h\nu} {}^*[Ru^{III}(L)_2(L^-)]^{2+}$$
$$(d\pi)^6(L\pi^*)^0 \qquad\qquad (d\pi)^5(L\pi^*)^1$$

$$[Ru^{II}(L)_2(L^-)]^+ + [Co^{III}(bpy)_3]^{3+}$$

$$\nearrow k_{red}$$

$$[Ru^{III}(L)_2(L^-)]^{2+} + [Co^{II}(bpy)_3]^{2+}$$

$$\searrow k_{ox}$$

$$[Ru^{III}(L)_3]^{3+} + [Co^{I}(bpy)_3]^+$$

In the first case the electron transfer is from $Co(bpy)_3^{2+}$ to $^*RuL_3^{2+}$ to produce $Co(bpy)_3^{3+}$ and RuL_3^+ while in the second case the electron transfer is from $^*RuL_3^{2+}$ to $Co(bpy)_3^{2+}$ to yield RuL_3^{3+} and $Co(bpy)_3^+$. The driving forces for these two reactions can be estimated from the relevant reduction potentials.

	E^o,V	ΔE^o,V
reductive pathway:		
$^*RuL_3^{2+} + e^- = RuL_3^+$	+0.67	0.32
$Co(bpy)_3^{3+} + e^- = Co(bpy)_3^{2+}$	+0.35	
oxidative pathway:		
$Co(bpy)_3^{2+} + e^- = Co(bpy)_3^+$	-0.89	0.12
$RuL_3^{3+} + e^- = {}^*RuL_3^{2+}$	-1.01	

The driving force is larger for the reductive pathway and favors this pathway by a factor of ~ 50 (C. V. Krishnan, C. Creutz, D. Mahajan, H. A. Schwarz, and N. Sutin, Israel J. Chem., 22, in press.) On this basis it might be tempting to conclude that the oxidative quenching pathway may be neglected in this system. This conclusion is not justified since it ignores any difference in the intrinsic barriers for the two reactions: these barriers are very different for the two reactions largely because of the difference in the nuclear configurations of the cobalt couples involved. This difference is reflected in their self-exchange rates.

	Δd_o,Å	k_{ex}, $M^{-1}s^{-1}$
reductive pathway:		
$Co(bpy)_3^{2+} + Co(bpy)_3^{3+}$ $(d\pi)^5(d\sigma^*)^2$ $(d\pi)^6$	0.19	20
oxidative pathway:		
$Co(bpy)_3^{+} + Co(bpy)_3^{2+}$ $(d\pi)^6(d\sigma^*)^2$ $(d\pi)^5(d\sigma^*)^2$	-0.02	$\geqslant 10^8$

The difference in the self-exchange rates of the two cobalt couples favors the oxidative pathway by a factor of ~ 300. (For a further discussion of the above and other self-exchange rates, see B. S. Brunschwig, C. Creutz, D. H. Macartney, T.-K. Sham, and N. Sutin, Faraday Discuss. Chem. Soc., No 74, in press). Evidently the difference in the intrinsic barriers is large enough to compensate for the less favorable driving force for the oxidative pathway. As a result the latter pathway can compete favorably with the reductive pathway.

This system illustrates the importance of both the thermodynamic and intrinsic barriers in determining the direction of electron transfer within a given reactant pair. In addition, systems such as the one considered here in which the oxidative and reductive pathways possess comparable rate constants afford an opportunity of controlling or switching the direction of electron transfer by modifying one of the barriers.

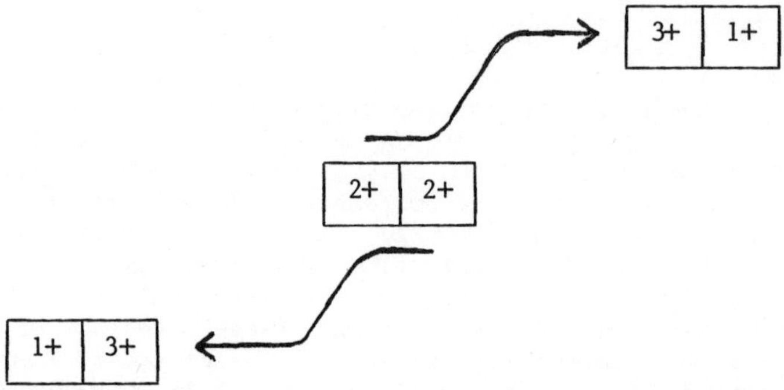

This may find application in biological and other systems. One way in which the effective thermodynamic barrier can be modified is through the movement of a charged group near one of the reactants: since the charge distribution following electron transfer differs in the two cases, a given pair of products can be preferentially stabilized (or destabilized) through interaction with nearby charged groups. Consequently the movement of such groups prior to electron transfer, for example through a protein configuration change or through the diffusion of ions, could change the direction or rate of the electron transfer in suitable systems.

9

Metalloproteins and Electron Transfer

S. K. CHAPMAN, D. M. DAVIES, A. D. WATSON, and A. G. SYKES

The University, Newcastle-upon-Tyne, Department of Inorganic Chemistry, Newcastle-upon-Tyne NE1 7RU England

Studies on 1:1 electron-transfer reactions of metalloproteins with inorganic complexes are, in a number of cases, at a stage where the site or sites on the protein at which electron transfer occurs can be specified. Results obtained for the blue Cu protein plastocyanin are considered here in detail as illustrative of different approaches yielding relevant information. The reduction of plastocyanin PCu(II) with cytochrome c(II) is also considered as an example of a protein-protein reaction. It has been possible in the latter to define the reaction sites of the two proteins with respect to each other at the time of electron transfer.

Metalloproteins fall into three main structure categories depending on whether the active site consists of a single coordinated metal atom, a metal-porphyrin unit, or metal atoms in a cluster arrangement. In the context of electron-transfer metalloproteins, the blue Cu proteins, cytochromes, and ferredoxins respectively are examples of these different structure types. Attention will be confined here mainly to a discussion of the reactivity of the blue Cu protein plastocyanin. Reactions of cytochrome c are also considered, with brief mention of the [2Fe-2S] ferredoxin, and high potential Fe/S protein [HIPIP].

It is timely to review the reactivity of plastocyanin in the light of recent aqueous solution studies, and the elegant structural work of Freeman and colleagues on both the PCu(I) and PCu(II) forms (1,2). Plastocyanin now ranks alongside cytochrome c (3) as the electron-transfer metalloprotein for which there is most structural information.

The aim in solution studies on metalloprotein is to be able to say more about intermolecular electron transfer processes, first of all by studying outer-sphere reactions with simple inorganic complexes as redox partners. With the information (and experience) gained it is then possible to turn to protein-protein reactions, where each reactant has its own complexities

0097-6156/83/0211-0177$06.25/0

and reactions are correspondingly more difficult to assess. A fuller understanding of reactions involving physiological protein partners is the ultimate objective. In addition to the active site chemistry of each protein, the identification of a site (or sites) on the protein surface at which electron transfer occurs, the nature and specificity of such functional sites, the distance over which electrons are transferred, and the manner in which electrons are transferred (intervening groups?) are relevant. Multimetal proteins such as cytochrome oxidase with more than one active site are not considered here. Intramolecular electron transfer is an additional feature with such proteins.

Function of Plastocyanin. Plastocyanin (M.W.10,500, 99 amino-acid residues) occurs in all higher plants as well as green and blue-green algae (104 amino acids) (4). It has a relatively well defined function in photosynthetic electron transfer and acts as an oxidant for membrane bound cytochrome f (M.W.33,000), and as a reductant for P700 which is the double chlorophyll pigment of photosystem 1. The Cu in plastocyanin has a Cu(II)/Cu(I) reduction potential of 370 mV at pH 7, which is between that of cytochrome f (360 mV) and P700 (520 mV). From the amino-acid composition cytochrome f is estimated to have a charge of -30 at pH 7 (5).

Structure of Plastocyanin. The structure of PCu(II) is now known to 1.6Å resolution, (2). The protein contains a single Cu atom with a distorted tetrahedral coordination geometry. It is coordinated to the imidazole N_δ atoms of two histidine residues (37 and 87), the thiolate S-atom of cystein 84, and the thioether S-atom of methionine 92, Figure 1, which has come to be regarded as the normal view. Metal-ligand distances at the Cu(II) site are indicated in Figure 2. The upper edge of the imidazole of His 87 is level with the molecular boundary, and the Cu is at this point at its closest to, and only 6Å from, the surface. Viewed from above the Cu is situated in a pocket, the walls and rim of which are lined with conserved hydrophobic groups. Solvent H_2O does not have access to the Cu site. The matching tetrahedral geometries for PCu(II) and PCu(I) provides an excellent example of the entatic state of Vallee and Williams (6).

From 13 completed amino-acid sequences and 54 partial sequences (>40 residues) of plastocyanins from higher plants it appears that sixty residues are invariant and 7 are conservatively substituted (2,7). With three algal plastocyanins included there are 39 invariant or conservatively substituted groups. It is believed that the same structural features apply to the whole family, and that highly conserved residues are an indication of functional sites on the protein surface. The upper hydrophobic and right-hand-side surfaces are believed to be particularly relevant in this context, the latter including four consecutive

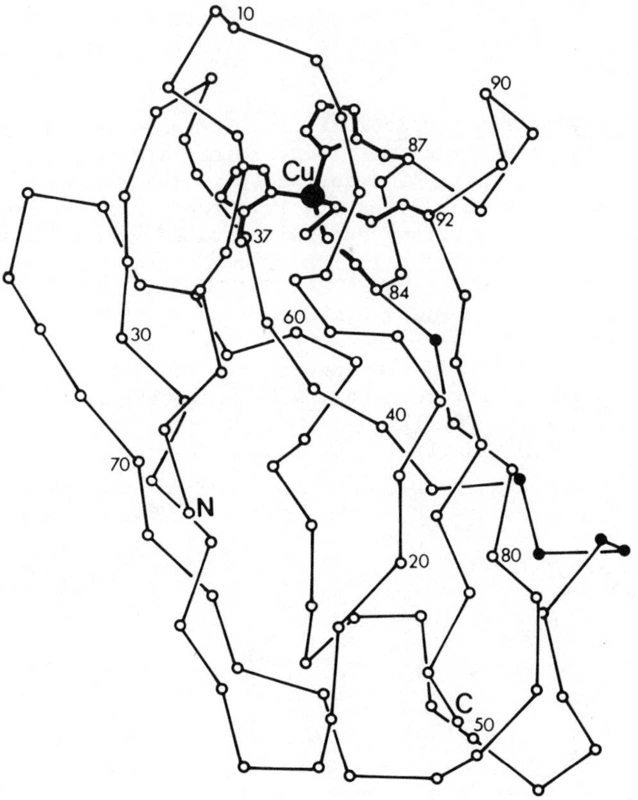

Figure 1. Structure of plastocyanin (2) showing the positions of α-carbon atoms of amino acid residues. The active site and positions of the conserved (plant) negative patch (42–45) and Tyr 83 are indicated (●).

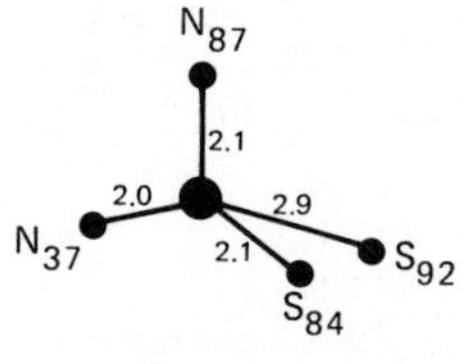

Figure 2. The Cu active site of plastocyanin PCu(II) (2).

negatively charged amino acids (which form a negative patch 42-45) as well as Tyr-83. Plastocyanin from the blue-green algal source Anabaena variabilis has no negative patch (only 42 is acidic) (4). The extra amino acids (104 compared to 99) are incorporated in lower sections of the structure remote from the active site.

Rate Constants and Reactivity. Electron-transfer reactions of plastocyanin (and other metalloproteins) are so efficient that only a narrow range of redox partners (having small driving force) can be employed. Rates are invariably in the stopped-flow range, Table I. Unless otherwise stated parsley plastocyanin

Table I

Summary of reduction potentials and rate constants for reactions with plastocyanin PCu(I) and PCu(II) at 25°C, pH 7-8, I = 0.10 M(NaCl).

	$E^\ominus$ (mV)	k_{oxid}[a] (M^{-1} s^{-1})	k_{red}[b] (M^{-1} s^{-1})
Cytochrome c(III)/(II)	260		1.45×10^6
$Ru(NH_3)_5py^{3+,2+}$	273	4.2×10^4 [c]	4.2×10^5
PCu(II)/PCu(I)	370		
$Co(phen)_3^{3+,2+}$	370	3.0×10^3 [d]	
$Co(bipy)_3^{3+,2+}$	370	313	
$Co(dipic)_2^{-,2-}$	400	510	
$Fe(CN)_6^{3-,4-}$	410	9.4×10^4	1.9×10^4

[a]As oxidant for PCu(I)
[b]As reductant for PCu(II)
[c]At pH 5.8, I = 0.10M ($NaClO_4$)
[d]Value obtained at low [$Co(phen)_3^{3+}$] before limiting kinetics effective.

is used in studies described. Reactions as in (1)

$$PCu(I) + Co(phen)_3^{3+} \longrightarrow PCu(II) + Co(phen)_3^{2+} \qquad (1)$$

are monitored at the PCu(II) peak at 597nm (ε 4500 M^{-1} cm^{-1}) with the inorganic redox partner in >10-fold excess. With cytochrome c (M.W. 12,500, 104 residues) the Fe(II) form is readily monitored at 417nm (Δε 4 x 10^5 M^{-1} cm^{-1}) and the reaction is best studied with PCu(II) in large excess, (2).

$$\text{Cyt c(II)} + PCu(II) \longrightarrow \text{Cyt c(III)} + PCu(I) \qquad (2)$$

Ionic strengths are generally adjusted to I = 0.10M (NaCl). A range of buffers (~ 10^{-2}M) has been employed with satisfactory agreement in overlap regions. Phosphate is known to associate with cytochrome c and was therefore avoided in all such studies. Care is required when studying reactions which are thermodynamically unfavourable, since it is necessary to have high concentrations of redox partner to ensure that the reaction proceeds to completion (>95%). If this condition is not met incorrect interpretation can result unless a more rigorous kinetic treatment is employed (8). Simple first-order kinetics (k_{obs}) apply in very many cases, (3),

$$\text{Rate} = k\,[P]\,[C] \qquad (3)$$

i.e. $k_{obs} = k[C]$, where [P] and [C] are protein and complex concentrations respectively. Typical rate constants, k, are listed in Table I. Attempts to apply Marcus correlations give a wide range of rate constants (50 to 3 x 10^6 M^{-1} s^{-1}) for the PCu(I) + PCu(II) self-exchange (9).

Limiting Kinetics. In some instances first-order rate constants (k_{obs}) give a less than first-order dependence on the reactant in large excess, and (3) no longer applies. Instead, in for example the $Co(phen)_3{}^{3+}$ oxidation of PCu(I), (4) holds,

$$k_{obs} = \frac{K\,k_{et}\,[C]}{1 + K\,[C]} \qquad (4)$$

with K and k_{et} as defined in (5) and (6).

$$P + C \underset{}{\overset{K}{\rightleftharpoons}} P,C \qquad (5)$$

$$P,C \xrightarrow{k_{et}} \text{products} \qquad (6)$$

Such behaviour is identified by curvature when k_{obs} is graphed against [C], Figure 3, (10). Both K and k_{obs} can be obtained from linear plots of $(k_{obs})^{-1}$ against $[C]^{-1}$. If K is small so that K[C]<<1 then the kinetics conform to (3). Under such circumstances it is still appropriate to think in terms of a two-step process with $k = K\,k_{et}$. Identification of (5) - (6) with the implication of long duration or 'sticky' collisions is important in that it could present a means by which low probability (long distance?) electron transfer could occur. For reactions involving large protein-molecules electron transfer over long distances may be a necessity.

A number of examples of limiting kinetics have been reported for reactions of [2Fe-2S] and 2[4Fe-4S] ferredoxins with inorganic complexes (11). Recent stopped-flow work has not however confirmed limiting kinetics for the reaction of azurin, ACu(I) +

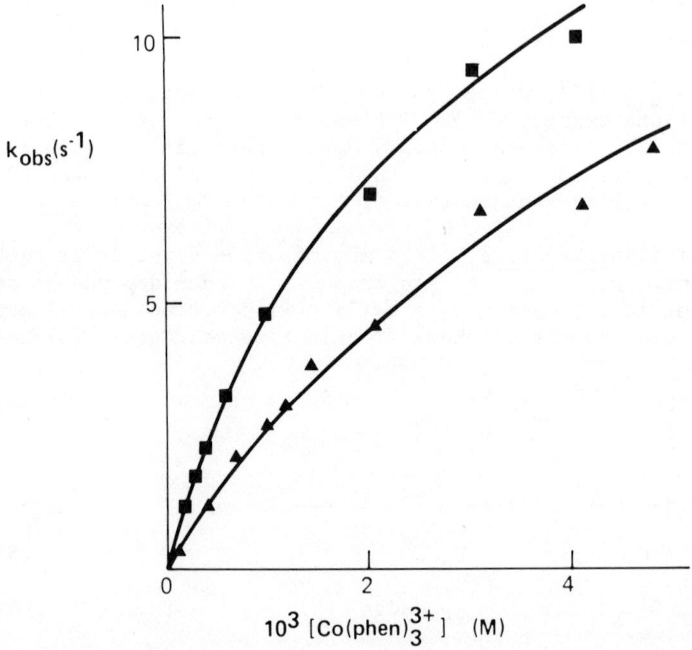

Figure 3. Dependence of first-order rate constants k_{obs} (25 °C) vs. [Co(phen)$_3^{3+}$] for the oxidation of plastocyanin PCu(I). Conditions: pH, 7.5 (phos); and I, 0.10 M *(NaCl). Key: ■, spinach; and ▲, parsley. (Reproduced from Ref. 10. Copyright 1978, American Chemical Society.)*

$Fe(CN)_6^{3-}$, and for the reaction $Fe(CN)_6^{4-}$ + PCu(II) at pH 7, and a requirement for (4) would now seem to be that the overall charges on the two reactants are of opposite sign. Clearly $k = K\,k_{et}$ applies since $\Delta H^{\ddagger}$ for k in the ACu(I) + $Fe(CN)_6^{3-}$ reaction is negative implying a composite term with ΔH for K having a negative value. From the amino-acid composition charges on ACu(I) and PCu(II) are estimated to be -1 and -7 respectively at pH 7. The ionic strength at which a kinetic study is carried out can of course have an influence on behaviour and interpretation. Since each reactant has associated with it an ionic atmosphere of opposite charge, K will become smaller as the ionic strength increases. Limiting kinetics (4), is less likely therefore at I = 0.50 M, (12), and conversely more likely to be detectable at I < 0.10 M. To test for limiting kinetics relatively high concentrations of C ($\sim 3 \times 10^{-3}$ M) are sometimes required, and replacement of a 1:1 electrolyte (NaCl) by e.g. a 3:1 electrolyte (the reactant) to maintain constant I may contribute. It is difficult to allow for such an effect (if any), so that when K is small as in the PCu(I) + $Co(phen)_3^{3+}$ reaction there will be some uncertainty attached to the magnitude of K. However other work described below in which 3+, 4+ and 5+ complexes associate more strongly with PCu(I) provide a self-consistant pattern of behaviour which leaves little doubt that an effect is present.

It should also be mentioned that (5) - (6) is not a unique interpretation for (4). An alternative is the so-called 'dead-end' mechanism (7) - (8)

$$P + C \underset{}{\overset{K}{\rightleftharpoons}} P,\ C \qquad (7)$$

$$P + C \xrightarrow{k'_{et}} \text{products} \qquad (8)$$

where association of C occurs at another site on the protein to give non-reactive P, C. Some sort of conformational change is envisaged to account for this 'switching-off' in reactivity. No evidence has yet been obtained in support of such a mechanism in the present context, and it is unlikely that it has general applicability. NMR measurements for example provide no support for a conformational change of PCu(I) on association with Cr(III) complexes (13). Moreover it has in one case been demonstrated that k_{et} in (4) is not dependent on ionic strength, consistant with an intramolecular as opposed to intermolecular process (11). Although caution is required, particularly as isolated examples of (7) - (8) may exist, the invoking of such a mechanism seems to be a case of looking for greater complexity than may actually exist. A reasonable stance, and one which we have adopted, is that discussion should proceed in terms of (5) - (6) until evidence in support of (7) - (8) is obtained.

The Effect of pH on Reactivity. In spite of the different charges on oxidants $Co(phen)_3^{3+}$ and $Fe(CN)_6^{3-}$ and evidence that they use different reaction sites on PCu(I) (see below), remarkably similar pH profiles of rate constants are observed, Figure 4. Dependence on $[H^+]$ are described by (9),

$$k = \frac{k_o + k_H K_H [H^+]}{I + K_H [H^+]} \tag{9}$$

where the constants are as defined in (10) - (12).

$$P + H^+ \underset{}{\overset{K_H}{\rightleftharpoons}} H^+P \tag{10}$$

$$P + C \xrightarrow{k_o} \text{products} \tag{11}$$

$$H^+ P + C \xrightarrow{k_H} \text{products} \tag{12}$$

The above reactions give $K_H = 0$ and acid dissociation pK_a values (from K_H) of 6.1 for $Co(phen)_3^{3+}$ and 5.7 for $Fe(CN)_6^{3-}$ (10). Results obtained for the spinach PCu(I) + $Co(phen)_3^{3+}$ reaction (pK_a 5.7) are also included in Figure 4. The complete switch-off in reactivity led to the suggestion that protonation at or near the Cu site occurring (10). Certainly formation of planar three-coordinate Cu(I) was an attractive possibility. This has now been confirmed by crystallography where protonation of the His 87 of PCu(I), but not of PCu(II), has been demonstrated, Figure 5. Clarification is required however of an earlier report that pK_a's at 4.9 and <4.5 from NMR correspond respectively to histidine ligand dissociations (14). Also with the oxidant $Co(4,7\text{-DPSphen})_3^{3-}$, (15), it has now been demonstrated that proton switch -off occurs at a pH just <5. Other protonation effects are presumably modifying the reactivity of $Co(4,7\text{-DPS phen})_3^{3-}$ but this remains a puzzling result, and one which we do not fully understand. Protonation of the negative patch 42-45 could affect reactivity in the pH range investigated, particularly with $Co(phen)_3^{3+}$ (see below). However this cannot be the only effect since preliminary results with plastocyanin from Anabaena variabilis (no negative patch) give a similar switch-off in reactivity with $Co(phen)_3^{3+}$ to that indicated in Figure 4, (16).

No kinetic evidence for significant protonation of PCu(II) has previously been reported (10). With $Fe(CN)_6^{4-}$ as reductant effects are small, Figure 6. The contrasting behaviour of $Fe(CN)_6^{3-}$ and $Fe(CN)_6^{4-}$ in Figures 4 and 6 respectively carries the implication that $E^\ominus$ for the protein increases with decreasing pH. With $Ru(NH_3)_5py^{2+}$ and cytochrome c(II) (8+ charge) as reductants significant pH effects are observed, Figure 7, giving pK_a values of 4.95. A fit of rate constants for $Ru(NH_3)_5 py^{2+}$ to (9) gives k_H which is $\sim$ 40% of k_o. For cytochrome c a value $k_H = 0$ is obtained. To account for this behaviour

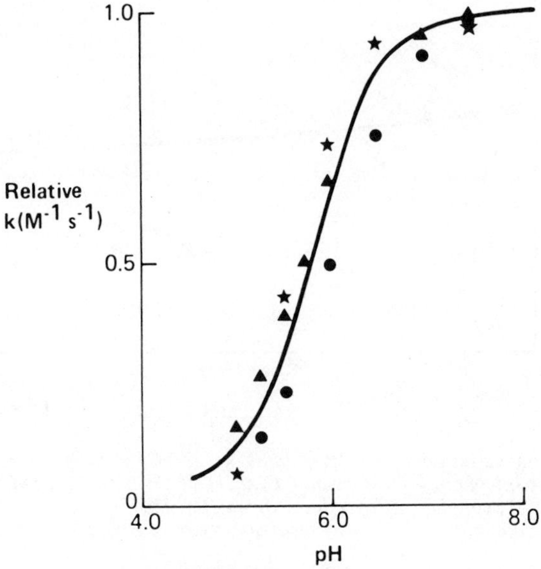

Figure 4. The variation of rate constants k($M^{-1}s^{-1}$) *at 25 °C (relative scale) for the oxidation of parsley PCu(I) with* $Fe(CN)_6^{3-}$ (▲) *and* $Co(phen)_3^{3+}$ (●), *and spinach PCu(I) with* $Co(phen)_3^{3+}$ (★) *[I = 0.10* M *(NaCl)]. (Reproduced from Ref. 10. Copyright 1978, American Chemical Society.)*

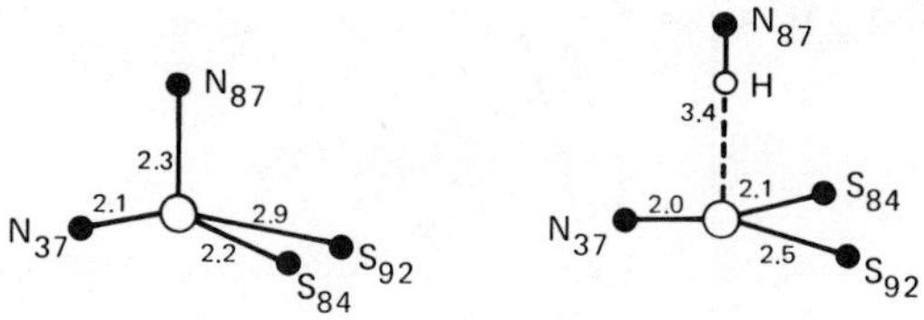

Figure 5. The Cu active site of plastocyanin PCu(I) at pH values > 7 (left) and < 4.5 (right) (2).

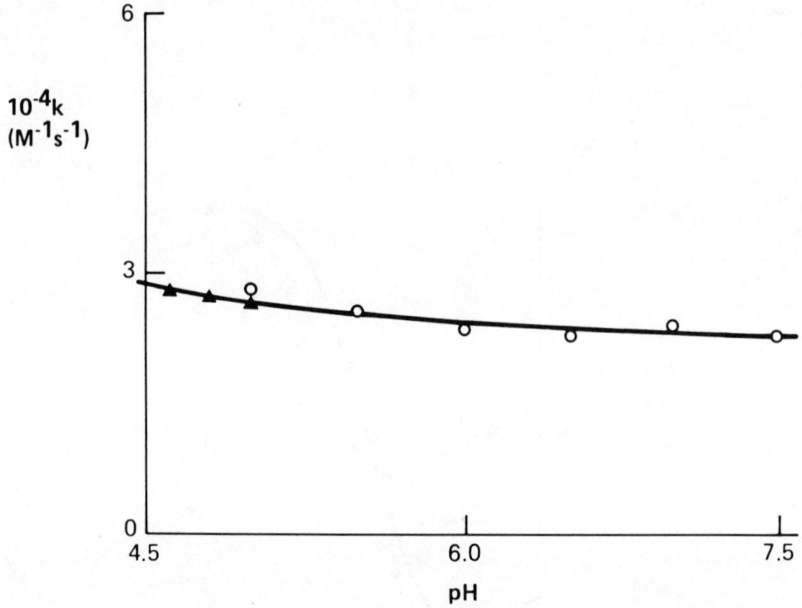

Figure 6. The variation of rate constants k *(25 °C) with pH for the reaction of* $Fe(CN)_6^{4-}$ *with parsley plastocyanin PCu(II) [I = 0.10* M *(NaCl)]. Key:* ▲*, acetate; and* ○*, cacodylate. (Reproduced from Ref. 10. Copyright 1978, American Chemical Society.)*

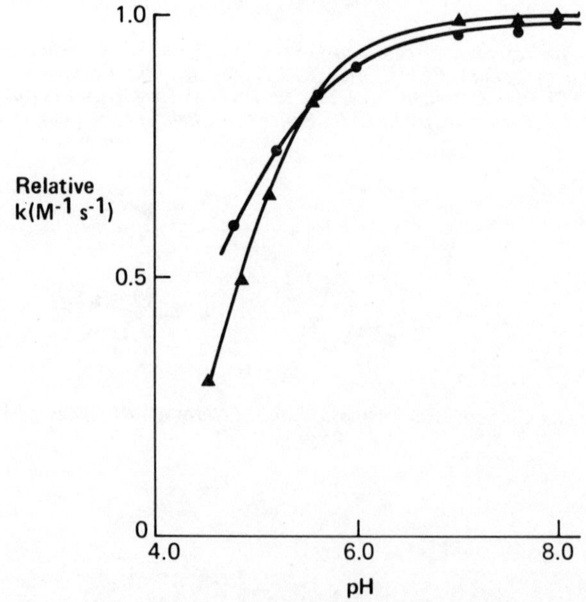

Figure 7. The variation of rate constants k *(25 °C) with pH for the reduction of parsley plastocyanin PCu(II) with* $Ru(NH_3)_5py^{2+}$ *(●) and cytochrome c(II) (▲) [I = 0.10* M *(NaCl)].*

protonation at the negative patch 42-45 is proposed. Therefore electron transfer at or near 42-45 is envisaged for these redox partners. With $Fe(CN)_6^{4-}$ the much smaller effect is consistent with reaction at a more remote site only slightly influenced by the overall change in charge attendant on protonation of for example the 42-45 patch.

<u>NMR Studies</u>

Imaginative use of high resolution ^1H NMR spectroscopy to examine the interaction of PCu(I) (3.5 mM) with redox inactive $Cr(CN)_6^{3-}$, $Cr(phen)_3^{3+}$ and $Cr(NH_3)_6^{3+}$ (0.2 - 1.2 mM), the first two of which are analogues of $Fe(CN)_6^{3-}$ and $Co(phen)_3^{3+}$ has been reported (17). The paramagnetic Cr(III) complexes give line-broadening effects which indicate preferred sites for association. Such experiments provide unequivocal evidence that $Cr(CN)_6^{3-}$ associates close to His 87 and near to the Cu site (6Å), whereas $Cr(phen)_3^{3+}$ and $Cr(NH_3)_6^{3+}$ associate at a site more distant from the Cu and close to Tyr 83. The negative patch 42-45 is close to Tyr 83. These results hold for the three plastocyanins from french beans, cucumber and parsley (17, 18). The conservation of amino-acid residues throughout all higher plant plastocyanins in the vicinity of His 87 and Tyr 83 is particularly relevant in the context of these results.

<u>Blocking Effects.</u> Highly charged redox inactive complexes $Cr(phen)_3^{3+}$, $Co(NH_3)_6^{3+}$, $Pt(NH_3)_6^{4+}$ and $(NH_3)_5Co.NH_2.Co(NH_3)_5^{5+}$ block the reaction of PCu(I) + $Co(phen)_3^{3+}$. Experiments with $Pt(NH_3)_6^{4+}$ are at pH 5.8 to avoid ammine ligand acid dissociation (pK_a 7.1). Blocking by $Cr(phen)_3^{3+}$ is an important link with NMR experiments (as above) indicating that the region close to Tyr 83 is used for electron transfer. More extensive association (K_B) of the ammine complexes with PCu(I) is observed, and these have been explored more fully therefore. As can be seen by inspection of Figure 8 blocking by different complexes (B) is incomplete. A satisfactory explanation is provided by the mechanisms (13) - (15) in which the adduct PCu(I), B retains some reactivity with $Co(phen)_3^{3+}$. The corresponding rate law

$$PCu(I) + B \underset{}{\overset{K_B}{\rightleftharpoons}} PCu(I), B \tag{13}$$

$$PCu(I) + C \xrightarrow{k} \text{products} \tag{14}$$

$$PCu(I), B + C \xrightarrow{k_B} \text{products} \tag{15}$$

dependence is as in (16), where k_{app} is the rate constant at a

$$(k - k_{app}) = \frac{k - k_B K_B [B]}{1 + K_B [B]} \tag{16}$$

particular value of [B]. Values of K_B are listed in Table II.

Table II

Reactions of Parsley Plastocyanin PCu(I). A Comparison of Protein-Complex Association Constants (25°C) at pH 7.5, I = 0.10 M(NaCl).

Complex	Association Constant (M^{-1})
$Co(phen)_3^{3+}$	167[a]
$Cr(phen)_3^{3+}$	176[b,c]
$Co(NH_3)_6^{3+}$	580[b]
$Pt(NH_3)_6^{4+}$	5×10^3 [b,d]
	1.6×10^4 [b,e]
$(NH_3)_5Co.NH_2.Co(NH_3)_5^{5+}$	1.6×10^4 [b]

[a]Redox active association K as in (5)
[b]Redox inactive association K_B as in (13)
[c]Complete blocking assumed
[d]$Pt(NH_3)_6^{4+}$ has acid dissociation pK_a = 7.1, the 3+ c.b. form is dominant at pH 7.5.
[e]pH 5.8

The possibility that (15) corresponds to unblocked reaction of $Co(phen)_3^{3+}$ at the His 87 site has been considered, but is discounted because (eventual) complete blocking of the Tyr 83 site by the 3+, 4+, 5+ complexes would be expected to give behaviour as in Figure 9. This is not observed. Accordingly a self-consistent interpretation along the following lines is possible. Ammine complexes associate strongly in the region of the 42-45 site, which influences the reaction of $Co(phen)_3^{3+}$ with the protein at a site near to the Tyr 83. An additional feature is that the conserved negatively charged Glu 59 residue is close to Tyr 83. By invoking two adjacent sites (or one large site) in this way partial blocking of $Co(phen)_3^{3+}$ can be explained.

Other redox partners $Co(bipy)_3^{3+}$ (oxidant) and $Ru(NH_3)_5py^{2+}$ (reductant) are likewise partially blocked by $Pt(NH_3)_6^{4+}$. Interestingly the reaction of cytochrome c(II) with PCu(II) is also blocked by $Pt(NH_3)_6^{4+}$, thus identifying this as a site for electron transfer with cytochrome c. This observation is consistant with a preliminary report of NMR results (19). The blocking is in fact more extensive than that observed with the above complexes, which is reasonable in view of the larger size of cytochrome c. Reaction with the negatively charged dipicolinate oxidant, $Co(dipic)_2^-$, was similarly investigated, where separate association of the oxidant with $Pt(NH_3)_6^{4+}$ can be

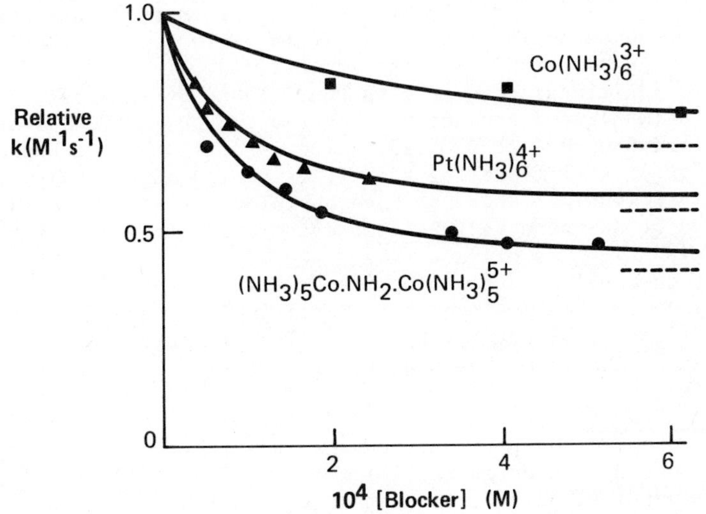

Figure 8. The blocking effect (25 °C) of redox inactive complexes on the reaction of parsley plastocyanin PCu(I) + $Co(phen)_3^{3+}$. Rate constants were determined at pH 7.5 (for ■ and ●) and pH 5.8 (for ▲) [I = 0.10 M (NaCl)].

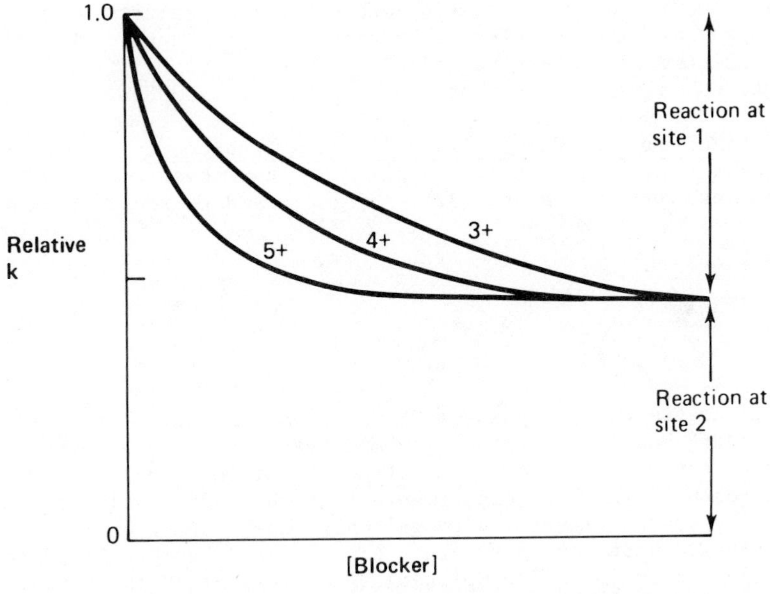

Figure 9. Behavior expected if there are two separate functional sites on the protein for electron transfer. One site (Site 1) is blocked directly by redox inactive $3^+ < 4^+ < 5^+$ complexes, the other (Site 2) is not blocked.

assumed negligible. No blocking effect (or enhancement) in rate was observed on addition of $Pt(NH_3)_6^{4+}$. This implies that as with $Fe(CN)_6^{3-}$, the His 87 site is utilized.

Two additional effects with PCu(II) are of interest. Firstly K_B for $Pt(NH_3)_6^{4+}$ decreases with decreasing pH, 16500 M^{-1} at pH 5.8 and 6500 M^{-1} at pH 5.45, which is consistant with protonation at 42-45 influencing the effectiveness of $Pt(NH_3)_6^{4+}$. Secondly $Pt(NH_3)_6^{4+}$ does not have any blocking effect on the reaction of Ananaena variabilis plastocyanin (no negative patch) with $Co(phen)_3^{3+}$ as oxidant (16).

It is concluded that blocking by association of the $Pt(NH_3)_6^{4+}$ at the 42-45 patch is a particularly sensitive test for utilization of the Tyr 83 site. The aromatic ring of Tyr 83 is located 10A from the Cu at the active site.

Reaction with Cr(III) Modified Plastocyanin. From thermolysin proteolysis experiments Farver and Pecht (20) have concluded that reduction of PCu(II) with labelled $Cr(H_2O)_6^{2+}$ (1:1 mole amounts) at pH 7 gives a product in which Cr(III) is attached to the peptide chain 40-49. Coordination of the Cr at one or two carboxylates in the 42-45 patch is favoured. It has now been demonstrated that rate constants (25°C) for the reaction of PCu(I).Cr(III) + $Co(phen)_3^{3+}$ are decreased by ~16%. The effect is of similar magnitude to that observed for blocking by the 3+ redox inactive $Co(NH_3)_6^{3+}$ (~30% decrease). Use of Cr(III) modified protein has no effect on the reaction with $Fe(CN)_6^{3-}$ as oxidant. These observations (21) support the belief that positive and negative redox partners utilize different functional sites on the protein for electron transfer.

Reaction with Chemically Modified Cytochrome c. Chemically modified (CDNP) cytochrome c derivatives have been prepared by Margoliash and colleagues (22). Lysine residues react as in (17),

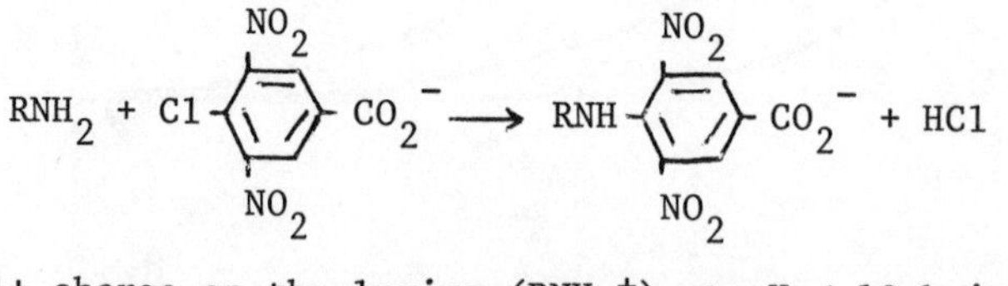

$$RNH_2 + Cl\text{-}\langle\text{ring}\rangle\text{-}CO_2^- \longrightarrow RNH\text{-}\langle\text{ring}\rangle\text{-}CO_2^- + HCl \qquad (17)$$

the 1+ charge on the lysine (RNH_3^+) at pH < 10 being replaced by a charge of 1-. Eight singly modified derivations have been used to investigate the effect of modification on reactivity. It is assumed in these experiments that the modification closest to the electron-transfer site will have most effect on rate constants. Rate constants are enhanced for 3+ and retarded for 3- redox partners. With $Co(phen)_3^{3+}$ and $Fe(CN)_6^{3-}$ as oxidants it has been demonstrated that both react at the exposed heme edge of cytochrome c (23). The exposed heme edge is also relevant with PCu(II) as oxidant, Table III. With all three oxidants

Table III

Second-order rate constants (25°C) for the oxidation of singly modified CDNP-Lysine Cytochrome c Derivatives by $Co(phen)_3^{3+}$ (k_{Co}), $Fe(CN)_6^{3-}$ (k_{Fe}) and PCu(II) (k_{Cu}) at pH 7.5, I = 0.10 M (NaCl).

Modification	$10^{-3}k_{Co}$	$10^{-6}k_{Fe}$	$10^{-5}k_{Cu}$
Native	1.5	12	15
Lys-7	2.1	10	12
Lys-13	3.6	3.9	7.5
Lys-25	2.8	7.8	8.6
Lys-27	11	5.1	8.1
Lys-60	1.7	13	15
Lys-72	4.4	3.3	13
Lys-86	1.7	7.7	8.6
Lys-87	1.9	9.1	9.0

use of lysine-60 modified cytochrome c (modification on the reverse side from the exposed heme edge) has no effect on reactivity. It can also be concluded that there is a preferred orientation of PCu(I) to the exposed edge, Figure 10. Since in cytochrome c the porphyrin ring delocalises electron density from the Fe(II), it can be argued that the 10Å separating Tyr 83 from the Cu is relevant as far as the electron-transfer process is concerned. Whether overlap of the aromatic Tyr 83 ring with the porphyrin, or indeed with the aromatic ligands of $Co(phen)_3^{3+}$ and $Ru(NH_3)_5py^{2+}$ (<u>24</u>), is a requirement remains to be demonstrated.

<u>Other Studies.</u> Experiments in which rate constant pH profiles, blocking effects of redox inactive complexes as well as the effect of Cr(III) modification should now be possible enabling sites on plastocyanin used by cytochrome f and P700 to be specified (<u>25</u>,<u>26</u>).

Rate constants for the oxidation of the negatively charge high potential Fe/S protein from <u>Chromatium Vinosum</u> with PCu(II) do not exhibit any dependence on pH 5.0 - 8.5 which suggests that the His 87 site is being used in this case.

Similar approaches to those described herein are already underway with [2Fe-2S] ferredoxins. From NMR line-broadening studies Markley has demonstrated that redox inactive $Cr(NH_3)_6^{3+}$ preferentially associates at a specific site close to Tyr 83 (<u>13</u>,<u>27</u>). This particular group is a relatively long way from the [2Fe-2S] cluster and the result is surprising since conserved or partially conserved negative patches at 67-69 and 94-96 are nearer to the active site and might have been expected

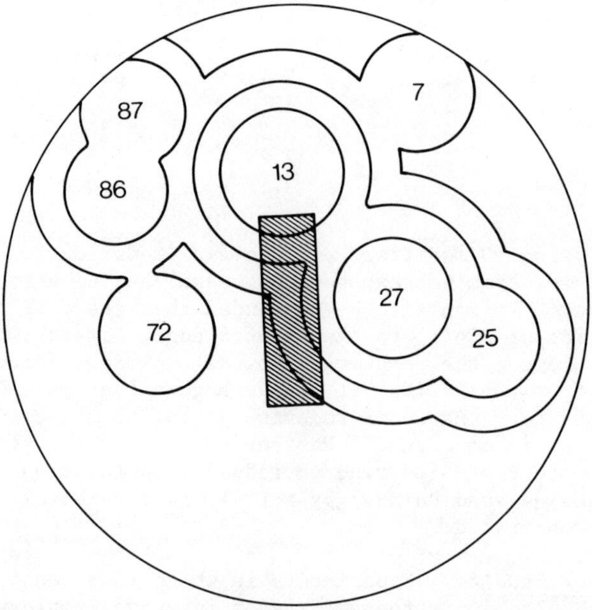

Figure 10. View of cytochrome c with positions of exposed heme edge (block) and lysine modifications (sequence numbers for α-carbons) shown. The smaller circles indicate the relative effectiveness of modifications on rate constants for the reaction of Cyt c(II) + PCu(II) (see *Table III*). *Preferential interaction with PCu(II) in the direction 25, 27, 13, 87 is indicated.*

to be involved. The effect of $Cr(NH_3)_6^{3+}$ on the [2Fe-2S] + $Co(NH_3)_6^{3+}$ reaction (100% blocking is observed) and related studies suggests that a single functional site on [2Fe-2S] is relevant (28). A comparison of reactivity patterns observed with [2Fe-2S] from parsley and algal *Spirulina plastensis* is also of interest in order to test the effect of amino-acid variations (35%). With oxidants $Co(NH_3)_6^{3+}$, $Co(acac)_3$, and $Co(edta)^-$ very similar kinetic behaviour is observed, and it is concluded that identical reaction sites must be involved on the two proteins.

Acknowledgments

It is a pleasure to acknowledge stimulating discussions with Professors H.C. Freeman and E. Margoliash. We are grateful to the U.K. Science Research Council for post-doctoral research assistantships (DMD and ADW) and a research studentship (SKC).

Literature Cited

1. Coleman, P.M.; Freeman, H.C.; Guss, J.M.; Murata, M.; Norris, V.A.; Ramshaw, J.A.M.; Venkatappe, M.P.; *Nature (London)*, 1978, *272*, 319.
2. Freeman, H.C. in "Coordination Chemistry-21"; Laurent, J.P. Ed.; Pergamon, Oxford 1981; p29.
3. Swanson, R.; Trus, B.L.; Mandel, N.; Mandel, G.; Kallai, O.B.; Dickerson, R.E.; *J. Biol. Chem.*, 1977, *252*, 759.
4. Aitken, A.; *Biochem. J.*; 1975, *149*, 675.
5. Gray, J.C.; *Eur. J. Biochem*, 1978, *82*, 133.
6. Vallee, B.L.; Williams, R.J.P.; *Proc. Nat. Acad. Sci. US.A.* 1968, *59*, 498.
7. Boulter, D.; Peacock, D.; Guise, A.; Gleaves, J.T.; Estabrook, G.; *Phytochem.*, 1979, *18*, 603.
8. Butler, J.; Davies, D.M.; Sykes, A.G.; *J. Inorg. Biochem.*, 1981, *15*, 41.
9. Lappin, A.G., "Metal Ions in Biological Systems", H. Sigal ed, M. Dekker (N.Y.), 1981, 13.
10. Segal, M.G.; Sykes, A.G.; *J. Amer. Chem. Soc.*, 1978, *100*, 4585.
11. Armstrong, F.A.; Henderson, R.A.; Sykes, A.G.; *J. Amer. Chem. Soc.*, 1979, *101*, 6912 and 1980, *102*, 6545.
12. Holwerda, R.A.; Knaff, D.B.; Gray, H.B.; Clemmer, J.D.; Crawley, R.; Smith, J.M.; Mauk, A.G.; *J. Amer. Chem. Soc.*, 1980, *102*, 1142.
13. Wright, P.E.; quoted in reference 2.
14. Markley, J.L.; Ulrich, E.L.; Berg, S.P.; Krogman, D.W.; *Biochem.*, 1975, *14*, 4428.
15. Lappin, A.G.; Segal, M.G.; Weatherburn, D.C.; Sykes, A.G.; *J. Amer. Chem. Soc.*, 1979, *101*, 2297.

16. Sisley, M.J.; Sykes, A.G.; work in progress.
17. Cookson, D.J.; Hayes, M.T.; Wright, P.E.; Biochem. Biophys. Acta.; 1980, 591, 162.
18. Handford, P.M.; Hill, H.A.O.; Lee, R.W.-K.; Henderson, R.A.; Sykes, A.G.; J. Inorg. Biochem.; 1980, 13, 83.
19. Cookson, D.J.; Wright, P.E.; quoted in reference 17.
20. Farver, O.; Pecht, I.; Proc. Nat. Acad. Sci. U.S.A.; 1981.
21. Adzamli, I.K.; Chapman, S.K.; Knox, C.V.; Sykes, A.G.; to be published.
22. Osheroff, N.; Brautigan, D.L.; Margoliash, E.; J. Biol. Chem.; 1980, 255, 8245, and references therein.
23. Butler, J.; Davies, D.M.; Sykes, A.G.; Koppenol, W.H.; Osheroff, N.; Margoliash, E.; J. Amer. Chem. Soc.; 1981, 103, 469, and unpublished work.
24. Cummins, D.; Gray, H.B.; J. Amer. Chem. Soc.; 1977, 99, 5158.
25. Niwa, S.; Ishikawa, H.; Nikai, S.; Takabe, T.; J. Biochem.; 1980, 88, 1177.
26. Beoku-Betts, D.; Sykes, A.G.; unpublished work.
27. Chan. T.-M.; Ulrich. E.L.; Markley, J.L.;"Photosynthesis II, Electron Transport and Photophospharylation", ed. G. Akoyunoglou, 1981, Balaban Int. Science Services, Pa.; p697.
28. Adzamli, I.K.; Ong, H.W.-K.; Sykes, A.G.; Biochem. J.; 1982, 203, 669.
29. Adzamli, I.K.; Petrou, A.; Sykes, A.G.; Rao, K.K.; Hall, D.O.; to be published.

RECEIVED December 10, 1982

Discussion

N. Sutin, Brookhaven National Laboratory: The 16% decrease in the rate of oxidation of reduced plastocyanin by $Co(phen)_3^{3+}$ resulting from the attachment of the chromium(III) label seems rather small if the chromium is indeed bound at or near the protein site used for electron transfer to $Co(phen)_3^{3+}$.

A.G. Sykes: I agree with Dr. Sutin that a 16% effect is small. Recent work has however shown that in our experiments the Cr binds at two or more sites on the plastocyanin (analyses confirm that there is attachment of one Cr per molecule of protein), and it is necessary therefore for us to elaborate on the original Farver and Pecht results. Our evidence is based on detailed kinetic studies with redox partners such as $Co(phen)_3^{3+}$ when kinetic plots are found to be biphasic. It is concluded that the Cr attached at one of the sites has little or no effect on reactivity. On further reaction of PCu(II).Cr(III) with a second Cr^{2+} a product PCu(I).2Cr(III) is obtained and a much larger 70% effect on the rate of reduction of $Co(phen)_3^{3+}$ is observed. With this further modification the kinetics are close to being uniphasic. The implication would seem to be that there is after attachment of a second Cr complete modification at the 42-45 patch (Farver and Pecht's original assignment), with the other Cr attached at a more remote (non-influential) site. Also relevant is the observation that PCu(I), PCu(I).Cr(III), and PCu(I).2Cr(III) all react with $Fe(CN)_6^{3-}$ at the same rate. This confirms that $Co(phen)_3^{3+}$ and $Fe(CN)_6^{3-}$ react at different sites and that the Cr modification procedure provides a means of identifying the site used by $Co(phen)_3^{3+}$.

H.B. Gray, California Institute of Technology: The rate of $Co(phen)_3^{3+}$ oxidation of the chromium derivatives of the reduced blue copper protein is surprisingly close to the rate for the native protein, in my view. Could you comment further on what you think this result means in terms of the proposed interaction sites in the various mechanisms?

A.G. Sykes: As already indicated (answer to Dr. Sutin) recent results have helped clarify the situation considerably. A 70% rather than a 16% effect of Cr(III) bond at the 42-45 patch is certainly to be regarded as substantial. That this falls short of 100% blocking suggests that the Cr is located close to rather than at the site on plastocyanin used for electron transfer. Redox partners such as $Co(phen)_3^{3+}$ probably need to approach more closely to the Cu active site than the 42-45 patch, and electron transfer at a site in the region stretching from Tyr83 and the upper-right-hand corner of the molecule (Figure 1) would be perfectly consistent with existing information.

C.H. Brubaker, Michigan State University: In the case of the cytochromes, it has been proposed that electron transfer from the iron porphyrin may involve the pi system of the porphyrin and even nearby aromatic rings. Do you think that a similar thing may happen in the case of the reaction between these copper(I) plastocyanins and the chromium(III)? You seem to favor the idea that the important factor is that the Cr(III) be at a site that is reasonably close to the copper center.

A.G. Sykes: The involvement of aromatic rings in electron-transfer reactions of metalloproteins is an attractive idea. There are two possible effects to consider. A single ring might function simply as a mediator for electron transfer between two metal centres, or when such rings are present on both reactants stacking may occur making electron transfer easier. In the case of plastocyanin highly conserved residues such as Tyr83 which are close to the 42-45 patch, could well function in one or other of these ways. Unfortunately evidence in support of such detailed electron transfer mechanisms is at present lacking. The porphyrin of cytochrome c certainly has a specific function in electron transfer, but this is rather a special case since the metal is coordinated at the centre of the ring system.

D.R. McMillin, Purdue University: In addition to the charge effects discussed by Professor Sykes, I would like to add that structural effects may help determine electron transfer reactions between biological partners. A case in point is the reaction between cytochrome C_{551} and azurin where, in order to explain the observed kinetics, reactive and unreactive forms of azurin have been proposed to exist in solution (1). The two forms differ with respect to the state of protonation of histidine-35 and, it is supposed, with respect to conformation as well. In fact, the ^1H nmr spectra shown in the Figure provide direct evidence that the nickel(II) derivative of azurin does exist in two different conformations, which interconvert slowly on the nmr time-scale, depending on the state of protonation of the His^{35} residue (2). As pointed out by Silvestrini *et al.*, such effects could play a role in coordinating the flow of electrons and protons to the terminal acceptor *in vivo*.

(1) Silvestrini, M.C.; Brunori, M.; Wilson, M.T.; Darley-Usmar, V.M. J. Inorg. Biochem. 1981, 14, 327-338.

(2) Blaszak, J.A.; Ulrich, E.L.; Markley, J.L.; McMillin, D.R., submitted for publication.

A.G. Sykes: Professor McMillin does right to draw attention to the effect of H^+ on reactions of azurin, which are somewhat different to those observed for plastocyanin.

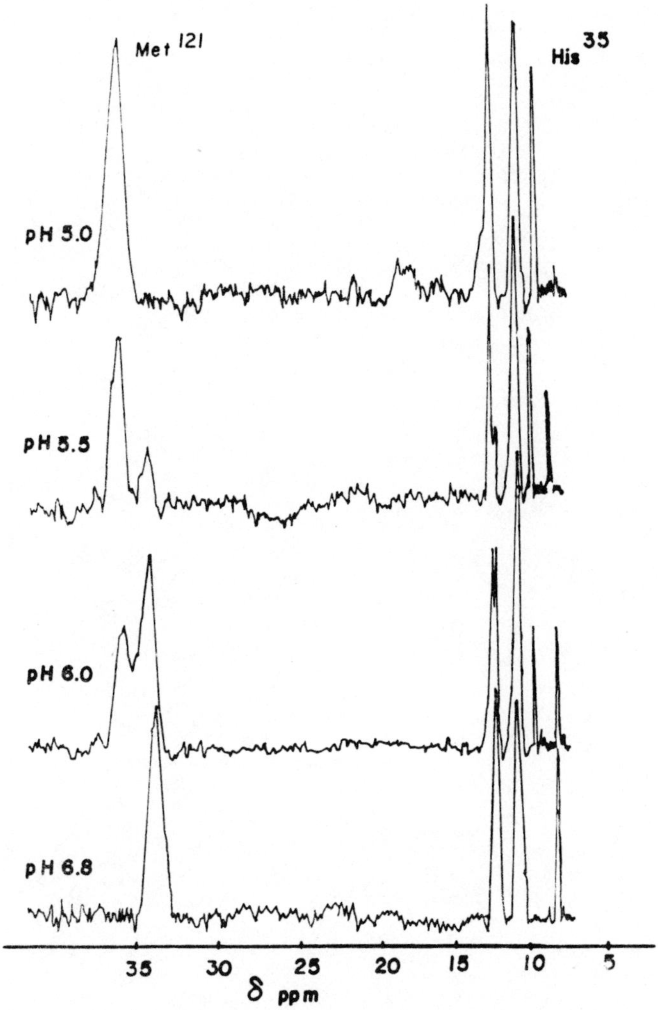

Figure. The pH titration of the C2-H proton of His[35] and the methyl protons of the copper ligand Met[83]. The discontinuous nature of the spectral shifts establish the existence of two structures which interconvert slowly on the NMR time-scale. Note that the same pK_a *is involved, namely that one associated with the protonation of His[35]. (The spectra were obtained with a Nicolet 8.5T NMR spectrometer, supported by NIH grant RR01077.)*

NONCLASSICAL COORDINATION COMPOUNDS

10

Nonclassical Coordination Compounds

JOHN P. FACKLER, JR. and JOHN D. BASIL

Case Western Reserve University, Case Institute of Technology, Department of Chemistry, Cleveland, OH 44106

The topic "Non-Classical Coordination Compounds" is troubling since it implies some knowledge of the meaning of both "classical" and "non-classical". To a young inorganic chemist reviewing some β-diketone work of ours, performed in the '60's, the term "classical" apparently meant work done a number of years ago. Obviously work being done today by modern young chemists is, therefore, "non-classical".

Rather than accept the above definition, the fourth edition of "Advanced Inorganic Chemistry" by F. A. Cotton and G. Wilkinson (1) was consulted to see what it contains relevant to the "non-classical" behavior of metal-metal bonded compounds. Page 980 of this text presents an interesting illustration. H. Schmidbaur, in some beautiful work involving coordination of methyl ylides to metals, demonstrated that the dinuclear gold(I) ylide complex (Figure 1), reacts with dichlorine to form a dichloride which formally contains Au(II). As described below, there is a single bond between the two Au atoms. However, a remarkable non-classical species appears to be formed upon reaction with $NaBPh_4$. This cation is formulated as a Au(II) compound containing only two ligands coordinated to each metal atom. As any good inorganic chemist knows, a d^9 metal ion such as Au(II) must be non-classical when it is only two coordinate. Even if a Au-Au bond is added, the species remains non-classical. Consequently, a careful examination of the system has been undertaken. After considerable study of the system and published results from Schmidbaur's laboratory (2), it has been concluded that this ion is so non-classical that it is NON-EXISTENT.

What Does Exist?

As indicated in Figure 2, Schmidbaur and his students have reported (3) some very interesting reactions of the dinuclear Au(I) ylide complexes. Not only do these species oxidatively add halogen atoms to form Au(II) species (Fig. 2;3) a second mole of

0097-6156/83/0211-0201$06.00/0

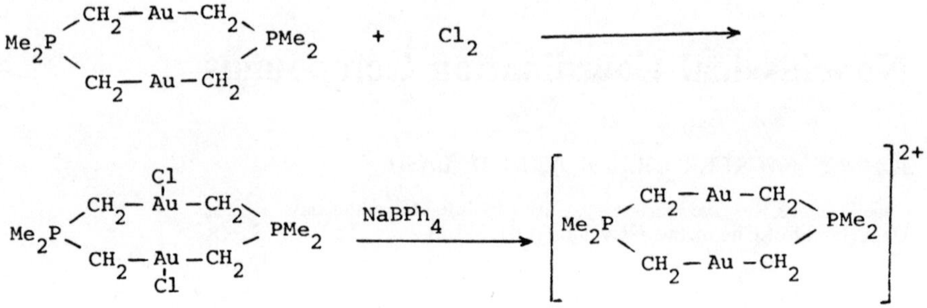

Figure 1. The remarkable "nonclassical" dinuclear gold(II) ion.

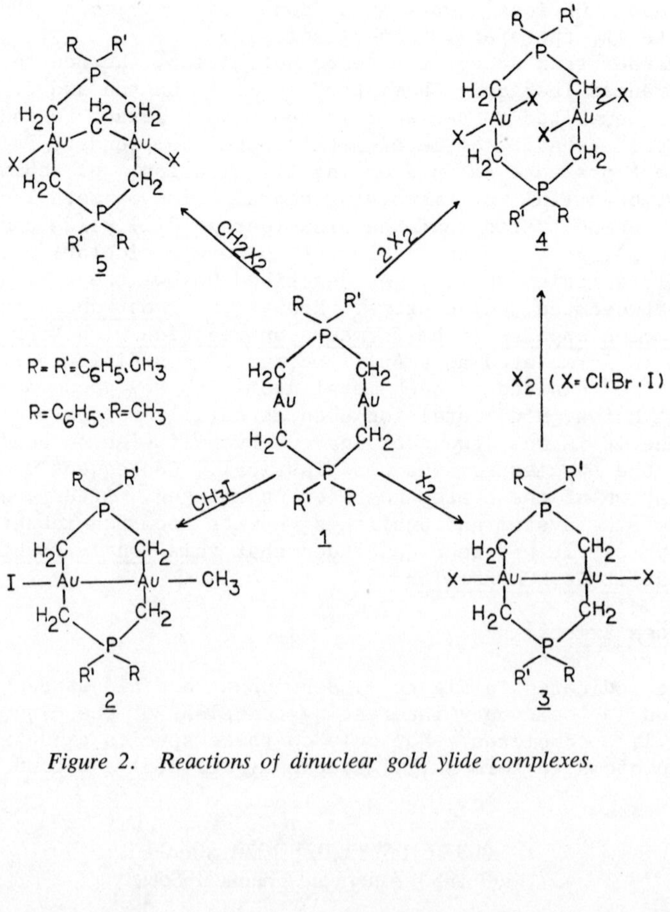

Figure 2. Reactions of dinuclear gold ylide complexes.

halogen will produce a dinuclear Au(III) species. Such a compound has not been subjected to a structural examination yet, but the dichloride species (Fig. 2;4) appears to be isolable as a pure compound. The oxidative addition of methylene dihalides leads (4) to bridged methylene species analogous to the types of compounds reported by Balch and coworkers (5) and by Puddephatt and coworkers (6) for the palladium and platinum A-frame complexes. Schmidbaur also reported (7) that methyliodide oxidatively adds to the Au(I) complex to form a species which contains methyl on one end and iodide on the other (Fig. 2;2). This interesting compound has been the subject of an intensive study in our own laboratory.

Methyliodide Oxidative Addition

First attempts to synthesize the methyliodide oxidative addition product in our laboratory led to crystalline materials which were found by X-ray crystallographically to be a mixture of the Au(I) starting material and the Au(II) diiodide species (8). As indicated in Figure 3, the Au(I) dimer shows a normal non-bonding metal-metal distance and a typical linear geometry about each metal atom. On the other hand, the diiodide displays an essentially linear I-Au-Au-I axis with a metal-metal bond of approximately 2.66Å. Note that the Au iodide distance is approximately 2.7Å long, Figure 4. The coordination geometry about each Au(II) atom is not unusual for a d^9 metal. The puzzling thing was that the syntheses of the methyliodide oxidative addition product appeared to be non-reproducible.

After several years of effort it has been possible to reproduce Schmidbaur's addition of methyliodide (8). The problem appears to be that the methyliodide product is extremely photosensitive in solution. It decomposes with the production of methyl radicals and the ultimate formation of a mixture of the Au(I), and Au(II) diiodide products. The X-ray crystal structure of the methyliodide product is presented in Figure 5. Two features are to be noted. Firstly, the metal-metal distance lengthens only a small amount, 0.04Å, but the Au iodide distance lengthens by approximately 0.20Å. The structural trans effect caused by the methyl group is not attenuated by the Au-Au bond.

A qualitative, symmetry based molecular diagram for the dinuclear system is presented in Figure 6. Note that the only empty orbital associated with the planar D_{2h} framework is the orbital which is antibonding both in the Au-Au interaction and in Au-halide interaction. Thus, one is left to conclude that a metal-metal single bond exists. Furthermore, it is clear that, as the metal-halide bond strength increases, the metal-metal bond strength decreases. An extended Hückel calculation on the related $Rh_2H_4Cl_2^{6-}$ has been reported by Hoffman and Hoffmann (9) As a function of metal-metal distance, it is found that the energies of the b_{2u}^* levels decrease strongly as the metal-metal distance is increased. On the other hand, the two a_g levels both increase in

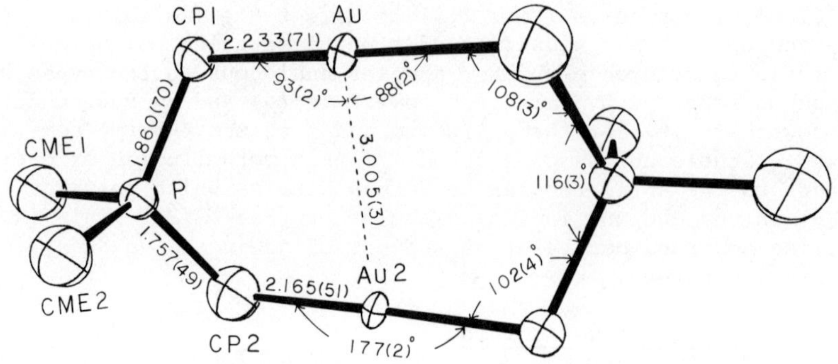

Figure 3. The x-ray structure of $Au_2[(CH_2)_2P(CH_3)_2]_2$.

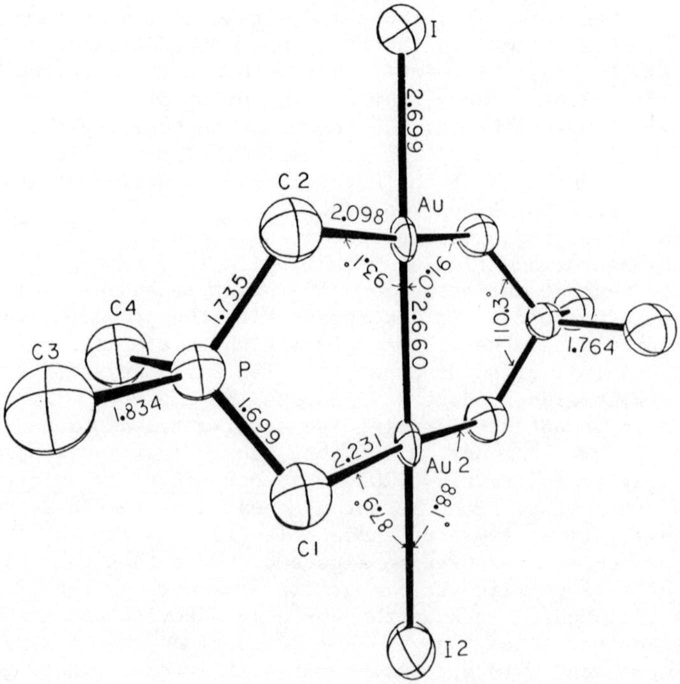

Figure 4. The x-ray structure of $Au_2[(CH_2)_2P(CH_3)_2]_2I_2$.

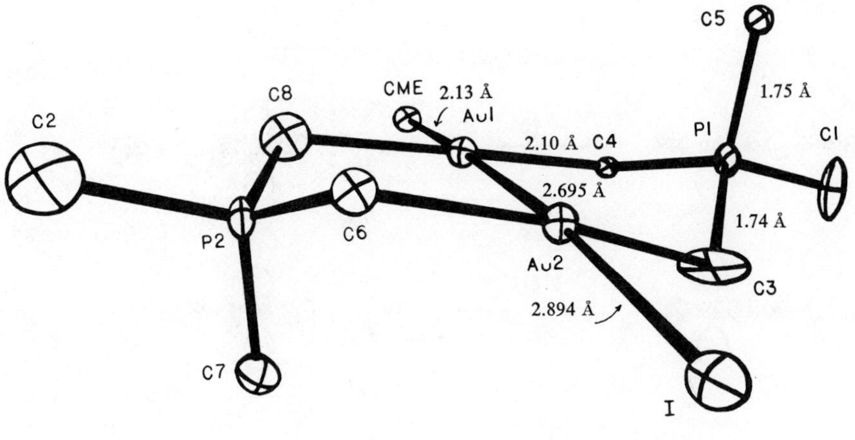

Figure 5. The x-ray structure of $Au_2[(CH_2)P(CH_3)_2]_2CH_3I$.

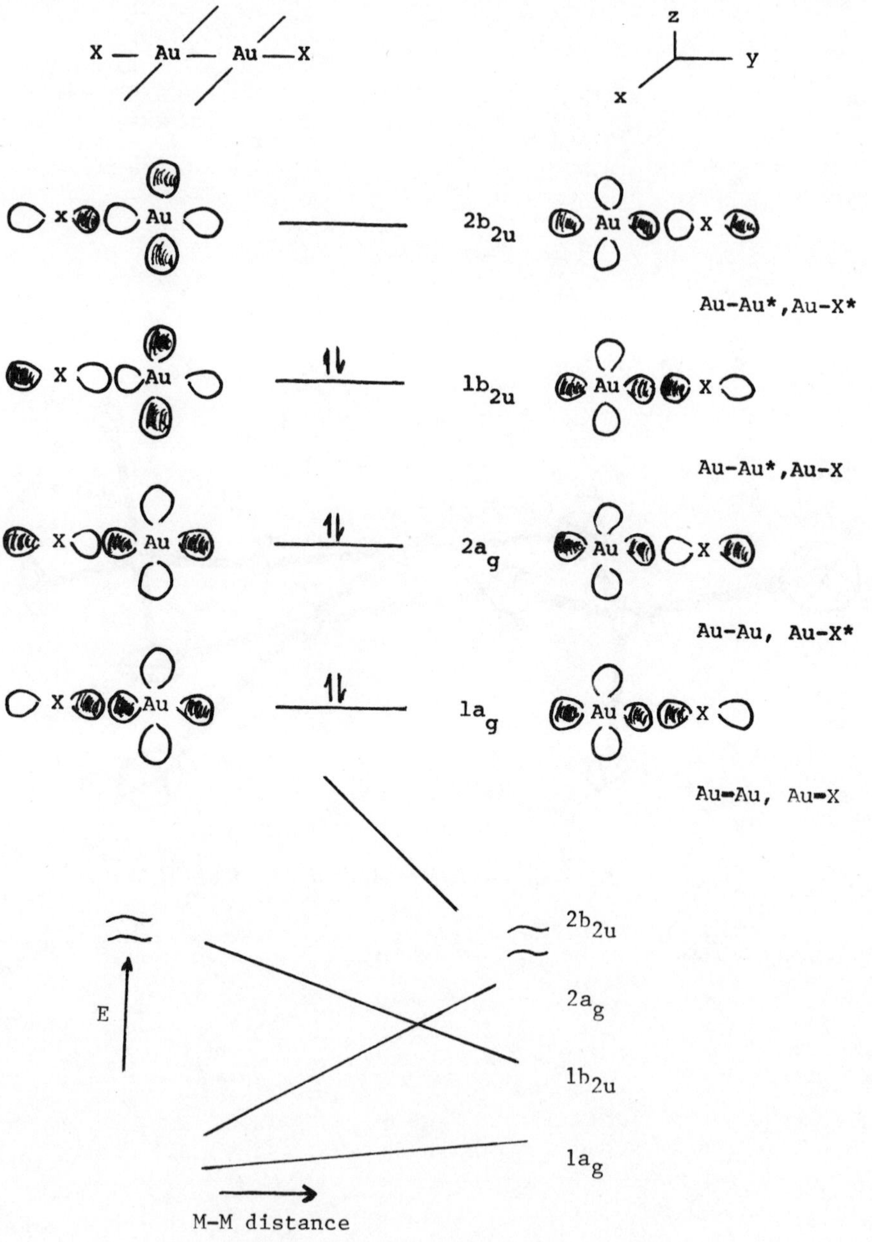

Figure 6. Orbital diagram for dinuclear gold complexes.

energy as the distance is lengthened. At some distance there, indeed, can be a crossover between the $2a_g$ level and the $1b_{2u}$ level. This work suggests that, as additional electron density goes into the linear sigma bonding system through the formation of covalently bonded ligands, such as produced by the addition of two methyl groups, the metal-metal bond may be destroyed.

Examination of the interesting reaction (Fig. 7; 2a,b to 4) suggests that, indeed, such a bond rupturing transformation can be observed. Schmidbaur and his coworkers (10) indicate that the methyliodide product and the Au(III) dibromide species react with methyllithium to produce a compound which appears to be the asymmetrical Au(III)-Au(I) dinuclear species 4. Spectroscopic evi-

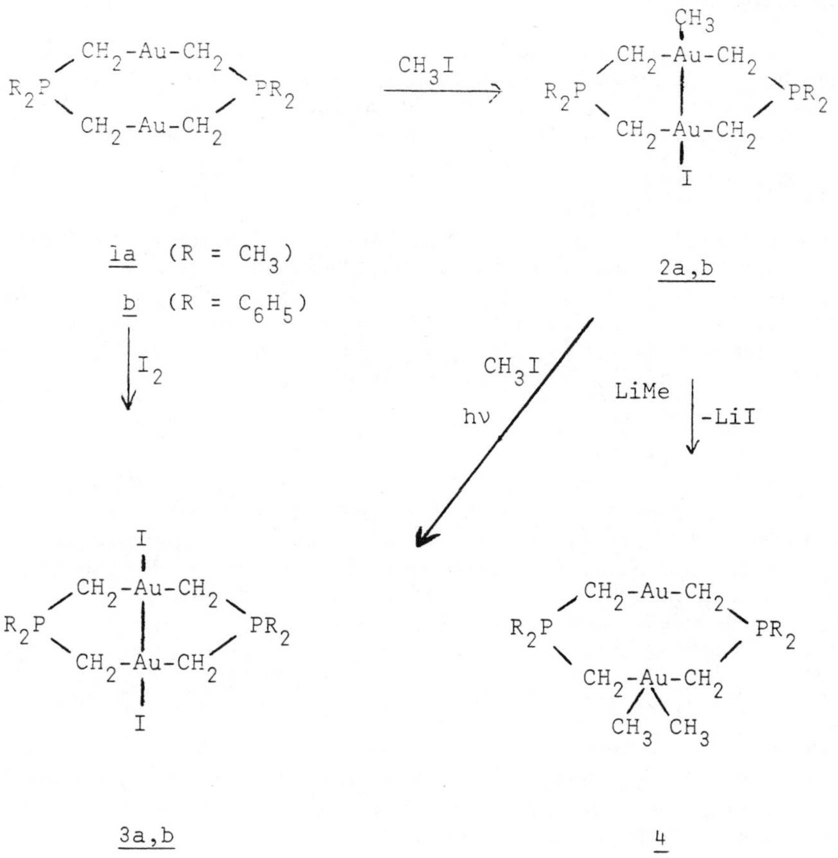

Figure 7. Organometallic reactions of dinuclear gold complexes.

dence in our laboratories confirms this result,(11), although to date no crystalline material has been isolated for X-ray structural analysis. The extreme photosensitivity in solution and chemical instability have made it difficult to achieve structural results to date. Hopefully, before the beginning of the 21st century, this problem will be sorted out, and the compound no longer will be "non-classical."

In summary, it can be concluded that "non-classical coordination compounds" are chemical compounds not explained by the textbooks. As the papers in this ACS symposium series "Inorganic Chemistry: Towards the 21st Century" are examined, the reader will find much that is "non-classical" today but will be "classical" by the 21st century.

Acknowledgments

Structural work reported here was performed by Dr. H. W. Chen and Dr. C. Paparizos. The National Science Foundation NSF-CHE 80-13141 and the donors of the Petroleum Research Fund as administered by the American Chemical Society have supported this work.

Literature Cited

(1) Cotton, F. A.; Wilkinson, G. "Advanced Inorganic Chemistry", John Wiley and Sons: New York, 1980; 980.
(2) Schmidbaur, H. Acc. Chem. Res. 1975, 8, 62.
(3) Schmidbaur, H.; Mandel, J. R.; Bassett, I.-M.; Blaschke, G.; Zimmer-Gasser, B. Chem. Ber. 1981, 114, 433; Schmidbaur, H.; Wohlleben, A.; Schubert, U.; Frank, A.; Huttner, G. Chem. Ber. 1977, 110, 2751-2757.
(4) Jandik, P.; Schubert, U.; Schmidbaur, H. Angew. Chem. Suppl. 1982, 1-12.
(5) Balch, A. L.; Hunt, C. T.; Lee, C.-L.; Olmstead, M. M.; Farr, J. P. J. Am. Chem. Soc. 1981, 103, 3764-3772.
(6) Brown, M. P.; Puddephatt, R. J.; Rashidi, M.; Manoihovic-Muir, L.; Muir, K. W.; Solomon, T.; Seddon, K. R. Inorg. Chim Acta 1977, 23, 633; Cooper, S. J.; Brown, M. P.; Puddephatt, R. J. Inorg. Chem. 1981, 20, 1374-1377.
(7) Schmidbaur, H.; Franke, R. Inorg. Chim. Acta 1975, 13, 85-89,
(8) Fackler, Jr., J. P.; Basil, J. D. Organomet. 1982, 1, 781-873.
(9) Hoffman, D. M.; Hoffmann, R. Inorg. Chem. 1981, 20, 3543-3555.
(10) Schmidbaur, H.; Slawisch, A. "Gmelin Handbuch der Anorganisthen Chemie. Au Organogold Compounds", 8th ed.; Springer-Verlag: Berlin/Heidelberg, 1980; 264.
(11) Basil, J. D. unpublished results.

RECEIVED September 16, 1982

11

High- and Low-Valence Metal Cluster Compounds: A Comparison

F. ALBERT COTTON

Texas A&M University, Department of Chemistry, College Station, TX 77843

The characteristic differences between metal cluster compounds having the metal atoms in low formal oxidation states and those having them in high formal oxidation states are reviewed critically and analytically.

There is now not only a great number but also a great variety of metal atom cluster compounds. In this essay I should like to discuss the differences between those that have metal atoms in a relatively high mean oxidation state (+2 to +4, and even, in rare cases, a bit higher) and those with metal atoms in oxidation states in the range -1 to +1. To keep the discussion within reasonable limits I shall restrict it almost exclusively to clusters consisting of only two or three metal atoms. I eschew the pedantic assertion that two atoms do not a cluster make.

For convenience and brevity in the discussion the abbreviations HVC and LVC for high-valent cluster and low-valent cluster, respectively, will be adopted.

To illustrate the distinction between the two types of cluster compounds the examples shown in Fig. 1 may be examined.

While the HVC/LVC dichotomy has rather routinely been taken for granted in the past, it has not been the subject of an explanatory discussion. There are five points of comparison and/or contrast to which I should like to draw attention; four are empirical and one theoretical.

Ligand Preferences. The LVC's are electron rich and in order to exist in stable compounds they require ligands, such as CO, that are weak donors and good π acceptors. On the other hand the HVC clusters are not attractive to such ligands nor do they require them for stability. In this respect, there is a direct parallel to mononuclear compounds where M° prefers CO, RNC or similarly π-acidic ligands while the M^{n+} (n = 2-4) ions do not generally form CO complexes.

0097-6156/83/0211-0209$06.00/0

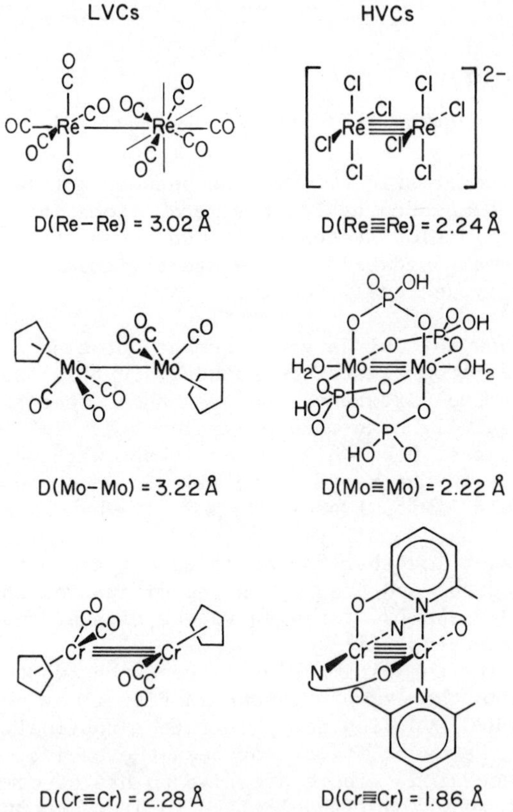

Figure 1a. Representative dinuclear cluster species of the high- and low-valent types.

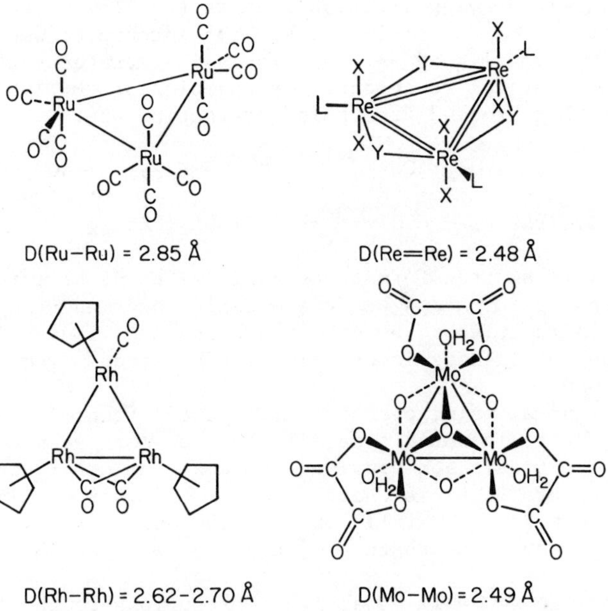

Figure 1b. Representative trinuclear cluster species of the high- and low-valent types.

Generally speaking, HVC's are found with virtually the entire gamut of ordinary ligands, such as halide ions, SCN^-, amines, phosphines, sulfate, carboxylates, etc., and they are unstable towards strong π-acceptors. The LVC's rarely bind such ligands and are not infrequently found surrounded entirely by CO groups or similar π-accepting ligands.

Clusters that contain M-M multiple bonds present a particularly interesting situation with respect to the interaction between π-acceptor ligands and the cluster itself. In most cases the cluster is actually rendered unstable by π-acceptor ligands, though there are a few exceptions to be discussed below. The general phenomenon is presumed to be due to the action of such ligands in withdrawing electron density from the π and δ components of the M-M multiple bonds. The matchup between M-M bonding orbitals and ligand π* orbitals is illustrated schematically in Fig. 2. Several representative examples of chemical reactions that proceed from this interaction are:

$$Mo_2(O_2CCH_3)_4 + 14CH_3NC \longrightarrow 2Mo(CNCH_3)_7^{2+} + 4CH_3CO_2^-$$

$$Mo_2Cl_4(PEt_3)_4 + 6CO \longrightarrow 2Mo(CO)_3Cl_2(PEt_3)_2$$

Presumably these transformations begin with one or more simple ligand replacement steps but these early intermediates are too unstable or reactive to be detected because of the action of the π-acceptor ligands in draining off electrons from the M-M π and δ bonds.

The general principle here is that electrons in M-M π and/or δ bonding orbitals would generally be even more stable in M-CO or M-CNR π orbitals and provided there are no other countervailing factors, that is where they go. It is recognized that this argument does not fully explain the above reactions since it does not cover the rupture of the σ as well as the π and δ bonds.

Among the few well-established exceptions to the generalization that M-M multiple bonds are unstable in the presence of strongly π-accepting ligands are the following:

(1) $Os_3(CO)_{10}H_2$
(2) $(\eta^5\text{-}C_5H_5)_2M_2(CO)_4$; M = Cr, Mo, W
(3) $Fe_2(CO)_6(Me_3CC{\equiv}CCMe_3)$
(4) $(\eta^5\text{-}C_5H_5)_2Fe_2(NO)_2$
(5) (cyclo-$R_4C_4)_2Fe_2(CO)_3$
(6) $Mo_2(OR)_6CO$

For (1) it is not actually clear that there is a multiple bond. The molecule is commonly depicted as in Fig. 3(a). From structural studies it is known that the Os-Os' and Os-Os" distances average 2.815(3)Å and Os'-Os" is 2.680(2)Å. This would seem to be prima facie evidence that the Os'-Os" bond has a higher order than the other two, whose characterization as single bonds is not likely to be controversial. However, the representation shown in Fig. 3(a) is not likely to be a true representation of the bonding. If the lines drawn between the Os-Os' and Os-Os" pairs of metal atoms are, as usual, meant to represent electron-pair bonds then there seems to be a superfluity of such lines in the region of Os', Os" and the two hydrogen atoms. For one thing, the hydrogen atoms will not form two H-Os single bonds; the best each of them can do is to participate in an Os H Os three-center/two electron bonding situation, of the kind familiar in boron hydrides, and which we might here also represent by the usual symbol

H
Os⌒Os

Thus, we might represent $Os_3(CO)_{10}H_2$ as in Fig. 3(b). However, this is still open to possible question since it leaves only four electrons each on Os' and Os" for Os-CO π-backbonding, which may not be enough. Moreover, a bond length decrease of only 0.14Å on going from a double to a single bond seems surprisingly small. I think it is likely that the Os'-Os" bond order is considerably less than 2 with a substantial fraction of the electron density needed to make a full double bond being withdrawn from the Os'-Os" region to strengthen the Os-CO π bonding. Thus, the molecule might most accurately (or least inaccurately) be represented as in Fig. 3(c).

For compounds (2) and (3) there is no reason to question the assignment of triple bonds and a double bond, respectively. The structure of the Cr compound of type (2) has been shown in Fig. 1. The structure of (3) is shown in Fig. 4(a). Such bonds are consistent with the attainment of 18-electron configurations. The reason why these bonds are not destabilized by the terminal CO groups present appears to be quite simple: sufficient metal d electrons are available to enter into M-CO π bonding without taking any from the M≡M or M=M bonds. For each metal atom in a molecule of type (2) there are, in addition to the six electrons in the M≡M bond, the six involved in the $M-C_5H_5$ bonding and the four in M-CO σ bonding, two more electrons that can participate in M-CO π bonding. Two electrons per two CO groups may be barely sufficient, but in these molecules we also have an unusual positioning of the CO groups that may enable each of them to donate electron density to the metal atom other than the one to which they are bound by the carbon atom.

In the case of (3) there are, on each metal atom, six electrons not required for Fe=Fe, Fe-acetylene or Fe-CO σ bonding. This situation contrasts sharply with that in, for example, an $Mo_2X_8^{4-}$ species, where there are no electrons in metal orbitals other than those in the Mo≣Mo bond and the Mo-X σ bonds.

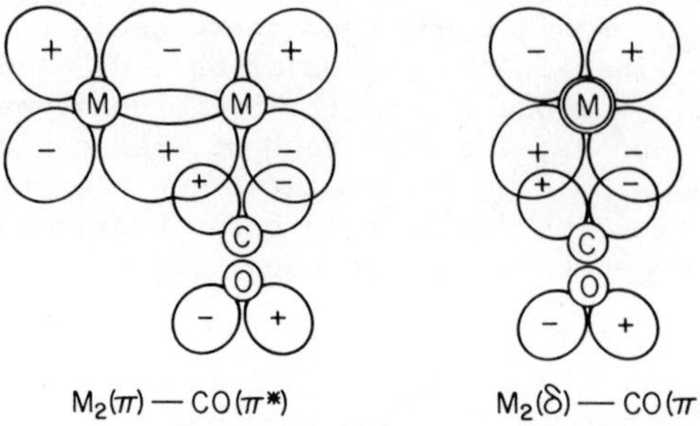

Figure 2. Diagrams of the overlaps between M–M π- or δ-orbitals and the π-orbitals of carbon monoxide.*

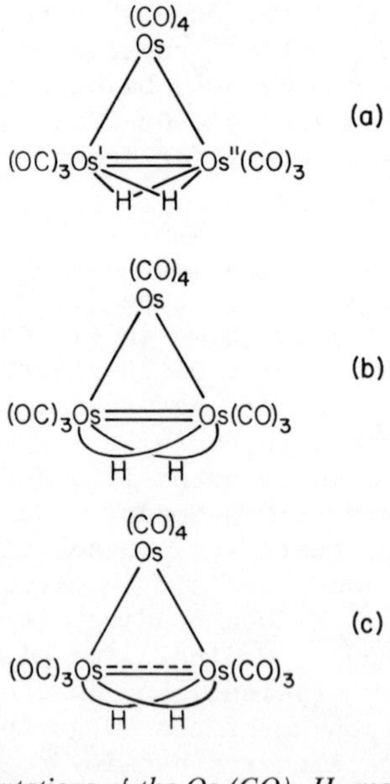

Figure 3. Three representations of the $Os_3(CO)_{10}H_2$ molecule. Key: a, a common, but incorrect one; b, a formally correct one; and c, a more realistic one.

In compounds (4) and (5), whose structures are shown in Fig. 3, there are also electrons available for M-NO or M-CO π bonding after all other bonds have been provided for. In (4) there are four such electrons and in (5) there are six. It may also be that the bridging CO or NO groups require relatively little (or even no) back-donation of metal electrons (although this is an unsettled question), but even if they do, in (4) and (5) there is one electron pair per bridging ligand, which would be quite sufficient even for a terminal CO or NO ligand.

Compound (6) has the structure shown in Fig. 4(d). It formally has 14-electron configurations about each metal atom and all fourteen electrons are involved in the bonds represented by lines in the structural drawing. Either the μ_2-CO group makes no appreciable demand for π electron density or there may be an indirect feeding of electrons from the lone-pair orbitals of the alkoxide groups to μ_2-CO. Alkoxide groups are known to be π-electron donors.

Metal Preferences. LVC's are formed mainly by transition metals to the right in the periodic table (especially elements in Group 8). This is in part due to the availability of d electrons that can be used in back-donation to the π-accepting ligands. Moreover, the formation of LVC's is not particularly "row-sensitive" by which I mean that the first-transition-series metals, Fe, Co and Ni, tend to form most of the same cluster compounds as their congeners, Ru, Rh, Pd and Os, Ir, Pt.

HVC's are formed mainly by the early transition elements, especially those in groups 6 and 7. Their stabilities are markedly "row-sensitive," with the light metals showing in general very little tendency to form clusters or multiple M-M bonds in higher oxidation states.

It is also notable that very large clusters are entirely in the province of the LVCs as far as we know at present. The HVCs include some octahedral clusters such as those in $[Ta_6Cl_{12}]^{2+}$ and $[Mo_6Cl_8]^{4+}$ derivatives, but there is nothing to rival the gigantic, metal-like rhodium clusters such as $[Rh_{14}(CO)_{25}]^{4-}$ and even larger ones such as $[Pt_{38}(CO)_{44}H_x]^{2-}$, in which the metal atoms adopt "packing" arrangements quite similar to those found in the metallic elements themselves.

M-M Bond Lengths. As a general rule, the M-M bonds are longer in the LVC compounds than in the HVC compounds. The values found in Fig. 1 are representative of this trend. To emphasize the point, some further discussion may be given. For HVC Mo-Mo single bonds, we have various values depending on structural details. In the $[Mo_3O_4(O_2CR)_3(H_2O)_3]^+$ type species, the distances are about 2.50Å but in the $[Mo_3(\mu_3\text{-}O)_2(O_2CR)_6\text{-}(H_2O)_3]^{2+}$ clusters, where the metal coordination numbers are higher they are about 2.75Å. In both cases the oxidation number

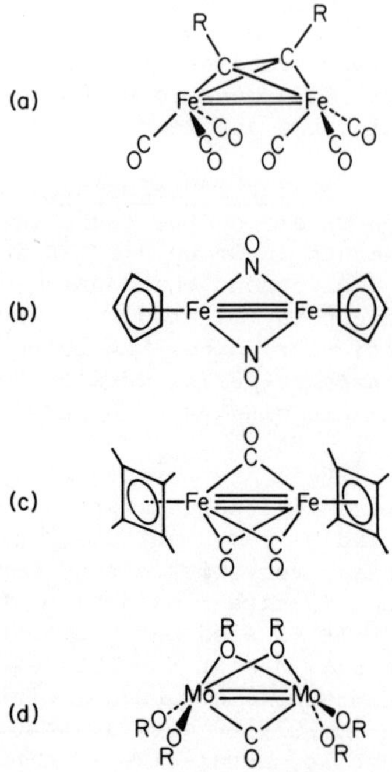

Figure 4. Some molecules containing both M–M multiple bonds and π-acceptor ligands.

number is +4. However, in LVC clusters of nearby elements, such as $Ru_3(CO)_{12}$ we have distances >2.80Å and the Mo-Mo single bond in $(\eta^2\text{-}C_5H_5)_2Mo_2(CO)_6$ has the astonishing length of 3.22Å. Among multiple bonds, similar trends are found. For Mo≡Mo triple bonds we have a length of 2.22Å for the HVC species $[Mo_2(HPO_4)_4]^{2-}$ but 2.45Å in the LVC species $(\eta^2\text{-}C_5H_5)_2Mo_2(CO)_4$. It is appropriate to stress that metal-metal bonds show far more complex behavior than those between main-group atoms and simple bond-length/bond-order relationships do not exist except in special cases. Nevertheless, the general trend of longer bonds in the LVCs as compared to the HVCs is unmistakable.

Any attempt to explain this observation must be speculative, but it is likely that two interrelated factors are largely responsible. In the HVCs the bonding is due to overlap of relatively compact metal *d* orbitals for which overlap (and hence bond strength) increases markedly with close approach of the metal atoms. In the LVCs the M-M bonding is largely due to overlap of *s*, *p* or sp^n hybrid orbitals, which are more diffuse and give best overlap at relatively long distances. The *d* orbitals in LVCs are mainly involved in π back-bonding to the ligands such as CO.

Lack of HVC-LVC Interconversions. A very mundane but sound reason for considering the LVCs and HVCs as separate categories is that members of the two classes are not often - if ever - interconvertible. One cannot start with a particular cluster of metal atoms, M_n, present in a compound of one class and by an oxidative or reductive procedure, in which the necessary ligand replacements would also have to take place, arrive at an M_n cluster compound of the other class. It is possible that one or a few examples of this might be found (or perhaps I have overlooked one already known) but at present this is not and does not seem likely to become a significant facet of metal atom cluster chemistry. I would, however, be delighted to have this judgement proven wrong, since such interconversions would afford valuable new synthetic routes.

Relative Importance of d, s and p Orbitals in M-M Bonding. Under the first four headings we have examined some facts and observations concerning the differences betwen the HVC and LVC classes of cluster compounds. We turn now to a theoretical point that underlies many of these differences, namely, the valence orbitals employed by the metal atoms in forming bonds.

For metal atoms in a formal state of oxidation of zero, or thereabouts, the energy gaps between *nd* and the (*n*+1)*s*, (*n*+1)*p* orbitals is fairly small and all three types of orbitals can play comparable roles in bond formation. Hybridization concepts are therefore useful in dealing with compounds containing LVCs. However, as the metal atom becomes more highly ionized, the energy gap between the (*n*+1)*s* and (*n*+1)*p* orbitals on the one hand and the nd orbitals on the other increases. At least for metal-metal bonding the former become much less important and in qualitative work can be neglected.

One of the consequences of this is that LVC compounds nearly always attain, or approach, the 18-electron configuration, whereas in HVC compounds, the number of electrons often falls far short of this. Thus, in $(\eta^5\text{-}C_5H_5)(OC)_2Mo{\equiv}Mo(CO)_2(\eta^5\text{-}C_5H_5)$ we have 18e configurations while in the $X_3Mo{\equiv}MoX_3$ (X = OR, NR_2, R) the metal atoms have only 12e configurations. Similarly in $[Re_2Cl_8]^{2-}$, $[Re_2Cl_4(PR_3)_4]$ and $[Mo_2(HPO_4)_4]^{2-}$, with quadruple, triple and triple bonds, respectively and metal oxidation states of +3, +2 and +3, respectively, the electron configurations about the metal atoms are 16, 18 and 14, respectively. In $Mo_6Cl_8{}^{4+}$ each metal atom has only 16 electrons although they weakly attach an additional ligand to give a total of 18 electrons, but in $Nb_6Cl_{12}{}^{2+}$ there are eight fewer electrons in the cluster. However, there are also HVC compounds in which formal 18e configurations are attained, as in $[Re_3Cl_{12}]^{3-}$ or $[Mo_3O_2(O_2CCH_3)_6(H_2O)_3]^{2+}$. The latter type, however, can be oxidized to lose one or two electrons while remaining intact.

For the purpose of explaining the bonding in cluster compounds it is at times crucial to keep in mind the fact that full participation of all (n+1)s and (n+1)p orbitals, on an equal footing with the nd orbitals, is not likely to be a correct assumption for HVC compounds even though it generally is so for LVC compounds. For example, in the earliest attempts by this author to account for the entire electronic structure of some compounds containing M-M quadruple bonds, and other compounds related to them, it was suggested that low-lying non-bonding σ orbitals formed by s and/or p_z orbitals might be available to hold up to four electrons beyond those occupying the σ, π and δ bonding orbitals. This led to the suggestion that in dirhodium species of the type $Rh_2(O_2CR)_4$ the Rh-Rh bond order would be 3, whereas, without such nonbonding orbitals, four more electrons would have to occupy antibonding orbitals, thereby reducing the bond order to 1. Chemical and structural evidence was indecisive, but an SCF-Xα-SW calculation showed convincingly that the 5s and 5p orbitals do not give rise to such nonbonding orbitals and that the Rh-Rh bond order is best regarded as 1.

For the triple bonds in $X_3Mo{\equiv}MoX_3$ and $X_3W{\equiv}WX_3$ species, the role of s and p orbitals is again one that requires judicious consideration in connection with the question of the barrier to internal rotation. The ligands themselves, of course, favor a staggered configuration, but the question is whether the triple bond itself imposes any barrier and, if so, which conformation (staggered or eclipsed) it favors. Reasonably rigorous calculations by either the SCF-Xα-SW or the Hartree-Foch method give a picture of the triple bond in which it consists of one σ and two π components, each formed from essentially pure d orbitals;

the predicted barrier to rotation is essentially zero in both calculations. On the other hand, from a frontier orbital treatment, in which the frontier orbitals are assumed to be $\underline{d}^2\underline{sp}^3$ octahedral hybrids, a preference for the eclipsed conformation was inferred, as of course it must be since the three "banana" bonds that constitute the triple bond should be strongest when the frontier orbitals point directly at one another. To the extent, however, that one reduces the p orbital participation this barrier will be reduced, and if there is not any significant amount of p orbital participation it will vanish. While the type of frontier orbital approach used has been quite successful in dealing with compounds containing metal atoms in very low formal oxidation states, it is questionable whether it is suitable for the metal atoms in an $Mo^{III}{}_2$ moiety. From a philosophical point of view the prediction that "with small ligands d^3-d^3 L_3MML_3 dimers will be eclipsed" is no meaningful scientific prediction at all since it is not falsifiable by any possible experiment. No matter what ligands are chosen, if (as I expect) the configuration remains staggered, the "predictors" can simply say: "Your ligands are still too big."

Concluding Remarks

The existence of two classes of metal atom cluster compounds is a fact of Nature. Like many such facts it is not neatly delineated; there are many blurred boundaries, few quantitative relationships, and exceptions to most if not all generalizations concerning it. Despite this, the way we recognize the difference, use it, and try to account for it is a good example of why chemistry is both less exact and more interesting (to me) than physics and mathematics. We chemists are forced to tackle far more complex and "messy" problems than workers in these other fields and, in our own way, I think we make a good job of it.

The sources of the facts and structural data cited and discussed above are multifareous. Rather than provide the usual reference list that might run to a great length, suffice it to say that all of the pertinent primary literature can be located through Advanced Inorganic Chemistry by Cotton and Wilkinson, 4th Edition, John Wiley and Sons, N. Y. 1980 and Multiple Bonds Between Metal Atoms by Cotton and Walton, John Wiley and Sons, N. Y. 1982.

Acknowledgment

I thank the National Science Foundation whose support of my research on metal-to-metal bonds over the years has made this article possible.

RECEIVED August 3, 1982

12

Metal–Metal Quadruple Bonds

Direct Experimental Determination of the Bonding Contribution of a δ-Orbital Electron

DENNIS L. LICHTENBERGER

University of Arizona, Department of Chemistry, Tucson, AZ 85721

The newly-developed capability to observe metal-metal vibrational fine structure in the valence ionizations of quadruply bonded dimers is illustrated for the delta-bond ionization of $Mo_2(O_2CCH_3)_4$. Observation of this structure provides direct information on the bonding influence of an electron in a delta-bonding orbital by showing the significant changes in metal-metal force constant and bond distance that occur when that electron is removed.

I wish to describe the first application of a significant new experimental capability to a rather "classic" question about a "classic" molecule. The beauty of the story is that direct and unique information is provided by the technique, and the explanation is short and simple.

The "classic" molecule is $Mo_2(O_2CCH_3)_4$, which is an important representative member of di-metal molecules containing a quadruple bond. The occupation of the delta-bonding orbital, which completes formation of the quadruple bond, is a special feature of these molecules. The classic question is the following: To what extent does an electron in the delta-bonding orbital contribute to the total bond strength and force constant between the two metal centers?

The obvious approach to answering this question is to remove an electron from this orbital and observe the effect on, for example, the metal-metal stretching frequency or metal-metal bond distance. Of course, removal of an electron from the delta bonding orbital creates a positive molecular ion for which determination of these properties may not be possible using normal techniques. In those cases where the ion is sufficiently stable that these properties can be measured, the meaning of the information may be clouded by changes in intermolecular interactions or other internal factors.

One simple method of taking the electron from the molecule

0097-6156/83/0211-0221$06.00/0

is photoelectron spectroscopy. Advantages of this technique are that the molecule is in the gas phase, so that solution or solid-state effects are non-existent, and the technique is fast, so that subsequent molecular occurrences are generally not observed. Also, this is the best technique for measuring accurate ionization energies. Professor Cotton discussed the importance of ionization energies to understanding bonding interactions in his talk. I wish to stress that these ionizations can also provide a direct measure of both the metal-metal stretching frequency and the equilibrium bond distance in the positive ion. This information is obtained if the vibrational fine structure comprising the ionization band envelope is observed.

The possibility of obtaining this direct information has not been discussed previously in this context because vibrational fine structure is not generally observed in the ionizations of molecules of this size, and has never before been observed for any transition metal-metal vibrational mode. Our breakthrough in demonstrating that this fine structure can be observed has followed from several developments of our instrumentation. The details of these developments have recently been published along with our report of the first observations of metal-ligand vibrational fine structure in the ionizations of metal carbonyls (1).

Figure 1 displays the ionization band of $Mo_2(O_2CCH_3)_4$ corresponding to loss of one electron from the delta-bonding orbital ($^2B_{2g}$ positive ion state). The fine structure due to the totally symmetric metal-metal vibrational mode levels in the positive ion is clearly observed. As Figure 1 shows, the band is well represented by an evenly spaced progression of vibrational components. No anharmonicity is detected. The progression has a skewed Gaussian intensity profile as expected for excitation to a potential well with a displaced equilibrium bond distance. This is illustrated in Figure 2.

The spacing between the vibrational components gives a metal-metal stretching frequency in the positive ion of 360(10) cm^{-1}, which is considerably less than the 406 cm^{-1} metal-metal stretching frequency in the neutral molecule. Thus, the force constant between the metals has decreased with removal of an electron from the delta-bonding orbital. Franck-Condon analysis of the progression shows that the metal-metal bond distance in this positive ion is about 0.17 Å longer than in the neutral molecule. This also shows that an electron in the delta-bonding orbital has a substantial influence on the strength of the metal-metal interaction.

Additional insight is obtained if these results are compared with the related absorption experiments in which an electron from the delta-bonding orbital is excited to the delta-antibonding orbital (2). The pertinent data is summarized in the Table. The state obtained by δ ionization has a greater formal bond order than the state obtained by $\delta \rightarrow \delta^*$ excitation, but has a weaker metal-metal force constant and a longer metal-metal bond. It is

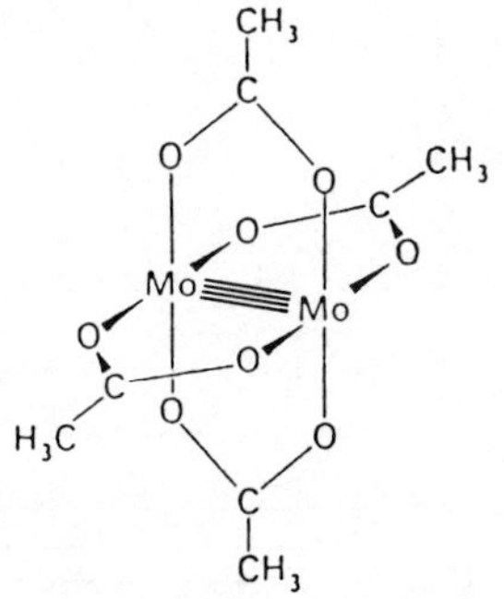

structure of $Mo_2(O_2CCH_3)_4$

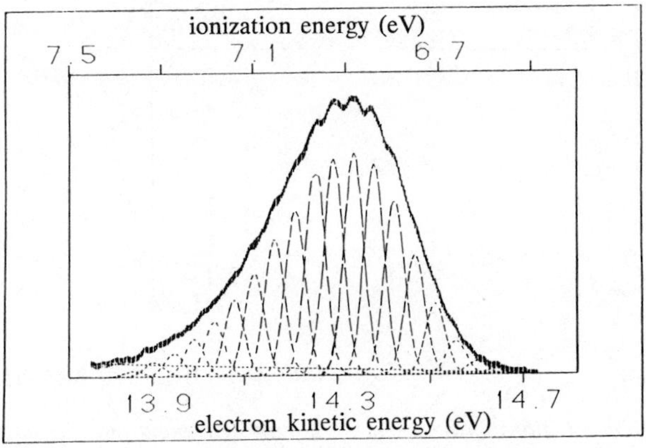

Figure 1. The δ ionization band ($^2B_{2g}$ positive ion state) of $Mo_2(O_2CCH_3)_4$ fit with equally spaced symmetric vibrational components.

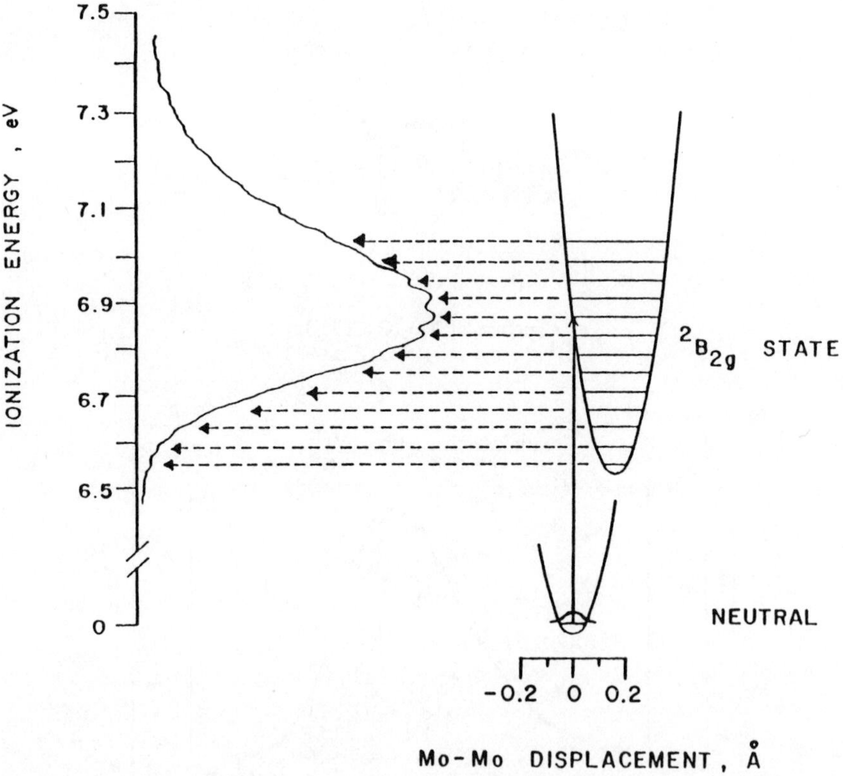

Figure 2. The relationship between the δ ionization band and the potential wells of the ground state and the displaced $^2B_{2g}$ positive ion state.

TABLE. Related data for three electronic perturbations of $Mo_2(O_2CCH_3)_4$.

state	formal bond order	Mo-Mo stretch	Mo-Mo distance
neutral	4.0	406 cm^{-1}	2.093 Å
δ-δ*	3.0	370(5)	2.20
δ ionization	3.5	360(10)	2.26(1)

important to note that, in addition to the change in formal bond order, the ionization process also changes the formal oxidation state of the metal centers. The increase in positive charge at the metal centers will contract the metal d orbitals and reduce their overlap. This will decrease the bonding ability of the remaining delta-bonding electron, as well as that of the pi-bonding electrons and quite possibly also the sigma-bonding electrons. This effect of positive charge is seldom discussed for vibrational fine structure in photoelectron spectroscopy, but has been discussed for quadruply bonded metal complexes (3).

The important point to remember is that an electron in the delta-bonding orbital of $Mo_2(O_2CCH_3)_4$ has a substantial influence on the strength of the metal-metal interaction. This influence is directly evidenced by the metal-metal vibrational fine structure observed with ionization from the delta orbital, which shows a lowering of the metal-metal stretching frequency and a lengthening of the equilibrium metal-metal bond distance.

Acknowledgment

These experiments were conducted by Mr. Charles H. Blevins II. We wish to thank the Department of Energy, contract DE-AC02-80ER10746 and the University of Arizona for partial support of this work. D.L.L. is an Alfred P. Sloan Fellow.

Literature Cited

1. Hubbard, J.L.; Lichtenberger, D.L., J. Am. Chem. Soc., 1982, 104, 2132.
2. Martin, D.S.; Newman, R.A.; Fanwick, P.E., Inorg. Chem., 1979, 18, 2511.
3. Bursten, B.E.; Cotton, F.A., Farad. Disc. Roy. Soc. Chem., 1980, 14, 180.

RECEIVED September 16, 1982

13

Higher Nuclearity Carbonyl Clusters

BRIAN T. HEATON

University of Kent, Chemical Laboratory, Canterbury CT2 7NH England

Professor Cotton's studies have considerably clarified our understanding of di- and tri-nuclear metal-metal bonded compounds. For higher nuclearity clusters, rationalisation of structures, bonding, reactivity etc., must be much more tenuous because of the increased number of variables (metal-metal, metal-ligand, steric effects) now present. However, I would briefly like to present a few trends which seem to be emerging in this area.

Substitution Sites in Tetra- and Penta-Nuclear Clusters

Apart from nitrile-substituted derivatives of $[Ir_4(CO)_{12}]$ (1, 2) which retain the non-carbonyl-bridged structure of $[Ir_4(CO)_{12}]$, (3) all other derivatives of $[Ir_4(CO)_{12}]$, both in the solid state and in solution, are based on the C_{3v}-structure of $[M_4(CO)_{12}]$[(M = Co, (4-7) Rh (4, 8)](Fig. 1). This structure, with three edge-bridging carbonyls, is better able to dissipate the increased nuclear charge induced on carbonyl substitution. Three possible isomers could result by carbonyl replacement at the apical, radial or axial site but the better back-bonding ability favours carbonyl replacement on the basal metal atom and the substituent is found exclusively in the axial site, eg. $[Ir_4(CO)_{11}X]^-$ (X = Br, (9) CO_2Me, (10) $[Ir_4(CO)_{11}]^-$ (11)), $[Ir_4(CO)_{11}L]$ (L = PR_3 (12)).

Further occupancy of axial sites occurs with small ligands, eg. $[Ir_4(CO)_{10}H_2]^{2-}$, (13) but minimisation of steric effects becomes important for larger ligands and the substitution pattern shown in Fig. 2 is preferred. (12, 14)

The reason for preferential axial site occupancy by monodentate ligands is probably related to results of recent CNDO calculations on $[Co_4(CO)_{12}]$, (15) which show that the axial carbonyls are least involved in back-bonding. Extrapolation to Rh_4- and Ir_4-derivatives seems reasonable and finds some support from n.m.r. measurements, which show that the ^{13}C n.m.r. chemical shift of the axial carbonyl is always at higher field than the radial carbonyl in both rhodium (16) and iridium derivatives. (12)

0097-6156/83/0211-0227$06.00/0

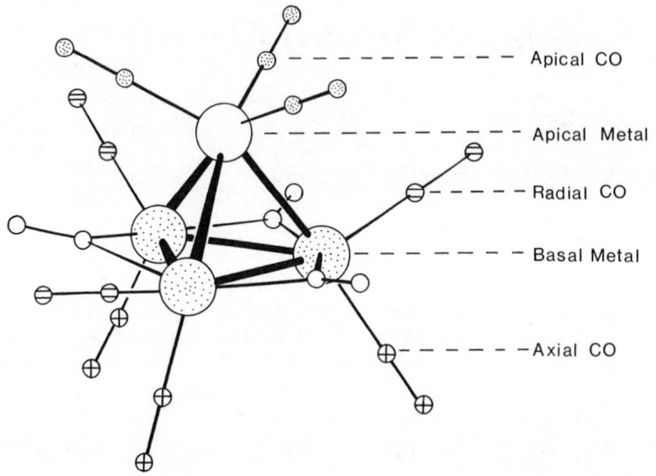

Figure 1. A representation of the C_{3v} structure of $[M_4(CO)_{12}]$ where M = Co, Rh.

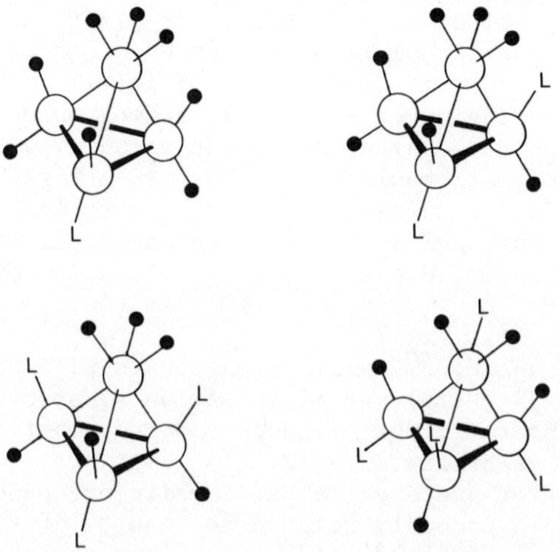

Figure 2. Isomers formed on ligand (L = PR_3, $P(OPh)_3$) substitution in $[M_4(CO)_{12}]$, (M = Rh, Ir). Key: ●, terminal CO; and ■, bridging CO.

Related substitution patterns are observed in tetranuclear cobalt and rhodium clusters. Thus, the small ligand, $P(OMe)_3$(L), occupies axial sites in $[Co_4(CO)_{12-x}L_x]$ (x = 1,2) (17) whereas steric effects become important with $P(OPh)_3$ and the isomers shown in Fig. 2 are obtained with tetrarhodium derivatives. (18, 19).

However, there is some deviation from these substitution patterns, especially with cobalt clusters. Thus, iodide in $[Co_4(CO)_{11}I]^-$ is predominantly in an apical site, although there is some substitution (2%) on the basal cobalt, (20) and $P(CH_2CH{:}CH_2)Me_2$(L) in $[Co_4(CO)_{10}L_2]$ occupies one axial and a *trans*-apical site. The reasons for these discrepancies are not clear but are probably associated with cobalt being both smaller and less electronegative than rhodium or iridium.

For the substituted penta-nuclear cluster, $[Rh_5(CO)_{14}I]^{2-}$, the iodide is on an apical rhodium, which is also coordinated to four other carbonyls (Fig. 3a). (22) Equalisation of charge over the cluster is probably the reason for iodide being on the apical rhodium since this arrangement then readily allows each of the other four rhodium atoms to also be associated with four carbonyls. It should be noted that the *iodide substituted apical rhodium is essentially octahedrally coordinated* to two terminal carbonyls, two edge-bridging carbonyls, iodide and the unbridged Rh_{eq}-Rh_{ap} bond. This group could exhibit isomerism in which the iodide is either *trans* to a bridging carbonyl (Fig. 3a) or *trans* to the unbridged rhodium-rhodium bond (Fig. 3b); only the former is found. This substituent site occupancy should be compared with the preferred axial site occupancy, *trans* to the Rh_{ap}-Rh_{bas} bond, in Fig. 1, (*vide supra*); in the latter case, there are no *trans* OC-M-CO groups available for substitution.

The structure of $[M'M_4(CO)_{15}]^{2-}$ (M' = Ru, M = Ir; (23) M' = Fe, M = Rh (24)) is related to $[Rh_5(CO)_{14}I]^{2-}$ since the iodide substituted Rh_{ap} atom is replaced by the electron rich metal, M', which is then associated with more carbonyls than the other metals, M.

Capping of Triangular- or Square-Metal Faces

Addition of either nucleophilic or electrophilic metallic species can result in the capping of triangular- or square-metal faces in carbonyl clusters. These redox reactions provide high yield syntheses of higher nuclearity clusters and somewhat resemble surface reconstruction on metals. With a few examples, I would like to show how there is often minimal rearrangement of the carbonyl skeleton.

The 90 electron trigonal prismatic cluster, $[Rh_6C(CO)_{15}]^{2-}$, is electron rich and seems to behave as though there is a pair of electrons pointing out of each triangular face. Each metal atom is surrounded by five groups (1 terminal CO, 3 bridging CO's and the interstitial carbide) in an essentially *square pyramidal* array. Octahedral coordination of the metal is accomplished by occupancy of the sixth site, which is above the trigonal face. Thus addition of electrophiles, $[Cu(NCMe)]^+$, (25) $[M(PEt_3)]^+$ (M = Ag, Au), (26) H^+ (27), results in the formation of either mono- or bis-adducts but perhaps more surprising are the structures of the adducts formed on incremental addition of Ag^+, which first acts as a linear bridge between staggered Rh_6-trigonal prismatic units and then progressive capping of the terminal trigonal faces occurs, resulting in breakdown of the polymer to a monomeric unit (see Table).

In contrast, nucleophilic addition to a Rh_3-face in $[Rh_6(CO)_{15}]^{2-}$ occurs on reaction with either $[Rh(CO)_4]^-$ or $[Ni(CO)_4]$ to give the iso-structural $[Rh_7(CO)_{16}]^{3-}$ (28) and $[NiRh_6(CO)_{16}]^{2-}$ (29) clusters respectively, (Fig. 4). In this case, three terminal carbonyls in $[Rh_6(CO)_{15}]^{2-}$ become edge-bridging to the capping atom with little other change of the carbonyl polyhedron.

The square-face in $[Rh_9E(CO)_{21}]^{2-}$ (E = P, As) (30) undergoes nucleophilic attack with $[Rh(CO)_4]^-$ to give $[Rh_{10}E(CO)_{22}]^{3-}$ (31, 32); the edge-bridging carbonyls on this square-face become edge-bridging to the capping atom but again there is little other reorganisation of the carbonyl polyhedron (Fig. 5).

In $[Rh_{13}(CO)_{24}H_{5-x}]^{x-}$ (x = 2, 3, 4), it has been shown by X-ray crystallography (33-35) that there are four pentagonal holes on the surface of the carbonyl polyhedron. It seems as though these holes are present because of a preference for all the edge-bridging carbonyls to be coplanar with the Rh_3-plane incorporating the carbonyl bridged edge and the interstitial rhodium atom; the terminal carbonyls then adopt positions which minimise steric interactions. Protonation can be used to successively convert the mono-hydride to the tetra-hydride cluster and low temperature ^1H-$[^{103}Rh]$ n.m.r. measurements have shown that, when x = 2, three of these pentagonal holes are occupied by hydrogen. (36) Similarly, electrophilic additions with $[Rh(CO)_2(NCMe)_2]^+$ results in capping of the Rh_4- square-faces below the pentagonal holes with minimal reorganisation of both the metallic and carbonyl polyhedra (Fig. 6). (37-39)

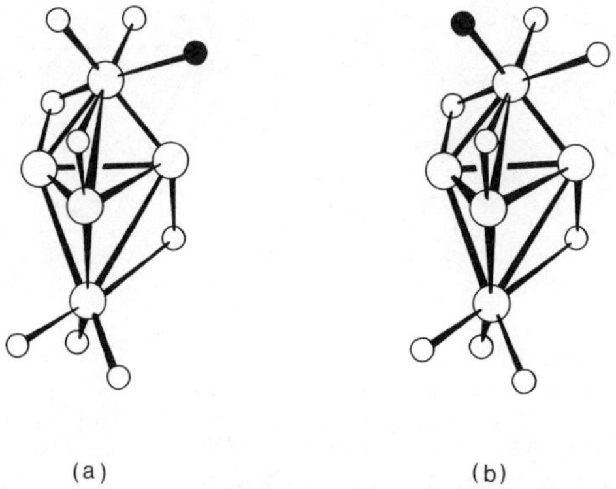

Figure 3. Possible isomers of $[Rh_5(CO)_{14}I]^{2-}$ that involve apical iodide substitution. Each equatorial rhodium has one terminal CO and each equatorial edge is carbonyl bridged; all these COs have been omitted for clarity. Key: ○, CO; ●, I; a, observed; and b, not observed.

Table

Adducts formed via addition of Ag^+ to trigonal faces of $[Rh_6C(CO)_{15}]^{2-}$, ($M = [Rh_6C(CO)_{15}]^{2-}$).

Adduct	$M:Ag^+$
$[M\ Ag\ M]^{3-}$	6:3
$[M\ Ag\ M\ Ag\ M]^{4-}$	6:4
$[M\ Ag]_n^{n-}$	6:6
$[Ag\ M\ Ag\ M\ Ag\ M\ Ag]^{2-}$	6:8
$[Ag\ M\ Ag\ M\ Ag]^-$	6:9
$[Ag\ M\ Ag]$	6:12

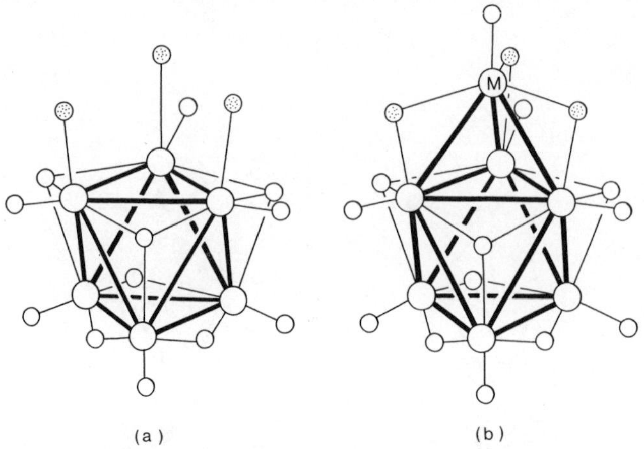

Figure 4. Capping of a triangular Rh_3-face in $[Rh_6(CO)_{15}]^{2-}$ (a) to give $[M\ Rh_6(CO)_{16}]^{n-}$ (b). (M = Ni, n = 2; M = Rh, n = 3.) The terminal COs (◎) in a become edge-bridging to the capping atom (M) in b.

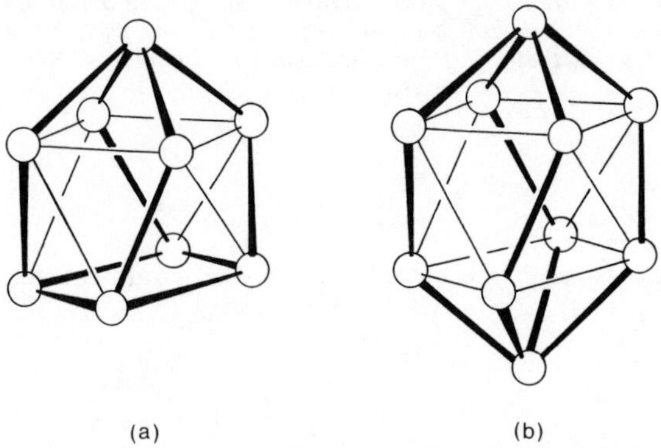

Figure 5. Capping of a square Rh_4-face in $[Rh_9E(CO)_{21}]^{2-}$ (a) to give $[Rh_{10}E(CO)_{22}]^{3-}$ (b). (E = P, As.) Each rhodium has one terminal carbonyl that, together with the interstitial atom, E, has been omitted for clarity. (■ = μ − CO.)

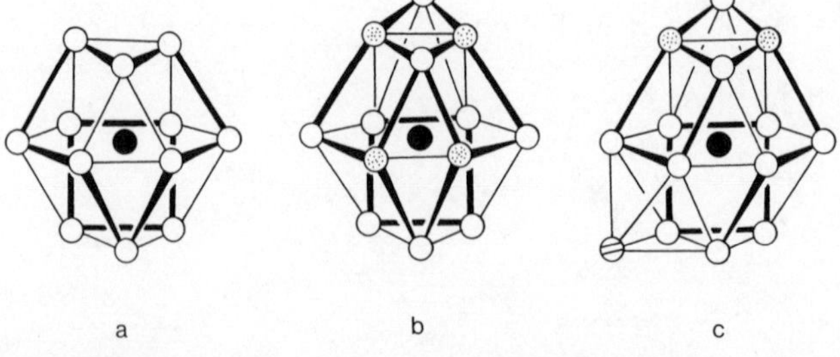

Figure 6. Capping of square Rh_4-faces in $[Rh_{13}(CO)_{24}]^{5-}$ (a) to give $[Rh_{14}(CO)_{25}]^{4-}$ (b) and $[Rh_{15}(CO)_{27}]^{3-}$ (c). Key: ○, Rh; ○, Rh–CO; ○, $Rh(CO)_2$; and ●, interstitial Rh.

Acknowledgments

I would like to acknowledge close collaboration with the group of cluster chemists in Milan and financial support from S.E.R.C.

Literature Cited

1. Shapley, J. R.; Stuntz, G. F.; Churchill, M. R.; Hutchinson, J. P. J. Chem. Soc. Chem. Commun. 1977, 219; Churchill, M. R.; Hutchinson, J. P. Inorg. Chem. 1979, 18, 2451.
2. Stuntz, G. F.; Shapley, J. R. J. Organomet. Chem. 1981, 213, 289.
3. Churchill, M. R.; Hutchinson, J. P. Inorg. Chem. 1978, 17, 3528.
4. Wei, C. H. Inorg. Chem. 1969, 8, 2384.
5. Carr, F. H.; Cotton, F. A.; Frenz, B. A. 1976, 15, 3580.
6. Cohen, M. A.; Kidd, D. R.; Brown, T. L. J. Am. Chem. Soc. 1975, 97, 4408.
7. Aime, S.; Osella, D.; Milone, L.; Hawkes, G. E.; Randall, E. W.; J. Am. Chem. Soc. 1981, 103, 5920.
8. Evans, J.; Johnson, B. F. G.; Lewis, J.; Matheson, T. W.; Norton, J. J. Chem. Soc. Dalton, 1978, 626.
9. Chini, P.; Ciani, G.; Garlaschelli, L.; Manassero, M.; Martinengo, S.; Sironi, A. J. Organomet. Chem. 1978, 152, C35.
10. Garlaschelli, L.; Martinengo, S.; Chini, P.; Canziani, F.; Bau, R. J. Organomet. Chem. 1981, 213, 379.
11. DeMartin, F.; Manassero, M.; Sansoni, M.; Garlaschelli, L.; Raimond, C.; Martinengo, S.; Canziani, F. J. Chem. Soc. Chem. Commun. 1981, 528.
12. Stuntz, G. F.; Shapley, J. R. J. Am. Chem. Soc. 1977, 99, 607.
13. Ciani, G.; Manassero, M.; Albano, V. G.; Canziani, F.; Giordano, G.; Martinengo, S.; Chini, P. J. Organomet. Chem. 1978, 150, C17.
14. Albano, V. G.; Bellon, P. L.; Scatturin, V. J. Chem. Soc. Chem. Commun. 1967, 730.
15. Freund, H. J.; Hohlneicher, G. Theor. Chimica Acta 1979, 51, 145.
16. Heaton, B. T.; Della Pergola, R.; Strona, L.; Smith, D. O. J. Chem. Soc. Dalton, submitted for publication.
17. Darensbourg, D. J.; Incorvia, M. J. Inorg. Chem. 1980, 20, 1911.
18. Ciani, G.; Garlaschelli, L.; Manassero, M.; Sartorelli, U.; Albano, V. G. J. Organomet. Chem. 1977, 129, C25.
19. Heaton, B. T.; Longhetti, L.; Mingos, D. M. P.; Briant, C. E.; Minshall, P. C.; Theobald, B. R. C.; Garlaschelli, L.; Sartorelli, U. J. Organomet. Chem. 1981, 213, 333.
20. Albano, V. G.; Braga, D.; Longoni, G.; Campanella, S.; Ceriotti, A.; Chini, P. J. Chem. Soc. Dalton 1980, 1820.

21. Keller, E.; Vahrenkamp, H. Chem. Ber. 1981, 114, 1111.
22. Martinengo, S.; Ciani, G.; Sironi, A. J. Chem. Soc. Chem. Commun. 1979, 1059.
23. Fumagalli, A.; Koetzle, T. F.; Takusagawa, J. Organomet. Chem. 1981, 213, 365.
24. Ceriotti, A.; Longoni, G.; Manassero, M.; Sansoni, M.; Della Pergola, R.; Heaton, B. T.; Smith, D. O. J. Chem. Soc. Chem. Commun., submitted for publication.
25. Albano, V. G.; Braga, D.; Martinengo, S.; Chini, P.; Sansoni, M.; Strumolo, D. J. Chem. Soc. Dalton, 1980, 52.
26. Heaton, B. T.; Strona, L.; Braga, D.; Albano, V. G.; Martinengo, S., unpublished results.
27. Heaton, B. T.; Strona, L.; Martinengo, S.; Strumolo, D.; Goodfellow, R. J.; Sadler, I. H. J. Chem. Soc. Dalton, in press.
28. Albano, V. G.; Bellon, P. L.; Ciani, G. J. Chem. Soc. Chem. Commun. 1969, 1024.
29. Fumagalli, A.; Longoni, G.; Chini, P.; Albinati, A.; Brückner, S. J. Organomet. Chem. 1980, 202, 329.
30. Vidal, J. L.; Schoening, R. C.; Pruett, R. L.; and Walker, W. E. Inorg. Chem. 1979, 18, 129.
31. Vidal, J. L. Inorg. Chem. 1981, 20, 243.
32. Vidal, J. L.; Walker, W. E.; Schoening, R. C. Inorg. Chem. 1981, 20, 238.
33. Albano, V. G.; Ceriotti, A.; Chini, P.; Ciani, G.; Martinengo, S.; Anker, W. M. J. Chem. Soc. Chem. Commun. 1975, 859.
34. Albano, V. G.; Ciani, G.; Martinengo, S.; Sironi, A. J. Chem. Soc. Dalton, 1979, 978.
35. Ciani, G.; Sironi, A.; Martinengo, S. J. Chem. Soc. Dalton, 1981, 519.
36. Heaton, B. T.; Goodfellow, R. J.; Martinengo, S., unpublished results.
37. Martinengo, S.; Ciani, G.; Sironi, A.; Chini, P. J. Am. Chem. Soc. 1978, 100, 7096.
38. Ciani, G.; Sironi, A.; Martinengo, S. J. Organomet. Chem. 1980, 192, C42.
39. Vidal, J. L.; Schoening, R. C. Inorg. Chem. 1981, 20, 265.

RECEIVED October 13, 1982

14

Resonance Raman Spectroscopy as a Complement to Other Techniques for Making Electronic Band Assignments for $(Re_2X_8)^{2-}$ Ions

ROBIN J. H. CLARK and MARTIN J. STEAD

University College London, Christopher Ingold Laboratories,
20 Gordon Street, London WC1H 0AJ England

The resonance Raman (RR) spectra of the $[Re_2X_8]^{2-}$ ions, X = F, Cl, Br or I, obtained at resonance with the $\delta^* \leftarrow \delta$, $\delta^* \leftarrow (X)\pi$ and $\pi^* \leftarrow \pi$ transitions are strikingly different from one another, and allow confirmation of the electronic band assignments in a new and convincing way.

Some years ago we demonstrated that very intense RR spectra of the $[Mo_2Cl_8]^{4-}$, $[Mo_2Br_8]^{4-}$, $[Re_2Cl_8]^{2-}$ and $[Re_2Br_8]^{2-}$ ions (1,2,3) could be obtained at resonance with the lowest electronic bands of the ions. These spectra (A-term RR spectra) took the form of long progressions in the ν_1 mode, the ReRe stretching mode, implying that (a) the resonant electronic transition is electric dipole allowed and that (b) in the excited state the ions suffer a substantial change to the metal-metal bond length. These results allowed the resonant electronic transition to be assigned to the $^1A_{2u} \leftarrow {}^1A_{1g}$, $\delta^* \leftarrow \delta$ transition since, on excitation to the δ^* state, the metal-metal bond order would be reduced from four to three with consequential elongation of the metal-metal bond.

Two features have encouraged us to study the RR spectra of these ions in greater detail. First, the recent syntheses of the $[Re_2F_8]^{2-}$ and $[Re_2I_8]^{2-}$ ions completed the tetrad of $[Re_2X_8]^{2-}$ ions, (4,5,6) and second, the development of new lasing dyes in the blue (stilbene 1 and 3) and red/infrared (LD 700) together with the availability of u.v. lines (Ar^{2+}, Kr^{2+}) now permit relatively easy coverage of the wide excitation range 330-800 nm. Thus the possibility emerges of irradiating, for each ion, within the contour of each of the electronic bands with $\lambda > 330$ nm in turn, thus opening the way to probing the nature of each excited state of each ion.

The electronic spectra of the ions are shown in Figure 1, and the RR spectra of the ions at resonance with their lowest electronic bands in Figure 2. For each ion, long progressions in

0097-6156/83/0211-0235$06.00/0

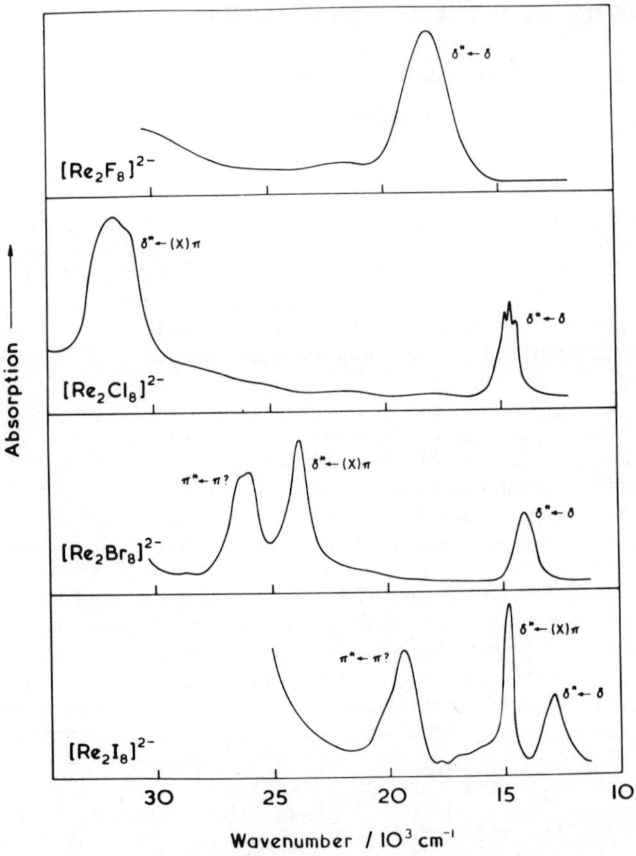

Figure 1. Electronic spectra of the complexes $[(n\text{-}C_4H_9)_4N]_2[Re_2X_8]$ at ~ 14 K in the region of their $\delta^* \leftarrow \delta$, $\delta^* \leftarrow \pi(X)$ and $\pi^* \leftarrow \pi$ (?) transitions ($Cs[BF_4]$, KCl, KBr, and CsI discs for X = F, Cl, Br, or I, respectively).

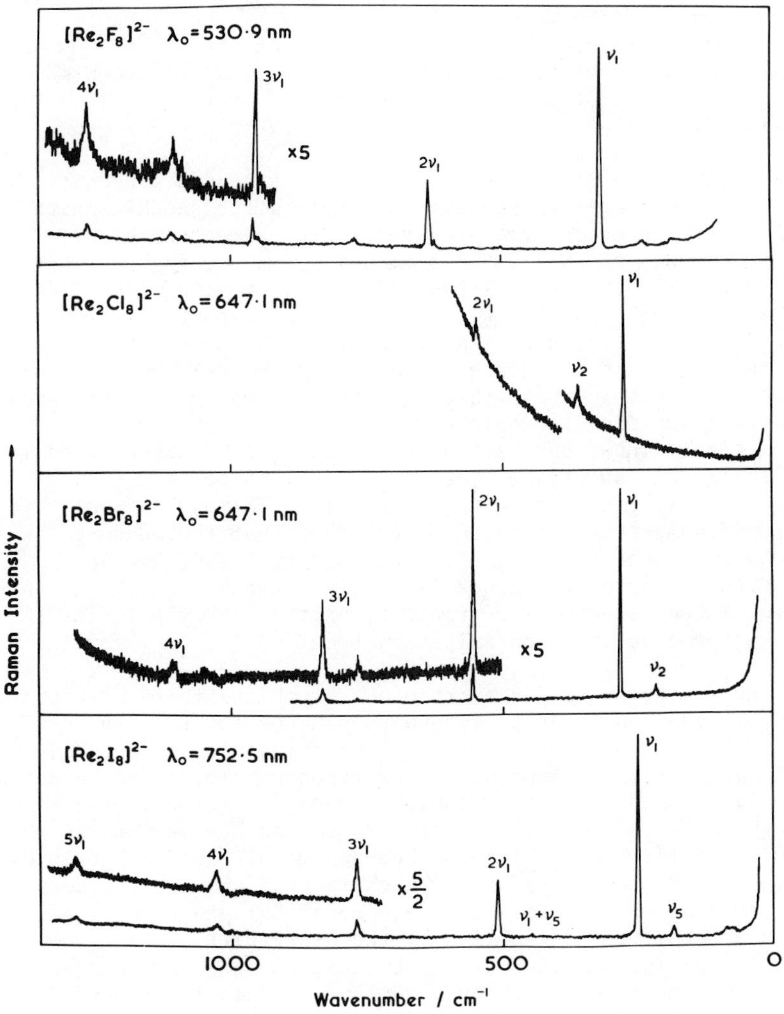

Figure 2. Resonance Raman spectra of the complexes $[(n\text{-}C_4H_9)_4N]_2[Re_2X_8]$ *at* ~ 80 *K in the region of their* $\delta^* \leftarrow \delta$ *transitions.*

ν_1 are observed (318, 275, 276 and 257 cm^{-1} for X = F, Cl, Br and I, respectively), clearly indicating the similar nature of the lowest electronic transition in each case and (by comparison with earlier work) the assignment of the lowest band to the $\delta^* \leftarrow \delta$, $^1A_{2u} \leftarrow {}^1A_{1g}$, transition. This conclusion is substantiated by measurements of the depolarization ratios of the ν_1 bands at resonance. These have been measured for the $[Re_2F_8]^{2-}$, $[Re_2Cl_8]^{2-}$ and $[Re_2Br_8]^{2-}$ ions and are found to be 1/3, a situation which can only arise if the resonant electronic transition is z-polarized, as the $\delta^* \leftarrow \delta$ transition (in the D_{4h} point group) must be.

Irradiation within the contour of the second electronic transition of each ion produces entirely different RR spectra in each case. These spectra (Figure 3) are characterised by resonance enhancement to the ν_2 mode and its overtones, where ν_2 is the totally symmetric metal-halogen stretching mode (624, 362, 211 and 152 cm^{-1} for X = F, Cl, Br and I, respectively). Thus the principal structural change undergone by each ion on excitation is, in this case, along the metal-halogen coordinate, a result in keeping with that expected in consequence of a non-bonding-to-antibonding halogen-to-metal charge-transfer transition. Thus, the assignment of the second strong band in the electronic spectra of the $[Re_2X_8]^{2-}$ ions to the $b_{1u}(\delta^*) \leftarrow (X)e_g(\pi)$, $^1E_u \leftarrow {}^1A_{1g}$, transition follows naturally. This assignment is confirmed by the fact that the measured depolarization ratio of the ν_2 band of the $[Re_2Cl_8]^{2-}$ and $[Re_2Br_8]^{2-}$ ions at resonance is 1/8, a situation which can only arise if the resonant electronic transition is xy polarized. This is precisely the polarization required (in D_{4h}) for the $^1E_u \leftarrow {}^1A_{1g}$ transition.

RR spectra obtained at resonance with the third electronic band of each ion leads to resonance enhancement to both the ν_1 and ν_2 bands (and their overtones) to comparable extents. Although no firm assignment of the resonant electronic band can be made in this case, the results are consistent with the assignment $e_g(\pi^*) \leftarrow e_u(\pi)$. This transition has been shown, by XαSCF calculations (7-9) to be between metal-based orbitals with a considerable amount of halogen character, and thus structural changes along both metal-metal and metal-halogen bond lengths would be expected, as implied by the RR results.

In conclusion, it can be seen that RR spectroscopy can be a valuable complement to other techniques for making electronic band assignments for these ions, and that the technique should have wide applicability to related studies of other inorganic species.

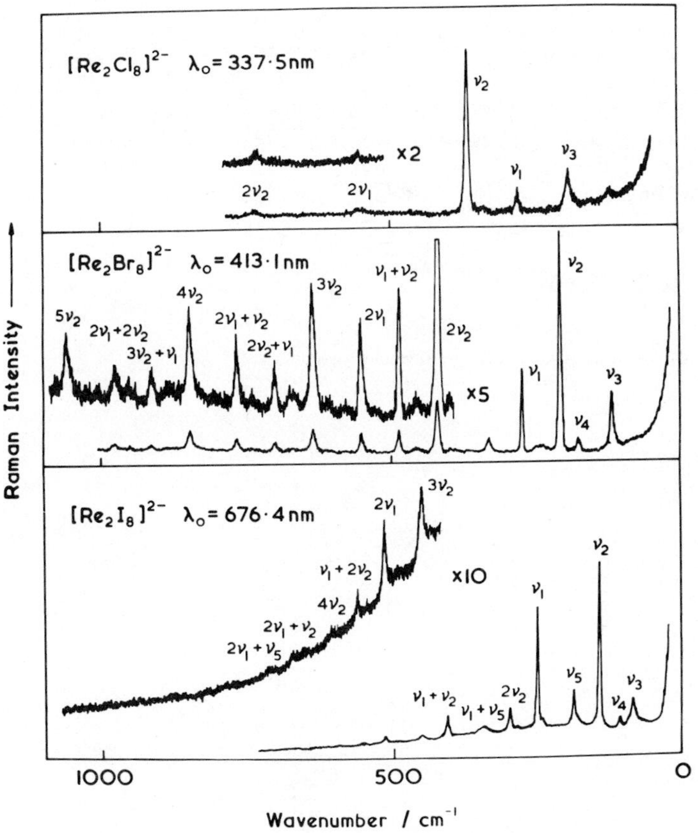

Figure 3. Resonance Raman spectra of the complexes $[(n\text{-}C_4H_9)_4N]_2[Re_2X_8]$ *at* $\sim$ *80 K in the region of their* $\delta^* \leftarrow \pi(X)$ *transitions.*

Literature Cited

1. Clark, R.J.H.; Franks, M.L. J. Amer. Chem. Soc. 1975, 97, 2691.
2. Clark, R.J.H.; Franks, M.L. J. Amer. Chem. Soc. 1976, 98, 2763.
3. Clark, R.J.H.; D'Urso, N.R. J. Amer. Chem. Soc. 1978, 100, 3088.
4. Preetz, W.; Rudzik, L. Angew. Chem. 1979, 91, 159.
5. Preetz, W.; Peters, G.; Rudzik, L. Z. Naturforsch. 1979, 34b, 1240.
6. Glicksman, H.D.; Walton, R.A. Inorg. Chem. 1978, 17, 3197.
7. Mortola, A.P.; Moskowitz, J.W.; Rösch, N.; Cowman, C.D.; Gray, H.B. Chem. Phys. Letters 1975, 32, 283.
8. Cotton, F.A. J. Mol. Struct. 1980, 59, 97.
9. Hay, P.J. J. Amer. Chem. Soc., to be published.

RECEIVED October 12, 1982

15

Synthetic Strategy for Interconverting High- and Low-Valence Clusters

M. DAVID CURTIS

University of Michigan, Department of Chemistry, Ann Arbor, MI 48109

Professor Cotton has pointed out some interesting differences between high valence clusters (HVC) and low valence clusters (LVC). I would like to comment on his observation that there appear to be few or no synthetic procedures which convert an HVC to an LVC or vice versa. Since the ligands in an HVC are typically halide or oxide ions (good π-donors), while those in an LVC are typically good π-acceptors, e.g. carbonyl, the synthetic problem reduces to the exchange of one ligand type for another without disrupting the cluster framework.

From that viewpoint, a synthetic strategy for converting HVC's to LVC's or vice versa would start with introducing ligands which can stabilize both high and low valent metals. Are there any such ligands? Experience would suggest that cyclopentadienyl and sulfur ligands are quite comfortable bonded to metals over a very wide range of formal oxidation states. These ligands are relatively moderate electron donors and relatively poor acceptors. Therefore, in concert with good π-acceptors, low oxidation states are possible, but in the absence of π-acceptors, the higher oxidation states are straightforwardly stabilized by increased electron donation from Cp or S.

A few examples illustrate this point. $Cp_2Mo_2(CO)_4$(Mo≡Mo) (1) reacts with S_8 or propylene sulfide to give the cluster, $Cp_3Mo_3(CO)_6S^+$ (2).[1] Although the direct conversion from 2 has not yet been demonstrated, the more highly oxidized cluster $Cp_3Mo_3S_4^+$ (3) possesses the same Mo_3S-tetrahedrane type skeleton.[2] Thus, in the 1 → 2 → 3 conversion the molybdenum oxidation state goes from +1 → +2 → +4 and at some point presumably crosses the border between the land of LVC and the kingdom of HVC.

As a second example, $Cp_2Mo_2(CO)_6$ (4) reacts with sulfur to give $Cp_2Mo_2S_4$ (5) which can be converted into complexes of the type, $Cp_2Mo_2(S)_2(SR)_2$ (6) and $Cp_2Mo_2(SR)_4$ (7).[3] The oxidation state of Mo in the series 4 → 5 → 6 → 7 changes in the order 1 → 5 → 4 → 3, again spanning both the LVC and HVC classifications.

0097-6156/83/0211-0241$06.00/0

The synergic effect of the Cp and S-ligand combination is dramatically shown by the changes in the Mo-S and Mo-C distances as the oxidation state of Mo is increased. CpMo complexes in which the oxidation state of Mo is I or II have Mo-C (Cp) distances in the range from 2.28 - 2.37Å.[4] In the Mo^{V} dimers this range is 2.35 - 2.47Å. In $Cp_2Mo_2(SMe)_4$ (Mo^{III}) the Mo-S distance is 2.42, but shrinks to 2.30 in the Mo(V) dimers. Thus, as the oxidation state of the Mo increases, the Mo-S distances decrease as expected, but the Mo-Cp distances tend to increase! This observation suggests that Cp is moderately effective at accepting electrons from low-valent metals via $d \rightarrow \pi^*$ bonding; but as the metal is oxidized, the $d \rightarrow \pi^*$ bonding is lost and the sulfur ligands take on the burden of increased electron donation to the metal via both σ- and π- donation. The M-S distances decrease, and the loss of $d \rightarrow \pi^*$-bonding (combined with the increasing trans-influence of the shorter M-S bonds) causes the M-Cp distances to increase.

In conclusion, it appears that the π-C_5H_5 ligand in concert with a polarizable donor ligand, e.g. S^{-2} (or possibly Se^{-2}, I^-, etc.), satisfies the requirements of our synthesis stratagem for interconversions of high and low valence clusters.

Literature Cited

1. M. D. Curtis and W. M. Butler, J.C.S. Chem. Commun., 1980, 998.

2. P. J. Vergamini, H. Vahrenkamp, and L. F. Dahl, J. Am. Chem. Soc., 93, 6327 (1971).

3. M. Rakowski DuBois, D. L. DuBois, M. C. VanDerveer, and R. C. Haltiwanger, Inorg. Chem., 20, 3064 (1981).

4. The range in Mo-C(Cp) distances reflects the electronic asymmetry about the molybdenum caused by the nature and spatial arrangement of the other ligands bonded to the metal. In other words, these variations are a manifestation of the trans-influence.

5. N. G. Connelly and L. F. Dahl, J. Am. Chem. Soc., 92, 7470 (1970).

RECEIVED October 12, 1982

16

Metal Alkoxides: Models for Metal Oxides

MALCOLM H. CHISHOLM

Indiana University, Department of Chemistry, Bloomington, IN 47405

In the spirit of this symposium, I should like to invite discussion of my proposal that metal alkoxides (1) may act as models for metal oxides in their reactions with a wide variety of hydrocarbons and small unsaturated molecules. Since metal oxides provide the most versatile class of heterogeneous catalysts used in the petrochemical industry, studies of metal alkoxides, which are hydrocarbon soluble, could shed light on metal oxide catalyzed reactions and furthermore might yield a new generation of homogeneous catalysts. The fact that metal oxide catalysts are effective for reactions such as olefin-hydrogenation, -isomerization, -polymerization and -metathesis implies that metal-hydrogen and metal-carbon bonds are involved, perhaps in an analogous manner to those which are well documented in organometallic chemistry (2). This by itself is an interesting thought since metal oxides seem far removed from the now classical organometallic compounds which abound with ligands such as tertiary phosphines, carbonyls, π-bonded cyclopentadienes and arenes, etc. How can a metal-oxide environment compete with the sophisticated ligand systems that the contemporary organometallic chemist has available for catalyst design? Apparently, metal-oxide systems compete very effectively and certainly in some instances there are no known homogeneous analogues. One can think of CO/H_2 activation in Fischer-Tropsch chemistry as just one system in which the metal oxide catalysts win hands down over homogeneous metal-carbonyl chemistry (3). Let us speculate what special requirements a metal oxide environment or surface may provide. Factors which come to my mind include (i) unusual coordination numbers, geometries and electronic configurations for metal atoms; (ii) oxo ligands are strong π-donors, which contrast with most ligands commonly used in organometallic chemistry which are π-acceptors; (iii) oxo ligands may act as terminal six-electron ($M \overset{+}{\equiv} O$) or four-electron (M=O) ligands (4) or may act as bridging (μ_2-, μ_3-, μ_4-, μ_5- and μ_6-) ligands and changes from one form to another may readily occur in response to the addition or elimination of

0097-6156/83/0211-0243$07.25/0

substrate molecules; and (iv) for many transition elements in their lower, middle and even penultimate oxidation states, metal-metal bonding may be important. Of course, in mixed metal oxide systems, there can be important contributions from two or more different metal atoms acting collectively to bring about activation that would not be possible in a homometallic system. The idea of heterobimetallic activation of small but "tough" molecules such as CO has gained much attention in homogeneous organometallic chemistry within the last few years (3) and in view of the mixed metal nature of many oxide catalysts it would seem equally important in heterogeneous systems.

Let us first examine some of the structural and electronic relationships that exist for metal oxides and metal alkoxides.

Structural and Electronic Analogies between Metal-Oxides and Alkoxides

I shall take the simple view that most metal oxide structures are derivatives of a closest packed O^{2-} lattice with the metal ions occupying tetrahedral or octahedral holes in a manner which is principally determined by size, charge (and hence stoichiometry) and d^n configuration (5). The presence of d electrons can lead to pronounced crystal field effects or metal-metal bonding. The latter can lead to clustering of metal atoms within the lattice with large distortions from idealized (ionic) geometries.

In 1958, Bradley (6) proposed a structural theory for metal alkoxides $M(OR)_n$ based on the desire of metal ions to achieve their preferred coordination number and geometry by oligomerization involving the formation of alkoxide bridges (μ_2 or μ_3). For compounds of formula MOR, octahedral and tetrahedral geometries are not attainable, but the cubane-like $M_4(OR)_4$ structure which affords the maximum coordination number, 3, is often found, e.g. M = Tl and Na (1). For $M(OR)_2$ compounds, a tetrahedral geometry is possible for a tetramer and an octahedral geometry can be envisaged for an infinite polymer. For $M(OR)_3$ compounds, tetrahedral geometries and octahedral geometries are easily obtained by oligomerization. One structural characterization of $Al(O\text{-}i\text{-}Pr)_3$ reveals a central octahedrally coordinated Al(III) ion surrounded by three tetrahedral Al(III) ions. (See I below.) $Cr(O\text{-}i\text{-}Pr)_3$, on the other hand, only exists in a polymeric form which is totally insoluble in hydrocarbon solvents. Based on spectroscopic and magnetic data, $[Cr(O\text{-}i\text{-}Pr)_3]_n$ is believed to have Cr(III) in only octahedral environments. The difference between Al(III) and Cr(III) is interesting and is readily understandable when one recognizes the marked propensity for the d^3 ion to adopt octahedral geometries to the exclusion of others - a fact which is generally reconciled with crystal field stabilization energies (5).

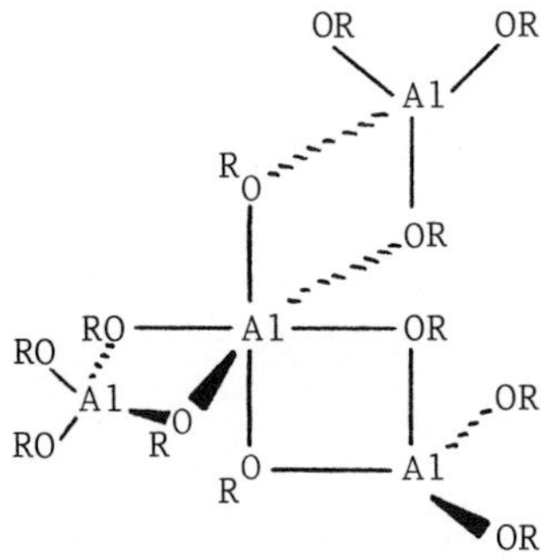

I

The structure of $[Ti(OEt)_4]_4$ reported by Ibers (7) in 1963 was of historic significance since it provided the first structural test of Bradley's theory for a compound of formula $M(OR)_4$. Each titanium atom achieves an octahedral environment, through the agency of four doubly bridging and two triply bridging alkoxy groups, as is shown in II.

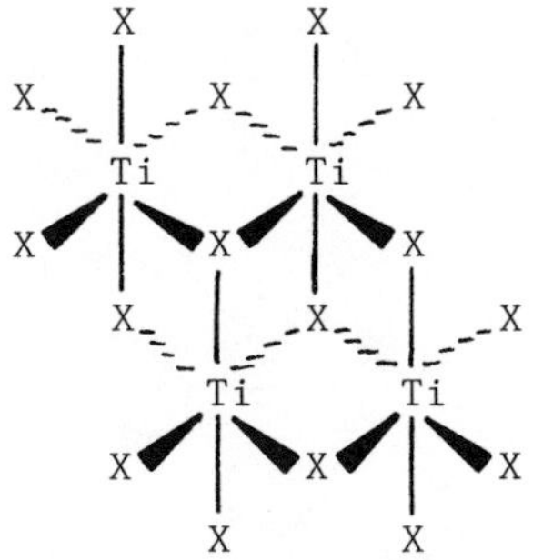

II

Compounds of formula $M(OR)_5$ can readily attain an octahedral geometry by dimerization to give two edge-sharing octahedra $[M(\mu_2\text{-}OR)(OR)_4]_2$ and this has been verified by the structural characterization of $[Nb(OMe)_5]_2$ (1).

In 1967, Bradley (8) wrote further about metal oxide -alkoxide (trialkylsiloxide) polymers. He noted that these compounds $[MO_x(OR)_{(y-2x)}]_n$ are interesting polymeric compounds which bridge the gap between the oligomeric alkoxides $[M(OR)_y]_m$ and the macromolecular metal oxides $[M_2O_y]\alpha$. Similarly, the metal oxide trialkylsilyloxides $[MO_x(OSiR_3)_{y-2x}]_n$ may be regarded as intermediate between the oligomeric metal trialkylsilyloxides $[M(OSiR_3)_y]_m$ and the mineral silicate macromolecules.

The central Ti_7O_{24} unit in $Ti_7O_4(OEt)_{20}$ (9), and the Nb_8O_{30} unit in $Nb_8O_{10}(OEt)_{20}$ (10) are shown in Figure 1 and reveal a striking similarity to both the $[Ti(OEt)_4]_4$ structure and the structures of isopoly and heteropolyacids of molybdenum, tungsten and vanadium.

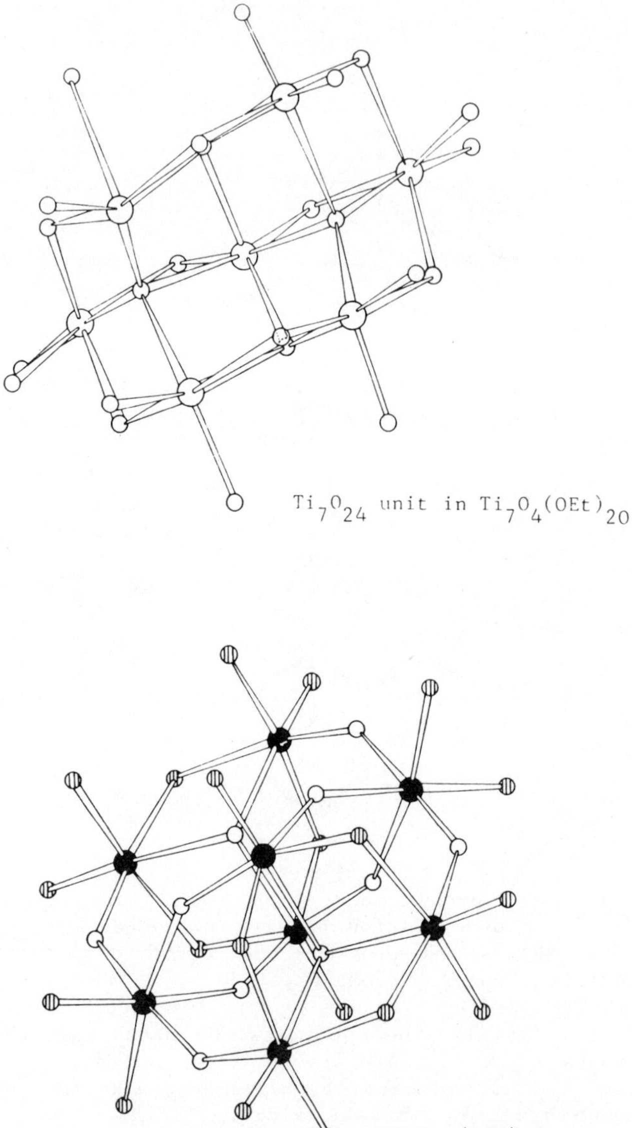

Figure 1. The Ti_7O_{24} unit in $Ti_7O_4(OEt)_{20}$ (top) and the Nb_8O_{30} unit in Nb_8O_{10}-$(OEt)_{20}$ (bottom). Key to top: o, oxygen; and ○, Ti. Key to bottom: ●, Nb; ○, oxygen in Nb–O–Nb bridge; and ◍, ethoxide oxygen.

Recently some interesting analogies have been found between metal-oxides and alkoxides where the d^n configuration is important in leading to strong metal-metal bonds. For example, the $[Ti(OEt)_4]_4$ type of structure involving four fused octahedra to give a centrosymmetric M_4 rhomboheral unit is also found for $W_4(OR)_{16}$, where R = Me or Et (11,12). Here the eight electrons from the W(4+) ions are used in cluster bonding. The same basic geometry is found in the mixed alkoxy -oxo molybdenum compound $Mo_4(=O)_4(\mu_2\text{-}O)_2(\mu_3\text{-}O)_2(\mu_2\text{-}O\text{-}i\text{-}Pr)_2(O\text{-}i\text{-}Pr)_2(py)_4$ (12) and in ternary metal oxides $Ag_8W_4O_{16}$ (13) and $Ba_{1\ 14}Mo_8O_{16}$ (14). The structure of the latter is particularly interesting: (1) There are infinite chains of metal cluster units bound together in such a way that tunnels are constructed for the Ba^{2+} ions. (2) The clusters are of two types, though both have the connectivity $Mo_4O_2O_{8/2}O_{6/3}$. Both share the basic M_4X_{16} geometry found in $M_4(OEt)_{16}$ (M = Ti and W), but differ with respect to M-M distances. In one cluster unit, there are roughly five equally short Mo-Mo distances, 2.58 Å (averaged), while in the other there are two longer Mo-Mo distances, 2.85 Å (averaged). In keeping with the stoichiometry $Ba_{1.14}Mo_8O_{16}$, it is reasonable to assign the regular and elongated clusters as having ten and eight cluster bonding electrons, respectively. A particularly interesting comparison of M-M distances in these M_4X_{16} containing units can be made as a function of the number of cluster bonding electrons 0, 4, 8 and 10. See Table I. Cotton and Fang (15) recently carried out calculations which suggested that the lengthening of two M-M bonds in the eight electron cluster $W_4(OEt)_{16}$ results from a 2nd order Jahn-Teller distortion which accompanies removal of two electrons from the regular C_{2h} rhombohedral M_4 geometry. The latter accommodates ten cluster electrons in effectively five M-M bonds. McCarley (14) noted in describing the Mo_4 clusters found in $Ba_{1.14}Mo_8O_{16}$ that these were metal-metal bonded adaptations of the well known hollandite structure and closely related to the cluster found for $CsNb_4Cl_{11}$.

Triangulo Mo_3 and W_3 units are common in discrete coordination clusters (16) and ternary oxides (19,20,21) and may accommodate five, six, seven and eight cluster bonding electrons (12). In the mixed oxo alkoxides of formula $Mo_3(O)(OR)_{10}$ where R = CH_2-t-Bu and i-Pr, the Mo-Mo distance is 2.535 Å (averaged) (23), which may be compared with Mo-Mo = 2.524(2) Å (averaged) in $Zn_2Mo_3O_8$ (19,20). The coordination about the molybdenum atoms is similar, but not the same for these triangulo complexes as is shown in Figure 2. In both cases, the molybdenum atoms are surrounded by six oxygen ligands in a distorted octahedral manner. One can imagine that the $Mo_3(\mu_3\text{-}O)(\mu_3\text{-}OR)(\mu_2\text{-}OR)_3(OR)_6$ structure could be converted to the $Mo_3(\mu_3\text{-}O)(\mu_2\text{-}O)_3(O)_9$ structure by a reaction in which the μ_3-OR ligand becomes a terminal OR group to one molybdenum and two new terminal bonds are formed, one to each of the other two molybdenum atoms. This has

Table I. M–M Distances (Å) Found in Compounds which are Structurally Related to $W_4(OEt)_{16}$.[a]

Compound	M(2)–M(1)'	M(1)–M(2)	M(1)–M(1)'	Number of M_4 Cluster Electrons	Ref.
$[Ti(OEt)_4]_4$	3.34	3.50	3.42	0	b
$Ag_8W_4O_{16}$	3.32	3.23	3.49	0	c
$Mo_4O_8(OPr^i)_4(py)_4$	3.47	2.60	3.22	4	d
$W_4(OEt)_{16}$	2.65	2.94	2.76	8	d
$Ba_{1.14}Mo_8O_{16}$	2.54	2.84	2.56	8	e
$Ba_{1.14}Mo_8O_{16}$	2.61	2.57	2.58	10	e

(a) Distances are quoted to ±0.01 Å; the labelling scheme for M(1), M(2) and M(1)' is shown below and is such that M(1) and M(2) have, respectively, two and three terminal groups. (b) Ibers, J.A. Nature 1963, 197, 686. (c) Skarstad, P.M.; Geller, S. Mat. Res. Bull. 1975, 10, 791. (d) Ref. 12. (e) McCarley, R.E.; Luly, M.H.; Ryan, T.R.; Torardi, C.C. ACS Symp. Series 1981, 155, 41.

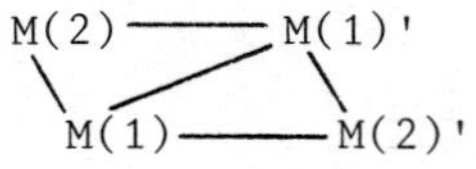

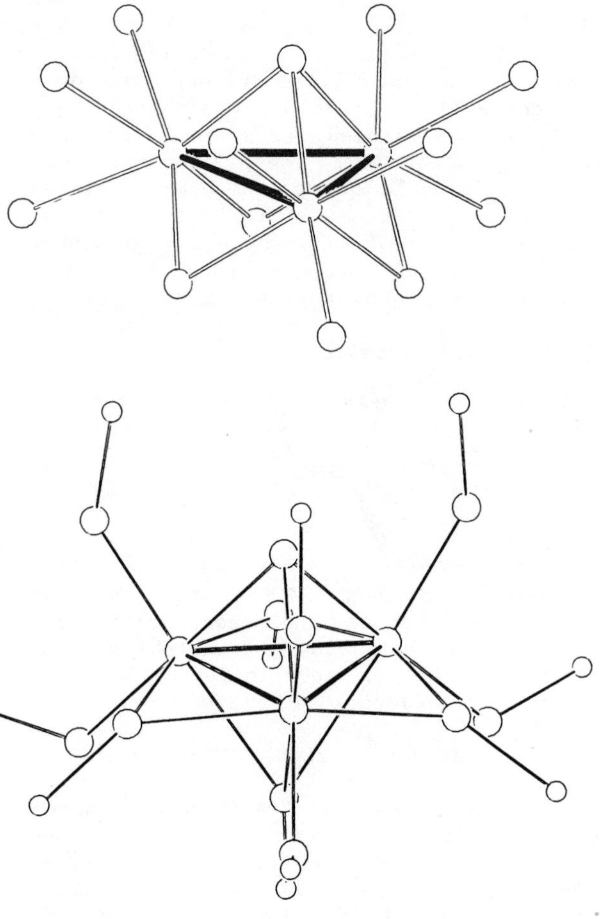

Figure 2. The Mo_3O_{13} unit found in $Zn_2Mo_3O_8$ (top), and the $Mo_3O(OC)_{10}$ unit found in $Mo_3O(OR)_{10}$ compounds ($R = CHMe_2$ and CH_2CMe_3) (bottom).

not yet been achieved chemically, but a hypothetical reaction is easily envisaged: $Mo_3(\mu_3\text{-}X)_2(\mu_2\text{-}X)_3(X)_6 + 2X \rightarrow Mo_3(\mu_3\text{-}X)(\mu_2\text{-}X)_3(X)_9$. Thus, the $Mo_3(\mu_3\text{-}O)(\mu_3\text{-}OR)(\mu_2\text{-}OR)_3(OR)_6$ structure may be viewed as coordinatively unsaturated with respect to that found in the presence of a single capping (μ_3) ligand.

M-M multiple bonding has long been known in metal oxide structures. The first Mo=Mo bond was seen in one crystalline form of MoO_2 which has a distorted rutile structure wherein the Mo(4+) ions occupy adjacent octahedral holes throughout the lattice (24). The octahedra are distorted because of the short Mo-Mo distances 2.51 Å. $La_4Re_2O_{10}$ has a fluorite type structure in which O^{2-} is substituted for F^- and four of the five Ca^{2+} sites are occupied by La^{3+} ions. The remaining Ca^{2+} site is occupied by an $(Re{\equiv}Re)^{8+}$ unit with an Re-Re distance 2.259(1) Å (25).

In the alkoxides $M_2(OR)_6$, where M = Mo and R = t-Bu, i-Pr and CH_2-t-Bu and M = W and R = t-Bu, there are unbridged M≡M bonds (26). The Mo_2O_6 geometry has approximate D_{3d} symmetry, as shown in III below, and in this way a $Mo_2{}^{6+}$ unit occupies a distorted O_6 octahedral cavity.

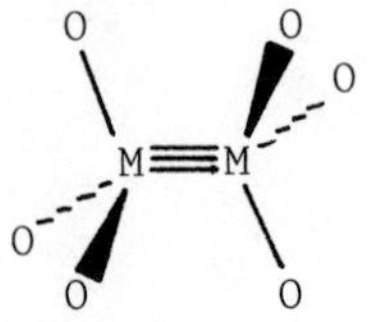

III

The compounds $M_2(OR)_6$ (M≡M), where M = Mo and W, are coordinatively unsaturated and react reversibly with donor ligands such as amines (27,28) and phosphines (29) according to eq. 1. The position of equilibrium appears to be largely determined by steric factors associated with R and L.

$$M_2(OR)_6 + 2L \rightleftharpoons M_2(OR)_6L_2 \quad (1)$$

In the solid state, the $M_2(OR)_6L_2$ compounds have four coordinated metal atoms united by M≡M bonds. The four ligands coordinated to each metal atom lie roughly in a square plane as is shown in Figure 3.

I shall mention just two further alkoxide structures which I believe are noteworthy with respect to metal oxide structures. The first is $W_4(\mu\text{-}H)_2(O\text{-}i\text{-}Pr)_{14}$ (30) whose central $W_4H_2O_{14}$ skeleton is shown in Figure 4. This molecule may be viewed as a dimer of $W_2(\mu\text{-}H)(O\text{-}i\text{-}Pr)_7$. The average oxidation state of tungsten is +4 and evidently the eight electrons available for M-M bonding are used to form two localized double bonds: W(1)-W(2) = W(1)'-W(2)' = 2.45 Å and W(1)-W(1)' = 3.41 Å. This may be viewed as a model for a section of a reduced tungsten oxide. Through the agency of bridging oxygens and

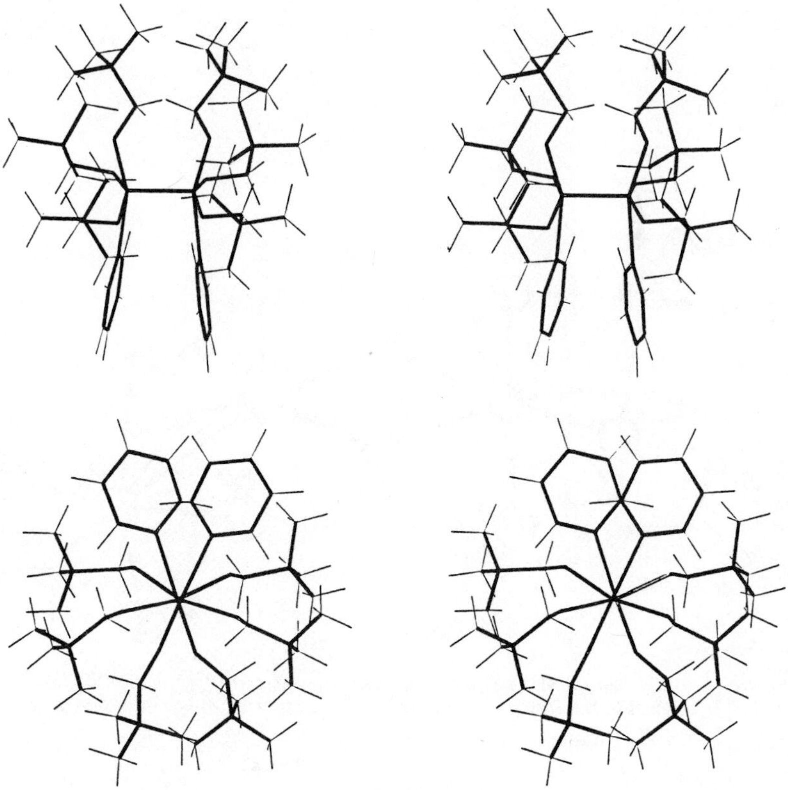

Figure 3. Two stereo stick views of the $Mo_2(OCH_2CMe_3)_6(py)_2$ molecule viewed perpendicular to the Mo–Mo bond (top) and down the Mo–Mo bond (bottom).

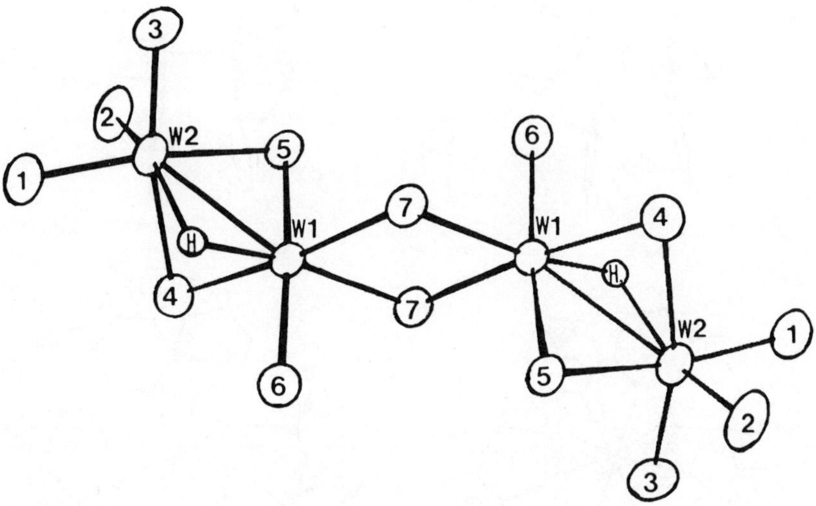

Figure 4. The central $W_4(\mu\text{-}H)_2O_{14}$ skeleton of the centrosymmetric $W_4(\mu\text{-}H)_2$-$(O\text{-i-}Pr)_{14}$ molecule emphasizing the octahedral geometries of the tungsten atoms.

hydrides, tungsten attains an octahedral environment and the d^n electrons are used to form metal-metal bonds. Formally, the outer tungsten atoms are in oxidation state +4½, while the inner ones are +3½.

The structure of the centrosymmetric molecule Mo_6O_{10}-(O-i-Pr)$_{12}$ (31) has an S-chain of six molybdenum atoms and is shown in Figure 5. Here there are both six and five coordinate molybdenum atoms and terminal and bridging oxo and alkoxide ligands. The average oxidation state for molybdenum is 5.33 which leaves four electrons available for metal-metal bonding. These are evidently used to form two localized single Mo-Mo bonds (2.585(1) Å). An interesting feature not previously seen in metal alkoxide structures is the presence of "semi-bridging" RO ligands. One of the alkoxy ligands on each of the six-coordinate terminal molybdenum atoms is positioned beneath the basal plane of its five coordinate neighboring molybdenum atom. The Mo---O distance is 2.88(1) Å, much too long for a regular bridging distance, but much too short to be viewed as non-bonding. Furthermore, the regular Mo-OR bond is lengthened and the Mo-O-C angle of this terminal OR group indicates that there is significant interaction between the oxygen lone pair and the neighboring molybdenum atom. The structure suggests incipient bond formation. The facile interconversion of terminal and bridging ligands (O and OR) is a dominant feature of the chemistry of these compounds. In solution, $W_4(\mu\text{-H})_2(\text{O-i-Pr})_{14}$ and $Mo_6O_{10}(\text{O-i-Pr})_{12}$ are fluxional on the nmr time-scale and both molecules are coordinatively unsaturated.

Oxygen-to-Metal π-Bonding

Certain general trends in M-O bond distances can be correlated with bond multiplicity and the degree of oxygen-to-metal π-bonding. A comparison of M-O bond distances for structurally related compounds having a $M_4(\mu_3\text{-X})_2(\mu_2\text{-X})_4X_{10}$ unit is given in Table II. For M = Mo and W in closely related compounds, differences in M-L distance are negligible.

1. The terminal M-O distances follow the order $Ba_{1.14}Mo_8$-O_{16} > RO-M in both $W_4(OEt)_{16}$ and $Mo_4O_8(\text{O-i-Pr})_4(py)_4$ > $Ag_8W_4O_{16}$ > O≡Mo in $Mo_4O_8(\text{O-i-Pr})_8(py)_4$. This trend is readily understood in terms of the relative degrees of O-to-M π-bonding in the series and places terminal RO ligands somewhere between the infinite chain oxygen donor ligands found in the ternary molybdenum oxide and the terminal oxo ligands found in $Ag_8W_4O_{16}$, which approximates to $W_4O_{16}^{8-}$, with weak coordination to Ag^+ ions.

2. M-μ_2-O and M-μ_3-O distances are slightly shorter for oxo ligands than for alkoxy ligands. This general observation may, however, be masked by other factors. For example, the M-μ_3-O distances in $Ag_8W_4O_{16}$ and $W_4(OEt)_{16}$ are comparable but longer than those in McCarley's compounds. These deviations

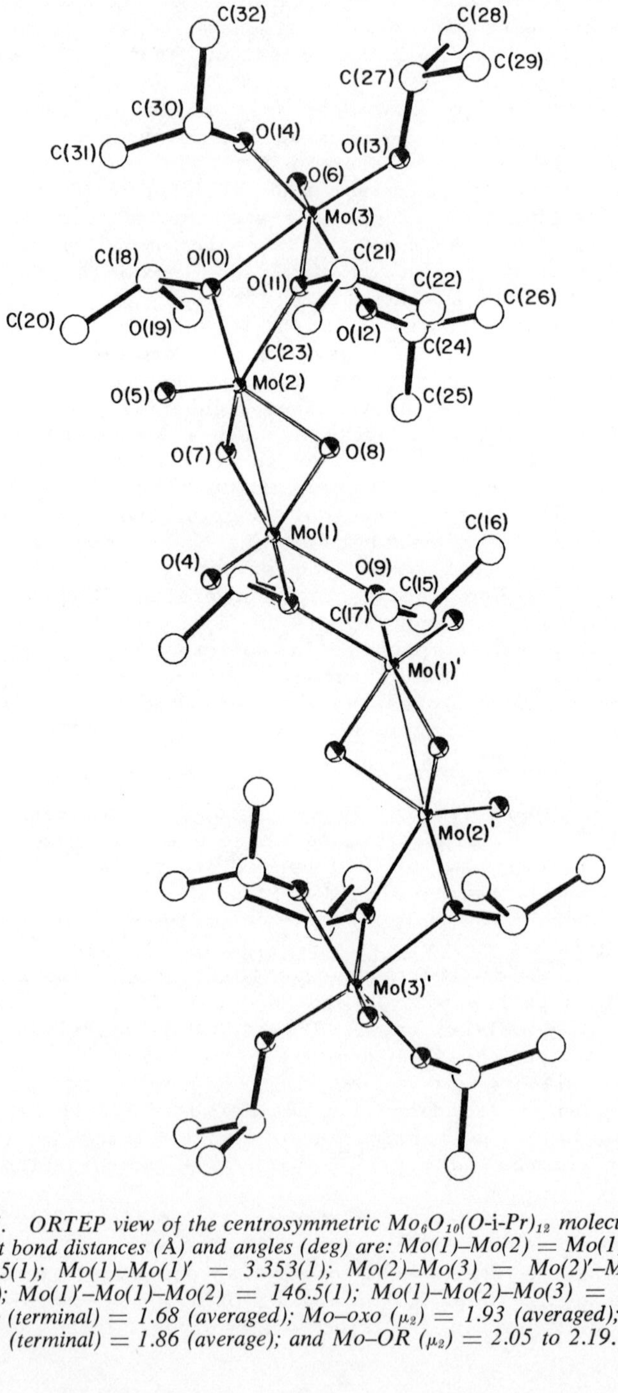

Figure 5. ORTEP view of the centrosymmetric $Mo_6O_{10}(O\text{-i-}Pr)_{12}$ molecule. Some pertinent bond distances (Å) and angles (deg) are: Mo(1)–Mo(2) = Mo(1)′–Mo(2)′ = 2.585(1); Mo(1)–Mo(1)′ = 3.353(1); Mo(2)–Mo(3) = Mo(2)′–Mo(3)′ = 3.285(1); Mo(1)′–Mo(1)–Mo(2) = 146.5(1); Mo(1)–Mo(2)–Mo(3) = 134.3(1); Mo–oxo (terminal) = 1.68 (averaged); Mo–oxo (μ_2) = 1.93 (averaged); Mo–OR (terminal) = 1.86 (average); and Mo–OR (μ_2) = 2.05 to 2.19.

Table II. Comparison of Metal-to-Oxygen Bond Distances (Å) in Compounds which are Structurally Related to $W_4(OEt)_{16}$.[a]

Compound	M-O Terminal	$M\text{-}O\mu_2$	$M\text{-}O\mu_3$
$W_4(OEt)_{16}$[b]	1.96 1.90 1.94 1.93 1.98	2.03 2.02 2.08 2.03	2.17 2.16 2.20
$Ag_8W_4O_{16}$[c]	1.80 1.73 1.77 1.79 1.83	1.86 2.20 1.87 2.14	2.11 2.15 2.22
$Mo_4O_8(OPr^i)_4(py)_4$[b]	1.68 (=O) 1.70 (=O) 1.94 (OR)	1.95 (-O-) 1.94 (-O-) 2.12 (OR) 2.24 (OR)	1.98 2.18 2.04
10 electron Mo_4 cluster in $Ba_{1.14}Mo_8O_{16}$[d]	2.10, 2.08 2.21, 2.09 2.06, 1.99 2.11, 2.14 2.03, 2.06	2.05, 2.01 2.06, 2.05 1.93, 1.92 2.08, 1.95	2.06, 2.10 2.03, 2.07 2.03, 2.09
8 electron Mo_4 cluster in $Ba_{1.14}Mo_8O_{16}$[d]	1.94, 1.85 2.06, 2.10 2.10, 2.12 2.09, 2.00 2.11, 2.14	1.94, 1.91 2.00, 1.99 2.10, 2.00 2.10, 2.04	2.05, 2.12 1.99, 2.12 1.99, 2.08

(a) Distances quoted to ±0.01 Å. (b) Ref. 12. (c) Skarstad, P.M.; Geller, S. Mat. Res. Bull. 1975, 10, 791. (d) McCarley, R.E.; Luly, M.H.; Ryan, T.R.; Torardi, C.C. ACS Symp. Series 1981, 155, 41.

from the general rule, $d_{M-\mu-OR} > d_{M-\mu-O}$, may be accounted for by considerations of trans influences (32) operating within the octahedral units. The triply bridging oxo groups in $Ag_8W_4O_{16}$ are trans to terminal W-O bonds which are short (1.79 Å average) because of multiple bond character. An examination of the three $Mo-\mu_3-O$ distances in $Mo_4O_8(O-i-Pr)_4(py)_4$ is particularly informative since the bonds are trans to three different ligands, namely oxo, isopropoxy and pyridine, which have different trans influences. A similar large asymmetry occurs in the $M-\mu_2-O$ distances in $Ag_8W_4O_{16}$ and $Mo_4O_8(O-i-Pr)_4(py)_4$ and may be traced to the different trans influences of terminal oxo and bridging oxo ligands.

What influence will strong π-donor ligands have in organometallic chemistry? This question is one which cannot be answered reliably at this time. The stabilization of unusual coordination numbers and geometries by π-donor ligands in compounds such as $Mo(O-t-Bu)_2(py)_2(CO)_2$ and $Mo(CO)_2(S_2CNR_2)_2$ has attracted the attention of Hoffmann (33) and Templeton (34) and their coworkers. We have noted that RO π-donors may produce anomalous properties in other ligands which are coordinated to the same metal. For example, in $Mo(CO)_2(O-t-Bu)_2(py)_2$ (35), the carbonyl stretching frequencies are anomalously low for carbonyl groups bonded to Mo(2+), ν(CO) = 1906 and 1776 cm^{-} and, in $Mo(O-i-Pr)_2(bpy)_2$, the 2,2'-bipyridyl ligands appear partially reduced from X-ray and Raman studies (36). Qualitatively, both of these observations may be rationalized in terms of RO-to-M π-bonding. Strong π-donor ligands raise the energy of the $t_{2g}{}^4$ electrons and, thus, enhance metal backbonding to π*-acceptor ligands, even to ligands such as 2,2'-bipyridine which do not usually behave as π-acceptor ligands in the ground state in transition metal coordination compounds.

Reactions of Metal-Metal Bonded Alkoxides of Molybdenum and Tungsten

Having established structural and electronic analogies between metal oxides and alkoxides of molybdenum and tungsten, the key remaining feature to be examined is the reactivity patterns of the metal-alkoxides. Metal-metal bonds provide both a source and a returning place for electrons in oxidative-addition and reductive elimination reactions. Stepwise transformations of M-M bond order, from 3 to 4 (37,38), 3 to 2 and 1 (39) have now been documented. The alkoxides $M_2(OR)_6$ (M≡M) are coordinatively unsaturated, as is evident from their facile reversible reactions with donor ligands, eq. 1, and are readily oxidized in addition reactions of the type shown in equations 2 (39) and 3 (39).

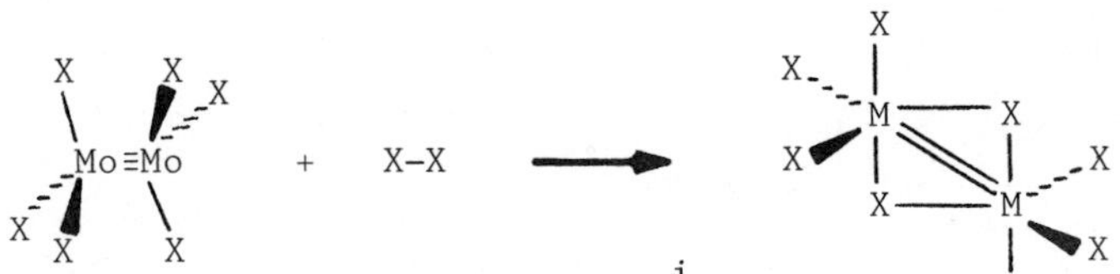

$X = OPr^i$ (2)

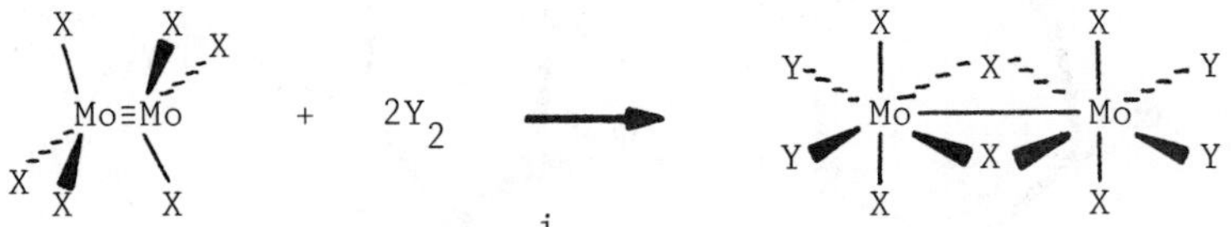

$X = OPr^i$; Y = Cl, Br (3)

Upon oxidation, alkoxy bridges seem to be invariably formed and this may be viewed as an internal Lewis base association reaction in response to the increased Lewis acidity of the metal ions which accompanies the changes $(M{\equiv}M)^{6+} \rightarrow (M{=}M)^{8+} \rightarrow (M{-}M)^{10+}$. The structural changes which take the ethane-like $X_3M{\equiv}MX_3$ unit to two fused trigonal bipyramids, sharing a common equatorial-axial edge, to two face- or edge-shared octahedra are fascinating.

The $M_2(OR)_6$ compounds provide a good source of electrons to ligands that are capable of being reduced upon coordination. For example, Ph_2CN_2, reacts with $Mo_2(O\text{-}i\text{-}Pr)_6$ in the presence of pyridine to give the adduct $Mo_2(O\text{-}i\text{-}Pr)_6(N_2CPh_2)_2(py)_2$ (40) in which the diphenyldiazomethane ligand may be viewed as a $2e^-$ ligand, $Mo{=}N{-}N{=}CPh_2$ and the dimolybdenum unit may be viewed as $(Mo{-}Mo)^{10+}$. See Figure 6.

The four electrons involved in the two Mo-Mo bonds in $Mo_6O_{10}(O\text{-}i\text{-}Pr)_{12}$ (Figure 5) are reactive to molecular oxygen: $Mo_6O_{10}(O\text{-}i\text{-}Pr)_{12} + O_2 \rightarrow 6/n[MoO_2(O\text{-}i\text{-}Pr)_2]_n$. The structure of $[MoO_2(O\text{-}i\text{-}Pr)_2]_n$ is not presently known, but the 2,2'-bipyridine (bpy) adduct $MoO_2(O\text{-}i\text{-}Pr)_2(bpy)$, which is obtained by addition of bpy to a hydrocarbon solution of $[MoO_2(O\text{-}i\text{-}Pr)_2]_n$ has been structurally characterized and shown to have the expected octahedral geometry with cis di-oxo ligands (41).

Carbon monoxide reacts with $M_2(OR)_6$ compounds and the first step has been shown to involve the reversible formation of $M_2(OR)_6(\mu\text{-}CO)$. A structural characterization of $Mo_2(O\text{-}t\text{-}Bu)_6\text{-}(\mu\text{-}CO)$ reveals an interesting square based pyramidal geometry for each molybdenum atom (42). See Figure 7. The two halves of the molecule are joined by a common apical bridging CO ligand, a pair of basal bridging OR ligands and formally a Mo=Mo bond. Closely related compounds $M_2(OR)_6L_2(\mu\text{-}CO)$ have been isolated for R = i-Pr and CH_2-t-Bu where L is a donor ligand such as an amine (43). The structures of the $M_2(O\text{-}i\text{-}Pr)_6(py)_2(\mu\text{-}CO)$ compounds (M = Mo, W) have been determined by X-ray studies and

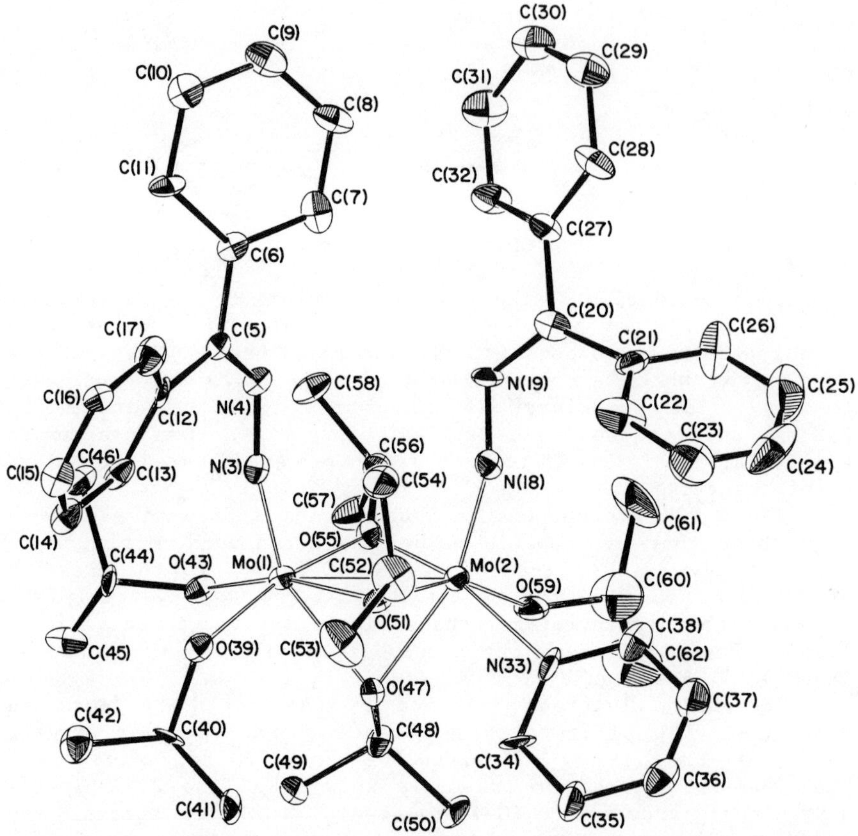

Figure 6. An ORTEP view of the Mo_2(O-i-Pr)$_6$(N_2CPh_2)$_2$(py) molecule. Some pertinent distances (Å) and angles (deg) are: Mo–Mo = 2.661(2); Mo(1)–O(39) = 1.96(1); –O(43) = 1.96(1); –O(47) = 2.24(1); –O(51) = 2.11(1); –O(55) = 2.13(1); –N(3) = 1.76(1); Mo(2)–O(47) = 2.16(1); –O(51) = 2.04(1); –O(55) = 2.04(1); –O(59) = 1.98(1); –N(18) = 1.78(1); –N(33) = 2.225(10); N(3)–N(4) = 1.30(1); N(4)–C(5) = 1.30(2); N(18)–N(19) = 1.30(1); N(19)–C(20) = 1.31(2); Mo(1)–N(3)–N(4) = 164(1)°; N(3)–N(4)–C(5) = 124(1)°; Mo(2)–N(18)–N(19) = 155(1)°; and N(18)–N(19)–C(20) = 122(1)°.

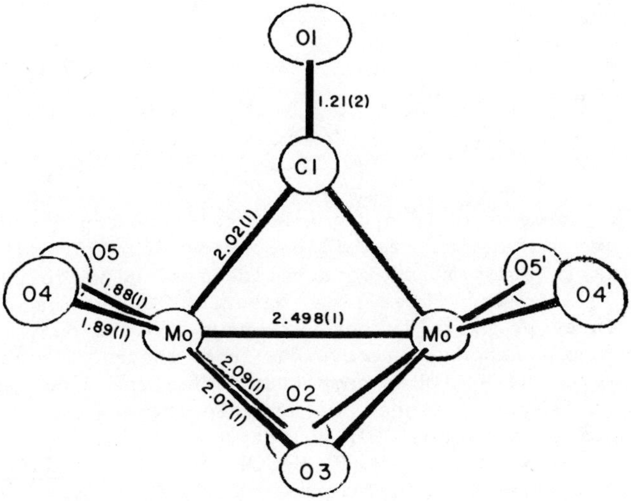

Figure 7. An ORTEP view of the central [$Mo_2O_6(CO)$] skeleton of the Mo_2-(O-t-$Bu)_6(\mu$-CO) molecule.

reveal a similarity to the $Mo_2(O\text{-}t\text{-}Bu)_6(\mu\text{-}CO)$ structure: the pyridine ligands complete the confacial bioctahedral geometry by forming bonds trans to the M–C bonds. Characteristic features of these compounds are their exceedingly low values of $\nu(CO)$, ca. 1650 (Mo) and 1550 cm^{-1} (W) and low carbonyl carbon chemical shifts, ca. 325 ppm (Mo) and 315 ppm (W) downfield from Me_4Si. These low values, which are unprecidented for neutral bridging (μ_2–) carbonyl compounds, imply a great reduction in C–O bond order. In a formal sense, the $Mo_2(OR)_6(\mu\text{-}CO)(py)_2$ compounds are inorganic analogues of cyclopropenones and it is possible to envisage a significant contribution from the resonance form IV, shown below, which emphasizes the analogy with a bridging oxy alkylidyne ligand. Bridging alkylidyne carbon resonances in $[(Me_3SiCH_2)_2W(\mu_2\text{-}CSiMe_3)]_2$ are found at 353 ppm (44).

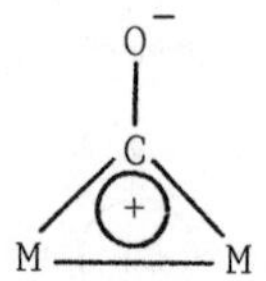

IV

$Mo_2(OR)_6$ compounds in hydrocarbon solvents rapidly polymerize acetylene to a black metallic-looking form of polyacetylene. Propyne is polymerized to a yellow powder, while but-2-yne yields a gelatinous rubber-like material (45). The detailed nature of these polymers is not yet known and the only molybdenum containing compounds recovered from these polymerization reactions were the $Mo_2(OR)_6$ compounds. When the reactions were carried out in the presence of pyridine/hexane solvent mixtures, simple adducts $Mo_2(OR)_6(py)_2(ac)$ were isolated for R = i-Pr and CH_2-t-Bu, and ac = HCCH, MeCCH and MeCCMe (45,46).

The structure of $Mo_2(O\text{-}i\text{-}Pr)_6(py)_2(\mu\text{-}C_2H_2)$ is shown in Figure 8. The central M_2C_2 unit is typical of those commonly found in dinuclear organometallic complexes, e.g. $Co_2(CO)_6(RCCR)$ (47), $Cp_2M_2(CO)_4(HCCH)$ where M = Mo (48) and W (49) and $(COD)_2Ni_2(RCCR)$ (50). The acetylenic C–C distance (1.368(6) Å) in the $Mo_2(OR)_6(py)_2(HCCH)$ compound is slightly longer than that in ethylene, 1.337(3) Å (51). The acetylenic distance in $W_2(O\text{-}i\text{-}Pr)_6(py)_2(HCCH)$, which is isostructural with the molybdenum compound, is even longer, 1.413(19) Å. Again, we see the ability of the $M_2(OR)_6$ compounds to donate electron density to π-acceptor ligands with the order W > Mo. In a formal sense, this can be viewed as an oxidative-addition reaction accompanied, as usual, by a return to octahedral geometries through the agency of alkoxy bridges and, in this instance, a bridging alkyne ligand.

The neopentoxy compound $Mo_2(ONe)_6(py)_2(\mu\text{-}C_2H_2)$, which can be prepared by the addition of one equivalent of acetylene to $Mo_2(ONe)_6(py)_2$, reacts further with another equivalent of acety-

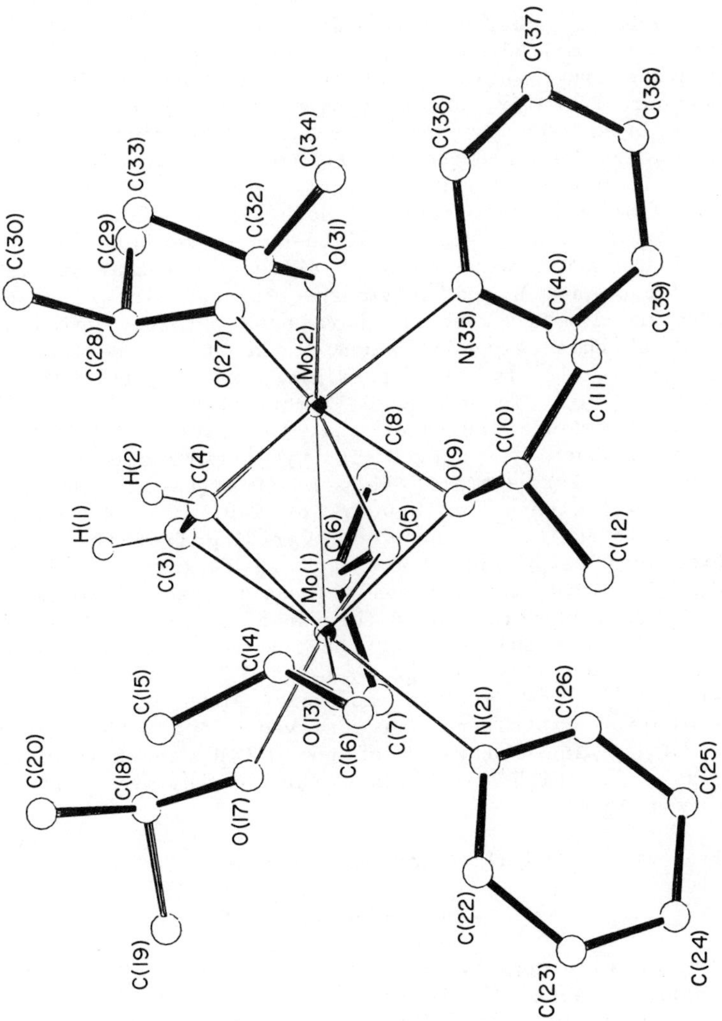

Figure 8. An ORTEP view of the $Mo_2(O\text{-}i\text{-}Pr)_6(\mu\text{-}C_2H_2)(py)_2$ molecule. Pertinent bond distances (Å) are: Mo(1)–Mo(2) = 2.554(1); Mo–O (terminal) = 1.94 (averaged); Mo–O (μ_2) = 2.14; Mo–N = 2.31 (averaged); Mo–C = 2.09 (averaged); and C–C (μ-acetylene) = 1.368(6).

lene to yield $Mo_2(ONe)_6(\mu-C_4H_4)(py)$ (45,46). An ORTEP view of this interesting molecule is shown in Figure 9. The formation of a $M_2(\mu-C_4R_4)$ unit by the coupling of two acetylenes at a dimetal center is a well recognized feature of organometallic chemistry and was first seen in 1961 with the structural characterization of $(CO)_6Fe_2(\mu-C_4Me_2(OH)_2)$, a product obtained from the reaction between but-2-yne and alkaline solutions of ironhydroxycarbonyl: $MeCCMe{\cdot}H_2Fe_2(CO)_8$ (53). In $Mo_2(ONe)_6$-$(\mu-C_4H_4)(py)$, one molybdenum atom is incorporated in a metallacyclopentadiene ring, while the other is π-bonded to the diene in a η^4-manner. Formally, one molybdenum is in oxidation state +4½ and the other +3½. In any event, the dimolybdenum center has been oxidized from $(Mo{\equiv}Mo)^{6+}$ to $(Mo{=}Mo)^{8+}$ and we see a return to the confacial bioctahedral geometry in which the $\mu-C_4H_4$ ligand occupies two sites of the bridging face.

These $\mu-C_2R_2$ and $\mu-C_4H_4$ molybdenum alkoxides are very soluble in hydrocarbon solvents which allows their characterization in solution by ^1H and ^{13}C nmr spectroscopy. In all cases, low temperature limiting spectra have been obtained which are consistent with those expected based on the structures found in the solid state. It is also possible, by nmr studies, to elucidate their role in alkyne oligomerization reactions and all available evidence indicates that they are not active in polymerization, but are active in cyclotrimerization which yields benzenes. For example, when $Mo_2(ONe)_6(\mu-C_4H_4)(py)$ was allowed to react with ca. 20 equiv. of C_2D_2 in a sealed nmr tube, the ^1H intensity of the signal arising from the $\mu-C_4H_4$ ligand decreased as a signal due to $C_6H_4D_2$ grew. Similarly, when $Mo_2(O\text{-}i\text{-}Pr)_6(py)_2(\mu-C_2H_2)$ was allowed to react with ca. 20 equiv. of C_2D_2, a benzene proton resonance appeared and, within the limits of ^1H nmr integration, the intensity of this signal corresponded to the loss of the $\mu_2-C_2H_2$ signal. When $Mo_2(O\text{-}i\text{-}Pr)_6(py)_2(\mu\text{-}MeCCH)$ was allowed to react with HCCH, ca. 20 equiv., formation of toluene was observed along with Mo (O-i-$Pr)_6(py)_2(\mu-C_2H_2)$ and the ratio of the $\mu-C_2\underline{H}_2$ and toluene-$C\underline{H}_3$ signals was 2:3, which establishes the stoichiometry of the reaction shown in eq. 4.

$$Mo_2(O\text{-}i\text{-}Pr)_6(py)_2(\mu\text{-}MeC_2H) + 3HCCH \longrightarrow Mo_2(O\text{-}i\text{-}Pr)_6(py)_2(\mu\text{-}C_2H_2) + C_6H_5Me \quad (4)$$

The species responsible for alkyne polymerization, which is kinetically more facile than cyclotrimerization since only a small fraction of the added alkyne is converted to benzenes, is not yet known. Carbene-metal complexes, both mononuclear (54) and binuclear (μ_2-CR_2) complexes (55,56), have been shown to act as alkyne polymerization initiators and several years ago it was shown that terminal alkynes and alcohols can react to give alkoxycarbene ligands (57). As yet, we have no evidence

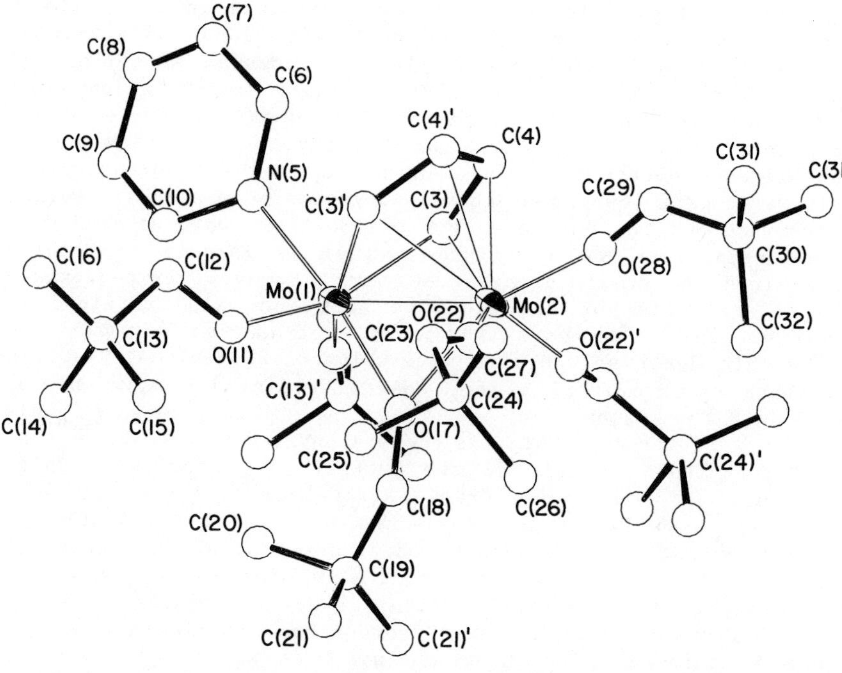

Figure 9. An ORTEP view of the $Mo_2(ONe)_6(\mu\text{-}C_4H_4)(py)$ molecule. Some pertinent distances (Å) are: Mo(1)–Mo(2) = 2.69(1); Mo–O (terminal) = 1.92 (averaged); Mo–O (μ_2) = 2.15; Mo–N = 2.15(1); Mo(1)–C(3) = 2.12(2); Mo(2)–C(3) = 2.39(2); Mo(2)–C(4) = 2.34(2); C(3)–C(4) = 1.47(3); and C(4)–C(4)' = 1.44(4).

that carbene ligands are formed in these reactions, but they remain high on the list of possibly active species present in these reactions.

$W_4(\mu\text{-}H)_2(O\text{-}i\text{-}Pr)_{14}$ (30) is one of the very few transition metal hydrido alkoxides that have been claimed in the literature (58,59). It is the only one to be fully structurally characterized by an X-ray study and, though in the solid state each tungsten is in a distorted octahedral environment (Figure 4), it behaves in solution as a coordinatively unsaturated molecule. It is fluxional on the ^1H nmr time-scale in toluene-d_8. Even at the lowest temperatures accessible in that solvent there is one time averaged type of O-i-Pr group and a hydride resonance at δ = 7.87 ppm flanked by tungsten satellites of roughly one-fifth intensity due to coupling to ^{183}W which has I = ½ and 14.4% natural abundance. The appearance of only one coupling constant, $J_{^{183}W-^1H}$ = 96 Hz, and the intensity of the satellites indicate that the hydride ligand sees two equivalent (time-averaged) tungsten atoms. Cryoscopic molecular weight determinations in benzene gave M = 1480 ± 80, which showed that the tetranuclear nature of the complex is largely or totally maintained in non-coordinating solvents. However, in p-dioxane, the molecular weight was close to half the value obtained in benzene, which suggests that the tetranuclear complex is cleaved in donor solvents to give solvated $W_2(\mu\text{-}H)(O\text{-}i\text{-}Pr)_7$. In the mass spectrometer, a very weak molecular ion was detectable, but the major ion, by field desorption, was $W_2(\mu\text{-}H)(O\text{-}i\text{-}Pr)_7^+$. All of this is consistent with the view that cleavage of one or both of the central alkoxy bridges, which are long (and weak) being trans to the hydrido ligand and span the W(1) to W(1)' atoms which are non-bonded (W(1)-W(1)' = 3.407(1) Å), creates a vacant coordination site. This molecule thus provides an interesting opportunity for the study of the reactivity of the bridging hydrido ligand. In a quick survey of its reactions with unsaturated hydrocarbons carried out in nmr tubes in toluene-d_8 solution, it has been found (30) to react rapidly with ethylene, allene and diphenyl acetylene. In each case, the Wμ-hydride resonance was lost. The reaction with ethylene is evidently reversible since attempts to isolate the presumed ethyl complex formed by insertion yield only $W_4(\mu\text{-}H)(O\text{-}i\text{-}Pr)_{14}$. Moreover, when $W_4(\mu\text{-}H)_2(O\text{-}i\text{-}Pr)_{14}$ was allowed to react with a large excess of $CH_2{=}CD_2$ in an nmr tube, the labels in the excess ethylene were rapidly scrambled and when $W_4(\mu\text{-}D)_2(O\text{-}i\text{-}Pr)_{14}$ was reacted with excess $CH_2{=}CH_2$, the $W_4(\mu\text{-}H)_2(O\text{-}i\text{-}Pr)_{14}$ compound was recovered. $W_4(\mu\text{-}H)_2(O\text{-}i\text{-}Pr)_{14}$ and 1-butene show no apparent reaction at room temperature in toluene-d_8, but at 60°C, 1-butene is selectively isomerized to cis 2-butene (60).

Concluding Remarks

1. The structural analogies between metal oxides, metal-alkoxides and metal-alkoxide-oxide polymers originally noted for d^0 metal systems can be extended to metal-metal bonded systems.

2. Metal-metal bonds in molybdenum and tungsten alkoxides provide a ready source of electrons for oxidative-addition reactions and addition reactions involving π-acidic ligands.

3. The structural properties and chemical reactivities of the M_2-μ-C_2R_2, -μ-C_4H_4, -μ-H groups have direct parallels with those of "classical" organometallic compounds.

4. The catalytic cycles that have been documented, namely alkyne cyclotrimerization and olefin isomerization, demonstrate that addition and elimination from dimetal centers can occur readily in the presence of metal-metal bonds and alkoxide ligands.

5. Alkoxide ligands, which are strong π-donors and may readily interchange terminal and bridging (μ_2 or μ_3) bonding sites, could play an important role in the future development of organometallic chemistry, particularly that of the early transition elements which may wish to increase their coordination numbers and valence shell electron configurations. Alkoxides of titanium, so-called organic titanates, in the presence of aluminum alkyls or alkylaluminum halides, have already been used for polymerization and copolymerization reactions of olefins and dienes, but this has attracted little mechanistic attention. Schrock and his coworkers (61) have shown that alkoxide ligands may be used to supress β-hydrogen elimination reactions in olefin metathesis reactions involving niobium and tantalum catalysts.

Finally, I should like to draw attention to the work of others who have provided models for metal oxides and their interactions with hydrocarbon and organometallic fragments. Specifically the work of Klemperer (62,63,64) and Knoth (65,66,-67) and their coworkers have provided us with discrete salts of the heteropolyanions to which a variety of organic/organometallic fragments have been adhered.

Acknowledgments

I thank the Chemical Science Division, Office of Basic Research, U.S. Department of Energy, the Office of Naval Research, and the National Science Foundation for their support of various aspects of the chemistry described, and also my many talented collaborators, cited in the references, who made the work possible.

Literature Cited

1. Bradley, D.C.; Mehrotra, R.C.; Gaur, P.D. "Metal Alkoxides"; Academic Press: London, New York, San Francisco, 1978.

2. Kochi, J.K. "Organometallic Mechanisms and Catalysis"; Academic Press: New York, 1974.
3. Catalytic Activation of Carbon Monoxide, ACS Sym. Ser. 1981, 152. Ford, P.C. Editor.
4. Nugent, W.A.; Haymore, B.L. Coord. Chem. Rev. 1980, 31, 123.
5. Cotton, F.A.; Wilkinson, G. "Advanced Inorganic Chemistry", John Wiley-Interscience, 4th Edit., 1980.
6. Bradley, D.C. Nature (London) 1958, 182, 1211.
7. Ibers, J.A. Nature (London) 1963, 197, 686.
8. Bradley, D.C. Coord. Chem. Rev. 1967, 2, 299.
9. Watenpaugh, K.; Caughlan, C.N. Chem. Commun. 1967, 76.
10. Bradley, D.C.; Hursthouse, M.B.; Rodesiler, P.F. Chem. Commun. 1968, 1112.
11. Chisholm, M.H.; Huffman, J.C.; Leonelli, J. J. Chem. Soc., Chem. Commun. 1981, 270.
12. Chisholm, M.H.; Huffman, J.C.; Kirkpatrick, C.C.; Leonelli, J.; Folting, K. J. Am. Chem. Soc. 1981, 103, 6093.
13. Skarstad, P.M.; Geller, S.; Mater. Res. Bull. 1975, 10, 791.
14. McCarley, R.E.; Ryan, T.R.; Torardi, C.C. ACS Symp. Ser. 1981, 155, 41.
15. Cotton, F.A.; Fang, A. J. Am. Chem. Soc. 1982, 104, 113.
16. Muller, A.; Jostes, R.; Cotton, F.A. Angew. Chem., Int. Ed. Engl. 1980, 19, 875.
17. Bino, A.; Cotton, F.A.; Dori, Z. J. Am. Chem. Soc. 1981, 103, 243.
18. Bino, A.; Cotton, F.A.; Dori, Z.; Kolthammer, B.W.S. J. Am. Chem. Soc. 1981, 103, 5779.
19. McCarroll, W.H.; Katz, L.; Ward, J. J. Am. Chem. Soc. 1957, 79, 5410.
20. Ansell, G.B.; Katz, L. Acta. Crystallogr. 1966, 21, 482.
21. McCarley, R.E.; Torardi, C.C. J. Solid State Chem. 1981, 7, 393.
22. Bursten, B.E.; Cotton, F.A.; Hall, M.B.; Najjar, R.C. Inorg. Chem. 1982, 21, 302.
23. Chisholm, M.H.; Folting, K.; Huffman, J.C.; Kirkpatrick, C.C. J. Am. Chem. Soc. 1981, 103, 5967.
24. Brandt, B.G.; Skapski, A.G. Acta. Chem. Scand. 1967, 21, 661.
25. Waltersson, K. Acta. Cryst. 1976, B32, 1485.
26. Chisholm, M.H.; Cotton, F.A.; Murillo, C.A.; Reichert, W.W. Inorg. Chem. 1977, 16, 1801.
27. Chisholm, M.H.; Cotton, F.A.; Extine, M.W.; Reichert, W.W. J. Am. Chem. Soc. 1978, 100, 153.
28. Akiyama, M.; Chisholm, M.H.; Cotton, F.A.; Extine, M.W.; Haitko, D.A.; Little, D.; Fanwick, P.E. Inorg. Chem. 1979, 18, 2266.
29. Chisholm, M.H.; et. al. Results to be published.
30. Akiyama, M.; Chisholm, M.H.; Cotton, F.A.; Extine, M.W.; Haitko, D.A.; Leonelli, J.; Little, D. J. Am. Chem. Soc. 1981, 103, 779.

31. Chisholm, M.H.; Huffman, J.C.; Kirkpatrick, C.C. J. Chem. Soc., Chem. Commun. 1982, 189.
32. Appleton, T.G.; Clark, H.C.; Manzer, L.M. Coord. Chem. Rev. 1973, 10, 335.
33. Kubacek, P.; Hoffmann, R. J. Am. Chem. Soc. 1981, 103, 4320.
34. Templeton, J.L.; Winston, P.B.; Ward, B.C. J. Am. Chem. Soc. 1981, 103, 7713.
35. Chisholm, M.H.; Huffman, J.C.; Kelly, R.L. J. Am. Chem. Soc. 1979, 101, 7615.
36. Chisholm, M.H.; Huffman, J.C.; Rothwell, I.P.; Bradley, P.G.; Kress, N.; Woodruff, W.H. J. Am. Chem. Soc. 1981, 103, 4945.
37. Chisholm, M.H.; Haitko, D.A. J. Am. Chem. Soc. 1979, 101, 6784.
38. Chisholm, M.H.; Chetcuti, M.J.; Haitko, D.A.; Huffman, J.C. J. Am. Chem. Soc. 1982, 104, 2138.
39. Chisholm, M.H.; Kirkpatrick, C.C.; Huffman, J.C. Inorg. Chem. 1981, 20, 871.
40. Chisholm, M.H.; Folting, K.; Huffman, J.C.; Ratermann, R.L. J. Chem. Soc., Chem. Commun. 1981, 1229.
41. Chisholm, M.H.; Huffman, J.C.; Kirkpatrick, C.C. Results to be published.
42. Chisholm, M.H.; Cotton, F.A; Extine, M.W.; Kelly, R.L. J. Am. Chem. Soc. 1979, 101, 7645.
43. Chisholm, M.H.; Huffman, J.C.; Leonelli, J.; Rothwell, I.P. J. Am. Chem. Soc., submitted.
44. Chisholm, M.H.; Cotton, F.A.; Extine, M.W.; Stults, B.R. Inorg. Chem. 1976, 15, 2252.
45. Chisholm, M.H.; Huffman, J.C.; Rothwell, I.P. J. Am. Chem. Soc. 1982, 104, xxxx.
46. Chisholm, M.H.; Huffman, J.C.; Rothwell, I.P. J. Am. Chem. Soc. 1981, 103, 4245.
47. Cotton, F.A.; Jamerson, J.D.; Stults, B.R. J. Am. Chem. Soc. 1976, 98, 1774.
48. Bailey, W.I. Jr.; Chisholm, M.H.; Cotton, F.A.; Rankel, L.A. J. Am. Chem. Soc. 1978, 100, 5764.
49. Ginley, D.S.; Bock, C.R.; Wrighton, M.S.; Fischer, B.; Tipton, D.L.; Bau, R. J. Organometal. Chem. 1978, 157, 41.
50. Muetterties, E.L.; Day, V.W.; Abdel-Meguid, S.S.; Debastini, S.; Thomas, M.G.; Pretzer, W.R. J. Am. Chem. Soc. 1976, 98, 8289.
51. Kutchitsu, K. J. Chem. Phys. 1966, 44, 906.
52. Chisholm, M.H.; Huffman, J.C.; Leonelli, J. Results to be published.
53. Hock, A.A.; Mills, O.S. Acta. Crystallogr. 1961, 14, 139.
54. Katz, T.J.; Lee, S.J. J. Am. Chem. Soc. 1980, 102, 422.
55. Knox, S.A.R.; Dyke, A.F.; Finnimore, S.R.; Naish, P.J.; Orpen, A.G.; Riding, G.H.; Taylor, G.E. ACS Symp. Ser. 1981, 155, 259.
56. Levisalles, J.; Rose-Munch, F.; Rudler, H.; Daran, J.C.; Dromzee, Y.; Jeannin, Y. J. Chem. Soc., Chem. Commun. 1981, 152.

57. Chisholm, M.H.; Clark, H.C. Acc. Chem. Res. 1973, 6, 202.
58. [Ti (OPh) H]: Flamini, A.; Cole-Hamilton, D.J.; Wilkinson, G. J. Chem. Soc., Dalton Trans. 1978, 454.
59. $[Ti_4(OEt)_{13}H]$: Sabo, S.; Gervais, D.C.R. Comptes Rend. Acad. Sci., Paris, Ser. C 1980, 291, 207.
60. Chisholm, M.H.; Leonelli, J. Results to be published.
61. Rocklage, S.M.; Fellmann, J.D.; Rupprecht, G.A.; Messerle, L.W.; Schrock, R.R. J. Am. Chem. Soc. 1981, 103, 1440.
62. Besecker, C.J.; Klemperer, W.G.; Day, V.W. J. Am. Chem. Soc. 1980, 102, 7598.
63. Day, V.W.; Fredrich, M.F.; Thompson, M.R.; Klemperer, W.G.; Liu, R.S.; Shum, W. J. Am. Chem. Soc. 1982, 103, 3597.
64. Day, V.W.; Fredrich, M.F.; Klemperer, W.G.; Liu, R.S. J. Am. Chem. Soc. 1979, 101, 491.
65. Knoth, W.H. J. Am. Chem. Soc. 1979, 101, 759.
66. Knoth, W.H. J. Am. Chem. Soc. 1979, 101, 2211.
67. Knoth, W.H.; Harlow, R.L. J. Am. Chem. Soc. 1982, 103, 4265.

RECEIVED July 26, 1982

Discussion

F. A. Cotton, Texas A&M University: In connection with the very nice results just presented by Dr. Chisholm, I would like to mention work done by Dr. Willi Schwotzer in my laboratory. He has carried out a related but different reaction, namely

$$W_2(OPr^i)_6py_2 + CO \rightarrow [W_2(OPr^i)_6(CO)py]_2$$

The product is a centrosymmetric "dimer of dimers" each half of which is $(py)(Pr^iO)_2W(\mu\text{-}OPr^i)(\mu\text{-}CO)W(OPr^i)_3$, with W-W = 2.654(1)Å. These are linked by bonds from the μ-CO oxygen atom of one to the pyridine-bearing W atom of the other, so that two

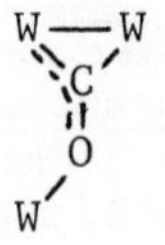

units are present. I know of no precedent for this type of $\eta^2\text{-}\mu_3$-CO group.

A.R. Siedle, 3M Central Research Laboratory: How may these novel polynuclear metal alkoxides be studied so as to contribute further to our understanding of chemical reactions which occur on the surfaces of bulk oxides?

M.H. Chisholm: No comment.

Dimolybdenum Versus Ditungsten Alkoxides

RICHARD A. WALTON

Purdue University, Department of Chemistry, West Lafayette, IN 47907

In the preceding paper, Malcolm Chisholm (1) has presented a cogent case for the modeling by metal alkoxides of certain aspects of the structural chemistry and reactivities of metal oxides. The focus of this work has been the dinuclear and polynuclear alkoxides of molybdenum and tungsten, an area of research which has also attracted our interest (2-4) and upon which I would now like to take this opportunity to comment.

In the case of the triply bonded unbridged alkoxides $M_2(OR)_6$, where M = Mo when R = t-Bu, i-Pr or CH_2-t-Bu, and M = W when R = t-Bu, (1, 5-7), the ditungsten derivative W_2(O-t-Bu)$_6$ is much less thermally stable than $Mo_2(OR)_6$. Furthermore, attempts to prepare other complexes of the type $W_2(OR)_6$, where R represents a bulky alkyl group, through the reactions of $W_2(NMe_2)_6$ with the appropriate alcohol leads to facile oxidation to tungsten(IV). This is shown, for example, in the conversion of $W_2(NMe_2)_6$ to W_4(O-i-Pr)$_{14}H_2$ (containing W=W bonds) (1,8), a reaction which corresponds formally to an oxidative addition. Another important example is the formation of tetranuclear $W_4(OR)_{16}$ upon reacting $W_2(NMe_2)_6$ with methanol or ethanol (1,9). In the case of the dimolybdenum $Mo_2(OR)_6$ complexes, oxidative additions require somewhat more potent reagents, as in the conversions of Mo_2(O-i-Pr)$_6$ to doubly bonded Mo_2(O-i-Pr)$_8$ by i-PrOO-i-Pr and to $Mo_2X_4(\mu$-O-i-Pr)$_2$(O-i-Pr)$_4$ by the halogens X_2 (1,10).

Qualitatively at least, the preceding observations hint at an increased stability of the higher valent ditungsten alkoxides over related dimolybdenum species, a trend which correlates with the increasing stability of higher oxidation states as a transition group is descended. Our own entry into this area stemmed from studies we made into the reactions between alcohol - HCl(g) mixtures and the quadruply bonded complexes $M_2(mhp)_4$, where M = Mo or W and mhp is the anion of 2-hydroxy-6-methylpyridine. In the presence of CsCl such reactions give $Cs_4Mo_2Cl_8$ and $Cs_3W_2Cl_9$ (2), reflecting the greater ease of oxidizing the $W_2{}^{4+}$ core. When the tertiary phosphines PEt_3 or P-n-Pr_3 are added in place of CsCl, each dimetal system is oxidized one step further, $Mo_2(mhp)_4$

0097-6156/83/0211-0269$06.00/0

affording $(R_3PH)_3Mo_2Cl_8H$ (11), a member of an already well characterized class of μ-hydrido bridged dimolybdenum(III) complexes (7), while $W_2(mhp)_4$ produces the dark green ditungsten(IV) alkoxides $W_2Cl_4(\mu\text{-}OR)_2(OR)_2(ROH)_2$(2-4). The latter molecules possess a W=W bond represented by a $\sigma^2\pi^2$ ground state electronic configuration (4). X-ray structure determination on the methoxide and ethoxide show that these complexes possess structure 1, the short

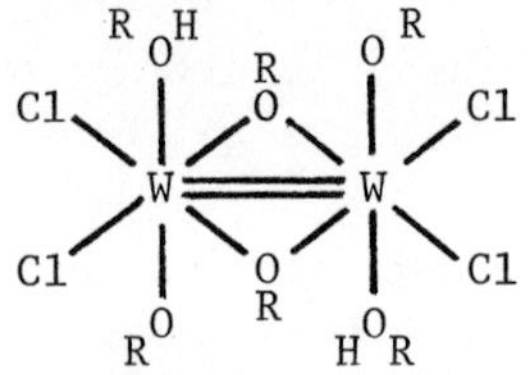

1

W-W distance (2.48Å) reflecting the double bond character. Oxidation of 1 by O_2, H_2O_2, NO_2 or Ag(I) salts gives dark red complexes (structure 2), which are the ditungsten(V) analogs of 1 (W-W dis-

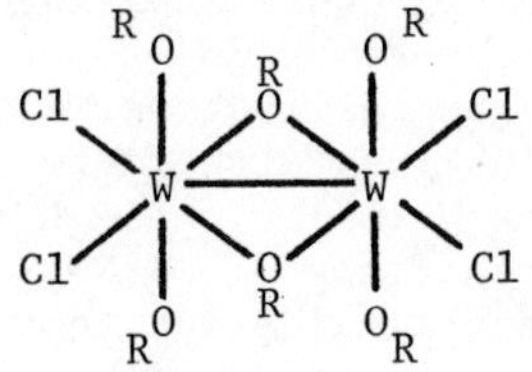

2

tance of 2.72Å)(3). In addition to the difference in W-W bond orders, 1 and 2 differ in that the former possesses two alkoxide and two alcohol ligands whereas in 2 there are now four terminal alkoxides. As a result, 2 does not possess the unsymmetrical hydrogen bond that exists between the adjacent alkoxide and alcohol ligands in 1, *viz.*

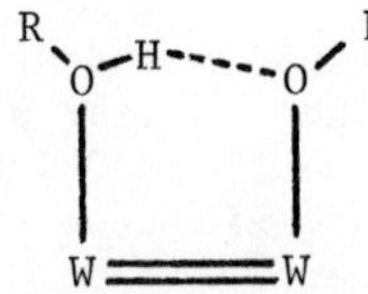

In contrast to the remarkable thermal and air stability of 1 and 2, the dimolybdenum(V) analogues of 2, $Mo_2X_4(\mu\text{-}O\text{-}i\text{-}Pr)_2(O\text{-}i\text{-}Pr)_4$ (X = Cl or Br) are thermally unstable and very moisture sensitive (10). Interestingly, molybdenum analogues of 1 are unknown although the closely related $Mo_2X_2(O\text{-}i\text{-}Pr)_6$ may be formed as a very unstable intermediate in the oxidation of $Mo_2(O\text{-}i\text{-}Pr)_6$ by halogens (10), and the dimolybdenum(IV) complex $Mo_2(O\text{-}i\text{-}Pr)_8$ is a very well characterized species (1,10). Whereas $Mo_2X_4(\mu\text{-}O\text{-}i\text{-}Pr)_2$-

(O-i-Pr)$_4$ decompose even at room temperature (to evolve isopropyl halides), high temperatures are required to decompose ditungsten complexes of types **1** and **2**, and preliminary experiments indicate that they may decompose by pathways different from the isopropoxide complexes of dimolybdenum(V). For example, the ditungsten isopropoxides of types **1** and **2** decompose at 160° in vacuum to give propene but little propyl chloride (12). In the case of the n-propoxide of ditungsten(IV), $W_2Cl_4(\mu\text{-O-}n\text{-Pr})_2(O\text{-}n\text{-Pr})_2(n\text{-PrOH})_2$, di-n-propylether is the principal volatile (12).

The doubly bonded complexes $W_2Cl_4(\mu\text{-OR})_2(OR)_2(ROH)_2$ are different from the fully substituted tungsten(IV) alkoxides $W_4(OR)_{16}$ (1,9), in that the latter 8-electron cluster exhibits a considerable degree of delocalized W-W bonding, the shortest W-W distance (2.65Å) being much longer than the W=W bond (2.48Å) in $W_2Cl_4(\mu\text{-OR})_2(OR)_2(ROH)_2$ (4).

The possibility of converting $W_2Cl_4(\mu\text{-OR})_2(OR)_2(ROH)_2$ to tetranuclear $W_4(OR)_{16}$ by substituting OR^- for Cl^- is being pursued (12). In the course of this work we have studied the alcohol exchange reactions of the ethoxide $W_2Cl_4(\mu\text{-OEt})_2(OEt)_2(EtOH)_2$. Complete exchange occurs in reactions with primary alcohols ROH (R = Me, n-Pr, n-Bu, n-Pent, n-Oct or i-Bu) to give $W_2Cl_4(\mu\text{-OR})_2(OR)_2(ROH)_2$ (12,13). On the other hand, the more sterically demanding secondary alcohols R'OH (R' = i-Pr, s-Bu or s-Pent) afford the mixed alkoxides $W_2Cl_4(\mu\text{-OEt})_2(OR')_2(R'OH)_2$ (12). X-ray structure determinations on the derivatives where R' = i-Pr or s-Pent (14) show that these complexes, while still possessing the same general type of structure as depicted in **1**, now contain a symmetrical hydrogen-bond, viz.

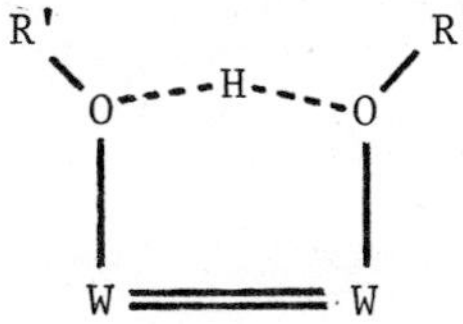

Thus the hydrogen-bonding is apparently dependent, amongst other factors, upon the nature of the alkyl substituents. Perhaps the more highly branched secondary alkyl substituents increase the hydrophobic environment about the hydrogen bond, thereby favoring a symmetrical interaction. In any event, it is apparent that this interaction is best described by a very broad, shallow potential function.

The hydrogen-bonding interactions within the complexes $W_2Cl_4(\mu\text{-OR})_2(OR)_2(ROH)_2$ and $W_2Cl_4(\mu\text{-OR})_2(OR')_2(R'OH)_2$ may provide the molecular analogues with which to model the structure and reactivities of transition metal oxide catalysts that possess surface hydroxyl groups. The thermal treatment which is often carried out in the pretreatment of metal oxides (leading to the loss of -OH groups and the evolution of H_2O) may have interesting analogies in the thermolysis of the ditungsten alkoxides.

Literature Cited

1. Chisholm, M. H. preceding paper in this symposium.
2. DeMarco, D.; Nimry, T.; Walton, R. A. Inorg. Chem., 1980, 19, 575.
3. Cotton, F. A.; DeMarco, D.; Kolthammer, B.W.S.; Walton, R. A. Inorg. Chem., 1981, 20, 3048.
4. Anderson, L. B.; Cotton, F. A.; DeMarco, D.; Fang, A.; Ilsley, W. H.; Kolthammer, B.W.S.; Walton, R. A. J. Am. Chem. Soc., 1981, 103, 5078.
5. Chisholm, M. H.; Cotton, F. A.; Murillo, C. A.; Reichert, W. W. Inorg. Chem., 1977, 16, 1801.
6. Akiyama, A.; Chisholm, M. H.; Cotton, F. A.; Extine, M. W.; Haitko, D. A.; Little, D.; Fanwick, P. E. Inorg. Chem., 1979, 18, 2266.
7. Cotton, F. A.; Walton, R. A. "Multiple Bonds Between Metal Atoms", J. Wiley and Sons, 1982, and references therein.
8. Akiyama, M.; Chisholm, M. H.; Cotton, F. A.; Extine, M. W.; Haitko, D. A.; Leonelli, J.; Little, D. J. Am. Chem. Soc., 1981, 103, 779.
9. Chisholm, M. H.; Huffman, J. C.; Kirkpatrick, C. C.; Leonelli, J.; Folting, K. J. Am. Chem. Soc., 1981, 103, 6093.
10. Chisholm, M. H.; Kirkpatrick, C. C.; Huffman, J. C. Inorg. Chem., 1981, 20, 871.
11. Pierce, J. L.; DeMarco, D.; Walton, R. A., submitted for publication.
12. DeMarco, D.; Walton, R. A. unpublished results.
13. Clark, P. W.; Wentworth, R.A.D. Inorg. Chem., 1969, 8, 1223.
14. Cotton, F. A.; DeMarco, D.; Walton, R. A. unpublished results.

RECEIVED October 8, 1982

18

Metal Clusters and Extended Metal–Metal Bonding in Metal Oxide Systems

ROBERT E. McCARLEY

Iowa State University, Ames Laboratory and Department of Chemistry, Ames, IA 50011

The results of recent investigations of the synthesis and structures of ternary and quaternary molybdenum oxide compounds having metal-metal bonded cluster units or infinite chains are presented. In all cases the average oxidation state of molybdenum is less than 4+, the lowest state previously known in structurally characterized oxide phases. New compounds containing discrete clusters are $LiZn_2Mo_3O_8$ and $Zn_3Mo_3O_8$, with triangular units, and $Ba_{1.14}Mo_8O_{16}$ with two variants of rhomboidal tetranuclear units, one having regular and the other distorted rhomboidal geometry. Infinite metal-metal bonded chains consisting of Mo_6 octahedral cluster units fused on opposite edges are found in the new compounds $NaMo_4O_6$, $Ba_5(Mo_4O_6)_8$, $Sc_{0.75}Zn_{1.25}Mo_4O_7$, and $Ti_{0.5}Zn_{1.5}Mo_4O_7$. The effects of the differing metal-based electron counts in the repeat units of the infinite chains among these compounds are discussed with reference to metal-metal bond order and structural variations within the units.

Metal clusters in metal oxide systems have not been well-characterized or abundantly investigated up to the present time. Only isolated examples of metal-metal bonded units in oxide lattices have appeared from time to time. It will be the thesis of this presentation to show that highly unusual structures determined by strong metal-metal bonding will be found in ternary and quaternary metal oxide systems, and that opportunities abound for creative work on the synthesis, theory and structure-property relationships of such compounds. Because of the well-known correlation of d-electron population and d-orbital radial extension with metal-metal bond formation,

0097-6156/83/0211-0273$06.00/0

we may expect that oxide systems containing the early 4d and 5d transition elements in reduced oxidation states will provide the most fertile area for new investigations. We emphasize here our recent work with the reduced ternary oxides of molybdenum, but expect that related chemistry will be observed for oxides of Nb, Ta, W, Tc and Re.

The most familiar examples of metal-metal bonding in oxide compounds are found in the dioxides, VO_2, NbO_2, MoO_2, WO_2 and α-ReO_2, all of which adopt low-symmetry variants of the rutile structure (1-4). In each case the metal atoms occur in chains of edge-shared MO_6 octahedra running parallel to the rutile c-axis. Displacement of each metal atom towards one and away from the other of its nearest-neighbor metal atoms along the chain accounts for the alternate short and long distances between metal atoms. The metal-metal bonds thus established may be regarded as single (VO_2, NbO_2) or double (MoO_2, WO_2 and α-ReO_2) (4). With the edge-shared bioctahedral arrangement of oxide ligands only 4 electrons may populate M-M bonding orbitals in the configuration $\sigma^2\pi^2$ (5). Additional electrons then populate nonbonding (M-M) or antibonding (M-O) orbitals of δ^* and δ symmetry with respect to the M-M bond axis (5,6). Thus, though 6 electrons reside in metal-centered orbitals, the Re-Re bond in α-ReO_2 is best regarded as a double bond, with two additional electrons in the δ^*, essentially nonbonding orbital. The compounds MoO_2 and WO_2 exhibit metallic conductivity because of overlap of the filled M-M π and empty M-O π^* valence bands, which then provide a partially filled conduction band (4). More recently metal-metal doublets with multiple bond character have been found in $La_4Re_6O_{19}$ (5), $La_3Os_2O_{10}$ (7) and $La_4Re_2O_{10}$ (8). In the first two cases the M-M bonds, established in an edge-shared bioctahedral arrangement, are at best considered as double bonds, with extra electrons consigned to M-O π^* delocalized orbitals. The doublets of Re atoms in $La_4Re_2O_{10}$ consist of Re_2O_8 units like that of $Re_2Cl_8^{2-}$, i.e. two ReO_4 fragments held together in an eclipsed configuration by a strong metal-metal bond. Here the Re atoms share 6 electrons in a triple bond with the configuration $\sigma^2\pi^4$, and the eclipsed conformation of the Re_2O_8 unit is imposed by the arrangement of oxygen atoms in the lattice.

We notice that in these examples of ternary oxides the oxygen/metal (O/M) ratio is high. This fosters small metal-metal bonded units because an octahedral arrangement of O atoms about each metal atom can be attained with a minimum of shared atoms, e.g. only two shared O atoms in the $Os_2O_{10}^{9-}$ units of $La_3Os_2O_{10}$. It is reasonable to expect that, as the O/M ratio is decreased, higher degrees of aggregation will ensue in order for the metal atom to attain higher coordination numbers. For a given electron/metal (e/M) ratio larger cluster units with M-M bonds of low bond order then will be favored over the smaller units with M-M bonds of higher order. The best known examples showing effects of this kind are found in metal halide systems, e.g. presumed dimers in WCl_4, trimers in Nb_3Cl_8 (9), and hexamers in Zr_6Cl_{12} (10).

Oxides with Trinuclear Clusters

The first compound containing a well-characterized cluster in a metal oxide system was $Zn_2Mo_3O_8$, reported in 1957 by McCarroll, Katz and Ward (11,12). In this compound the $Mo_3O_8^{4-}$ cluster units are constructed such that three MoO_6 octahedra share common edges, and share O atoms with adjacent cluster units as reflected in the connectivity formula $Zn_2Mo_3O_4O_{6/2}\,O_{3/3}$. Thus the true cluster unit is of the type M_3X_{13} as shown in Figure 1. Although the e/Mo ratio is the same as in MoO_2, all electrons are utilized in single bonds in the $Mo_3O_8^{4-}$ cluster as opposed to the double bond in the Mo_2 dimers of MoO_2. Possibly this difference is dictated by the higher O/total metal ratio in MoO_2 as opposed to $Zn_2Mo_3O_8$.

Other compounds containing the $Mo_3O_8^{4-}$ cluster units have been prepared as members of two series. The first series is of the type $M_2^{II}Mo_3O_8$ (11) with M^{II} = Zn, Cd, Mg, Mn, Fe, Co and Ni, and the second series is of the type $LiM^{III}Mo_3O_8$ (13, 14,15) with M^{III} = Ga, In, Sc, Y, and rare earth ions having atomic numbers greater than 62(Sm). The structural arrangement in crystals of the two series however is very similar. In each case one-half of the cations occupy tetrahedral sites and one-half occupy octahedral sites within a close-packed oxide lattice. Indicating these as M^o for ions in octahedral sites and M^t for those in tetrahedral sites, we have the formulas $M^tM^oMo_3O_8$ and $Li^tM^oMo_3O_8$ for the two series.

Our work was initiated on the reduced ternary molybdenum oxides with the thought that the metal cluster electron count (MCE) should be variable for the Mo_3O_8 cluster units. Based on Cotton's previous molecular orbital treatment of such clusters (16) it appeared that MCE's from 6 to 8 could be accommodated, but it was not clear whether the seventh and eighth electrons would occupy bonding or antibonding orbitals with respect to the M-M interactions. We thus set about to determine this from structural data on suitable compounds. The attempted replacement of Zn^{2+} with Sc^{3+} to secure the compound $Zn^tSc^oMo_3O_8$ was conducted via the reaction shown in equation 1.

$$2Sc_2O_3 + 4ZnO + 11MoO_2 + Mo \xrightarrow[1100^\circ]{} 4ZnScMo_3O_8 \qquad (1)$$

X-ray powder patterns showed that the product of this reaction is indeed isomorphous with $Zn_2Mo_3O_8$ and hence is the desired 7-electron cluster derivative. Unfortunately single crystals for a complete structure determination have not been obtained. Subsequent work (17) however showed that additional cations could

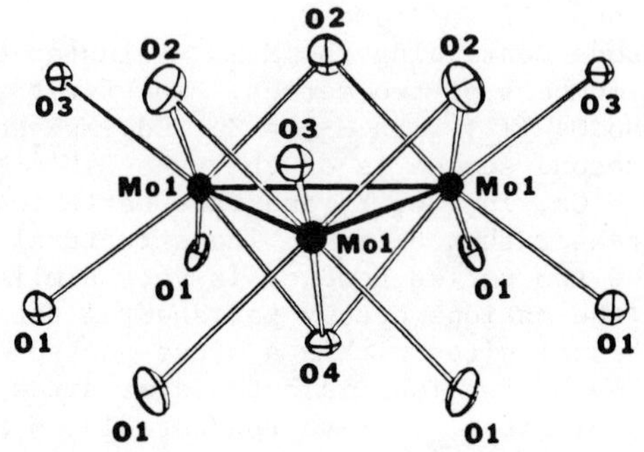

Figure 1. The M_3X_{13} cluster type as found in the oxides $Zn_2Mo_3O_8$, $LiZn_2Mo_3O_8$, and $Zn_3Mo_3O_8$. (See Table I for atom numbering scheme.)

be introduced into the lattice of a modified $Zn_2Mo_3O_8$ structure via high temperature reactions, equations 2 and 3.

$$Li_2MoO_4 + 4ZnO + 4MoO_2 + 1Mo \xrightarrow[1100°]{} 2LiZn_2Mo_3O_8 \quad (2)$$

$$6ZnO + 5MoO_2 + Mo \xrightarrow[1100°]{} 2Zn_3Mo_3O_8 \quad (3)$$

Structure determinations of $LiZn_2Mo_3O_8$ and $Zn_3Mo_3O_8$ (18) showed that these compounds are isomorphous but the pattern of oxide-layer stacking is different from that of $Zn_2Mo_3O_8$. Otherwise the structures are closely related and all contain the cluster units with the same connectivity, $[Mo_3O_4O_{6/2}O_{3/3}]^{n-}$. Thus we may compare structural data for $Zn_2Mo_3O_8$ (6MCE) with those for $LiZn_2Mo_3O_8$(7MCE) and $Zn_3Mo_3O_8$(8MCE) to discern the effects of adding successive electrons to the $Mo_3O_8^{4-}$ cluster unit. A comparison of Mo-Mo and Mo-O bond distances is given in Table I for these compounds. It is clear that addition of

Table I. Comparison of Mo-Mo and Mo-O bond distances (Å) in $Zn_2Mo_3O_8$, $LiZn_2Mo_3O_8$ and $Zn_3Mo_3O_8$.

Bond	$Zn_2Mo_3O_8$ [a]	$LiZn_2Mo_3O_8$	$Zn_3Mo_3O_8$
Mo1-Mo1	2.524(2)	2.578(1)	2.580(2)
Mo1-O1	2.058(10)	2.063(6)	2.100(9)
Mo1-O2	1.928(20)	2.003(8)	2.056(13)
Mo1-O3	2.128(30)	2.138(5)	2.160(8)
Mo1-O4	2.002(30)	2.079(7)	2.054(11)
Mo1-O (aver.)	2.017	2.058	2.088

[a] Data taken from reference (12).

electrons to the $Mo_3O_8^{4-}$ cluster causes a net elongation of both the Mo-Mo and Mo-O bonds. This may be taken as evidence that the seventh and eighth MCE's enter a molecular orbital which is antibonding with respect to both Mo-Mo and Mo-O interactions. Among the Mo-O bonds, those involving the three-edge-bridging O atoms are the most strongly affected. It therefore appears that the set of three d-orbitals not involved in either Mo-O or Mo-Mo σ-bonding mix primarily with the set of three $p\pi$ orbitals located on the edge-bridging O atoms to form a set of 3 bonding and 3 antibonding MO's. The bonding set (a_1 + e) is populated by electrons from the O atom lone pairs, and these probably lie below the Mo-Mo σ-bonding orbitals; the antibonding set (a_1^* + e^*) would then constitute the LUMO's of $Zn_2Mo_3O_8$. Magnetic susceptibility data for $LiZn_2Mo_3O_8$ showed Curie-Weiss behavior over the temperature range 95-298 K with a very large Weiss constant $\theta = -545°$. At 298 K, $\mu_{eff} = 1.18\ \mu_B$, a low value caused by strong antiferromagnetic coupling. Magnetic measurements on $Zn_3Mo_3O_8$, while less detailed, show only a weak paramagnetism, $\mu_{eff} = 0.6\ \mu_B$, probably indicative of Van Vleck paramagnetism. We thus infer that the electron spins are paired in $Zn_3Mo_3O_8$, and that the a_1^* level lies below the e^* level.

Oxides with Tetranuclear Clusters

In the structure refinement of $LiZn_2Mo_3O_8$ the Li^+ ions were not located because of their apparent scrambling with Zn^{2+} ions in both the tetrahedral and octahedral sites. An effort to overcome this difficulty was made with the attempted synthesis of $NaZn_2Mo_3O_8$, where it was expected that the Na^+ ions would be confined only to the octahedral sites. This attempted synthesis led instead to the formation of the new compound $NaMo_4O_6$ (19) which grew in the reaction mixture and on the wall of the molybdenum container as thin needles with silvery metallic luster. The serendipitous discovery of this compound has proven to be extremely important to our visions of possibilities for new metal-metal bonded structures in reduced oxide phases. In retrospect it is amazing that oxide phases containing molybdenum in oxidation states less than 4^+ were essentially unknown and certainly structurally uncharacterized. The existence of the previously mentioned series $M_2^{II}Mo_3O_8$ and $LiM^{III}Mo_3O_8$ should have been a tip-off to an extensive chemistry for metal-metal bonded molybdenum oxide systems. Indeed, subsequent work has revealed a plethora of new compounds all of which (where structure has been determined) feature strong metal-metal bonding in either discrete cluster units or extended chain arrays.

A chart of these structurally characterized compounds is given in Table II, showing the correlation of average oxidation state of the molybdenum with cluster formation or extended arrays. There are two notable effects which are pertinent to the nature of the metal-metal bonding, namely the electron/metal (e/M) ratio and the oxygen/metal (O/M) ratio. With other factors constant an increase in e/M ratio should promote more extensive M-M bonding, and a decrease in the O/M ratio should promote clusters of increasing nuclearity, progressing from dimers through larger clusters to systems with extended chains, sheets or 3-dimensional arrays (20, 21).

The interesting compound $Ba_{1.14}Mo_8O_{16}$ (17,22) contains tetranuclear rhomboidal cluster units, tied into infinite chains by Mo-O-Mo bonding as represented in the connectivity $Mo_4O_2O_{8/2}O_{6/3}$. The Ba^{2+} ions are stacked in sites along tunnels formed by weaving together the $[Mo_4O_8^{n-}]_\infty$ chains, as shown in Figure 2. In this low-symmetry, metal-metal bonded adaptation of the well-known hollandite structure the chains forming the four sides of each tunnel are related in pairs by $P\bar{1}$ symmetry.

Table II. Characteristics of metal-metal bonded ternary and quaternary molybdenum oxides.

Compound	O.N.[a]	$e/_{Mo}$ [b]	$O/_{Mo}$ [c]	Structural feature
$Ba_{1.14}Mo_8O_{16}$	3.72	2.28	2.0	rhomboidal clusters
$LiZn_2Mo_3O_8$	3.67	2.33	2.67	triangular clusters
$Zn_3Mo_3O_8$	3.33	2.67	2.67	triangular clusters
$NaMo_4O_6$	2.75	3.25	1.50	infinite chains
$Ba_5(Mo_4O_6)_8$	2.69	3.31	1.50	infinite chains
$Sc_{0.75}Zn_{1.25}Mo_4O_7$	2.31	3.69	1.75	infinite chains
$Ti_{0.5}Zn_{1.5}Mo_4O_7$	?	?	1.75	infinite chains

[a] Oxidation number (average) of molybdenum

[b] Average electron to metal ratio for metal-metal bonding

[c] Oxygen to molybdenum ratio

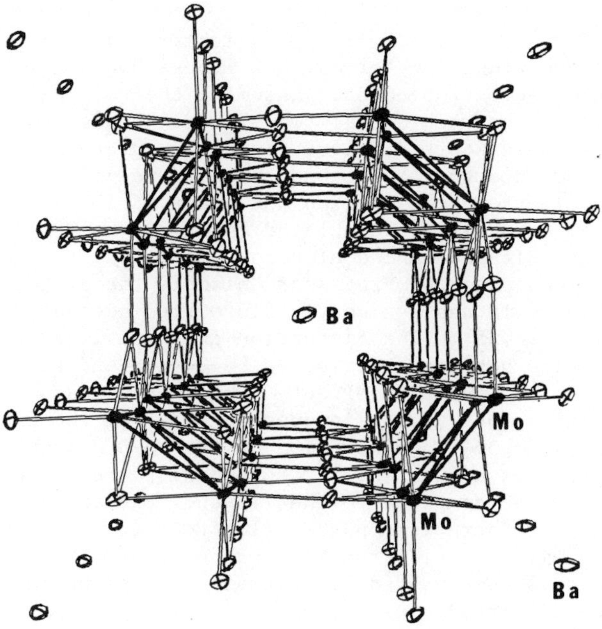

Figure 2. A three-dimensional perspective of the structure of $Ba_{1.14}Mo_8O_{16}$ as viewed down the unique axis parallel to the direction of chain growth and tunnel formation.

In one pair of chains the rhomboidal cluster units have five Mo-Mo bonds of nearly equal length (regular units), and in the other pair the (distorted) rhomboidal units have three short and two long Mo-Mo bonds. Figure 3 shows how these rhomboidal units are bound together within the infinite chains. A comparison of Mo-Mo and selected Mo-O bond distances for the regular and distorted cluster units is given in Table III.

Table III. Selected bond distances (Å) within the distorted and regular cluster units of $Ba_{1.14}Mo_8O_{16}$.

Atoms [a]		Distorted Cluster	Regular Cluster
Mo1-Mo2		2.847(1)	2.616(1)
Mo1-Mo2'		2.546(1)	2.578(1)
Mo2-Mo2'		2.560(1)	2.578(1)
Mo1-O3	edge	1.931(6)	1.936(6)
Mo1-O8	edge	1.894(6)	2.022(6)
Mo1-O4	edge	2.079(6)	2.143(6)
Mo1-O1	capping	2.082(6)	2.053(6)
Mo1-O2	capping	2.046(6)	2.055(6)
Mo1-O2'	capping	2.104(6)	2.095(6)

[a] See Figure 3 for atom numbering scheme.

Since the Mo1-Mo2 distance in the distorted unit corresponds to a Pauling bond order (23) of ca. 0.5 these bonds may be considered as one-electron bonds. Assuming the Mo1-Mo2' and Mo2-Mo2' bonds are each normal two-electron bonds, there are ca. 8 MCE's for the distorted units. In the regular units the Mo-Mo distances indicate that these bonds should be considered as normal two-electron bonds and an MCE count of 10 is obtained. Thus the compound may be formulated as $Ba_{1.14}(Mo_4O_8^{2-})(Mo_4O_8^{0.28-})$ to represent this unbalanced electron distribution. As noted in the previous paper by Professor Chisholm the distorted Mo_4O_8 cluster unit has a recently discovered molecular analog in $W_4(OEt)_{16}$ (24,25). The distortion in these 8-electron rhomboidal clusters has been attributed by Cotton and Fang (26) to a second order John-Teller effect. This seems reasonable for the molecular $W_4(OEt)_{16}$ but is questionable for the distorted unit in $Ba_{1.14}Mo_8O_{16}$. Examination of the Mo-O distances in Table III reveals a much shorter Mo1-O8 distance in the distorted cluster as compared to that in the regular cluster. Thus it appears that there is a push-pull effect at work, such that electron density removed from Mo1 through weakening of the Mo1-Mo2 bond is compensated by increased π-bonding in the Mo1-O8 bond of the distorted cluster.

It appears that other compounds having this structure type should be possible. In particular, by appropriate substitution of Ba^{2+} by cations of different charge the MCE count should be variable over some range, and conceivably even compounds having more than 10 MCE per cluster unit might be prepared, with ad-

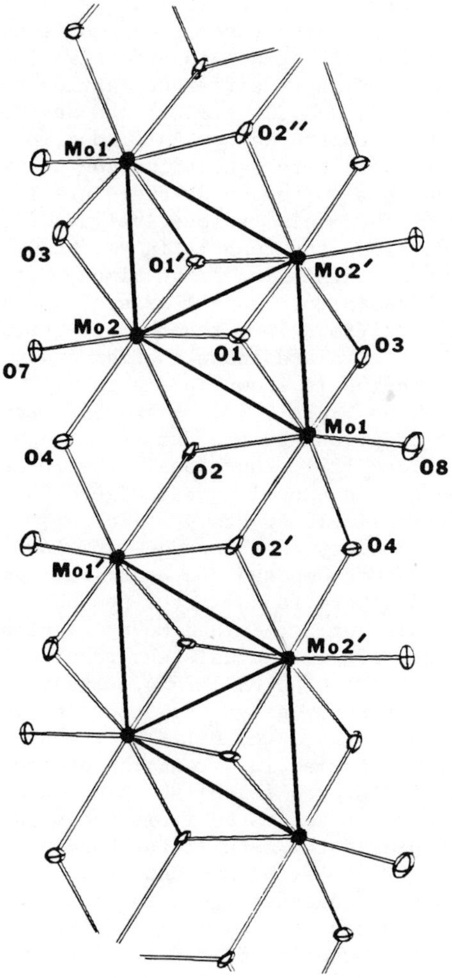

Figure 3. A view of one Mo_4O_8 cluster chain in $Ba_{1.14}Mo_8O_{16}$ which shows intrachain Mo–Mo and Mo–O bonding. (See Table III for atom numbering scheme.)

ditional electrons entering a conduction band. Work along this line is presently in progress in our laboratory.

Oxides with Extended Metal-Metal Bonding

The remaining compounds listed in Table II all adopt structures with infinite metal-metal bonded chains consisting of octahedral cluster units fused on opposite edges. However, because of the large difference in effective ionic radius of the cations concerned, very different lattice types are dictated. The compounds $NaMo_4O_6$ (19,22) and $Ba_5(Mo_4O_6)_8$ (17) adopt tunnel structures with the Na^+ or Ba^{2+} ions located in sites along the tunnels with 8-fold coordination by oxygen atoms. In the sodium and barium compounds there are 13 and 13.2 MCE's per Mo_4O_6 unit of the chain, respectively. Ordering of the cation vacancies along the tunnels in $Ba_5(Mo_4O_6)_8$ results in a super-lattice with a dimension eight times that of $NaMo_4O_6$ along the unique chain (or tunnel) axis. Views of the $NaMo_4O_6$ (tetragonal) and $Ba_5(Mo_4O_6)_8$ (orthorhombic) structures as viewed down the c-axes are shown in Figure 4. Construction of the $[Mo_4O_6]_\infty$ chains is shown in Figure 5.

The compounds $Sc_{0.75}Zn_{1.25}Mo_4O_7$ and $Ti_{0.5}Zn_{1.5}Mo_4O_7$ are the most recent additions to the family of reduced, ternary or quaternary oxides (27). Once again they were obtained as unexpected products in experiments designed for another purpose. Attempts to obtain single crystals of $ScZnMo_3O_8$ produced only microcrystalline powders in the range 1100-1400°C. At ∿1450°C the latter compound decomposed and beautiful gem-like crystals of the new compound were found in the product mixture. Analyses of several crystals for Sc, Zn and Mo by electron-microprobe x-ray fluorescence provided consistent results and the average atomic ratios Sc:Zn:Mo of 0.75:1.25:4. These data, combined with the subsequent x-ray structure determination, established the composition $Sc_{0.75}Zn_{1.25}Mo_4O_7$. The composition of crystals of $Ti_{0.5}Zn_{1.5}Mo_4O_7$, formed in a reaction between TiO_2, ZnO, MoO_2 and Mo at 1450°C, was established in like manner. Although these new compounds proved to be isomorphous, distinct differences were found in the metal-metal bonding characteristics, evidently caused by differing MCE counts in the repeat units of the infinite chains. Because the oxidation state of Ti in $Ti_{0.5}Zn_{1.5}Mo_4O_7$ has not been established, the assessment of the MCE count and interpretation of bonding differences with $Sc_{0.75}Zn_{1.25}Mo_4O_7$ is uncertain. For this reason only the structure and bonding features of the scandium compound will be described here.

A view down the c-axis (parallel to the chain axes) of the scandium compound is shown in Figure 6. It can be seen that the $[Mo_4O_7]_\infty$ chains are bound together through Mo-O-Mo bridge bonding to form layers which are, in turn, bound together through O-Sc-O and O-Zn-O bonding. The sites for these metal ions are of two types: tetrahedral sites occupied only by Zn^{2+} ions, and octahedral sites occupied by either Zn^{2+} or Sc^{3+}. In

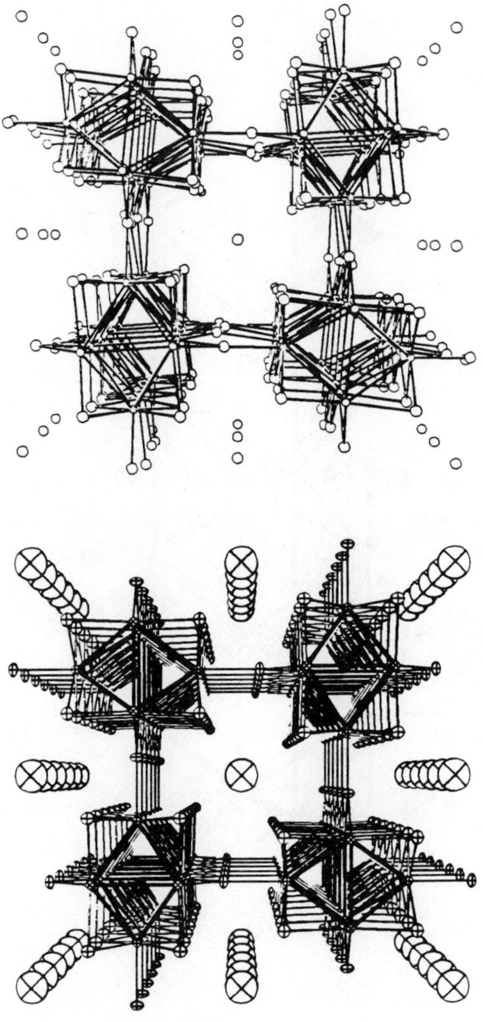

Figure 4. A three-dimensional perspective of the structures of $NaMo_4O_6$ (top) and $Ba_5(Mo_4O_6)_8$ (bottom) as viewed down the unique axis of chain growth and tunnel formation.

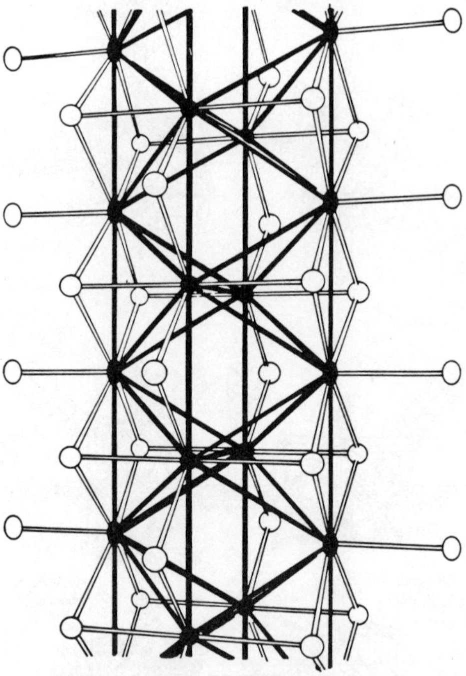

Figure 5. Segment of one $(Mo_4O_6^-)_\infty$ chain in $NaMo_4O_6$ which shows intrachain Mo–Mo and Mo–O bonding. Key: ●, Mo; and ○, oxygen.

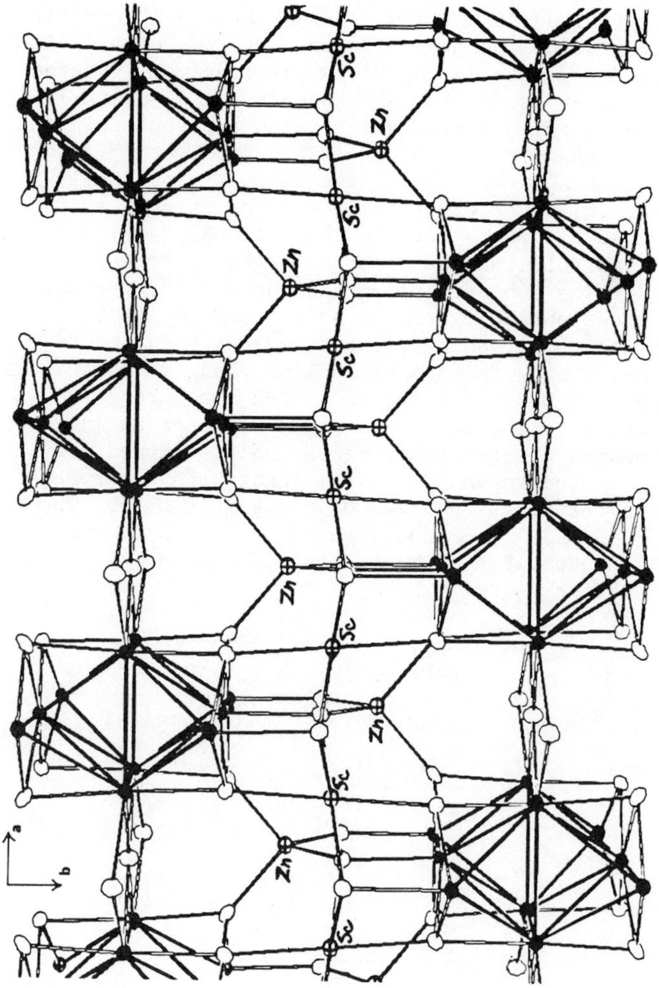

Figure 6. A view down the axis of $[Mo_4O_7^{4.75-}]_\infty$ chain growth in $Sc_{0.75}Zn_{1.25}Mo_4O_7$.

the octahedral sites the metal ions appear to be statistically disordered with 0.25 Zn^{2+} and 0.75 Sc^{3+} fractional occupation. Based on these considerations and the sharing of O-atoms along and between chains, the compound can be formulated as $Sc^{o}_{0.75}Zn^{o}_{0.25}Zn^{t}_{1.00}[(Mo_4O_{8/2}O_{2/2})O_2]$.

Electron transfer from the Sc and Zn to the Mo_4O_7 units of the chains results in 14.75 MCE per unit. The average formal oxidation state of 2.31+ for Mo in this compound is the lowest yet attained in a ternary oxide phase. A comparison of Mo-Mo bonding along the chains between $NaMo_4O_6$ with 13 MCE and $Sc_{0.75}Zn_{1.25}Mo_4O_7$ with 14.75 MCE is instructive. Pertinent Mo-Mo bond distances for these compounds are given in Table IV. The principal bonding difference between the chains in the two structures is that in $NaMo_4O_6$ the spacing between Mo atoms along the chain, both apex-apex and waist-waist, is perfectly regular. In contrast, while the waist-waist spacing is regular, the apex-apex distances alternate between short (2.625(2) Å) and long (3.139(2) Å) within the chains of $Sc_{0.75}Zn_{1.25}Mo_4O_7$, as shown in Figure 7. The sum of Pauling bond orders for the net of two apex-apex bonds per Mo_4O_6 repeat unit in $NaMo_4O_6$ is 0.773. For the net of one short and one long apex-apex bond per repeat unit in $Sc_{0.75}Zn_{1.25}Mo_4O_7$ the sum of Pauling bond orders is 1.092. This increase of 0.319 for the apex-apex bond order sum in going from the former to the latter compound dictates an increase of ca. 0.64 bonding electrons per repeat unit. Summed over all Mo-Mo bonds of the repeat unit, the total bond order for $NaMo_4O_6$ is 6.36, and for $Sc_{0.75}Zn_{1.25}Mo_4O_7$ the sum is 6.84. These

Table IV. A comparison of Mo-Mo bond distances (Å) in $NaMo_4O_6$ and $Sc_{0.75}Zn_{1.25}Mo_4O_7$.

	d (Mo-Mo)	
Bond	$NaMo_4O_6$	$Sc_{0.75}Zn_{1.25}Mo_4O_7$
(waist-waist)[a]	2.8618(2)	2.8854(5)
(waist-waist)[b]	2.753(3)	2.817(2)
apex-waist	2.780(2)	2.747(1) 2.782(1)
apex-apex	2.8618(2)	2.625(2) 3.139(2)

[a] Bond distance parallel to chain direction.

[b] Bond distance perpendicular to chain direction.

numbers strongly indicate that all MCE's participate in bonding interactions in $NaMo_4O_6$ and that the additional electrons required to form the chains in $Sc_{0.75}Zn_{1.25}Mo_4O_7$ further enhance the Mo-Mo bonding.

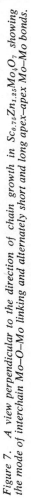

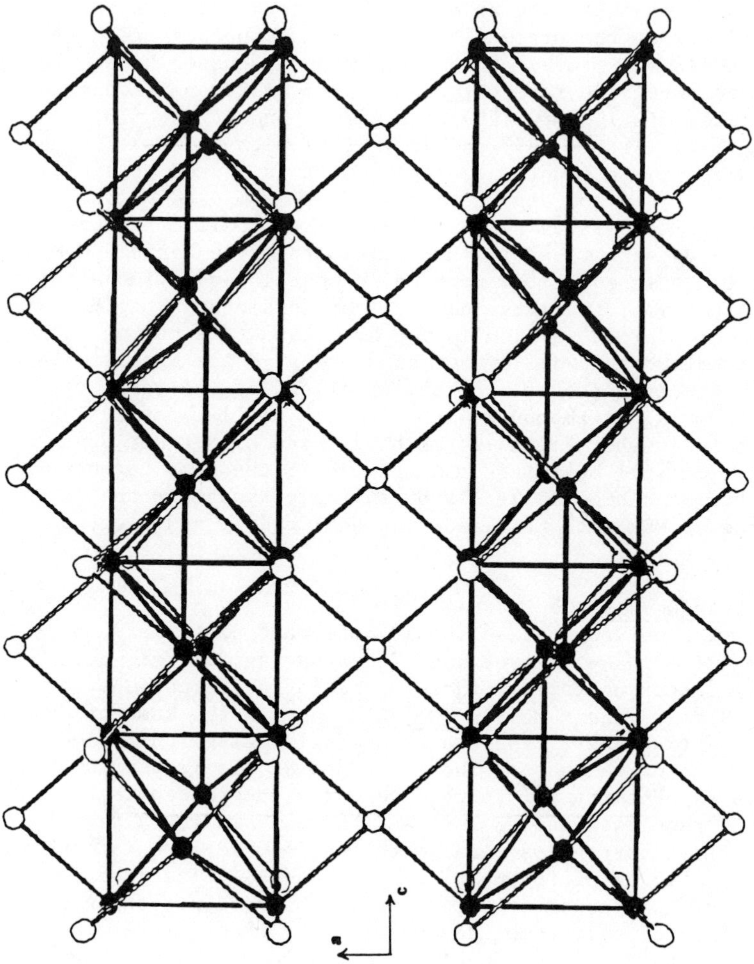

Figure 7. A view perpendicular to the direction of chain growth in $Sc_{0.75}Zn_{1.25}Mo_4O_7$ *showing the mode of interchain Mo–O–Mo linking and alternately short and long apex–apex Mo–Mo bonds.*

Several important questions then arise. What range of electron counts per repeat unit is permitted in chain structures of this type? What changes may be expected in the pattern of metal-metal bonding within the units as the electron count is varied across the permitted range? What is the detailed bond structure? How many discrete bands separated by forbidden gaps will occur? Answers to these and related questions are very important as guides to further synthetic work, i.e. what compounds are possible, and as a base to understand physical properties. At the present time several compounds are known having chain structures of the same type discussed here. These are found among the rare earth and group III transition metal subhalides, viz. Tb_2Br_3 (20) with 6 MCE; and Sc_5Cl_8 (28), Gd_5Br_8 (29), Tb_5Br_8 (29) and Er_4I_5 (30) with 7 MCE. Thus a wide range of MCE counts per repeat unit is indeed possible. The compounds at the low end of the range are composed of both metal and nonmetal atoms having relatively large radii. Because of the chain construction, which requires that the average M-M separation must equal the nonmetal-nonmetal separation along the chain direction, the large nonmetal atoms dictate long M-M distances. This condition is met by metal atoms having large radii forming bonds of low net bond order, hence metals with low valence electron counts. At the upper end of the range are the molybdenum oxide compounds with the much smaller nonmetal atoms, comparatively short metal-metal bonds, and higher net bond orders. It will be interesting to see if the gaps between these two extremes can be filled-in by suitable synthesis and requisite matching of MCE with metal and nonmetal radii, and electronic band structure.

Concluding Remarks

As a final comment we should note that no ternary molybdenum oxide phase has yet been found having discrete octahedral cluster units like those found in the famous Chevrel phases $M_xMo_6S_8$ and $M_xMo_6Se_8$ (31,32). Only in the case of $Mg_3Nb_6O_{11}$ (33) has a discrete octahedral cluster unit been found in an oxide phase. The existence of the latter however signals opportunity for further research, and taken together with the results reported here and those known for a few other metal oxides, extensive metal-metal bonded, metal oxide chemistry can be anticipated for Nb, Ta, W, Tc and Re. The great stability and unusual structure types observed for the known metal oxide cluster compounds should make them interesting candidates for the study and development of useful physical and chemical properties.

Acknowledgments

The author gratefully acknowledges the several coworkers upon whose work this article is based and whose contributions are indicated in the literature citations. The continuing support of the U.S. Department of Energy, Division of Basic Energy Sciences, for this research program is also gratefully acknowledged.

Literature Cited

1. Marinder, B.; Magneli, A. Acta Chem. Scand. 1957, 11, 1635.
2. Goodenough, J.B. "Magnetism and the Chemical Bond," Interscience Monographs in Chemistry, Inorganic Section, Vol. 1, F. A. Cotton, Ed., Interscience Division, John Wiley and Sons, Inc., New York, N. Y. 1963.
3. Goodenough, J.B. Bull. Soc. Chim. France 1965, 4, 1200.
4. Rogers, D.B.; Shannon, R.D.; Sleight, A.W.; Gillson, J.L. Inorg. Chem. 1969, 8, 841.
5. Sleight, T.P.; Hare, C.R.; Sleight, A.W. Mat. Res. Bull. 1968, 3, 437.
6. Skaik, S.; Hoffman, R. J. Am. Chem. Soc. 1980, 102, 1194.
7. Abraham, F.; Trehoux, J.; Thomas, D. J. Solid State Chem. 1979, 29, 73.
8. Waltersson, K. Acta Crystallogr. 1976, B32, 1485.
9. Schäfer, H.; von Schnering, H.G. Angew. Chem. 1964, 76, 833.
10. Imoto, H.; Corbett, J.D.; Cisar, A. Inorg. Chem. 1981, 20, 145.
11. McCarroll, W.H.; Katz, L.; Ward, R. J. Am. Chem. Soc. 1957, 79, 5410.
12. Ansell, G.B.; Katz, L. Acta Crystallogr. 1966, 21 482.
13. McCarroll, W.H. Inorg. Chem. 1977, 16, 3351.
14. Donohue, P.C.; Katz, L. Nature (London) 1964, 201, 180.
15. Kerner-Czeskleba, H.; Tourne, G. Bull. Soc. Chim. Fr. 1976, 729.
16. Cotton, F.A. Inorg. Chem. 1964, 3, 1217.
17. Torardi, C.C. and McCarley, R.E. J. Solid State Chem. 1981, 37, 393.
18. Torardi, C.C. and McCarley, R.E. to be published.
19. Torardi, C.C. and McCarley, R.E. J. Am. Chem. Soc. 1979, 101, 3963.
20. Simon, A. Angew. Chem. Int. Ed. Engl. 1981, 20, 1.
21. Corbett, J.D. Acc. Chem. Res. 1981, 14, 239.
22. McCarley, R.E.; Ryan, T.R.; Torardi, C.C. ACS Symp Ser. 1981, 155, 41.
23. Pauling, L. The Nature of the Chemical Bond, 3rd Ed. Cornell University Press, 1960, p.400.
24. Chisholm, M.H.; Huffman, J.C.; Leonelli, J. J. Chem. Soc., Chem. Commun. 1981, 270.
25. Chisholm, M.H.; Huffman, J.C.; Kirkpatrick, C.C.; Leonelli, J.; Folting, K. J. Am. Chem. Soc. 1981, 103, 6093.
26. Cotton, F.A.; Fang, A. J. Am. Chem. Soc. 1982, 104, 113.
27. Brough, L.F.; Carlin, R.T.; McCarley, R.E. Results to be published.
28. Poeppelmeier, K.R.; Corbett, J.D. J. Am. Chem. Soc. 1978, 100, 5039.
29. Mattausch, H.J.; Simon, A.; Eger, R. Rev. Chim. Miner. 1980, 17, 516.
30. Berroth, K.; Simon, A. J. Less-Common Met., to be published.

31 Chevrel, R.; Sergent, M.; Prigent, J. J. Solid State Chem. 1971, 3, 515.
32 Yvon, K. Current Topics Mat. Sci. 1978, 3, 55.
33. Marinder, B.-O. Chem. Scripta 1977, 11, 97.

RECEIVED September 13, 1982

Discussion

J.P. Fackler, Jr., Case Western Reserve University: Bob, I find your study to be most interesting. Has it been possible for you to isolate mixed O,S phases of the Chevrel type by oxygen substitution?

R. E. McCarley: No, we have done very little work along this line, although I think important results will be obtained from studies of mixed O, S phases. So far we have found no direct structural analogies between the reduced ternary molybdenum oxide phases and the various ternary sulfide or selenide phases of the Chevrel type.

A.R. Siedle, 3M Central Research Laboratory: If one of the extended structures described by Professor McCarley were truncated through a low Miller index plane, can one, following the appproach of Solomon, predict what metal orbitals would protrude from the surface so generated? Have ultraviolet photoelectron spectra been obtained on single crystals of any of these materials?

R. E. McCarley: In response to the first question, it should be easy to predict those metal orbitals which would protrude from the 001 and 002 faces of $NaMo_4O_6$, i.e. those faces normal to the direction of chain growth. However, up to the present time we have not considered such questions because of our generally poor understanding of bonding details in these extended systems.

In reference to the latter question, we have obtained UPS data on polycrystalline samples of $NaMo_4O_6$, but not on single crystals. These measurements indicate a relatively high density of states at the Fermi level, which is in agreement with the low resistivity and metallic character of the material.

19

Low-Valent Dimers of Tantalum and Tungsten

A. P. SATTELBERGER

University of Michigan, Department of Chemistry, Ann Arbor, MI 48104

I am not going to discuss early transition metal oxide or alkoxide chemistry here. Instead I would like to use the few minutes allotted me to describe our work with low valent dimers of tantalum and tungsten. Professor McCarley will no doubt see his influence on our work and forgive this digression from the topic of his excellent presentation.

The synthesis of the first quadruply bonded tungsten(II) carboxylate, $W_2(O_2CCF_3)_4$ (hereafter referred to as $W_2(TFA)_4$), was reported by us last year (1). The structure of its diglyme adduct, $W_2(TFA)_4 \cdot 2/3$ diglyme is shown in Figure 1. This particular view shows the partial contents of two unit cells and it emphasizes the tridentate nature of the axially coordinated polyether which "stiches" $W_2(TFA)_4$ units together in the solid state. We now have in hand several other adducts of $W_2(TFA)_4$ and one of these, $W_2(TFA)_4 \cdot 2PPh_3$ (an axial adduct), has been structurally characterized (2). Despite many attempts, X-ray quality crystals of unsolvated $W_2(TFA)_4$ have not been obtained.

The successful synthesis of $W_2(TFA)_4$ was the culmination of two years of hard work in our laboratory. In retrospect the synthesis now appears trivial (reaction 1); all of our more grandiose approaches failed. We have further streamlined reaction 1 by

$$W_2Cl_6(THF)_4 + 2Na/Hg + 4NaO_2CCF_3 \xrightarrow{THF} W_2(TFA)_4 + 6NaCl \qquad (1)$$

using polymeric WCl_4 as the starting material and $W_2(TFA)_4$ is usually obtained in $\sim$65% yield from a "one-pot" synthesis. Our efforts to extend this method to other carboxylates have only been successful in the case of $W_2(O_2CCMe_3)_4$. Sodium pivalate, like sodium trifluoroacetate, is slightly soluble in tetrahydrofuran and this factor appears to be crucial to the success of reaction 1. No quadruply bonded products were isolated in reactions with sodium acetate or sodium benzoate, both virtually insoluble in THF.

0097-6156/83/0211-0291$06.00/0

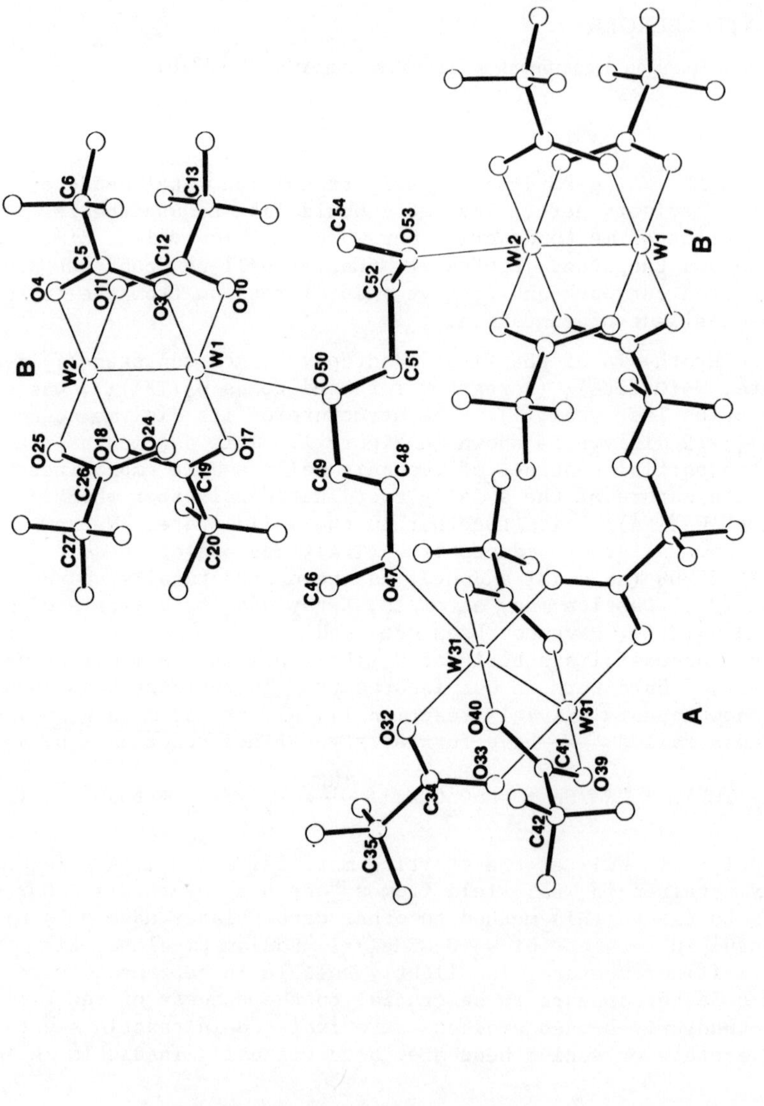

Figure 1. Molecular structure of $W_2(O_2CCF_3)_4 \cdot 2/3$ diglyme. Molecules A and B are in the same unit cell as the polyether. Molecule B′ is related to B by a unit translation. The symmetry related partner of B is not shown (1).

With $W_2(TFA)_4$ in hand and $Mo_2(TFA)_4$ already known and well characterized (3), we began to pursue comparative spectroscopic studies. The results of our IR and Raman experiments have already been presented in preliminary form (1). We have also measured the gas phase photoelectron spectra in collaboration with Professor G. M. Bancroft, Department of Chemistry, University of Western Ontario (4). The HeI spectra are shown in Figure 2. The spectrum of $Mo_2(TFA)_4$ (top) is in good agreement with that published previously by Green and coworkers (5). We also find two peaks below the onset of ligand ionization (ca. 12 eV). The peak at 8.76 has traditionally been assigned as the δ ionization (5,6). The assignment of the 10.46 eV band is another matter. SCF-Xα-SW results favor assigning this band as the π ionization (7), whereas ab initio calculations suggest that this band be assigned to overlapping π and σ ionizations (8,9,10). Cotton has documented all of the objections to this latter assignment (6). Now we turn to the $W_2(TFA)_4$ spectrum. Here we find three distinct bands in the region below 12 eV. These are at considerably lower binding energy than in the molybdenum dimer, a feature not inconsistent with the greater reactivity (e.g., toward O_2 or HCl) of $W_2(TFA)_4$ relative to $Mo_2(TFA)_4$ (1). We have assigned the lowest energy band at 7.39 eV as the δ ionization. The remaining two bands (9.01 and ca. 9.76 eV) are separated by 0.75 eV and this presents an assignment problem. It might be tempting to assign these as the spin-orbit components of the ^2E state of $W_2(TFA)_4^+$. However, 0.75 eV is outside the range expected from the spin-orbit interaction, ca. 0.3 to 0.6 eV (11) and we consider it unlikely, even taking 0.75 eV as an acceptable spin-orbit splitting, that the photoionization cross sections, width and structure of the two levels would be so different. We are therefore inclined to assign the bands at 9.01 and 9.76 eV as the π and σ ionizations, respectively. The opposite interpretation has been invoked to explain the PES spectrum of $W_2Cl_4(PMe_3)_4$ (12). Here the separation between the second and third bands in the spectrum in 0.4 eV, i.e., within the range expected from tungsten spin-orbit splitting but the band shapes are anomolous even considering mixing with the σ and δ levels.

Now I would like to leave tungsten chemistry and describe a few of our results in low valent tantalum chemistry. Our entry into this area began with the synthesis (reaction 2) and structural characterization (Figure 3) of the tantalum(III) dimer,

$$TaCl_5 + 2Na/Hg + 2PMe_3 \xrightarrow{PhCH_3} 0.5[TaCl_2(PMe_3)_2]_2(\mu\text{-}Cl)_2 + 2NaCl \qquad (2)$$

$[TaCl_2(PMe_3)_2]_2(\mu\text{-}Cl)_2$ (13). The short metal-metal bond length of 2.710(2)Å has been interpreted as a formal metal-metal double bond on the basis of molecular orbital arguments (13,14). An unusual feature of this complex is that it reacts with molecular

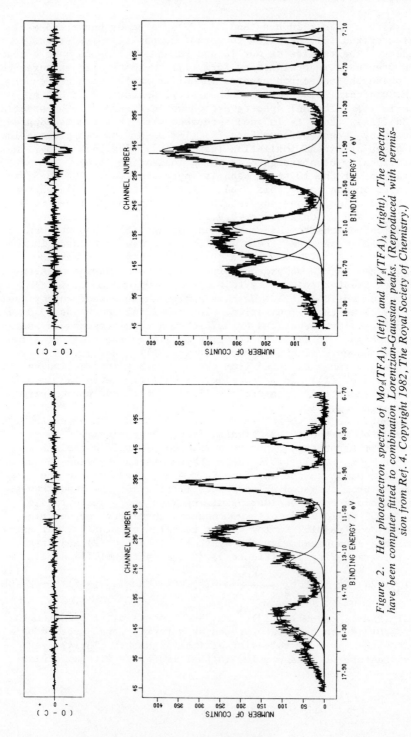

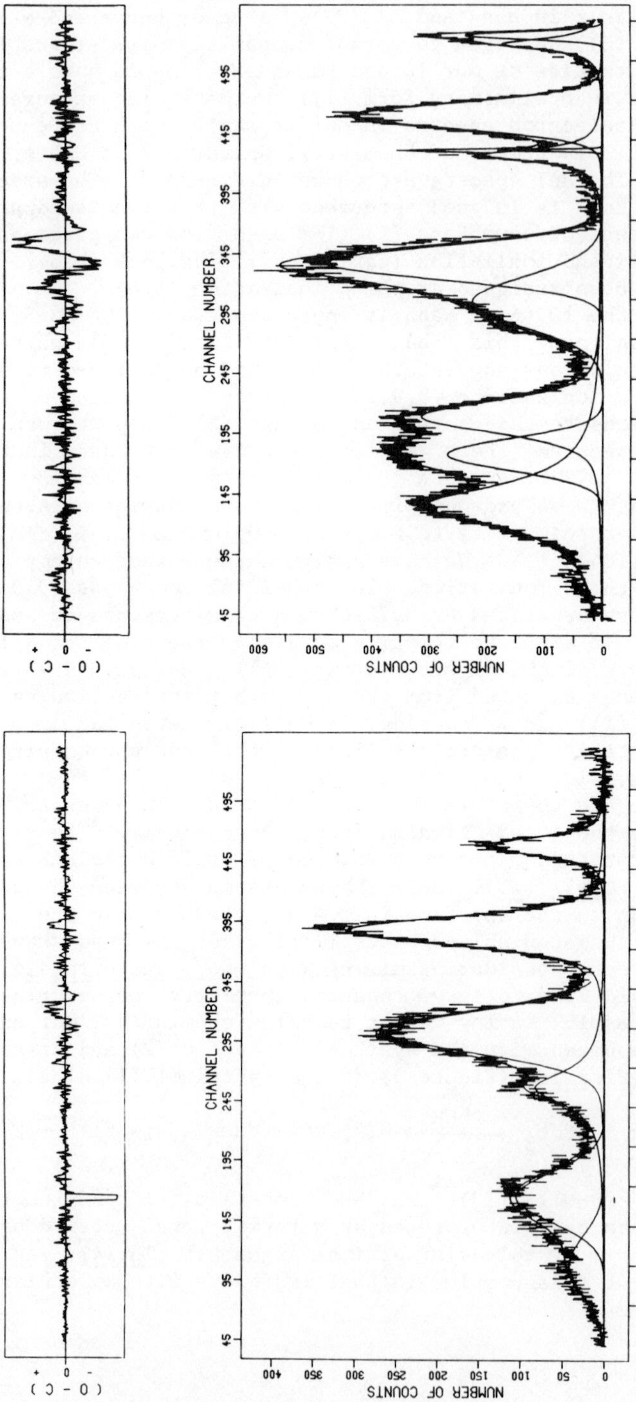

Figure 2. HeI photoelectron spectra of $Mo_2(TFA)_4$ (left) and $W_2(TFA)_4$ (right). The spectra have been computer fitted to combination Lorentzian-Gaussian peaks. (Reproduced with permission from Ref. 4. Copyright 1982, The Royal Society of Chemistry.)

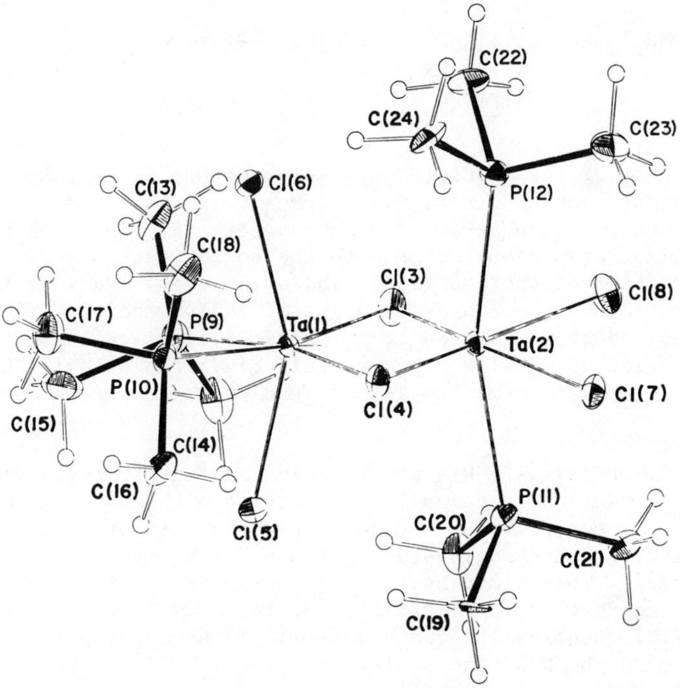

Figure 3. Molecular structure of $[TaCl_2(PMe_3)_2]_2(\mu\text{-}Cl)_2$ (13, 14).

hydrogen under very mild conditions (1 atm H_2, 25°C) to form the quadruply bridged tantalum(IV) dimer, $[TaCl_2(PMe_3)_2]_2(\mu\text{-Cl})_2(\mu\text{-H})_2$ (13,15). A view of this dimer looking down the Ta-Ta axis is shown in Figure 4. The bridging hydrogens were not located in the final difference Fourier of this X-ray structure but a battery of spectroscopic techniques (^1H and ^{31}P NMR and IR) places them in the cavity below the bridging chlorides (15). The $(\mu\text{-Cl})_2(\mu\text{-H})_2$ dimer offers a convenient entry back into tantalum(III) dimer chemistry (reaction 3) and the product, $[TaCl_2(PMe_3)_2]_2(\mu\text{-H})_2$, has been characterized by X-ray crystallography (16). This time we were fortunate enough to locate the bridging hydrides (Figure 5).

$$[TaCl_2(PMe_3)_2]_2(\mu\text{-Cl})_2(\mu\text{-H})_2 + 2Na/Hg \xrightarrow{\text{glyme}} [TaCl_2(PMe_3)_2]_2(\mu\text{-H}_2) + 2NaCl \qquad (3)$$

The geometry of the $Ta_2Cl_4(PMe_3)_4$ substructure is reminiscent of the quadruply bonded dimers $Mo_2Cl_4(PMe_3)_4$ and $W_2Cl_4(PMe_3)_4$ (17). The room temperature ^1H and ^{31}P NMR spectra (Figure 6) of the $(\mu\text{-H})_2$ dimer are inconsistent with the solid state structure. Rapid rotation of the end groups and/or bridging hydrides is required to account for apparent magnetic equivalencies. The molecule does, however, have built in pseudo-cylindrical symmetry, i.e., one set of metal π-type orbitals binds the bridging hydrides and the other set forms the π-component of the metal-metal double bond.

$[TaCl_2(PMe_3)_2]_2(\mu\text{-H})_2$ reacts with many substrates. Time permits me to mention only one - the reaction with molecular hydrogen. The latter provides the quadruply hydrogen bridged tantalum (IV) dimer, $[TaCl_2(PMe_3)_2]_2(\mu\text{-H})_4$ whose X-ray structure is shown in Figure 7. This dimer has crystallographically imposed D_{2d} symmetry and a very short (2.511(2)Å) tantalum-tantalum bond length (18). Unlike its tantalum(III) precursor, the $(\mu\text{-H})_4$ dimer is static on the NMR time scale (Figure 8). This is in keeping with the predictions of Hoffmann and coworkers on $(\mu\text{-H})_4$ dimers (19). All of the metal π-type orbitals are engaged in Ta-H-Ta bonding and a substantial rotational barrier arises from the overlap of orbitals of δ symmetry. This combination is shown below.

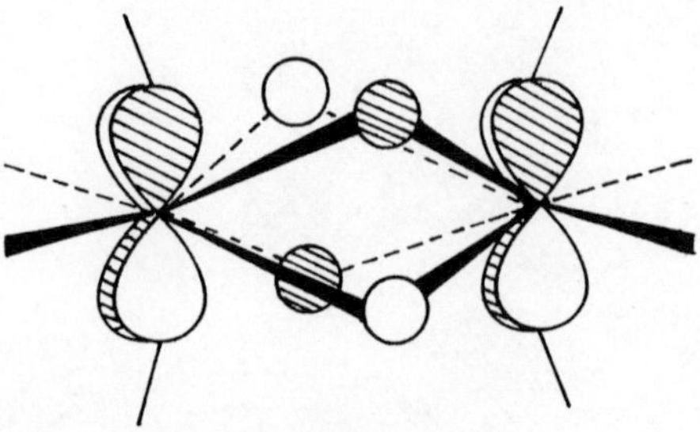

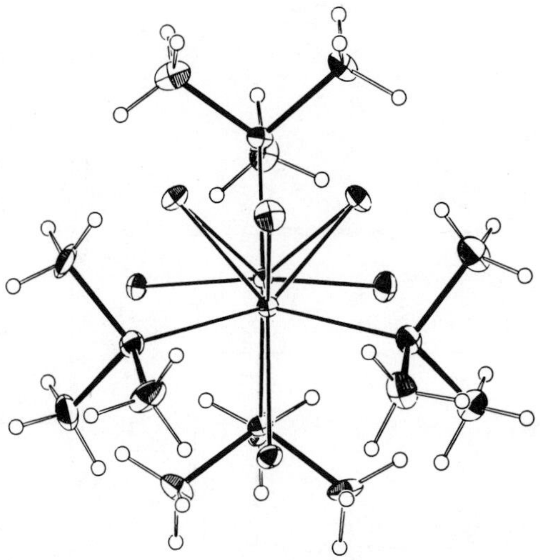

Figure 4. Molecular structure of $[TaCl_2(PMe_3)_2]_2(\mu\text{-}H)_2(\mu\text{-}Cl)_2$ (13, 15).

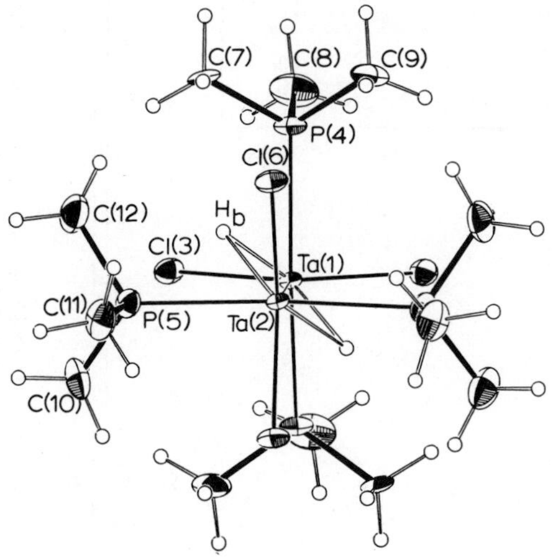

Figure 5. Molecular structure of $[TaCl_2(PMe_3)_2]_2(\mu\text{-}H)_2$ (16).

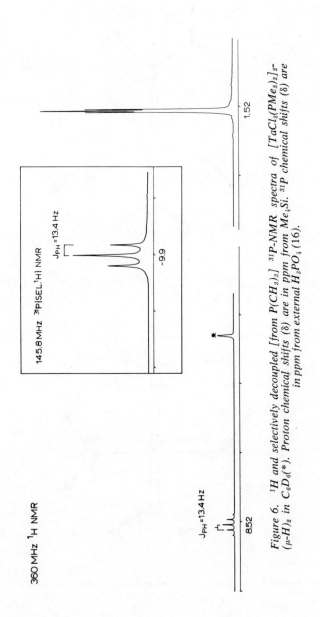

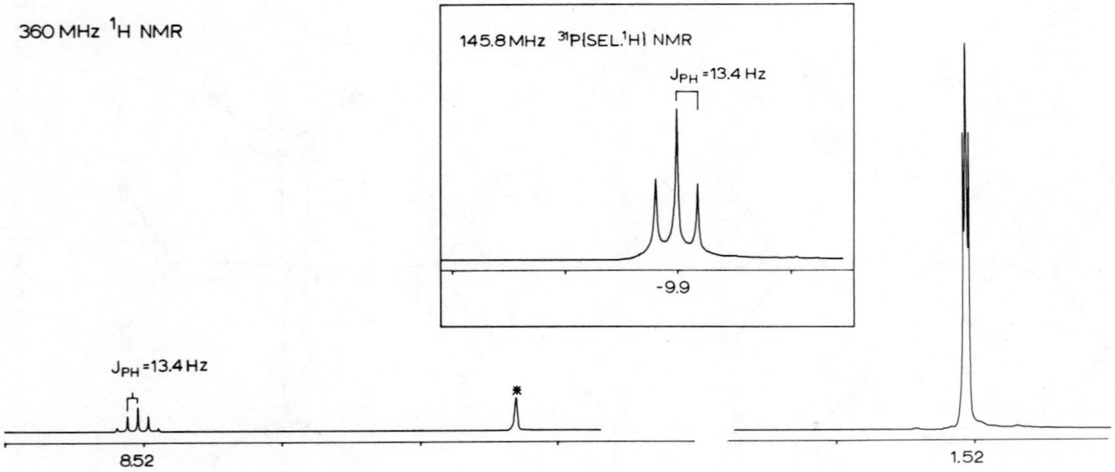

Figure 6. *^{1}H and selectively decoupled [from $P(CH_3)_3$] ^{31}P-NMR spectra of $[TaCl_2(PMe_3)_2]_2$-$(\mu$-$H)_2$ in C_6D_6(*). Proton chemical shifts (δ) are in ppm from Me_4Si. ^{31}P chemical shifts (δ) are in ppm from external H_3PO_4 (16).*

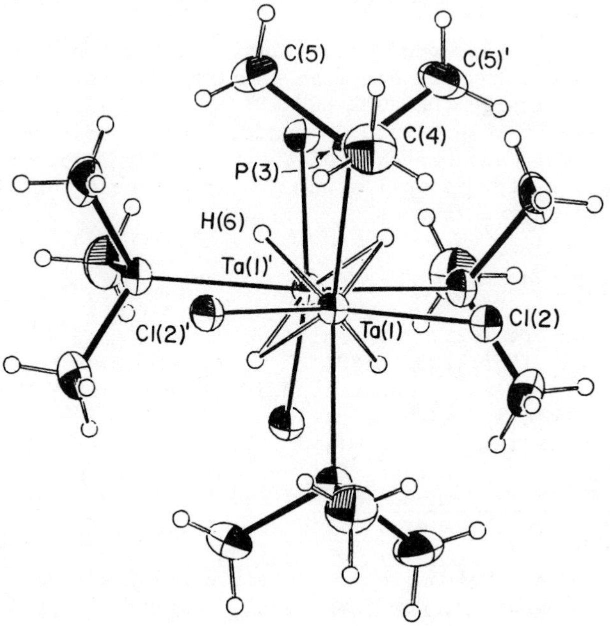

Figure 7. Molecular structure of $[TaCl_2(PMe_3)_2]_2(\mu\text{-}H)_4$ (18).

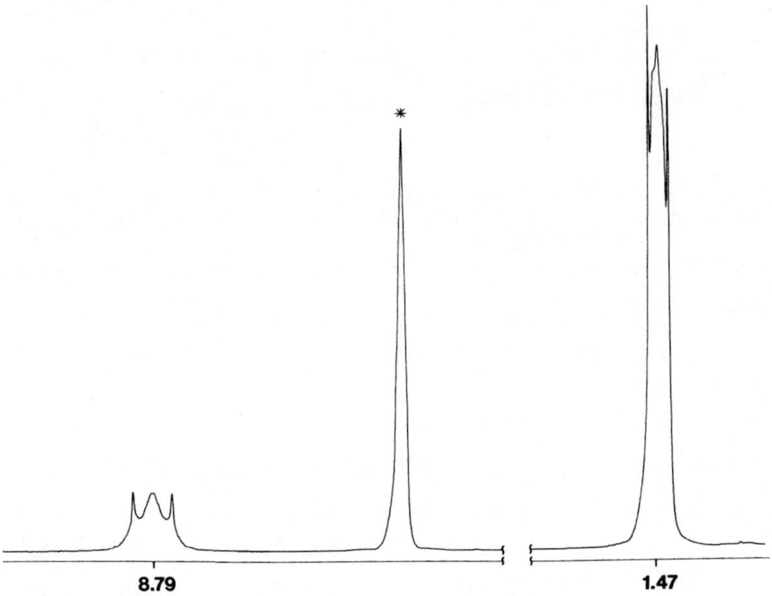

Figure 8. 360 MHz ^{1}H-NMR spectrum of $[TaCl_2(PMe_3)_2]_2(\mu\text{-}H)_4$ in C_6D_6(*). Chemical shifts (δ) are in ppm from Me_4Si. The amplitude of the δ8.79 signal (4 bridging hydrides) is twice that of the δ1.47 resonance (36 PMe_3 protons) (18).

Acknowledgments

I thank the Research Corporation, the Petroleum Research Fund administered by the American Chemical Society and the U.S. Department of Energy (DE-FG02-80ER10125) for their financial support. I also acknowledge my fruitful collaborations with Dr. John C. Huffman, Director of the Molecular Structure Center, Indiana University, as well as my own group whose contributions are cited in the references.

Literature Cited

1. Sattelberger, A.P.; McLaughlin, K.W.; Huffman, J.C. J. Am. Chem. Soc. 1981, 103, 2880.
2. Sattelberger, A.P., Santure, D.J.; McLaughlin, K.W.; Huffman, J.C., manuscript in preparation.
3. Cotton, F.A.; Norman, J.G., Jr. J. Coord. Chem. 1971, 1, 161.
4. Bancroft, G.M.; Pellach, E.; Sattelberger, A.P., McLaughlin, K.W. JCS Chem. Commun. 1982, xxxx.
5. Coleman, A.W.; Green, J.C.; Hayes, A.J.; Seddon, E.A.; Lloyd, D.R.; Niwa, Y. J. Chem. Soc. Dalton 1979, 1057.
6. Cotton, F.A.; Walton, R.A. "Multiple Bonds Between Metal Atoms" John Wiley, New York, 1982, Chapter 8, pp. 415-425.
7. Norman, J.C., Jr.; Kolari, H.J.; Gray, H.B.; Trogler, W.C. Inorg. Chem. 1977, 16, 987.
8. Benard, M. J. Am. Chem. Soc. 1978, 100, 2354.
9. Guest, M.F.; Hillier, I.H.; Garner, C.D. Chem. Phys. Lett. 1977, 48, 587.
10. Guest, M.F.; Garner, C.D.; Hillier, I.H.; Walton, I.B. J. Chem. Soc. Faraday II 1978, 74, 2092.
11. Bursten, B.E.; Cotton, F.A.; Cowley, A.H.; Hanson, B.E.; Lattman, M.; Stanley, G.G. J. Am. Chem. Soc. 1979, 101, 6244.
12. Cotton, F.A.; Hubbard, J.L.; Lichtenberger, D.L.; Shim, I. J. Am. Chem. Soc. 1982, 104, 679.
13. Sattelberger, A.P., Wilson, R.B., Jr., Huffman, J.C. J. Am. Chem. Soc. 1980, 102, 7911.
14. Sattelberger, A.P.; Wilson, R.B., Jr.; Huffman, J.C. Inorg. Chem. 1982, 21, 2392.
15. Sattelberger, A.P.; Wilson, R.B., Jr.; Huffman, J. C. Inorg. Chem. 1982, 21, xxxx.
16. Wilson, R.B., Jr.; Sattelberger, A.P.; Huffman, J.C. J. Am. Chem. Soc. 1982, 104, 858.
17. Cotton, F.A.; Extine, M.W.; Felthouse, T.R.; Kolthammer, B.W.S.; Lay, D.G. J. Am. Chem. Soc. 1981, 103, 4040.
18. Sattelberger, A.P.; Luetkens, M.L., Jr.; Wilson, R.B., Jr.; Huffman, J.C., manuscript in preparation.
19. Dedieu, A.; Albright, T.A.; Hoffman, R. J. Am. Chem. Soc. 1979, 101, 3141.

RECEIVED October 18, 1982

Discussion

I would like to make a few comments regarding the photoelectron spectra presented by Dr. Sattelberger. In referring to his spectrum of $W_2(O_2CCF_3)_4$, he has assigned the ionization band near 10 eV to be associated with removal of an electron from the sigma-bonding orbital between the two metal centers. This assignment is made primarily on the basis that the splitting of this band from the band near 9 eV, and the widely different cross-sections for these ionizations, are not consistent with spin-orbit effects on the predominantly metal pi bond ionization. The occurrence of the sigma ionization in this region would have some important implications. However, there is considerable information, both in the spectra of the compounds presented by Dr. Sattelberger, and from other related work, that indicate assignment of this band to the sigma ionization is not warranted. I will restrict my comments to the compounds discussed by Dr. Sattelberger.

First, inspection of the low energy ionization bands presented by Dr. Sattelberger for $W_2(O_2CCF_3)_4$ shows that the band near 10 eV is extremely narrow. If this band is assigned to the sigma bond ionization, then it follows that the change in metal-metal bond distance with removal of an electron from the sigma bond is very small and is much less than the change that occurs with removal of an electron from the delta bond. This point is related to the comments I made in my contribution following Professor Cotton's talk. A narrow band and negligible change in bond distance for a sigma ionization is unprecedented and would be quite remarkable.

Second, spin-orbit effects in a dimetal system must be treated rigorously before stating what is or is not reasonable for the ionization structure. One component of the spin-orbit split pi ionization will mix with the sigma framework of the molecule, while the other component will mix with the delta framework. These components will have different relaxation energies with ionization, different cross-sections for ionization, and different displacement of positive ion potential wells above the ground state of the molecule. A prediction of the magnitude of these differences is a major calculational challenge. In view of the present limit of our knowledge of the spin-orbit effects, interpretation of these ionizations as originating from the pi-bond electrons cannot be considered unreasonable.

Finally, the implication is made that the sigma ionization is also in the region of the pi ionization for the molybdenum dimers. We have recently carried out an examination of the pi ionization of $Mo_2(O_2CCH_3)_4$ in which we observe clear vibrational fine structure across the band. This observation is not expected if two different ionizations, the sigma and pi, are overlapped in the same envelope. Thus, I do not see any evidence at this stage that the sigma ionization is in the region of the pi ionization for these complexes.

20

Dynamic Processes of Metal Atoms and Small Metal Clusters in Solid Supports

GEOFFREY A. OZIN and S. A. MITCHELL

University of Toronto, Lash Miller Chemistry Laboratories,
Toronto, Ontario M5S 1A1 Canada

Major developments in the field of metal vapor synthesis and matrix isolation spectroscopy involving ground state metal atomic and cluster reagents over the past decade are briefly contemplated. A new direction which focuses attention on the chemistry and spectroscopy of these reagents in selected excited electronic states is the subject of this paper. In particular, matrix cage relaxation dynamics, following uv and visible excitation of atomic and small cluster guests of copper and silver in rare gas lattices will be examined. Reactivity patterns of electronically excited metal atomic reagents with methane and dioxygen will also be briefly described. Metal-support effects involving both ground and excited electronic states of the metal guest-cage unit will feature prominently in these discussions. These interactions play a critical role in determining reactivity patterns and relaxation processes of immobilized metal guests and clearly bear a relationship to metal-support effects known to be important in supported metal catalysts.

As the main theme of this meeting is to assess and consolidate past achievements in various key areas of inorganic/organometallic chemistry, with the objective of gazing deep and hard into the futuristic chemical crystal ball of the 21st century, the purpose of my presentation will be to focus attention on pivotal developments in the field of transition metal atom/metal cluster chemistry over the past decade and then to attempt to project and forecast some of the more promising directions that the area is likely to follow in the years ahead.

If one surveys the exciting growth period of the early seventies one cannot help but notice the natural but constrained subdivision of the field of metal vapor (MV) chemistry into a macroscale synthetic school, conducting experiments usually at 77-300K and a matrix scale spectroscopic school, working in the lower

0097-6156/83/0211-0303$07.25/0

temperature range of 4.2-300K (1). Recently, the two method ologies have been successfully merged to the great benefit of both. Currently, both solid state cocodensation and solution phase MV experiments of a combined synthetic/spectroscopic type are performed on a routine basis in our laboratory (2). Moreover, matrix and macroscale MVS equipment is now commercially available (3). These early studies were aimed at expoliting the often unique chemical reactivity of naked metal atoms, molecular metal clusters and colloidal metal particles in their ground electronic states, with each other to form well defined metal aggregates, as well as with organic ligands to produce novel reactive intermediates and products with varying degrees of stability over the cryogenic temperature to room temperature range.

The field of ground state MVS enjoyed profusive growth in the seventies and attracted a multidisciplinary audience of scientists. This was because of the wide ranging ramifications of the results in fields as diverse as nucleation theory, photographic and xerographic science, cluster and chemisorption model theory, catalysis by supported metal clusters, organometallic synthesis, homogeneous catalysis, organometal polymers, to name but a few (4).

Along with the elegant chemical achievements and impressive instrumental design, one also witnessed a rapid expansion in spectroscopic and kinetic techniques for characterizing matrix entrapped atomic/cluster metal vapor reaction products. Noteable amongst these pioneering experiments are SIMS (Michl (5)), MCD (Grinter(8), Schatz (7)), UPS (Jacobi (8)), EXAFS (Montano (9)), NMR (Michl (10)), ESR (Weltner, (11), Kasai (12), Lindsay (13)), Mössbauer (Barrett, Montano (14)), optical absorption and fluorescence (Kolb (15), Ozin (16)), Laser fluorescence, Raman, Resonance Raman (Bondybey (17), Nixon (18) Schulze (19), Gruen (20), Moskovits (21)).

Molecular electronic structure calculations of naked metal clusters, with nuclearities spanning the range from atom to bulk became the focus of intense scientific interest in the 70's (22). Experimental techniques, on the other hand, for fabricating and spectroscopically probing specific clusters developed more slowly. Presently, molecular beam and matrix isolation methods have emerged as the premier approaches for studying ligand-free metal clusters in the gas and condensed phases respectively. Each method displays advantages and limitations.

In the ideal collision free environment of a molecular beam, the properties of a metal cluster can be considered to be truely isolated from cluster-substrate effects. Therefore, spectroscopic methods that can selectively extract information from metal cluster beams hold great promise for illuminating diverse size dependent properties of aggregates of metal atoms in their equilibrium configuration (23).

Condensation of metal atoms and/or metal clusters, with or in rare gases at cryogenic temperatures, from either jets or beams represents the other main approach for observing nucleation phenomena from isolated metal atoms to bulk metal aggregates in a weakly interacting solid support (4,24). In this way, metal dispersion, thermal and photolytic behaviour of particular metal clusters may be observed. The spectra obtained are often of low resolution and observations reflect not only the properties of the guest but also their interaction with the host (25). The matrix method appears to be able to provide individual cluster properties up to about six atoms (4,24,26). Above this size, spectral overlap problems between different species in the matrix usually preclude an unambiguous determination of cluster nuclearity. Nevertheless, the important transition, atom to molecule to bulk and quantum size effects in small metal aggregates can be studied in this way (26,27).

Because the size regime of n=1-6 atoms is of great practical significance to the spectroscopic, chemical and catalytic properties of supported metal clusters in both weakly and strongly interacting environments (28), it is important to study very small metal clusters in various types of substrate as well as in the gas phase. In this way, one can hope to develop a scale of metal cluster-support effects (guest-host interactions) and evaluate the role that they play in diverse technological phenomena.

In the past few years, the field of metal atom/metal cluster chemistry has taken an interesting turn out of the realm of the ground electronic state into the world of the excited state. This promises to be an intriguing new phase and one with considerable potential for new scientific discovery and technological development.

The original swing of emphasis away from ground state metal atom/metal cluster chemistry can be traced to investigations of the photoprocesses of these reagents embedded in weakly interacting supports, mainly of the rare gas variety (24). Studies of this kind revealed that the interactions between the excited states but not the ground states of certain matrix entrapped metal guests with the surrounding cage, were not quite as innocent as might have been initially anticipated. In fact, these metal support effects were of sufficient magnitude to result in a range of hithertofore unobserved matrix relaxation processes, including metal atom photodiffusion and aggregation (30), metal cluster photofragmentation/fluorescence/cage-recombination (31), and metal cluster photoisomerization (32). These photoeffects were found to be exceptionally sensitive to the nature of the support and unveiled the existence and operation of surprisingly strong guest-host interactions.

It was as a result of investigations of the aforementioned kind that a new kind of excited state metal atom/metal cluster photoprocess was discovered, involving chemical reaction with the support itself (33). A prerequisit for the successful exploitation of this novel kind of chemistry, is a weakly interacting metal atom/metal cluster - cage ground electronic state. Only in this situation is it possible to prepare the metal reagents in the desired electronic state for investigation of their excited state reactivity patterns. This phase of development is very recent and may be considered to have opened up a new era of metal vapor chemistry, certainly worthy of consideration by chemists of the 21st century.

Early indicators emerging from studies of the excited state chemistry of metal atomic and cluster reagents point to an interesting and bright future for the field. Highlights over the past year include excited state metal atom agglomeration reactions (34), selective photoinsertion processes of metal atoms into the CH bonds of paraffinic hydrocarbons (33,35), the splitting of dihydrogen (36), dioxygen (37), water and ammonia (38) by photoexcited metal atoms, and photoinduced electron-transfer processes (37). A taste of the intriguing opportunities that this new field has to offer, can be appreciated from the recently discovered selective, room temperature, atmospheric pressure, photoheterogeneous dimerization of paraffins to higher saturated alkanes on a metal zeolite (39), an interesting new avenue in the quest for transforming natural gas ultimately to gasoline hydrocarbons.

For the remainder of this presentation, I will concentrate on some of the more interesting photophysical and photochemical properties recently observed for metal atomic and small cluster reagents entrapped in non-reactive and reactive solid substrates. Attention will be directed to matrix-cage relaxation energetics and dynamics following uv and visible photoexcitation of atomic and small cluster guests of copper and silver in rare gas, methane and dioxygen lattices. Metal-support interactions involving both ground and excited electronic states of the metal guest-cage unit will feature prominently throughout this discussion as they play a critical role in controlling the observed reactivity patterns and relaxation processes of the entrapped metal guests.

Results and Discussion

The major absorption and emission processes of gaseous Cu and Ag atoms in the uv-visible range are rather straightforwardly interpreted in terms of electronic transitions between an isotropic $nd^{10}(n+1)s^1, {}^2S_{1/2}$ ground state and $nd^{10}(n+1)p^1, {}^2P_{1/2,3/2}$ and $nd^9(n+1)s^2, {}^2D_{3/2,5/2}$ excited states (40). Only the ${}^2S_{1/2} \rightarrow {}^2P_{1/2,3/2}$ transitions are both parity and spin allowed although the ${}^2D_{5/2} \rightarrow {}^2S_{1/2}$ transition is seen in the emission spectra of Ag vapor (41). The ${}^2P_{1/2,3/2}$ terms are almost coincident for Cu and Ag atoms. Furthermore, the ${}^2D_{3/2,5/2}$ term of silver is almost equienergetic with the ${}^2P_{1/2,3/2}$ term, whereas this term for Cu lies well below that of the ${}^2P_{1/2,3/2}$ term. The spin orbit coupling constant for Ag($920.7 cm^{-1}$) is considerably larger than for Cu($248.4 cm^{-1}$).

When these atomic species are embedded in a solid rare gas, dramatic alterations in their spectroscopic and photolytic properties ensue which can be traced to specific guest-host interactions operating in their ground and excited states (30,31,34). These perturbations can be described in terms of attractive van der Waals interactions and short range repulsive forces due to overlap between charge clouds on the metal and matrix atoms (42). It is evident that the potential energy of interaction between a metal atom and a support will in general be different for distinct electronic states of the metal atom, since the polarizability, radial extent and symmetry of the electronic charge distribution will generally vary from one elctronic state to another.

One of the objectives of this paper is to evaluate the spectroscopic and photochemical consequences of the occurrence of markedly disparate guest-host interactions in the ground and optically excited states of Cu and Ag atoms, and some of their low nuclearity clusters, in rare gas as well as other supports. Original papers should be consulted for details.

In all discussions of metal support effects, whether weak or strong, a knowledge of the local structure and symmetry of the metal site is mandatory.For the solid rare gases, the available spectroscopic (43) and theoretical (44) evidence leans heavily towards substitutional incorporation into a tetradecahedral site with O_h symmetry, as the most probable location for entrapped Cu, Ag and Au atoms. In principle,matrix EXAFS measurements could place this proposal on a concrete footing. Comparison of the estimated van der Waals diameters of Cu, Ag, Au atoms (4.40,4.70, 4.60Å respectively) with the known substitutional site diameters for Ar, Kr, and Xe solids (3.83, 4.05, 4.41Å respectively) suggest that this would be a reasonable trapping site arrangement especially for Cu atoms in Xe, although a rather large mismatch in size would occur for Ag atoms in Ar matrices. It is expected that these comparisons are useful only as rough qualitative guidelines.

Atomic Silver in Rare Gas Supports

Consider the absorption spectra for atomically dispersed Ag in Ar, Kr, Xe (30) (Figure 1). Instead of displaying the gas phase spin-orbit doublet pattern with its characteristic 1:2 intensity ratio for the $^2S_{1/2}\rightarrow{}^2P_{1/2}$,$^2S_{1/2}\rightarrow{}^2P_{3/2}$ transitions, one observes three well resolved components (major blue site) with roughly equal intensities, with the splitting between the low energy component and the mean of the two high energy components being close to the gas phase spin-orbit splitting energy. Excitation of any one of these three lines results in rapid bleaching at equal rates with concurrent growth of Ag_n cluster absorptions straddling the Ag atomic bands (29,30). This phenomenon of photoaggregation has been found to be a sensitive function of the metal dispersion, nature and temperature of the support. During the photolysis of these Ag atom-rare gas films one cannot help but notice intense visible fluorescence (Figure 2). A time

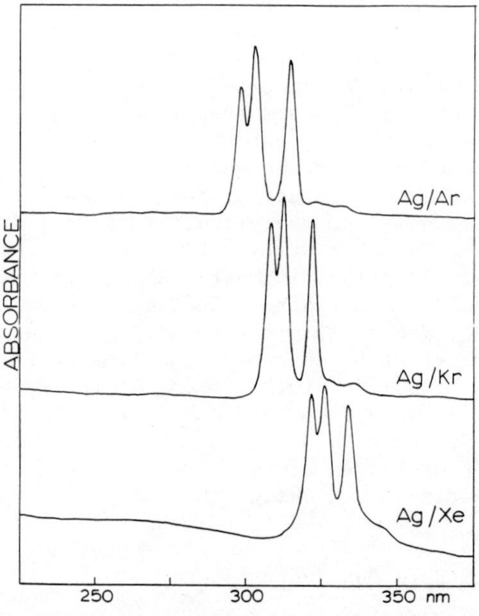

Figure 1. Optical absorption spectra of well-isolated Ag atoms in Ar, Kr, and Xe matrices (Ag/inert gas ~1/10[5]) at 10–12 K. (Reproduced from Ref. 30. Copyright 1980, American Chemical Society.)

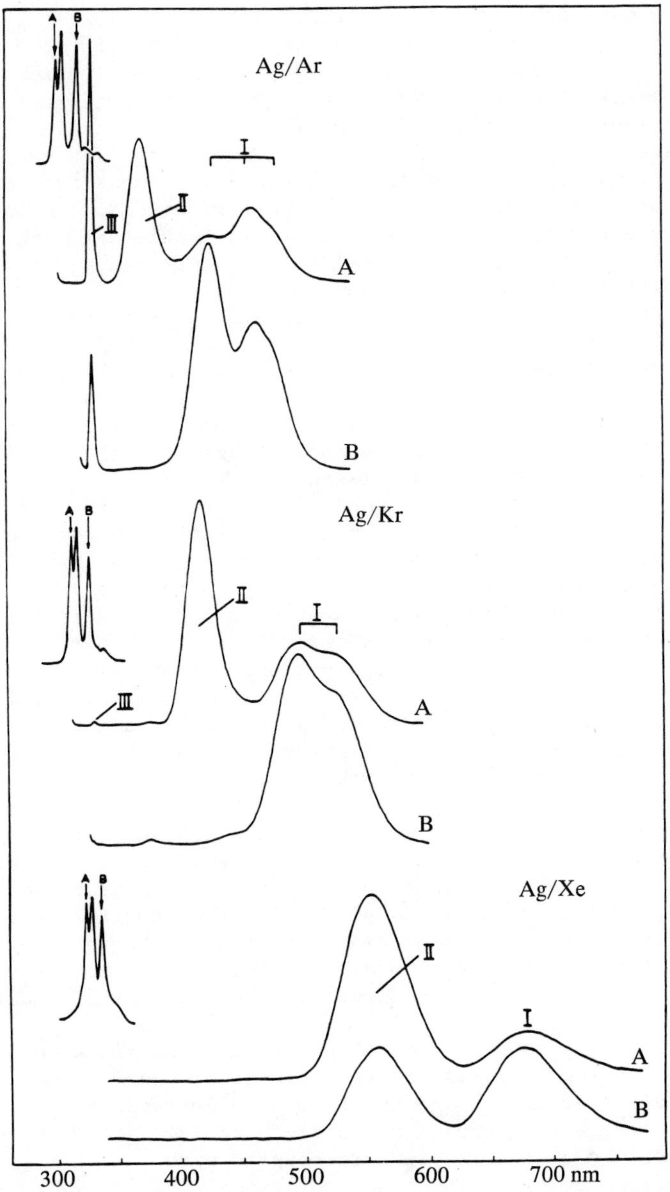

Figure 2. Comparison of the major emission bands of Ag atoms isolated in Ar, Kr, and Xe matrices. The ordinate represents emission intensity in arbitrary units, and the corresponding excitations are illustrated on the absorption spectra at the left. (Reproduced from Ref. 30. Copyright 1980, American Chemical Society.)

resolved study of the two main emissions for Ag/Xe using a pulsed N_2 laser, yielded corrected radiative lifetimes (∿12ns) within a factor of two of that of the $^2P_{3/2}$ level of free silver atoms (6.5±0.6ns at 328nm)(45). This suggests that the fluorescent states in question decay exclusively by the observed radiative pathways since lifetimes significantly shorter than the free atom values would be expected if concurrent non-radiative decay processes were important. The measurements also demonstrate that both fluorescent states were formed well within nano seconds following excitation of the silver atoms, consistent with the picture of a fast vibrational relaxation process and a relatively long radiative lifetime such that emission occurs after the matrix cage has attained its fully relaxed configuration.

This raises the issue of the occurrence of shifts between the matrix and the gas phase electronic transition energies. These can be understood in terms of differences between the interatomic potentials which apply for ground and electronically excited states of the metal atom in its matrix cage. If the net interaction energy between the metal atom and the matrix is not identical in the two states involved in an electronic transition then a matrix shift in the transition energy results as seen for Ag in Ar, Kr, and Xe. The interaction energy for the excited state of the metal atom refers in this context to the interaction where the configuration of the rare gas atoms about the metal atom is unchanged between the ground and excited states, that is the lattice is "frozen" during the electronic transition. Since the interatomic potentials change as a result of the electronic transition, there will be a tendency for the rare gas atoms to relax to a new equilibrium configuration about the excited metal atom. If the guest-host interactions are appreciably different between the ground and excited states, then the situation immediately following absorption is one in which the matrix cage is in a state of high vibrational potential energy. In the absence of very fast non-radiative electronic relaxation processes the system would tend to vibrationally relax and thus the matrix cage would attain the equilibrium configuration appropriate for the excited state of the metal atom. In this way part of the electronic energy is channelled directly into vibrational excitation of the host lattice.

It is useful to view optical absorption and emission processes in such a system in terms of transitions between distinct vibrational levels of the ground and excited electronic states of a metal atom-rare gas complex or quasi-molecule. Since the vibrational motions of the complex are coupled with the bulk lattice vibrations, a complicated pattern of closely spaced vibrational levels is involved and this results in the appearance of a smooth, structureless absorption profile (25). Thus the homogeneous width of the absorption band arises from a coupling between the electronic states of the metal atom and the host lattice vibrations, which is induced by the differences between the guest-host interactions in the ground and excited states of the metal atom. Inhomogeneous contributions to the band width most likely arise from the occurrence of a distribution of slightly different matrix trapping sites for the metal atom.

The fluorescence profiles of Ag in say Kr, basically shows two major emission bands around 415 and 510 nm which can be traced through their excitation dependence to originate from the three $^2S \rightarrow ^2P$ atomic components. Specifically, the lowest energy absroption band gives rise to only the lowest energy emission system, whereas the two higher energy absorption bands give rise to both emission band systems.

The observation of large Stokes shifts for the $^2P-^2S$ transition of entrapped Ag atoms indicates that the guest-host interactions are markedly different for the ^2S and ^2P states of this system and can be explained in terms of matrix cage relaxation effects.

An interesting way to visualize the origin of these shifts is to consider the ways in which twelve rare gas atoms in the symmetry of a substitutional site can approach a silver atom in its ground and excited states and to evaluate the van der Waals attractive forces and repulsive interactions that contribute to the binding energy of the cage-complex (46). It is well known that one electron wave functions for the $|j,m_j>$ spin-orbit levels arising from a np^1 configuration can be written in the following form (47):

$$|3/2, \pm 3/2> = (1/2)^{1/2} (p_x \pm ip_y)|\pm 1/2>$$
$$|3/2, \pm 1/2> = (2/3)^{1/2} p_z|\pm 1/2> + (1/6)^{1/2} (p_x \pm ip_y)|\mp 1/2>$$
$$|1/2, \pm 1/2> = -(1/3)^{1/2} p_z|\pm 1/2> + (1/3)^{1/2} (p_x \pm ip_y)|\mp 1/2>$$

where P_x, P_y and p_z refer to the real p-orbitals and $|\pm 1/2>$ is the spin function corresponding to $m_s = \pm 1/2$. Here $|3/2, \pm 3/2>$ and $|3/2, \pm 1/2>$ are the degenerate components of $^2P_{3/2}$ and $|1/2, \pm 1/2>$ corresponds to $^2P_{1/2}$, where each state is a Kramers doublet (48). Multiplication of these wave functions by their complex conjugates and integrating over the spin variable yields the following expressions for the associated charge distributions:

$$\rho_{3/2, \pm 3/2} = 1/2(p_x^2 + p_y^2)$$
$$\rho_{3/2, \pm 1/2} = 1/6(p_x^2 + p_y^2) + 2/3\, p_z^2$$
$$\rho_{1/2, \pm 1/2} = 1/3(p_x^2 + p_y^2 + p_z^2)$$

Thus the van der Waals stabilization should be maximized in the $^2P_{3/2} \pm 3/2$ state by allowing the rare gas atoms above and below the xy plane to approach the metal atom more closely than those lying in the xy plane. In terms of the substitutional site, this corresponds to an axial contraction along the z-axis, which should lead to a destabilization of the $^2P_{3/2,\pm 1/2}$ state because of the overlap between the p_z charge density on the metal atom and the charge clouds on the axial rare gas atom. An expansion along the z-axis would reverse this situation and cause the $^2P_{3/2,\pm 1/2}$ state to be stabilized. It can be seen in this way that an axial distortion from O_h symmetry gives rise to a splitting of the $^2P_{3/2}$ level into $^2P_{3/2,\pm 3/2}$ and $^2P_{3/2,\pm 1/2}$ components such that one of these components is stabilized by the distortion. These ideas can be put on a more solid footing by discussing the matrix cage relaxation process in the ^2P excited state of Ag in terms of a Jahn-Teller effect. It can be shown that an axial distortion is required by symmetry consideration. The above simply provides a physical picture of this requirement in terms of van der Waals stabilization of the silver atom-rare gas complex.

The aspect of the Jahn-Teller problem (49) which is of interest in the present context is the form of the nuclear potential energy surface for a ^2P electronic state in the AgX_{12} quasi-molecule. This state transforms as $^2T_{1u}$ in O_h symmetry. Orbital angular momentum is not quenched in a T_{1u} state. If spin-orbit coupling is very strong, as in the case of Ag, then the Jahn-Teller effect is of secondary importance and the $^2T_{1u}$ state splits into $^2G_{3/2u}$ and $^2E_{1/2u}$ spin-orbit components (correlating directly with the $^2P_{3/2}$ and $^2P_{1/2}$ states of the free atom) where the symmetry labels denote irreducible representations of the double group O_h' appropriate for handling half angular momenta states. In the limit of strong spin-orbit coupling (50), it is reasonable to consider Jahn-Teller interactions only in the $^2G_{3/2u}$ state rather than in the $^2T_{1u}$ term as a whole, which seems more appropriate for Cu (see later). The vibrational modes that can couple in first order with a $^2G_{3/2u}$ state are contained in the antisymmetric square $\{G_{3/2u}{}^2\} = A_{1g} + E_g + T_{2g}$. Coupling with e_g modes stabilizes tetragonal distortions from O_h symmetry whereas coupling with t_{2g} modes stabilizes trigonal distortions. These axial distortions remove the degeneracy of the two Kramers doublets corresponding to the $^2G_{3/2u}$ state. The solution of the static part of the Jahn-Teller problem involving a $G_{3/2u}$ electronic state interacting with e_g or t_{2g} vibrational modes is well known (51). On the basis of the description given earlier of the symmetry properties of the charge distribution associated with the degenerate sublevels of a $^2P_{3/2}$ state it is suggested that the strongest vibronic coupling should involve e_g stretching vibrations. The form of the tetragonal distortion which is expected to be stablilized by coupling with e_g stretching modes (52) is illustrated in Figure 3. The form of the potential energy surface (50,51) for this case in the simplest theory is also illustrated in Figure 3. The upper and lower branches of the potential energy surface labelled U_1 and U_2 represent the two electronic states which replace the original degenerate state.

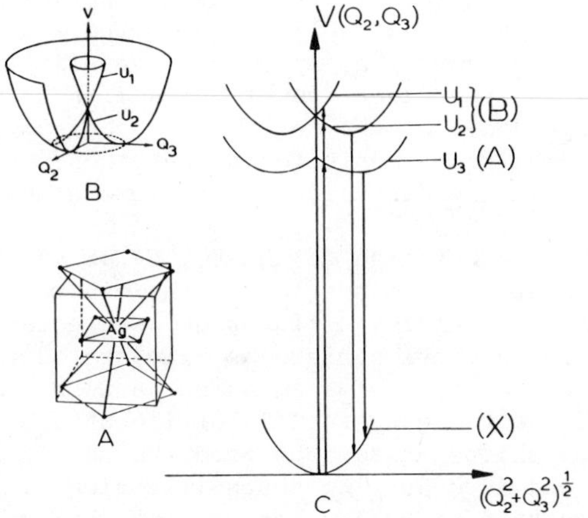

Figure 3. A: Tetragonal distortion of an MX_{12} tetradecahedral complex. B: Schematic potential energy surface for a doubly degenerate electronic state in O_h symmetry interacting with a doubly degenerate vibrational mode, neglecting anharmonicity effects. V *represents the nuclear potential energy and* Q_2 *and* Q_3 *represents the degenerate components of an* e_g *vibrational mode. C: Scheme of the ground and excited state potential energy surfaces for AgX_{12} cage complexes ($X = Ar$, Kr, or Xe). The absorption and emission processes are illustrated on this configuration coordinate diagram (*49–51*).*

The separation of the U_1 and U_2 branches at any point in the (Q_2, Q_3) coordinate space represents the vibronic splitting of the $G_{3/2u}$ state. If anharmonic terms are included in the nuclear potential then the energy surface looses its cylindrical character and retains only the three fold symmetry arising from the cubic symmetry of the Hamiltonian (53). In this case, the potential surface has three equivalent, trigonally disposed minima separated by saddle points, the minima corresponding to equivalent tetragonal distortions along the x,y,z directions of the molecule fixed cartesian coordinate system. The curve labelled U_3 represents the potential energy surface for the $^2P_{1/2}$ state in which account has been taken of the effects of $^2P_{3/2}$ - $^2P_{1/2}$ mixing by a second order perturbation treatment (53).

The configuration coordinate diagram described above can be used to interpret the major splittings in the absorption and emission spectra of entrapped Ag atoms, including the origin of the Stokes shifts, the direction of the shifts Xe>Kr>Ar, the temperature dependence of the absorption spectrum, fluorescence lifetimes and provides a basis for speculation concerning the origin of photoinduced diffusion and agglomeration effects described earlier (30).

Thus, optical excitation to the U_1, U_2 or U_3 levels should be followed by vibrational relaxation to the minimum point of the respective potential energy surfaces and subsequent radiative decay to the ground state surface as illustrated in Figure 3. The large Stokes shifts of the emission bands are seen to be the consequence of the tendency for distortion of the excited state complexes. The shifts arise both from the stabilization of the excited state and accompanying destabilization of the ground state and is expected to follow the order Xe>Kr>Ar as observed(30).

It is likely that the destabilization caused by producing ground state complexes in the relaxed excited state configuration, provides the driving force for photoinduced diffusion of the Ag atoms (29). Perhaps the anisotropic forces arising from the axial distortion of the AgX_{12} complex, causes the Ag atom in its ground electronic state to be ejected from its trapping site. The ejected silver atom could displace one of the surrounding rare gas atoms in such a way that an exchange of the original site positions occur. This would have the effect of moving the silver atom one intersite distance and restoring the original structural arrangement of the trapping site. As the Ag atom absorption profiles generally tend to retain their well defined structure during photoinduced aggregation experiments this suggests that the Ag atoms migrate between sites which are not very different in structure.

Atomic Copper in Rare Gas Supports

At first glance, the three fold splitting of the ^2P-^2S absorption band of Cu in Ar, Kr and Xe matrices might lead one to expect similar spectroscopic and photolytic properties to those described for Ag atoms (Figure 4). This is certainly not the case and differences can be traced to diminished importance of spin-orbit coupling effects and to the accessibility of the low lying $^2D_{3/2,5/2}$ term for Cu compared to Ag atoms (34). These changes translate experimentally into distinct relaxation mechanisms for the excited state cage complexes of Cu and Ag atoms, which manifest themselves for example, in different photo-aggregation kinetic behaviour (54). This can be seen in Figure 5 where the M/Kr ratio is lower by a factor of two for Ag compared to Cu.

If similar considerations described for Ag atoms are applied to Cu atoms in rare gas solids taking into account that the Jahn-Teller effect in the $^2T_{1u}$ excited electronic state dominates over that of spin-orbit coupling, then one can proceed with the analysis using the orbital wave functions representing the $^2T_{1u}$ state in O_h symmetry (49). By neglecting spin-orbit coupling and anharmonicity effects, the solution to the problem of $T_{1u} \otimes e_g$ vibronic coupling, is an excited state potential energy surface in (Q_2, Q_3) normal coordinate space which consists of three disjoint (mutually orthogonal) paraboloids (49,51). In this case, the $^2T_{1u} \leftarrow {}^2A_{1g}$ absorption band is expected to be a single Gaussian with a single relaxed emission band from the minimum of the upper state paraboloids. This scheme is not consistent with the spectral observations of a threefold splitting of the $^2P \leftarrow {}^2S$ absorption profile (34,55). On the other hand,for dominant $T_{1u} \otimes t_{2g}$ vibronic coupling, the upper state potential surface has been shown to consist of three sheets with four equivalent minima in (Q_4, Q_5, Q_6) space the latter being the normal coordinates spanning the t_{2g} representation (49,51).Such a potential energy hypersurface can be seen to yield a threefold structure in absorption with the expectancy of a complex emission profile. Thus the $T_{1u} \otimes t_{2g}$ coupling scheme can accurately describe the absorption profile of CuX_{12}, whereas a strong spin-orbit and weak vibronic coupling model would appear to be more appropriate for AgX_{12} (30,34,55).

Photoexcitation of Cu rare-gas films in the region of the $^2P \leftarrow {}^2S$ absorption band produces intense narrow emissions bands showing large spectral red shifts as seen in Figure 4,(34). In contrast to Ag, these emission profiles are insensitive to variations of the excitation wavelength within the threefold structure of the $^2P \leftarrow {}^2S$ absorption band. Simultaneous with the photolysis of any of the three $^2P \leftarrow {}^2S$ components, one observes gradual bleaching of all lines with concurrent formation of Cu_n where $n = 2 - 5$ (34,56). A further intriguing observation concerns the appearance of a weak structured emission near 420 nm for excitations centered on the secondary atomic site band of Cu in all three rare gas films (Figure 4), which has been found from independent studies of the absorption and emission spectra of matrix entrapped Cu_2 to arise from the A-state of Cu_2 (34).

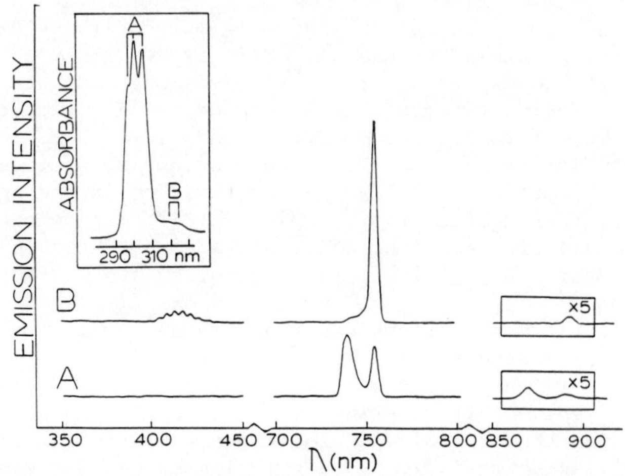

Figure 4. Fluorescence spectra of Cu atoms isolated in solid Ar at 12 K (Cu/Ar ~ 1/10[4]) uncorrected for instrumental factors. The corresponding excitation wavelengths are indicated on the absorption spectrum shown at the upper left. (Reproduced from Ref. 34. Copyright 1982, American Chemical Society.)

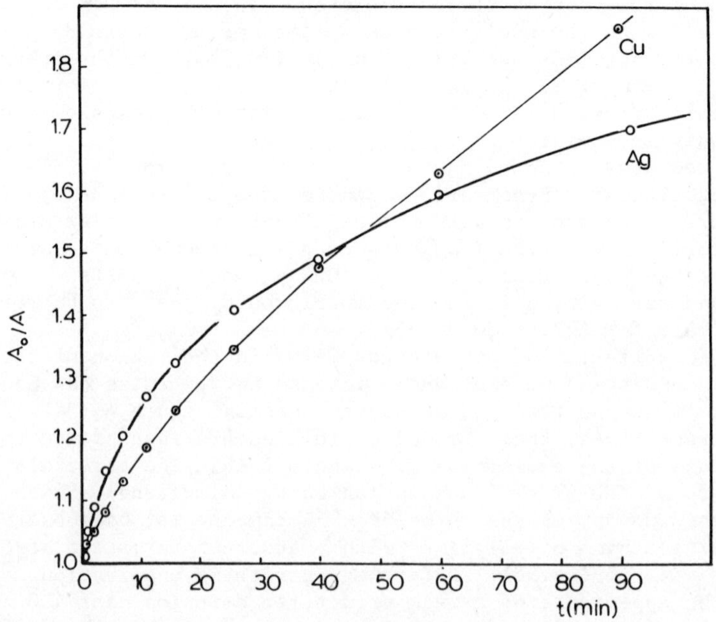

Figure 5. Agglomeration of Cu and Ag atoms in solid Kr at 10–12 K induced by $^2P \leftarrow {}^2S$ photoexcitation. Experimental conditions were identical except: Cu/Kr $\simeq 2/10^4$, and Ag/Kr $\simeq 1/10^4$ (54).

The major relaxation processes of $^2P \leftarrow {}^2S$ photoexcited Cu atoms in rare gas matrices are summarized in the following scheme (34).

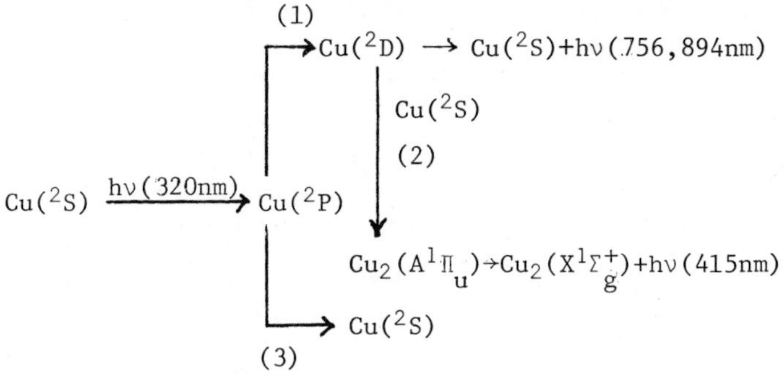

where the wavelengths in parenthesis apply to Ar. Here (1) and (3) represent non-radiative transitions which dissipate part, or all, respectively of the $^2P \leftarrow {}^2S$ excitation energy into the rare gas lattice. The occurrence of process (1) can be recognized directly from the fluorescence spectra since $^2P \leftarrow {}^2S$ photoexcitation is followed by $^2D \rightarrow {}^2S$ emission with no sign of $^2P \rightarrow {}^2D$ fluorescence. Similarly the absence of $^2P \rightarrow {}^2S$ fluorescence suggests that process (3) is important, since the quantum yield for $^2D \rightarrow {}^2S$ fluorescence is significantly less than unity as seen from comparing the fluorescence intensities for Cu atoms trapped in two different matrix sites, where the majority site (>90%) clearly shows a smaller quantrum yield for $^2D \rightarrow {}^2S$ fluorescence following $^2P \leftarrow {}^2S$ photoexcitation.

Although the details of these non-radiative transitions are not presently clear, it seems reasonable to suggest that the energy released into the matrix by processes (1) and/or (3) could account for the observed photo-induced diffusion and aggregation processes of Cu atoms in rare gas matrices (34,56). Related to this photoinduced diffusion effect is process (2) of the above scheme, which depicts an excited state dimerization reaction involving $Cu(^2D)+Cu(^2S)$ separated atoms, forming Cu_2 in an electronically excited state as seen by the observation of vibrationally relaxed A → X Cu_2 fluorescence following $^2P \leftarrow {}^2S$ Cu atomic excitation. The following points can be cited in support of the excited state reaction proposal. First the ^2D Cu atom excited state is likely to be quite long lived as a consequence of the forbidden nature of the optical transitions to the ground state. Second, the formation of these states should be accompanied by a release of considerable energy into the matrix as a result of the $^2P \rightarrow {}^2D$ non-radiative transition. This could result in "local" melting or matrix softening which would promote diffusion of the ^2D excited Cu atoms and allow for reactive encounters with nearby ground state ^2S Cu atoms. These considerations suggest that photoinduced diffusion and aggregation of Cu atoms in rare gas solids is promoted by $^2P \rightarrow {}^2S$ and $^2P \rightarrow {}^2D$ non-radiative transitions (34), in contradistinction to Ag atoms, which is believed to involve ^2P excited Ag atom-cage relaxation processes resulting in strong destabilization of ground state Ag atoms (30).

Dicopper and Disilver in Rare Gas Supports

Studies of the energetics and dynamics of Cu_2 and Ag_2 in rare gas solids have also been completed (31,34). The absorption and fluorescence spectra are similarly indicative of strong guest-host interactions in the low lying states of Cu_2 and Ag_2. Rather than presenting the spectroscopic and photolytic details, a summary of the observed radiative relaxation processes of visible and uv excited Cu_2 and Ag_2 in rare gas solids is shown below:

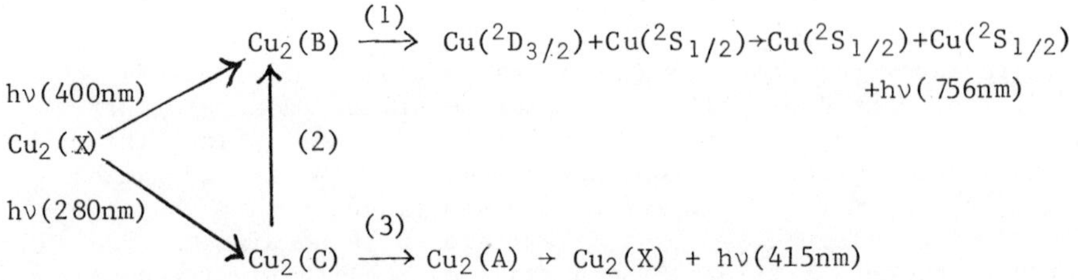

It is proposed that the B-state of Cu_2 (bound in the gas phase)(57) is sufficiently strongly destabilized in the matrix to the extent that it is unstable with respect to dissociation to $Cu(^2D_{3/2})$ + $Cu(^2S_{1/2})$ fragments following photoexcitation of Cu_2 from the ground state, process (1) in above scheme. The extent to which the dissociation actually occurs depends on the local dynamics following photoexcitation and the details of the Cu_2-rare gas potentials for the specific trapping site involved.

The absorption and fluorescence spectra of Ag_2 in rare gas solids are also clearly indicative of strong guest-host interactions involving the $A^1\Sigma_u^+$ and $C^1\Pi_u$ states of Ag_2 as summarized below:

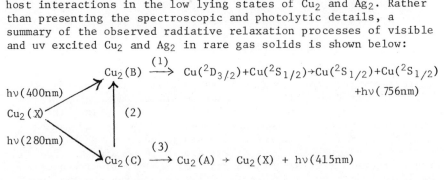

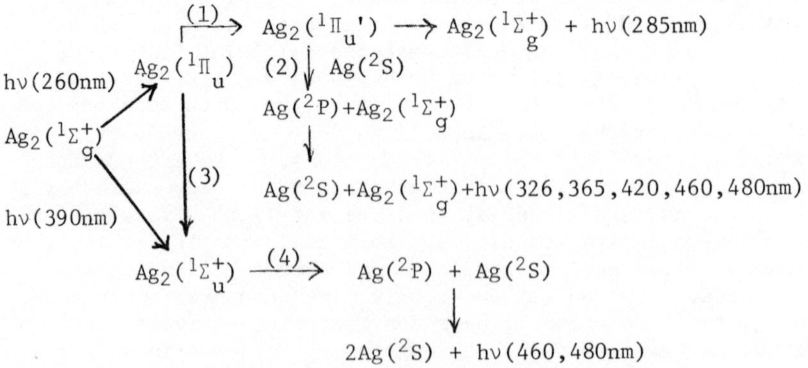

The emission spectrum produced by A←X Ag_2 excitation is interpreted in terms of an excited state Ag_2 dissociation involving strong stabilization of the $^2P+^2S$ Ag fragments, process (4) in the above scheme, by matrix cage relaxation effects as described for the ^2P excited state of Ag itself. Electronic relaxation of the C-state of Ag_2 is interpreted in terms of matrix cage relaxation, process (1), energy transfer to Ag atoms, process (2) and non-radiative decay to the A-state, process (3) followed by cage assisted photofragmentation, process (4).

Excited State Metal Atom Chemistry

Let us now turn our attention to the newly emerging field of excited state metal atom chemistry. The discovery of the excited state dimerization reaction (34):

$$Cu(^2D_{3/2})+Cu(^2S_{1/2}) \xrightarrow[12K]{\text{rare gas}} Cu_2(A^1\Pi_u)\rightarrow Cu_2(X^1\Sigma_g^+)+h\nu(415)$$

indicated that a prerequisit for performing condensed phase excited state metal atom chemistry was a weak interaction between the ground state metal atomic species and the matrix. This would allow for the preparation of the desired excited state metal atomic species, as illustrated in the recently discovered activation of CH_4 by photoexcited metal atoms (33):

$$Cu(^2S_{1/2})\{CH_4\}_{12} \xrightarrow[12K]{h\nu(320nm)} Cu(^2P_{1/2,3/2})\{CH_4\}_{12} \rightarrow CH_3CuH$$

weak ground state cage-complex; strong excited state cage-complex; photoinsertion

On the other hand to perform excited state metal atom chemistry in the M/CH_3Br system is not quite so straightforward (58):

$$\left.\begin{matrix}Fe\\Co\\Ni\end{matrix}\right\} \xrightarrow[12K]{CH_3Br} M\{BrCH_3\}_n \xrightarrow{h\nu} \text{no reaction}$$

$M\{BrCH_3\}_n$: highly coloured strong ground state charge transfer complex

because the metal atomic reagents tend to form strong ground state charge transfer complexes, thereby precluding the ability of achieving the desired metal atomic excited states. One way to surmount this difficulty is to work in dilute matrices as can be seen in the ground and excited state reactivity patterns of Cu atoms towards dioxygen summarized in the scheme below (37):

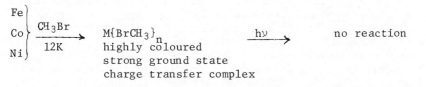

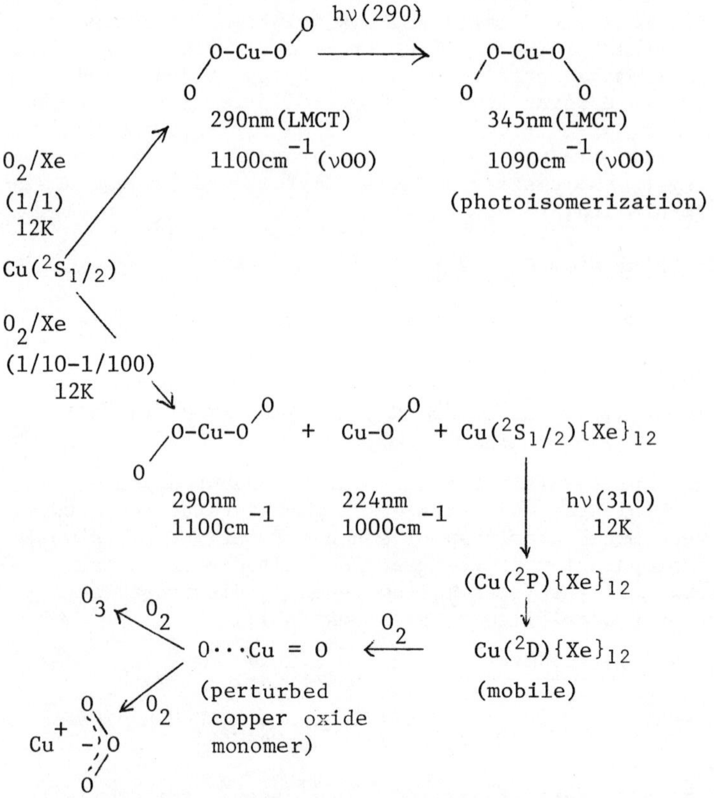

The above reaction scheme was established by a combination of uv-visible absorption and fluorescence, ir isotopic substitution, esr and kinetic measurements (37). The important point to note here is that in O_2 rich Xe matrices, ground state $Cu(^2S_{1/2})$ cannot avoid reactive encounters with O_2 to form $Cu(O_2)_2$ and $Cu(O_2)$ dioxygen complexes,whereas it is proposed that the formation of CuO, $Cu(O_3)$ and O_3 in dilute O_2/Xe matrices arises from the reaction of a long lived mobile excited state $Cu(^2D)$ with O_2. On the other hand the reactions of photoexcited $Ag(^2P)$ with O_2 are different (37), electron transfer being favoured to form $Ag^+O_2^-$. It is thought that this difference between the GS/ES reactivity of Cu and Ag atoms with O_2 originates in the accessibility of the reactive $Cu(^2D)$ state from the photolytically prepared $Cu(^2P)$ state which is not possible for $Ag(^2P)$ as illustrated in Figure 6.

Reactivity differences between $^2P \leftarrow {}^2S$ photoexcited Cu and Ag atoms have also been observed with CH_4 (59). For example,comparative matrix quenching kinetic measurements for Cu and Ag atoms in solid CH_4 show that the first order rate constant is considerably larger for Cu Fig.7. Detailed studies on the Cu system

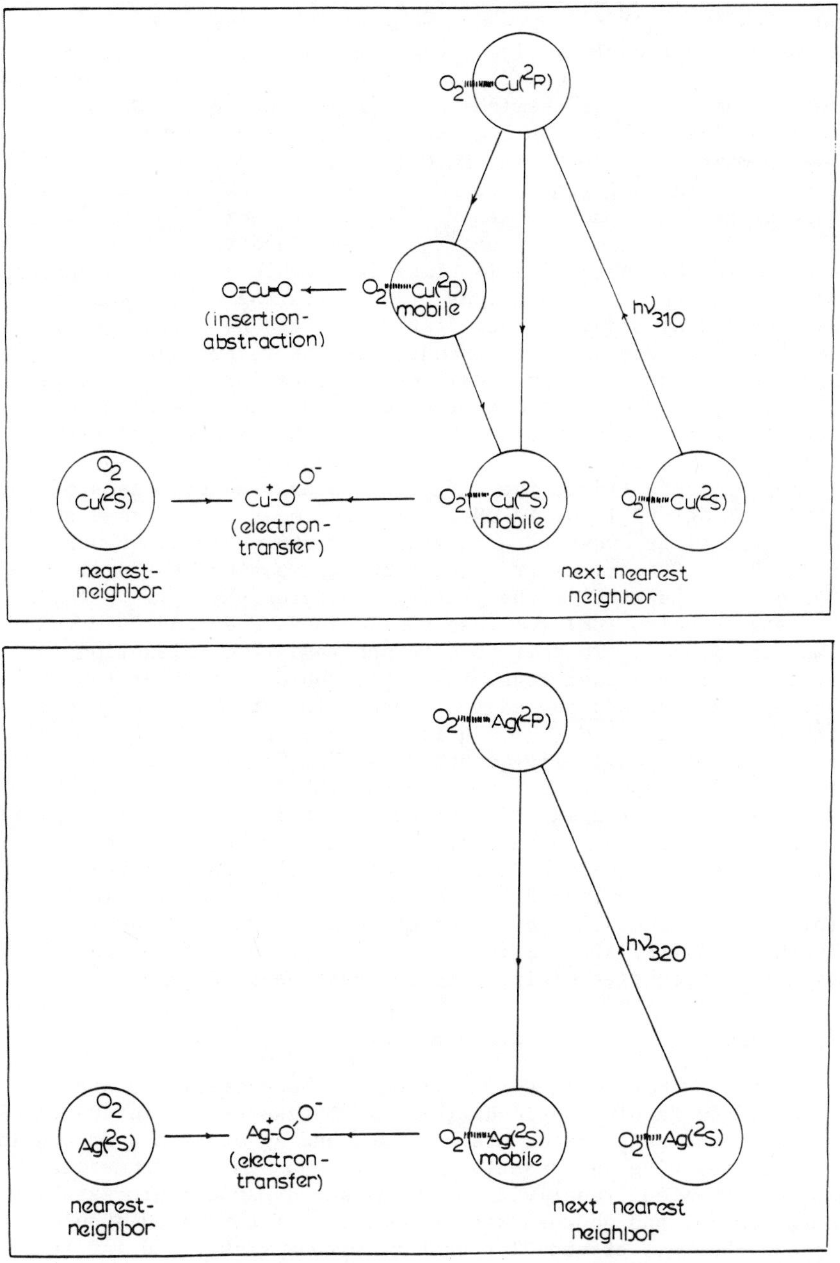

Figure 6. Reaction pathways for ground and excited state Cu and Ag atoms in O_2/Xe matrices. (Reproduced from Ref. 37. Copyright, American Chemical Society.)

(50) show that (i) the rate of chemical quenching by CH_4, CD_4 and CH_2D_2 are identical within experimental error,(2) the activation energy for the $Cu(^2P)/CH_4$ reaction is essentially zero,(3) the quantum yield for chemical quenching of $Cu(^2P)$ by CH_4 is close to unity,(4) isotopic scrambling is not observed in $Cu(^2P)/CH_4/CD_4$ mixtures and (5) in the $Cu(^2P)/CH_2D_2$ reaction the [CuH]/[CuD] ratio in the secondary photolysis channel of CH_2DCuD/CHD_2CuH is unity within experimental error. Collectively these results strongly support the direct insertion mechanism rather than the energetically less favourable electron transfer pathway. Furthermore, reactions of CH_4 with the $B^1\Sigma_u{}^+$ state of Cu_2 produced by 378nm photoexcitation of Cu_2/CH_4 solids (60), show only the formation of $CuH+CH_3$; CH_3CuH is not detected. These results provide some indirect evidence for excited state selectivity in the reactions of Cu with CH_4, that is the ^2P state inserts into a CH bond whereas the ^2D state abstracts H atoms. These ideas have been incorporated into Figure 8 which attempts to consolidate the different relaxation processes of $Cu(^2P)$ and $Ag(^2P)$ in solid CH_4. In essence it is proposed that the higher first order rate constant for the chemical quenching of $Cu(^2P)$ by CH_4 arises from the fact that a substitutionally incorporated Cu atom in the partially ordered phase II of solid CH_4 (stable below 20K) (61) can sample apex, edge and face configurations of CH_4.Thus if $^2P \rightarrow {}^2D$ non-radiative relaxation is favoured for apex positions, then this could lead to H-atom abstraction, whereas the edge or face orientation could lead to direct insertion of $Cu(^2P)$. In either case, photoexcited $Cu(^2P)$ is efficiently chemically quenched by CH_4. On the other hand,the ^2D state of Ag appears not to be accessible from the ^2P state,so cage relaxation of apex oriented CH_4 molecules around the $Ag(^2P)$ state could lead to ground state Ag in a relaxed CH_4 cage. This would cause photoinduced diffusion of $Ag(^2S)$ atoms for reasons similar to those proposed for the rare gases, resulting in Ag atom photoagglomeration as a competing pathway to CH_4 activation. Like Cu, a $Ag(^2P)$ atom lying on an edge or face of CH_4 is proposed to be reactive giving direct insertion to CH_3AgH.

Conclusion

From the above discussion, it should be possible to appreciate how extremely subtle differences in guest-host interactions in the ground and excited states of Cu and Ag atoms and dimers in both non-reactive and reactive supports can lead to dramatically distinct chemical reactivity patterns and dynamical processes. Photochemical and photophysical phenomena of this kind should provide chemists of the 21st century with a rich field for fundamental and applied research, offering considerable scope for experimental challenges and intellectual stimulation.

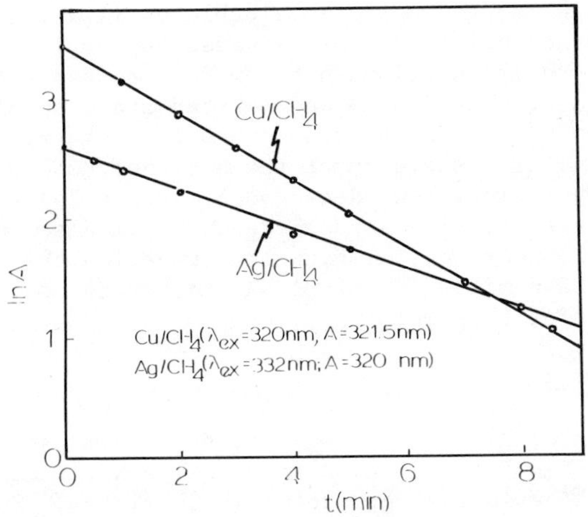

Figure 7. First-order kinetic plots for the quenching of $^2P \leftarrow {}^2S$ photoexcited Cu and Ag atoms in solid CH_4 at 10–12 K. Identical experimental conditions were used with $Cu/CH_4 \simeq Ag/CH_4 \simeq 1/10^4$. (Reproduced from Ref. 59. Copyright American Chemical Society.)

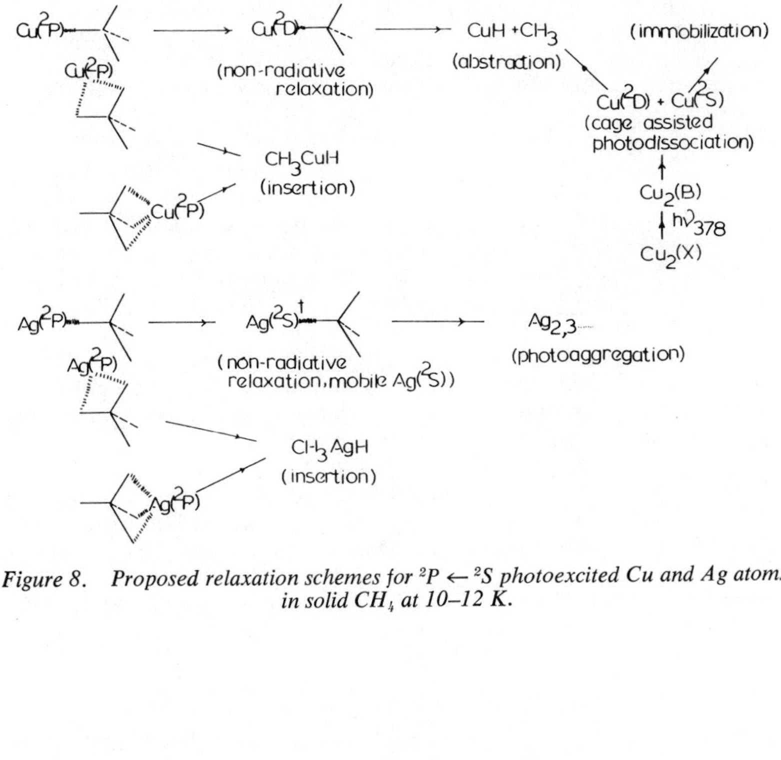

Figure 8. Proposed relaxation schemes for $^2P \leftarrow {}^2S$ photoexcited Cu and Ag atoms in solid CH_4 at 10–12 K.

Acknowledgments

I wish to acknowledge the invaluable assistance of my graduate students and colleagues whose names appear in the cited articles, for their contributions towards the understanding of the properties of both ground and excited state metal atomic and cluster reagents in reactive and non-reactive matrices. The generous financial assistance of the National Research Council of Canada's Operating, New Ideas and Strategic Energy Grant Programmes are gratefully acknowledged. I am also indebted to the Atkinson Foundation, the Bickell Foundation, the Connaught Fund and the Lash Miller Chemical Laboratory for support at various stages of this work.

Literature Cited

1. (a) Klabunde, K. J. "Chemistry of Free Atoms and Particles", Academic, New York, 1980; Blackborow, J.R.; Young, D., "Metal Vapour Synthesis in Organometallic Chemistry," Springer-Verlag, New York, 1979; Moskovits, M.; Ozin, G.A., "Cryochemistry", Wiley-Interscience, New York, 1976; Craddock, S.; Hinchcliffe, A.J., "Matrix Isolation", Cambridge University Press, London, 1975 (and references cited therein).
2. Ozin, G.A.; Francis, C.G., J. Mol. Struct., 1980, 59, 55; Ozin, G.A.; Francis, C.G.; Huber, H.X.; Andrews, M.P.; Nazar, L., J. Amer. Chem. Soc., 1981, 103, 2453; Ozin, G.A.; Francis, C.G.; Huber, H.X.; Nazar, L.F., Inorg. Chem., 1981, 20, 3635 (and references cited therein).
3. Torrovap Industries Inc., Toronto; Planar Industries Inc., Oxford.
4. Timms, P.L.; Turney, T.W., Adv. Organomet. Chem, 1977, 15, 53; Klabunde, K.J., Acc. Chem. Res., 1975, 8, 393, Skell, P.S.; Havel, J.J.; McGlinchy, M.J., ibid., 1973, 6, 97; Ozin, G.A., ibid., 1973, 6, 313; Ozin, G.A.; Power, W.J., Adv. Inorg. Chem. Radiochem., 1980, 23, 29; Ozin, G.A., Cat. Rev. Sci. Eng., 1977, 16, 191; Coord. Chem. Rev., 1979, 28, 117; J. Macromol. Sci., 1981, A16, 167; (and references cited therein).
5. Jonkman, H.T.; Michl, J., J. Chem.Soc. Chem. Comm., 1978, 751.
6. Grinter, R.; Armstrong, S.; Jayasooriya, U.A.; McCombie, J.; Norris, D.; Springall, J.P., Faraday Symp. Chem. Soc., 1980, 14, 94; Armstrong, S.; Grinter, R.; McCombie, J., J. Chem. Soc. Faraday Trans., 1981, 77, 123.
7. Mowery, R.L.; Miller, J.C.; Krausz, E.R.; Schatz, P.N.; Jacobs, S.M.; Andrews, L., J. Chem. Phys., 1979, 70, 3920.
8. Jacobi, K.; Schmeisser, D.; Kolb, D.M., Chem. Phys. Lett., 1980, 69, 113; J. Chem. Phys., 1981, 75, 5300.
9. Montano, P.A.; Shenoy, G.K., Solid State Comm., 1980, 35, 53.
10. Zilm, K.W.; Colin, R.T.; Grant, D.M.; Michl, J., J. Amer. Chem. Soc., 1978, 100, 8030.

11. Van Zee, R.J.; Baumann, C.A.; Weltner, W. Jr., J. Chem. Phys., 1981, 74, 6977 (and references cited therein).
12. Kasai, P.H.; McLeod, D. Jr., J. Amer. Chem. Soc., 1979, 101, 5860; Kasai, P.H., Accounts Chem. Res., 1971, 4, 329.
13. Thompson, G.A.; Lindsay, D.M., J. Chem. Phys., 1981, 74, 959; Thompson, G.A.; Tischler, F.; Garland, D.; Lindsay, D.M., Surf. Sci., 1981, 106, 408.
14. Montano,P.A.; Barrett, P.H.; Micklitz, H., Ber. Bunsenges. Phys. Chem., 1978, 82, 37.
15. Leutloff, D.; Kolb, D.M., Ber Bunsenges. Phys. Chem., 1979, 83, 666.
16. Ozin, G.A., Faraday Symp. Chem. Soc., 1980, 14, 7 (and references cited therein).
17. Bondybey, V.E.; English, J.H., J. Chem. Phys., 1981,74,6978; 1980, 73, 42 (and references cited therein).
18. Ahmed, F.; Nixon, E.R., J. Chem. Phys., 1979, 71, 3547.
19. Schulze, W.; Becker, H.U.; Minkwitz, R.; Mansel, K., Chem. Phys. Lett., 1978, 55, 59.
20. Pellin, M.J.; Foosnaes, T.; Gruen, D.M., J. Chem. Phys., 1981, 74, 5547.
21. Moskovits, M.; DiLella, D., J. Chem. Phys.,1980, 72, 2267; 1980, 73, 4917.
22. Demuynck,J.; Rohmer, M.M.; Strich, A.; Veillard,A., J. Chem. Phys., 1981, 75, 3443 (and references cited therein).
23. Proceedings of the Second International Meeting on Small Particles and Inorganic Clusters, Lausanne, Switzerland, 1980; Surf. Sci., 1981, 106, 1.
24. Moskovits, M.; Hulse, J.E., J. Chem. Phys., 1977, 67, 4271; 1977, 66, 3988.
25. Bondybey, V.E.; Brus, L.E., Adv. Chem. Phys., 1980, 36 , 269.
26. Dyson, W.; Montano, P.A., Solid State Comm.,1980, 33, 191.
27. Ozin, G.A., J. Amer. Chem. Soc., 1980, 102, 3301 (and references cited therein).
28. "Growth and Properties of Metal Clusters; Applications to Catalysis and the Photographic Process", Proc. 32nd Int. Meeting of the Soc. De Chim. Phys., Ed. J. Bourdon, Elsevier, New York, 1980.
29. Huber, H.; Ozin, G.A., Inorg. Chem., 1978, 17, 155. Klotzbücher, W.; Ozin, G.A.,J. Amer. Chem. Soc.,1978,100,2262; J. Mol. Catalysis, 1977, 3, 195; Mitchell,S.A.; Ozin, G.A., J. Amer. Chem. Soc., 1978, 100, 6776.
30. Mitchell, S.A.; Farrell, J.; Kenney-Wallace, G.A.; Ozin,G.A., J. Amer. Chem. Soc., 1980, 102, 7702.
31. Farrell, J.; Kenney-Wallace, G.A.; Mitchell, S.A.; Ozin, G.A., J. Amer. Chem. Soc., 1981, 103, 6030.
32. Ozin, G.A.; Huber, H.X.; Mitchell, S.A., Inorg. Chem., 1979, 18, 2932.
33. Mitchell,S.A.; Ozin,G.A.; Garcia-Prieto,J., J. Amer. Chem.Soc., 1981,103,1574; Margrave, J.; Billups,W.E.; Konarski, M.M.; Hauge,R.H.; Margrave,J.L., J. Amer. Chem. Soc.,1980,102,7393.

34. Mitchell, S.A.; Garcia-Prieto, J.; Ozin, G.A., J. Phys. Chem., 1982, 86, 473.
35. Mitchell, S.A.; Garcia-Prieto, J.; Ozin, G.A., Angew, Chem. Int. Ed., 1982, __, .
36. Mitchell, S.A.; Garcia-Prieto, J.; Ozin, G.A., Angew. Chem. Int. Ed., 1982, __, .
37. Mitchell, S.A.; Garcia-Prieto, J.; Ozin, G.A., J. Amer. Chem. Soc., (submitted for publication).
38. Billups, W.E.; Hauge, R.H.; Margrave, J.L., J. Amer. Chem. Soc., 1980, 102, 6005.
39. Hugues, F.; Ozin, G.A., U.S. Patent, Filed Aug. 1981; J. Amer. Chem. Soc. (submitted for publication).
40. Moore, C.E., Natl. Bur. Stand. (US), Circ., 1958, No.467, Vol. 2,3,
41. Shenstone, A.G., Phys. Rev., 1940, 57, 894.
42. McCarty, M.; Robinson, G.W.Jr., Mol. Phys., 1959, 2, 415.
43. Kasai, P.H.; McLeod, D. Jr., J.Chem. Phys., 1971, 55, 1566.
44. Forstmann, F.; Ossicini, S., J. Chem. Phys., 1980, 73, 5997.
45. Klose, J.Z., Astrophys. J., 1975, 198, 229
46. Ammeter, J.H.; Schlosnagle, D.C., J. Chem. Phys., 1973, 59, 4784.
47. Anderson, J.M., "Introduction to Quantum Chemistry", Benjamin, W.A., Inc. New York. 1969, Chapt.4.
48. Tinkham, M., "Group Theory and Quantum Mechanics", McGraw-Hill, New York, 1964, p.78.
49. Engleman, R., "The Jahn-Teller Effect in Molecules and Crystals", Wiley, New York, 1972.
50. Moran, P.R., Phys. Rev., 1965, 137, A1016; Fulton, T.A.; Fitchen, D.B., Phys. Rev., 1969, 179, 846.
51. Sturge, M.D., Solid State Phys., 1967, 20, 91.
52. Liehr, A.D., J. Phys. Chem., 1963, 67, 389.
53. Öpic, U.; Pryce, M.H.L., Proc. Roy. Soc., 1957, A238, 425.
54. Mitchell, S.A., "Photoprocesses of Metal Atoms and Clusters", Ph.D. Thesis, University of Toronto, 1982.
55. Moskovits, M.; Hulse, J.E., J. Phys. Chem., 1981, 35, 2904.
56. Ozin, G.A.; Huber, H.X.; McIntosh, D.; Mitchell, S.A.; Norman, J.G. Jr.; Noodleman, L., J. Amer. Chem. Soc., 1979, 101, 3504.
57. Klemen, B.; Lindkvist, S., Ark. Fys., 1954, 8, 333; 1955, 9, 385; J. Ruamps, Ann. Phys. (Paris), 1954, 4, 1111; Maheshwari, R.C., Indian J. Phys., 1963, 31, 368; Aslund, N.; Barrow, R.F.; Richards, W.G.; Travis, D.N., Ark. Fys., 1965, 30, 171.
58. Klabunde, K.J.; Tanaka, Y.; Davis, S.C., J. Amer. Chem. Soc., 1982, 104, 1013.
59. Mitchell, S.A.; Garcia-Prieto, J.; Ozin, G.A., J. Amer. Chem. Soc., (submitted for publication).
60. Mitchell, S.A.; Garcia-Prieto, J.; Ozin, G.A., Angew, Chem. Int. Ed., 1982, 798, .
61. W. Press. J. Chem. Phys., 1972, 56, 2597.

RECEIVED August 3, 1982

Discussion

R.L. Sweany, University of New Orleans: I was surprised at seeing your report of a ^2D Cu atom being able to abstract a hydrogen atom from methane, but, of course, the copper atom is "hot". I wonder if you see methyl take back its hydrogen atom after photolysis or does the radical pair collapse to give $HCuCH_3$?

G. A. Ozin: Responding to your first comment. The ^2D Cu atom is produced indirectly by 370-400 nm photofragmentation of Cu_2 entrapped in solid CH_4. In this photolysis we observe rapid bleaching of the Cu_2 absorptions, complete quenching of the ^2D emissions of Cu atoms, efficient photoproduction of ^2S Cu atoms and the observation of some CH_3 radicals with trace amounts of H atoms.

In contrast, the corresponding Cu_2 photolysis in the solid rare gases causes only a small net permanent photodissociation of Cu_2 and photodissociative yield of ^2S Cu atoms with accompanying ^2D Cu atom emissions being easily observed.

(1) Ozin, G.A.; Mitchell, S.A.; Garcia-Prieto, J.; J. Phys. Chem., 1982, 86, 473; Angew, Chem. Suppl., 1982, 798,

These observations indicate that ^2D Cu atoms react with CH_4 according to the scheme shown below:

$$Cu_2(X) \xrightarrow[CH_4;\ 12K]{370\text{-}400\ nm} Cu_2(B) \rightarrow Cu(^2D) + \underset{(trapped)}{Cu(^2S)}$$

$$Cu(^2D) \xrightarrow{CH_4,\ 12K} CH_3 + CuH$$

(abstraction)

to give mainly the abstraction products CH_3 + CuH, whereas ^2P Cu atoms produced directly yield the insertion product CH_3CuH which fragments via a secondary photolysis channel to give mainly CH_3+CuH. The ^2D Cu atom reaction with CH_4 described above is thermoneutral or slightly exothermic based on the known gas phase bond dissociation energy of CuH_1 (65.5 kcal mol^{-1}) and the energies of the $^2D_{5/2}$ (33.6 kcal mol^{-1}) and $^2D_{3/2}$ (38.0 kcal mol^{-1}) states respectively. Even so, the $Cu(^2D)/CH_4$ result is somewhat surprising and clearly suggests that further work is needed to elucidate the excited state reactivity patterns of the Cu atom CH_4 reaction. We are actively pursuing this line of research.

(2) Ozin, G.A.; Mitchell, S.A.; McIntosh, D.F.; Garcia-Prieto, J.; J. Amer. Chem. Soc., 1981, 103, 1574).

In response to your second enquiry, we do not have any evidence for a reaction of the photoproduced CH_3 radical with its hydrogen atom partner (or CH_4 itself) after photolysis. However, we do have data which depicts the presence of a CH_3 radical interacting with CuH (cf Pimentel and Tan's CH_3....LiI species) formed as a product of the $Cu(^2P)/CH_4$ reaction by either

$$Cu(^2P) + CH_4 \rightarrow CH_3,,,, CuH \text{ (abstraction)}$$

or

$$\{Cu(^2P) + CH_4 \rightarrow CH_3CuH \text{ (insertion)}$$
$$\{CH_3CuH \rightarrow CH_3,,,,CuH \text{ (secondary photolysis)}$$

Thermal annealing of the CH_3,,,CuH interacting pair at 30-35K leads to efficient generation of the insertion product CH_3CuH, which suggests that this reaction is to some degree exothermic.

21

Synthetic Inorganic Chemistry in Low Temperature Matrices

J. L. MARGRAVE, R. H. HAUGE, Z. K. ISMAIL, L. FREDIN, and W. E. BILLUPS

Rice University, Department of Chemistry, Houston, TX 77001

Simplicity and spontaneity are two of the attractive features of many metal atom reactions which justify the prediction that synthetic inorganic chemistry at low temperatures will be an important technique in the next two decades. Pimentel,[1] Skell,[2] Timms,[3] Green,[4] Klabunde,[5] Andrews,[6] Ozin,[7] and others have made significant contributions to this area.

For over 15 years we have conducted research utilizing metal atoms in low temperature spectroscopic and synthetic studies at Rice University.[8] Our synthetic work was started in the late 1960s with the work of Krishnan,[9] on lithium atom reactions with carbon monoxide, extended by Meier[10] in his studies of lithium atom reactions with water and ammonia and expanded over the next several years to include metal atom interactions with HF, H_2O, H_3N, H_4C, and their hundreds of organic analogs--RF, R_2O, ROH, R_3N, . . . H_3N, R_4C, R_3CH, etc.[11-15] A most exciting aspect of metal atom chemistry seems to lie in the photoassisted insertion processes by which one can produce molecules like HMF, HMOH, $HMNH_2$, $HMCH_3$, HNC_2H_5, etc. simply by co-depositing metal atoms with the potential reactant material and either through spontaneous reaction or through photochemically induced reactions achieving the formation of new molecular species. This approach has already been taken to a multigram preparative scale and there seems to be no reason why one cannot move on to preparation of industrially significant quantities of organometallic materials by this technique. Table I summarizes the studies of Fe_2 and

0097-6156/83/0211-0329$06.00/0

TABLE I

A. Reactions of Metal Atoms with H_2O and Isoelectronic Species

    (a) $M + HF \rightarrow [M:FH] \rightarrow HMF \rightarrow MF$

    (b) $M + H_2O \rightarrow [M:FH_2] \rightarrow HMOH \begin{smallmatrix} \nearrow MO \\ \searrow MOH \end{smallmatrix}$

    (c) $M + NH_3 \rightarrow [M:NH_3] \rightarrow HMNH_2 \begin{smallmatrix} \nearrow MNH \\ \searrow MNH \end{smallmatrix}$

    (d) $M + CH_4 \rightarrow [M \cdots CH_4] \rightarrow HMCH_3 \begin{smallmatrix} \nearrow MH \\ \searrow MCH_3 \end{smallmatrix}$

    (e) $M + ROR \rightarrow [M:OR_2] \rightarrow RMOR$

B. Reactions of Metal Atoms with CO and CO_2

    (a) $M + CO \rightarrow [M^+, CO^-] \rightarrow [M^+, (CO)_2^-] \rightarrow M_2C_2O_2$

    e.g. $\qquad\qquad\qquad\qquad\qquad\qquad\qquad [Li^+; {}^-OC \equiv CO^-; Li^+]$

    (b) $M + CO_2 \rightarrow [M^+, CO_2^-] \rightarrow [M^+, C_2O_4^-] \rightarrow M_2C_2O_4$

C. Reactions of Metal Atoms with Aliphatic & Cyclic Hydrocarbons

    (a) $M + C_2H_6 \xrightarrow{h\nu} HMC_2H_5$

    (b) $M + C_3H_8 \xrightarrow{h\nu} HMC_3H_7$

    (c) $Fe + cyclo\text{-}C_3H_6 \xrightarrow{h\nu}$ $\begin{smallmatrix} H_2C - CH_2 \\ | \qquad | \\ Fe - CH_2 \end{smallmatrix}$

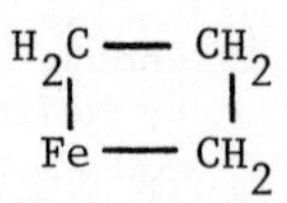

D. Reactions of Elemental Fluorine in Low-Temperature Matrices

    (a) $F_2 + UF_4 \xrightarrow{Al} UF_5 + UF_6$ (complete reaction)

    $\qquad\qquad\;\; \xrightarrow{CO} HF_4 + F_2$ (no reaction)

    (b) $F_2 + CH_4 \xrightarrow[10K]{or\ N_2} CH_4$ (no reaction) $\xrightarrow{h\nu} CF_4, CF_3H$, etc.

    (c) $F_2 + C_2H_4 \xrightarrow[10K]{Dark}$ (no reaction) $\xrightarrow[10K]{IR} C_2H_4F_2, C_2H_3F$, etc.

metal atom chemistry already performed at Rice University and some of the various molecules which can be prepared.

New formulas, new stoichiometries, new structures, new reactants, new reducing agents, new oxidizing agents, and new areas of chemical research will result from the studies of chemical reactions in low-temperature matrices.

ACKNOWLEDGMENTS

Studies of chemical reactions in low-temperature matrices have been supported by funding from the National Science Foundation, the U.S. Army Research Office, Durham, and from the Robert A. Welch Foundation.

Literature Cited

1. G.C. Pimentel, Pure Appl. Chem. 4, 61 (1962).
2. P.S. Skell, J.J. Havel and M.J. McGlinchey, Acc. Chem. Res., 6, 97 (1973).
3. P.L. Timms and T.W. Turney, Adv. Organomet. Chem., 15, 53 (1977).
4. M.L.H. Green, Journal of Organomet. Chem., 200, 119 (1980).
5. Chemistry of Free Atoms and Particles, K.J. Klabunde, Academic Press (1980).
6. L. Andrews, Ann. Rev. Phys. Chem. 22, 109 (1971).
7. Cryo Chemistry, M. Moskovits and G.A. Ozin, eds. (Wiley, N.Y., 1973).
8. J.W. Hastie, R.H. Hauge and J.L. Margrave, Spectroscopy in Inorganic Chemistry, Vol. 1, C.N.R. Rao and J.R. Ferro, eds. (Academic Press, 1970), pp. 57-106.
9. C.H. Krishnan, Ph.D. Thesis, Rice University (1975).
10. J.L. Margrave, P.F. Meier and R.H. Hauge, J. Am. Chem. Soc., 100, 2108 (1978).
11. (a) J.L. Margrave, W.E. Billups, M.M. Konarski and R.H. Hauge, Tetranedron Letters, 21, 3861-3864 (1980).
    (b) J.L. Margrave, W.E. Billups, M.M. Konarski and R.H. Hauge, J. Am. Chem. Soc., 102, 7393 (1980).
12. (a) J.L. Margrave, R.H. Hauge and J.W. Kauffman, J. Am. Chem. Soc., 102, 19, 6005-6011 (1980).
    (b) J.L. Margrave, R.H. Hauge, J.W. Kauffman and L. Fredin, Proc., Symp. on High Temperature Chemistry, ACS Advances in Chemistry Series, Ed. J.L. Gole (1981) pp. 363-376.
13. J.L. Margrave, M.A. Douglas and R.H. Hauge, Symp. on High Tempeature Chemistry, ACS Advances in Chemistry Series, Ed. J.L. Gole (1981) pp. 347-354.
14. Z.K. Ismail, R.H. Hauge, L. Fredin, J.W. Kauffman and J.L. Margrave, J. Chem. Phys., Aug. 15 (1982).
15. J.L. Margrave, R.H. Hauge, J.W. Kauffman, N.A. Rao, M.M. Konarski, J.P. Bell and W.E. Billups, J. Chem. Soc., Chem. Comm., 24, 1258-1260 (1981).

RECEIVED September 16, 1982

22

Heteroatom Cluster Compounds Incorporating Polyhedral Boranes as Ligands

NORMAN N. GREENWOOD

University of Leeds, Department of Inorganic and Structural Chemistry, Leeds LS2 9JT England

Synthetic routes to metalloboranes are briefly summarized. Two specific aspects of the field are selected for more detailed comment: (i) the occurrence of exopolyhedral cycloboronation of P-phenyl groups; (ii) oxidative cluster closure reactions. These topics are illustrated by reference to reactions of closo-$B_{10}H_{10}^{2-}$, nido-$B_9H_{12}^-$ and arachno-$B_9H_{14}^-$ with a variety of Ir^I complexes, and several novel metalloborane cluster geometries (as determined by X-ray diffraction analysis) are described, notably nido-[Ir(B_9H_{13})H(PPh_3)$_2$], iso-nido-[{IrC(OH)B_8H_6(OMe)}($C_6H_4PPh_2$)(PPh_3)], closo-[$Ir_2B_4H_2$)($C_6H_4PPh_2$)$_2$(CO)$_3$(PPh_3)], iso-closo-[($HIrB_9H_8$)($C_6H_4PPh_2$)(PPh_3)], arachno-[($HIrB_8H_{11}Cl$)(CO)(PMe_3)$_2$], nido-[(IrB_8H_{11})(CO)(PMe_3)$_2$], and closo-[($HIrB_8H_7Cl$)(PMe_3)$_2$]. The significance of these results is discussed.

Boron hydrides are generally regarded as electron deficient compounds in which the lack of electrons is compensated by forming multicentre bonds which lead to polyhedral cluster geometries of the molecules. Three principal series of boranes are now recognized: closo-$B_nH_n^{2-}$, nido-B_nH_{n+4} and arachno-B_nH_{n+6}. In addition, there are the hypho-B_nH_{n+8} and precloso-B_nH_n series (so far known only as derivatives) and also the general class of conjuncto-boranes B_nH_m formed by joining together various polyhedral fragments from the parent series. Most metals share with boron the property of having fewer valency electrons than orbitals available for bonding; accordingly, in the early 1960's, we initiated a research programme to investigate the consequences of considering metal atoms as honorary boron atoms, so that they could be incorporated as vertices within the polyhedral borane clusters. This idea has proved enormously fruitful and many groups on both sides of the Atlantic have made notable contributions to the burgeoning field of metalloborane cluster chemistry. As a

0097-6156/83/0211-0333$06.00/0

result, it is now known that most metals (except the most electropositive metals of Groups I, II, and III of the Periodic Table) can be so incorporated (1) and to date some 30 metals, having electronegativities in the range 1.5-2.2, have been shown to form polyhedral metalloborane clusters (cf. electronegativities of B 2.1, C 2.5). Equally significant has been the growing recognition that boranes and borane anions can act as excellent polyhapto ligands to appropriate metal centres (2). These new perceptions have greatly expanded the field of boron-hydride chemistry and have led to a deeper understanding of the factors influencing the structure, thermal stability, and chemical reactivity of these compounds.

Synthesis of Metalloboranes

Numerous routes have been devised for the synthesis of metalloboranes. Among the most useful of these are:

1. Coordination of a metal by a (polyhapto) borane ligand; examples of all modes from η^1 to η^6 are known.
2. Oxidative insertion of a metal into a borane cluster.
3. Partial degradation of a borane cluster in the presence of a metal complex.
4. Expansion, degradation, or modification of a preformed metalloborane cluster.

These routes are not mutually exclusive and many syntheses involve aspects of more than one category. For example reaction of arachno-$B_9H_{14}^-$ with Vaska's compound trans-$[Ir(CO)Cl(PPh_3)_2]$ gives, amongst other products, a small yield of the nido-iridadecaborane cluster $[Ir(B_9H_{13})H(PPh_3)_2]$, Figure 1, whereas its reaction with the isoelectronic platinum complex cis-$[PtCl_2PMe_2Ph)_2]$ gives a high yield of the monodeboronated product arachno-$[Pt(\eta^3\text{-}B_8H_{12})(PMe_2Ph)_2]$, Figure 2. This latter product can be deprotonated and reacted with a further mole of cis-$[PtCl_2(PMe_2Ph)_2]$ to yield the 10-vertex diplatinadecaborane cluster arachno-$[Pt_2(\eta^3,\eta^3\text{-}B_8H_{10})(PMe_2Ph)_4]$, Figure 3. Whilst 6-subrogated metallodecaboranes were previously known [see references cited in (1)], dimetalladecaboranes had not previously been reported. Mixed-metal clusters e.g. PtPd, and PtIr can also be prepared by this and related routes. Rather than attempt a general review of metallaborane chemistry and the use of boranes as ligands (which would necessarily be superficial in the space available) two specific topics will be selected in order to illustrate the advances that are at present being made and which point to some of the ways in which the field is likely to develop. Both topics involve borane cluster complexes with iridium in various oxidation states.

Exo-polyhedral Cycloboronation

Recent work by Janet Crook on the reaction of trans-[Ir(CO)Cl-

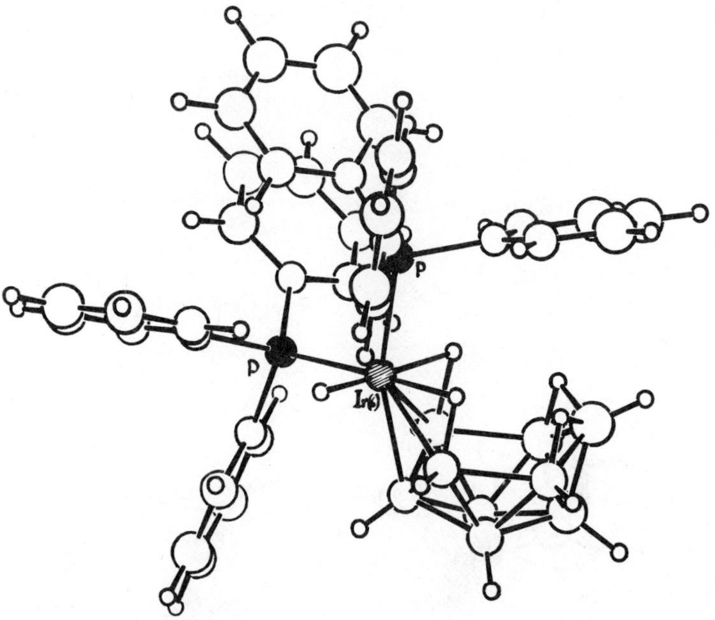

Figure 1. Structure of nido-*[Ir(B_9H_{13})H(PPh_3)$_2$] showing the subrogation of BH at position 6 by the isolobal unit {IrH(PPh_3)$_2$} (3).*

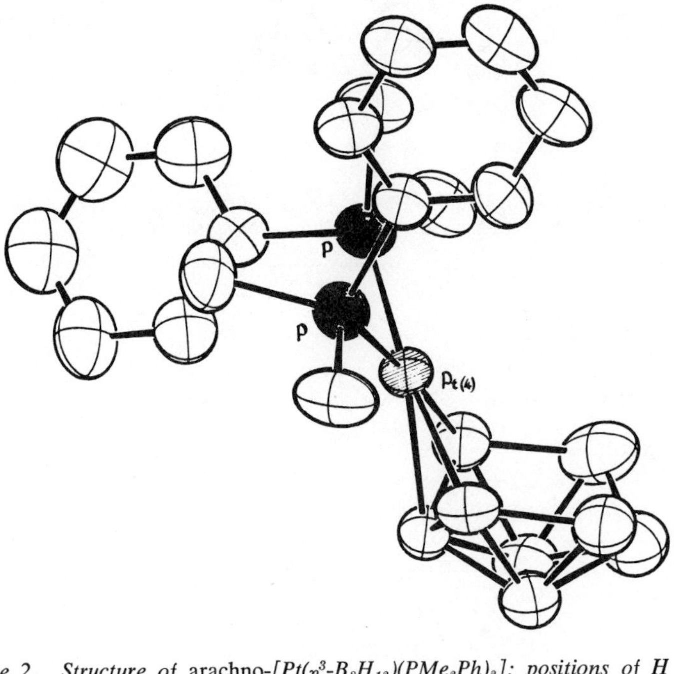

Figure 2. Structure of arachno-*[Pt(η^3-B_8H_{12})(PMe_2Ph)$_2$]; positions of H atoms are not determined (4).*

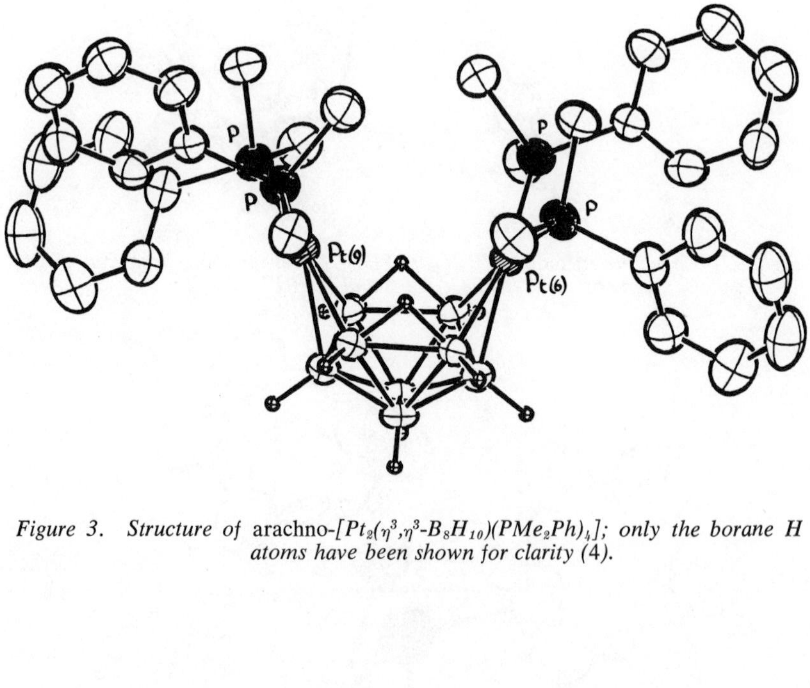

Figure 3. Structure of arachno-*[$Pt_2(\eta^3,\eta^3\text{-}B_8H_{10})(PMe_2Ph)_4$]; only the borane H atoms have been shown for clarity (4).*

$(PPh_3)_2]$ in boiling methanol with the normally unreactive closo-$B_{10}H_{10}^{2-}$ dianion led to the isolation of ten new metalla- and dimetalla-boranes, and the structures of eight of these have so far been unambiguously established, including the novel ten-vertex iso-nido-cluster $[\{IrC(OH)\overline{B_8H_6(OMe)\}(C}_6H_4PPh_2)(PPh_3)]$ shown in Figure 4 and the equally unexpected exo-bicyclic closo six-vertex dimetallahexaborane $[(Ir_2\overline{B_4H_2)(C}_6H_4PPh_2)_2(CO)_3(PPh_3)]$ shown in Figure 5. Of the numerous significant aspects of these two compounds (5, 6) a particularly interesting feature common to both is the occurrence of ortho-cycloboronation of one or more P-phenyl groups on the tertiary phosphine ligands. The mechanistic origin of this feature (which is unprecedented in metalloborane chemistry though well established in organometallic chemistry) was unclear. Further work suggested that it did not occur by straightforward elimination of H_2 from the acyclic analogues.

Jonathan Bould has now shown that ortho-cycloboronation occurs essentially quantitatively in certain reactions using $[IrCl(PPh_3)_3]$ as the source of iridium (7). Thus, as shown in Scheme 1 reaction with nido-$B_9H_{12}^-$ in CH_2Cl_2 solution at room temperature yields two isomeric forms of the yellow nido iridium(III) compound **1**, $[(HIr\overline{B_9H_{12})(C}_6H_4PPh_2)(PPh_3)]$ by the unusually clean reaction:

$$[IrCl(PPh_3)_3] + B_9H_{12}^- \longrightarrow [(HIr\overline{B_9H_{12})(C}_6H_4PPh_2)(PPh_3)] + Cl^- + PPh_3$$

The precise stoichiometry and high yield of this reaction suggest that a driving force of the ortho-cycloboronation in this case is the provision of the two additional cluster electrons required for the nido ten-vertex structure. These can be thought of as being provided by two H atoms (the ortho-H atom on the cyclo-boronating phenyl group and the terminal H atom on B(5)); the nett result is the formation of an Ir-H_t and an additional Ir-H_μ-B(5) bond. In fact, two isomers of the ortho-cycloboronated nido-iridadecaborane are formed which differ in the arrangement of the ligands about the pseudo-octahedral Ir(III) centre; the more labile, **1a**, can be quantitatively converted into the more stable, **1b**, by mild thermolysis at 65°C. At slightly higher temperatures (85°C) quantitative loss of $2H_2$ results in the smooth production of the novel bright orange-yellow iso-closo-iridium(V) compound, **2** (Scheme 1). It will be noted that the $\{IrB_9\}$ cluster (Figure 6) has the previously unobserved C_{3v} symmetry $\{IrB_3B_3B_3\}$ rather than the normal D_{4d} bicapped square antiprismatic arrangement of vertices. A formal electron count requires 22 electrons (2n + 2) for the closo-cluster bonding; of these 18 are supplied by the nine boron atoms, leaving 4 to be contributed by the Ir atom. This, together with the Ir-H terminal bond indicates that the compound can be regarded as a further example of the growing number of compounds in which the high oxidation state Ir(V) is stabilized by coordination to a 'soft' polyhedral borane ligand (5,7,8).

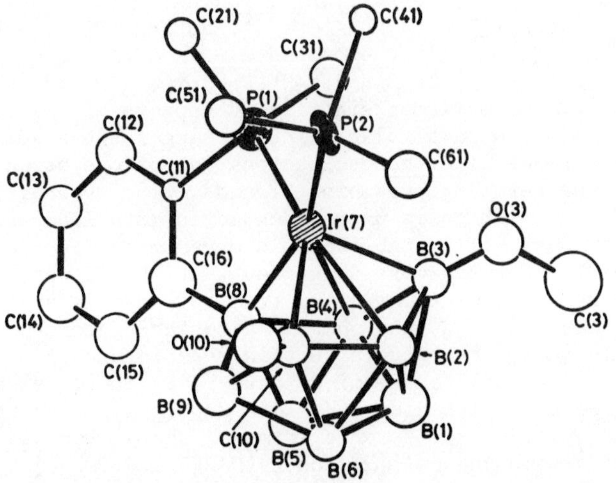

Figure 4. Structure of iso-nido-*[{IrC(OH)B₈H₆(OMEe)}(C₆H₄PPh₂)(PPh₃)] showing* ortho-*cycloboronation of a* P-*phenyl group (5).*

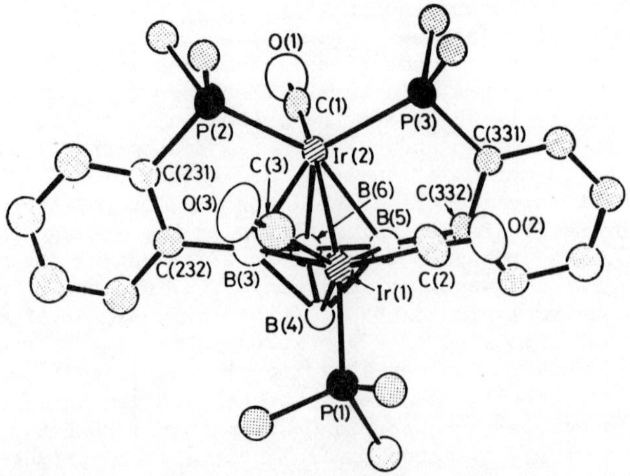

Figure 5. Structure of closo-*[(Ir₂B₄H₂)(C₆H₄PPh₂)₂(CO)₃(PPh₃)] with the* P-*phenyl groups omitted (apart from their* ipso *C atoms), but with the two* ortho-*cycloboronated* P-*phenylene groups retained (6).*

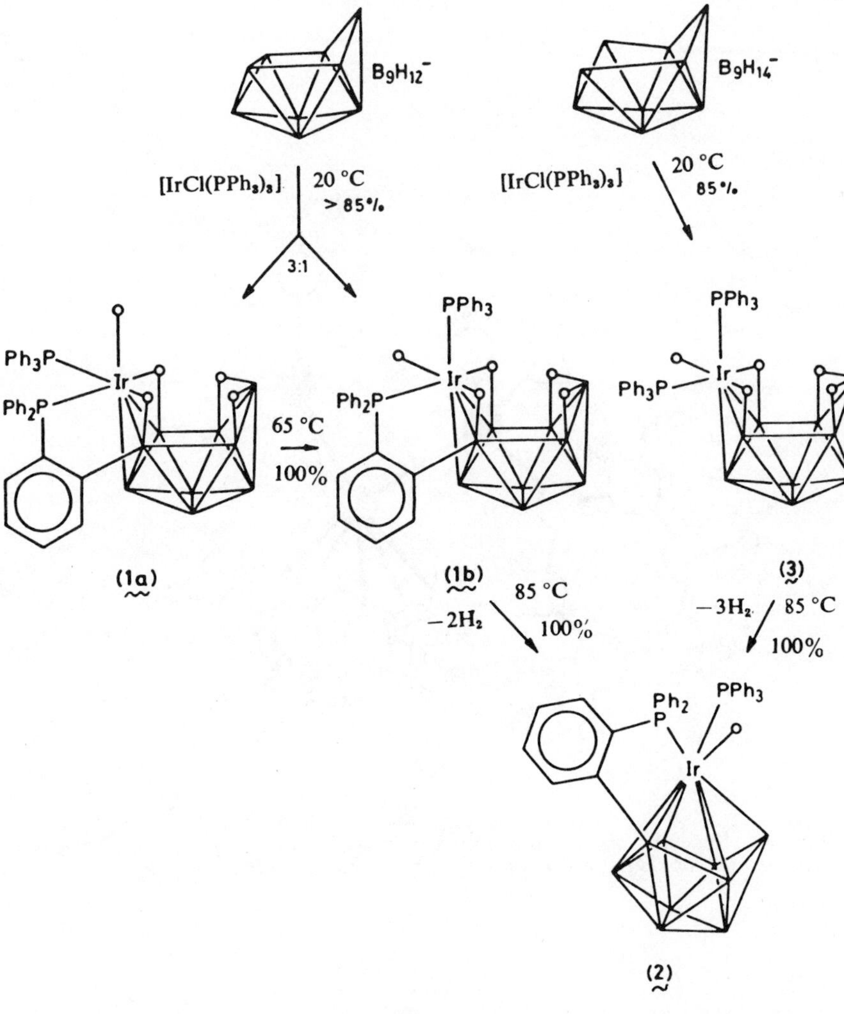

Scheme 1. *Reactions of* nido-$B_9H_{12}^-$ *and* arachno-$B_9H_{14}^-$ *with* $[IrCl(PPh_3)_3]$ *at room temperature, and the quantitative thermolytic dehydrogenation of the products at 85 °C to give* iso-closo-$[(HIrB_9H_8)(C_6H_4PPh_2)(PPh_3)]$. *(Reproduced with permission from Ref. 7. Copyright 1982, Royal Society of Chemistry.)*

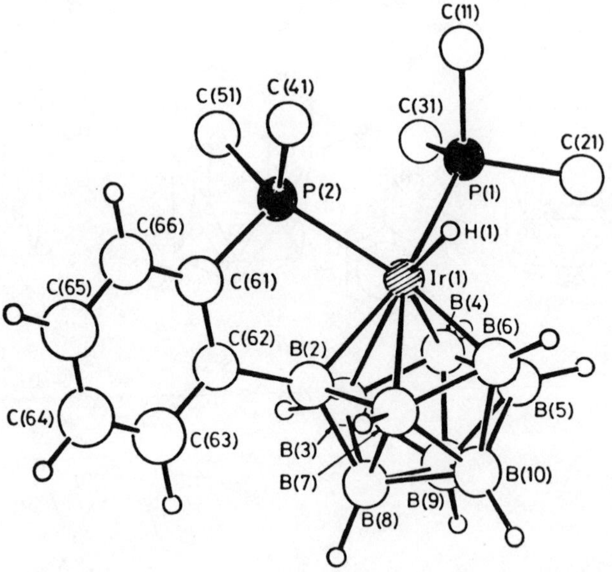

Figure 6. Structure of iso-closo-$[(HIrB_9H_8)(C_6H_4PPh_2)(PPh_3)]$ *with* P-*phenyl groups omitted for clarity (except for their* ipso *C atoms). All H atoms were located (7).*

Scheme 1 also shows that this ortho-cycloboronated iso-closo iridium(V) compound, **2**, can be formed directly and quantitatively by the mild thermolysis of compound **3** which is the acyclic analogue of the cycloboronated isomers **1a** and **1b**. Compound **3**, whose molecular structure has already been given in Figure 1, was formerly obtained (3) in <2% yield by reaction of arachno-$B_9H_{14}^-$ with trans-[Ir(CO)Cl(PPh$_3$)$_2$] but can now be obtained virtually quantitatively (7) by using the related [IrCl(PPh$_3$)$_3$] instead of Vaska's complex.

Oxidative Cluster Closure Reactions

In a related set of experiments the complete series of formal cluster oxidations from arachno → nido → closo has been effected by mild thermolysis of bis(trimethylphosphine) iridanonaboranes (8). Although the possibility of such reactions is implied by the structural classifications mentioned at the beginning of this lecture, and isolated examples are known as discussed above, the complete sequence has not previously been demonstrated. Jonathan Bould has shown that reaction of trans-[Ir(CO)Cl(PMe$_3$)$_2$] with nido-$B_9H_{12}^-$ affords a mixture of products amongst which are the colourless arachno-[(HIrB$_8$H$_{12}$)(CO)(PMe$_3$)$_2$] **4** and its monochloro derivative **5**. The crystal structure of this latter compound is shown in Figure 7. Mild thermolysis of **4** (and **5**) in hydrocarbon solutions effects a facile first-order dehydrogenation to the pale yellow nido-analogues **6** and **7** (Scheme 2 and Figure 8). Further heating at 135° results in loss of a further mole of H_2 and the CO to give a 45% yield of the poppy red closo-iridium(V) species **8** (Figure 9) though the process is now accompanied by some decomposition and degradation (which is even more severe for the non-chlorinated compound). Before discussing the mechanistic implications of these reactions it is worth noting the unique cluster geometry of the closo-IrB$_8$ species; this does not adopt the usual D_{3h} tricapped trigonal prismatic arrangement typified by closo-$B_9H_9^{2-}$ and its derivatives, but has an idealized C_{2v} cluster geometry which features an η^6-'inverted boat' coordination of the borane moiety to iridium: the 4 equivalent Ir-B(2,4,5,7) distances are 218 pm and the 2 Ir-B(3,6) are 231 pm. There is no significant bonding between B(2)-B(4) (307 pm) and B(5)-B(7) (304 pm) as would be required for tricapped trigonal prismatic geometry; the bonding B-B contacts are all within the normal range.

The mechanistic implications of these facile formal cluster oxidations arachno → nido → closo by nett loss of H_2 at moderate temperatures are considerable. The processes are accompanied by, and presumably assisted by a flexibility of coordination geometry about the Ir atom and also by its ready oxidation. The metal atom can be seen as a potential source of electrons for cluster bonding either by involving its lone pairs of electrons or by switching between Ir-H-B bridging and Ir-H terminal bonding

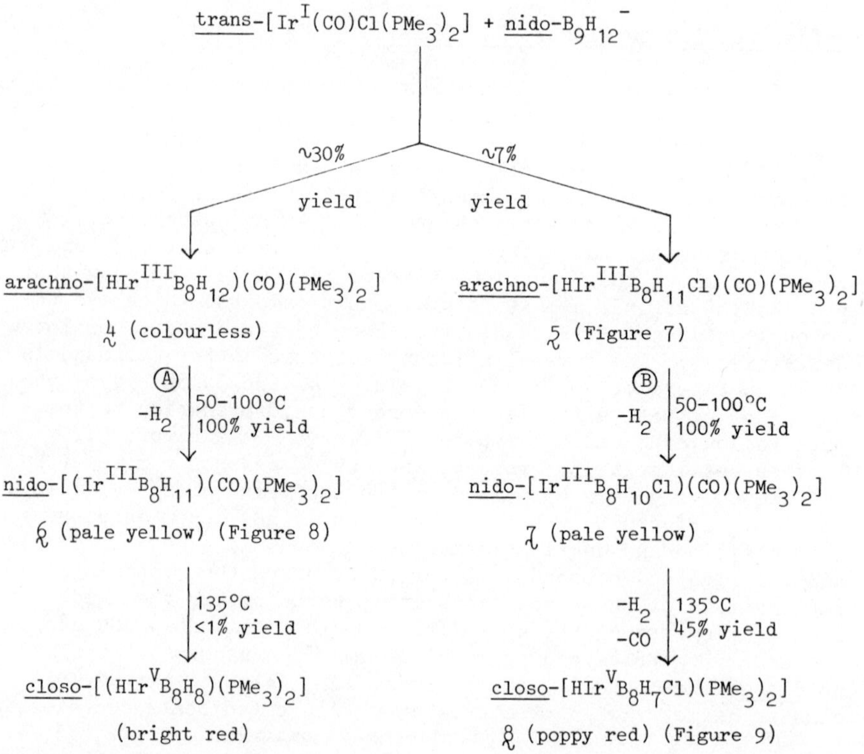

Scheme 2. Formal cluster oxidation reactions arachno → nido → closo. *Reactions labelled* Ⓐ *and* Ⓑ *follow first-order kinetics with activation parameters:* $\Delta H^{\ddagger}$, *127 and 122 kJ mol*$^{-1}$; *and* $\Delta S^{\ddagger}$, *35 and 13 J/K*$^{-1}$ *mol*$^{-1}$.

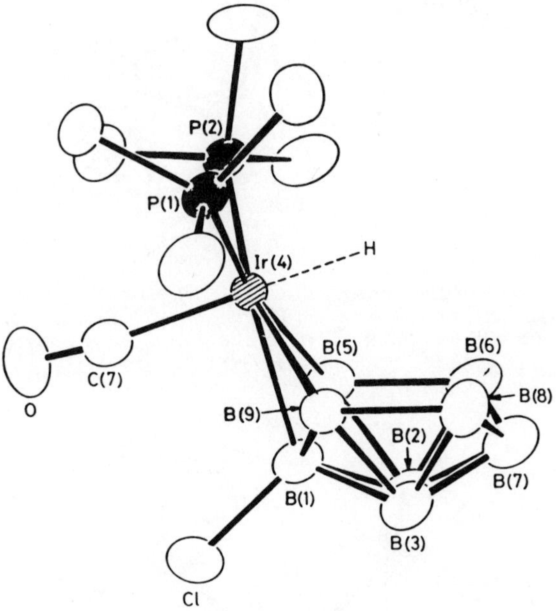

Figure 7. Structure of arachno-*[(HIrB$_8$H$_{11}$Cl)(CO)(PMe$_3$)$_2$]. The terminal H atom on Ir(4), mutually* cis *to P(1) and P(2), was established by NMR spectroscopy, which also showed the presence of an* exo-*terminal H atom on each B atom except B(1),* endo-*terminal H atoms on B(6) and B(8), and bridging H atoms between B(5,6) and B(8,9) (8).*

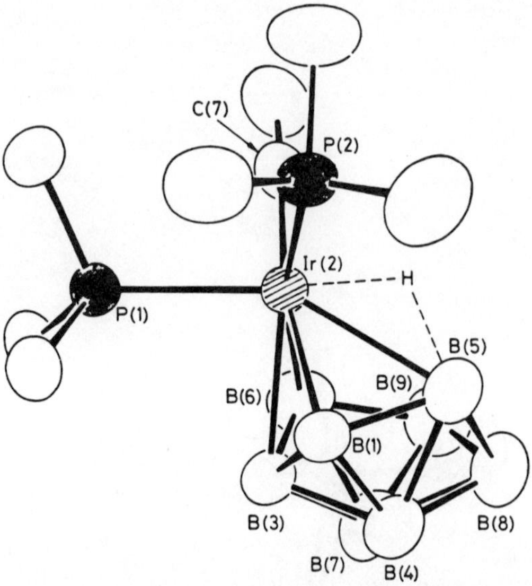

Figure 8. Structure of nido-*[IrB_8H_{11})(CO)($PMe_3)_2$]. ^1H-{^{11}B} NMR spectroscopy showed the presence of an* exo-*terminal H atom on each B atom and bridging H atoms between B(6,9) and B(8,9). There is also a bridging H atom Ir(2)–H–B(5)* trans *to P(1) (8).*

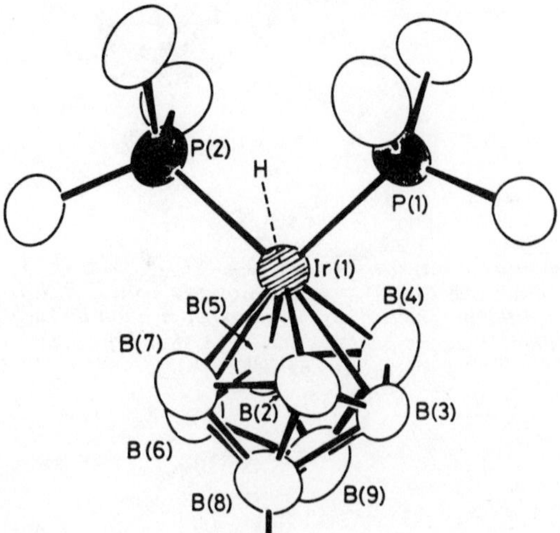

Figure 9. Structure of closo-*[($HIrB_8H_7Cl$)($PMe_3)_2$]. Selective ^1H-{^{11}B} NMR spectroscopy showed that each B atom except B(8) has one terminal H atom attached and that there is a terminal H atom on Ir(1) symmetrically* cis *to the two P atoms (8).*

pairs. Similar processes are known in organometallic systems, particularly those showing activity as hydrogenation catalysts. However, the processes occurring appear to be rather more complex than this simple picture suggests, since the positions of the Cl substituents in Figures 7 and 9 imply that some cluster rearrangement (such as a diamond-square-diamond process) is also taking place. One possible sequence which would account for the observed products is shown in Figure 10.

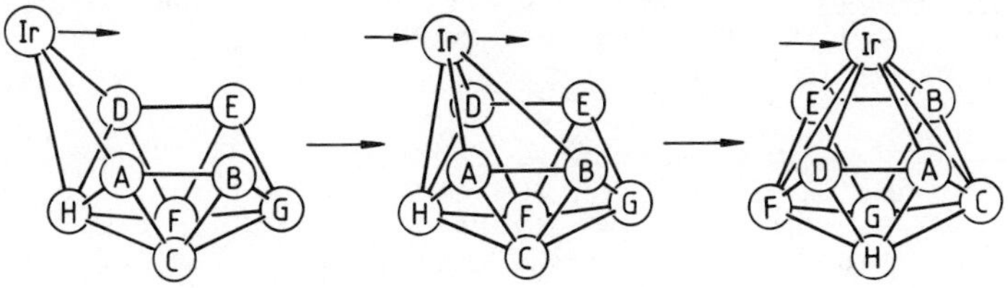

Figure 10. A possible sequence of skeletal rearrangements occurring during the arachno → nido → closo *cluster closure reactions discussed in the text.*

Acknowledgments

Dr J.D. Kennedy has been closely connected with all aspects of this work and it is a pleasure to acknowledge his stimulating collaboration. Crystal structures were obtained in collaboration with Dr W.S. McDonald. The work was supported by the U.K. Science and Engineering Research Council.

Literature Cited

1. Greenwood, N.N.; Kennedy, J.D. "Metal Interactions with Boron Clusters"; Grimes, R.N., Ed.; Plenum Press, New York, 1982; Chapter 2.
2. Greenwood, N.N. (Convenor); "Boranes as Ligands"; Micro-symposium 1, Proc. 19th International Conference of Coordination Chemistry, Prague, 1978; pages 69-88.
3. Boocock, S.K.; Bould, J.; Greenwood, N.N.; Kennedy, J.D.; McDonald, W.S. J. Chem. Soc. Dalton Trans. 1982, 713.
4. Boocock, S.K.; Greenwood, N.N.; Hails, M.J.; Kennedy, J.D.; McDonald, W.S. J. Chem. Soc. Dalton Trans. 1981, 1415.
5. Crook, J.E.; Greenwood, N.N.; Kennedy, J.D.; McDonald, W.S. J. Chem. Soc. Chem. Commun. 1981, 933.
6. Crook, J.E.; Greenwood, N.N.; Kennedy, J.D.; McDonald, W.S. J. Chem. Soc. Chem. Commun. 1982, 383.
7. Bould, J.; Greenwood, N.N.; Kennedy, J.D.; McDonald, W.S. J. Chem. Soc. Chem. Commun. 1982, 465.
8. Bould, J.; Crook, J.E.; Greenwood, N.N.; Kennedy, J.D.; McDonald, W.S. J. Chem. Soc. Chem. Commun. 1982, 346.

RECEIVED August 25, 1982

Discussion

W.L. Gladfelter, University of Minnesota: The thermolysis of arachno-[(CO)H$(PMe_3)_2(IrB_8H_{12})$], in which H_2 is evolved, is an example of a dinuclear reductive elimination. Have you observed any dependence of the rate on the PMe_3 concentration? Also, do the compounds undergo H/D exchange when placed in a D_2 atmosphere?

N.N. Greenwood: We have not yet done the experiments you mention, though we have them very much in mind.

T. Baker, DuPont Central Research: In reference to your so-called iso-closo iridaborane complex, $H(PPh_3)_2IrB_9H_9$, we have recently published two papers (1,2) dealing with isoelectronic, isostructural ten-vertex ruthenacarborane complexes and have demonstrated that these structures are related to the common closo bicapped square antiprismatic structure by the removal of two electrons (i.e. 2N skeletal electrons for an N vertex polyhedron). Such complexes have been referred to as hyper-closo to imply that the electronic unsaturation is not primarily metal-based (as in, for example, nido-$(PPh_3)_2RhC_2B_8H_{12}$ (3) or closo-$(PPh_3)ClRh(1,7\text{-}C_2B_9H_{11})$ (4), but is delocalized throughout the polyhedral network. The hyper-closo structures indicate that with the inclusion of a transition metal-containing vertex, capped-closo structures are not always the preferred polyhedral geometries.

(1) Jung, C.W.; Baker, R.T.; Knobler, C.B.; Hawthorne, M.F. J. Am. Chem. Soc. 1980, 102, 5782.

(2) Jung, C.W.; Baker, R.T.; Hawthorne, M.F. ibid 1981, 103, 810.

(3) Jung, C.W.; Hawthorne, M.F. ibid 1980, 102, 3024.

(4) Baker, R.T.; King III, R.E.; Long, J.A.; Marder, T.B.; Paxson, T.E.; Teller, R.G.; Hawthorne, M.F. ibid 1982, submitted for publication.

N.N. Greenwood: Whether your compounds are described as hyper-closo or iso-closo depends on the number of electrons assumed to be contributed by the metal atom to the cluster. If, as is generally assumed, ruthenium contributes two electrons to the cluster in compounds such as yours, then $\{RuC_2B_7\}$ has a closo 22e skeltal electron count (i.e. 2n+2) rather than a 20e hyper-closo count. The uncertainty concerning the most appropriate choice of formal oxidation state for metals in covalent compounds permeates the whole of organometallic chemistry and can only be resolved when experimental evidence from XPS and other techniques provides grounds for preferring one particular oxidation state rather than another. Fortunately this uncertainty does not normally extend to crystal structure determinations and, at least where these are available, we are both in agreement about the geometrical description of our respective compounds.

A.J. Carty, Guelph-Waterloo Centre: With regard to the rather interesting examples of orthocycloboronation of a triphenylphosphine ligand in the reaction of $[IrCl(PPh_3)_3]$ with nido-$B_9H_{12}^-$, do you have any information on the mechanisms of these reactions? It is thought, for example, that in the case of orthopalladation as in the reaction of $PdCl_4^{2-}$ with azobenzene, the mechanism involves an electrophilic attack by the metal on the arene ring. Could electrophilic attack by the borane fragment on an arene ring of iridium coordinated Ph_3P be a possible mechanism here? It is also interesting to note that $[IrCl(PPh_3)_3]$ itself undergoes intramolecular orthometallation, although under more severe conditions than used for orthocycloboronation.

N.N. Greenwood: We have not yet undertaken detailed mechanistic studies of the ortho-cyclo boronation reactions but the sequence you envisage has also seemed quite plausible to us.

23

Reactions of Iridium and Rhodium Complexes with the *arachno*-$B_4H_9^-$ Ion

SHELDON G. SHORE

The Ohio State University, Department of Chemistry, Columbus, OH 43210

Professor Greenwood has provided routes to metalloboranes which are derived from the noble metals Pt and Ir and the nido-$B_9H_{12}^-$, arachno-$B_9H_{14}^-$, and closo-$B_{10}H_{10}^{2-}$ ions. We wish to make note of some new reactions of Ir and Rh complexes with the arachno-$B_4H_9^-$ ion and compare these results with his observations of arachno-$B_9H_{14}^-$.

The following reaction studied by Dr. S. K. Boocock produced arachno-[Ir(B_4H_9)(CO)(PMe_2Ph)$_2$] in about 60% yield. Its structure (Figure 1) was determined by J. C. Huffman and K. Folting of the

$$[Ir(CO)Cl(PPh_3)_2] + K[B_4H_9] \longrightarrow [Ir(B_4H_9)(CO)(PMe_2Ph)_2] + KCl$$

Molecular Structure Center, Indiana University. This compound is an analogue of arachno-B_5H_{11} with Ir replacing the apical boron. It can also be considered to be an analogue of a metal-1,3-butadiene complex since the $B_4H_9^-$ unit is isoelectronic with 1,3-butadiene. A similar complex, arachno-[Ir(B_4H_9)(CO)(PMe_3)$_2$] has been isolated in about 1% yield from the reaction of trans-[Ir(CO)Cl(PMe_3)$_2$] with nido-$B_9H_{12}^-$(1).

The reaction of arachno-$B_4H_9^-$ cited above differs from those described by Professor Greenwood in that there is no oxidative transfer of hydrogen from B_4 to irridium. Such transfer occurs in the reaction of arachno-$B_9H_{14}^-$ with [IrCl(PPh_3)$_3$] and also with trans-[Ir(CO)Cl(PMe_3)$_2$] to give nido-[Ir(B_9H_{13})H(PPh_3)$_2$] in respective yields of 85% and 2%. Interestingly, Dr. M. A. Toft observed oxidative transfer of hydrogen from B_4 to the metal when he reacted [RhCl(PPH_3)$_3$] with K[B_4H_9]

$$[RhCl(PPh_3)_3] + K[B_4H_9] \longrightarrow [Rh(B_4H_8)H(PPh_3)_2] + KCl + PPh_3$$

The proposed structure of nido-[Rh(B_4H_8)H(PPh_3)$_2$] is shown in Figure 2. It is based upon ^1H, ^{11}B, ^{31}P NMR and IR spectra.

0097-6156/83/0211-0349$06.00/0

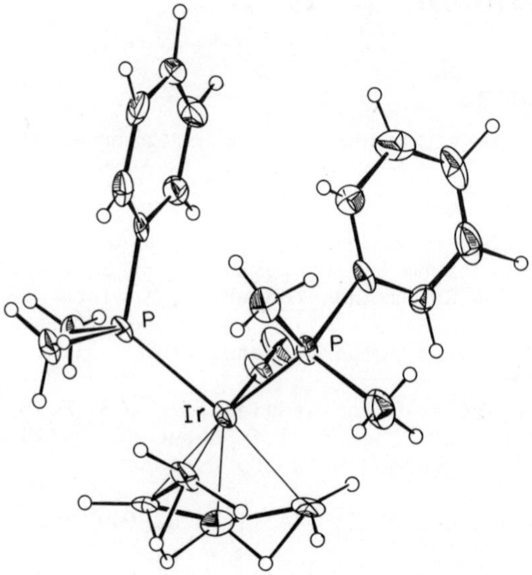

Figure 1. The molecular structure of arachno-*[IrB_4H_9)(CO)($PMe_2Ph)_2$].*

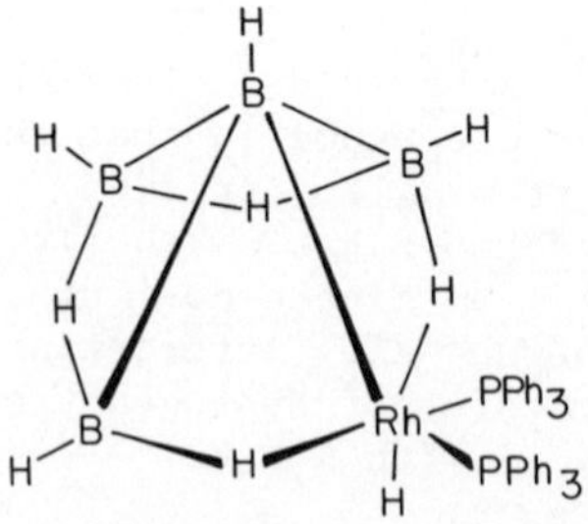

Figure 2. The proposed molecular structure of nido-*[$Rh(B_4H_8)H(PPh_3)_2$].*

Literature Cited

1. Bould, J., Greenwood, N. N., and Kennedy, J. D. J. Chem. Soc., Dalton, Trans., 1982, 481.

RECEIVED December 27, 1982

24

Platinacyclobutane Chemistry: Skeletal Isomerization, α-Elimination, and Ring Expansion Reactions

R. J. PUDDEPHATT

University of Western Ontario, Department of Chemistry, London, Ontario N6A 5B7 Canada

The rearrangement of platinacyclobutanes to alkene complexes or ylide complexes is shown to involve an initial 1,3-hydride shift (α-elimination), which may be preceded by skeletal isomerization. This isomerization can be used as a model for the bond shift mechanism of isomerization of alkanes by platinum metal, while the α-elimination also suggests a possible new mechanism for alkene polymerisation. New platinacyclobutanes with $-CH_2OSO_2Me$ substituents undergo solvolysis with ring expansion to platinacyclopentane derivatives, the first examples of metallacyclobutane to metallacyclopentane ring expansion. The mechanism, which may also involve preliminary skeletal isomerization, has been elucidated by use of isotopic labelling and kinetic studies.

Metallacyclobutanes have been proposed as intermediates in a number of catalytic reactions, and model studies with isolated transition metallacyclobutanes have played a large part in demonstrating the plausibility of the proposed mechanisms. Since the mechanisms of heterogeneously catalysed reactions are especially difficult to determine by direct study, model studies are particularly valuable. This article describes results which may be relevant to the mechanisms of isomerization of alkanes over metallic platinum by the bond shift process and of the oligomerization or polymerization of alkenes.

The proposed mechanism of the bond shift isomerization of neopentane is shown in Scheme I (1-3). There are now good models for each step in the proposed sequence, but no simple transition metal complex can accomplish all steps since there cannot be sufficient co-ordination sites. The first steps involve α,γ-dimetallation of the alkane, for which there are good precedents in both platinum and iridium chemistry (4, 5, 6). The

0097-6156/83/0211-0353$06.00/0

skeletal isomerization then occurs within the metallacyclobutane intermediate. Model studies show that this reaction is possible (equation 1) but do not positively show that the carbene-alkene intermediate shown in Scheme I is necessary (3, 7, 8).

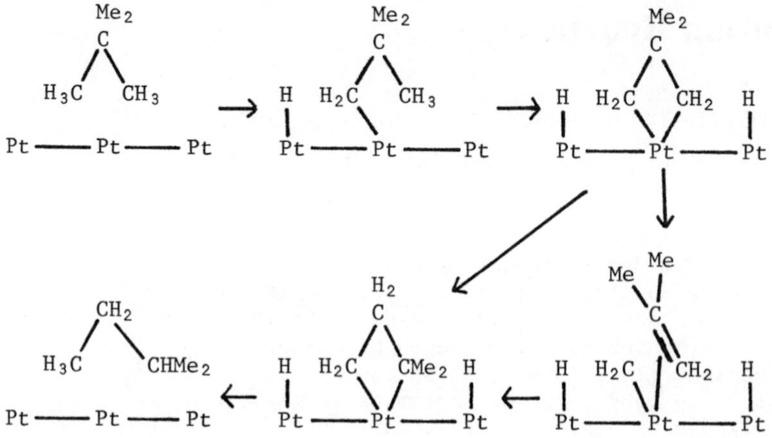

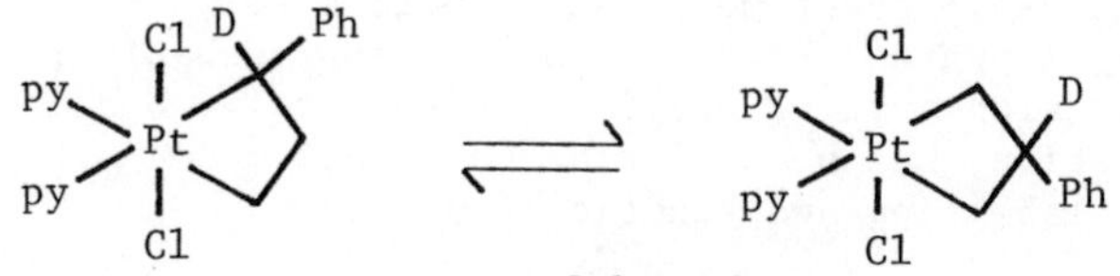

(1)

Scheme I

This skeletal isomerization is strongly retarded by the presence of free pyridine and is not observed when the monodentate pyridine ligands are replaced by bidentate ligands such as 2,2'-bipyridyl (7). This behavior is explained by the need to dissociate a pyridine ligand before the skeletal isomerization can occur, and can be useful in mechanistic investigations (*vide infra*).

Our first observations related to the particular skeletal isomerization of Scheme I were obtained in a study of steric effects of ligands on the stability of platinacyclobutanes (9). Three products could be obtained as shown in equation 2.

The isolated platinacyclobutane, (I), has the neopentane skeleton but the rearrangement products (II) and (III) have the isopentane skeleton. Low temperature NMR experiments, when L = 2,6-dimethylpyridine, showed that at -70°C the product was (I) but that this decomposed to (II) at -30°C and then (II) decomposed to

(III) at -10°C. The mechanism shown in Scheme II was therefore suggested, involving skeletal isomerization and α-elimination from the platinacyclobutane.

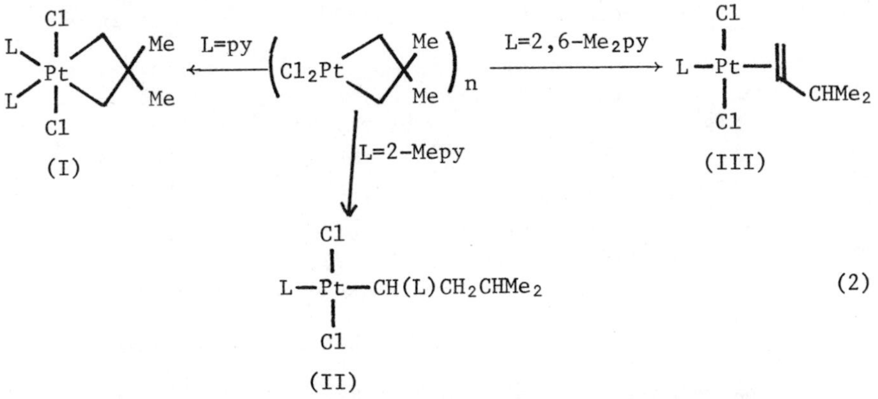

(2)

At this point it should be noted that this mechanism is unexpected. Simple platinum alkyls decompose by β-elimination whenever possible and there are no well-established examples of α-elimination [10]. All previous studies have indicated that metallacyclobutanes decompose by β-elimination, even for tantalum and titanium derivatives for which α-elimination is a frequent mechanism for decomposition of the simple alkyls [11, 12]. There is even a labelling study which appears to prove the β-elimination mechanism for decomposition of platinacyclobutanes (equation 3) [13].

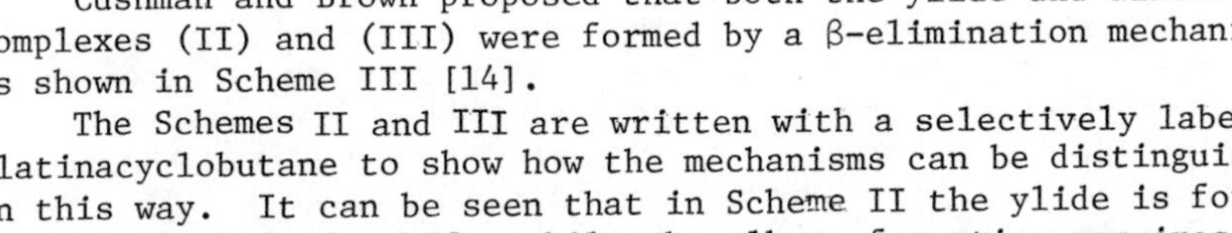

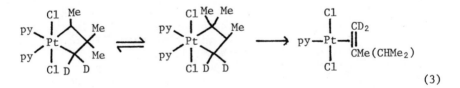

(3)

Cushman and Brown proposed that both the ylide and alkene complexes (II) and (III) were formed by a β-elimination mechanism as shown in Scheme III [14].

The Schemes II and III are written with a selectively labelled platinacyclobutane to show how the mechanisms can be distinguished in this way. It can be seen that in Scheme II the ylide is formed by a single hydride shift, while the alkene formation requires two hydride shifts. As a result the ylide CH group in (V) is adjacent to a CD_2 group and should therefore give a singlet in the ^1H NMR spectrum, while in the alkene complex products (VI) and (VII) the end groups are =CHD. In scheme III, on the other hand, the ylide is formed by a complex series of β-elimination-insertion steps and the ylide CH group in (XI) is adjacent to CHD group and should give a doublet in the ^{1}H NMR spectrum, while the alkene complexes (VIII) and (IX) are formed by a single hydride shift so that the end groups are $=CH_2$ and $=CD_2$. This will be a typical result for rearrangement of other selectively labelled platinacyclobutanes.

It was shown that the ylide formed from $[\{PtCl_2(CH_2CMe_2CD_2)\}_n]$ was (V) by the ^{1}H NMR spectrum, in the cases where L = 2,6-dimethylpyridine (at low temperature) or 2-methylpyridine. For example, the complex trans- $PtCl_2L(CHLCD_2CHMe_2)$, where L = 2,6-dimethylpyridine, gave a singlet in the ^{1}H NMR spectrum at -30°C [δ(CH)6.1 ppm, ^2J(PtH)108Hz]. Similar results were found for decomposition of the complex $[\{PtCl_2(CH_2CHPhCD_2)\}_n]$ to the ylide derivative $[PtCl_2L(CHLCD_2CH_2Ph)]$ (9), so that the formation of the ylides according to the α-elimination mechanism of Scheme II appears to be general.

It is, of course, still possible that alkene complexes are formed by β-elimination. This is most easily investigated using the platinacyclobutane $[\{PtCl_2(CD_2CHMeCHMe)\}_n]$. The predicted products according to the two opposing mechanisms are shown in equation (4).

In this case ylide complexes are not observed and therefore the reactions are very simple. When L = 2-methylpyridine or acetonitrile, the product was shown to be (XII) rather than (XIII). Complex (XII) could be characterised directly by ^{1}H and ^{13}C NMR spectroscopy or, more simply, treated with triphenylphosphine to release the alkene. Figure 1 shows the $^{13}C\{^1H\}$ NMR spectrum of the released alkene (together with 2-methylpyridine), which clearly shows 1:1:1 triplets for carbon atoms C^1 and C^4 due to coupling to deuterium as expected for the alkene from (XII) but not from (XIII). In addition, the $^2H\{^1H\}$ NMR spectrum shows approximately equal integration for deuterium at C^1 and at C^4 (15), and the ^{1}H NMR spectrum gives a doublet due to vicinal H-H coupling for the Me^5 signal. This alkene is therefore formed by the α-elimination mechanism of Scheme II.

This unexpected result promoted us to reinvestigate the decomposition of the platinacyclobutane $[\{PtCl_2(CD_2CMe_2CHMe)\}_n]$. In our hands, this gave as principal product on reaction with pyridine the complex $[PtCl_2(py)\{CHD{=}C(Me)CDMe_2\}]$ and not the product shown in equation (3). This reaction is complicated by a side reaction apparently involving β-elimination from one of the methyl substituents, but an analysis similar to that described above by ^{1}H, $^{13}C\{^1H\}$ and $^2H\{^1H\}$ NMR spectroscopy showed that the major product was formed by the α-elimination pathway (15).

Thus it seems that all platinacyclobutanes decompose by the α-elimination mechanism of Scheme II, and the reactions can be understood in terms of the following empirical rules.

(i) The skeletal isomerization is fast compared to the hydride shift reactions. As a result the ylides and alkene complexes frequently have a different carbon skeleton from the dominant platinacyclobutane (3, 9, 13-16).

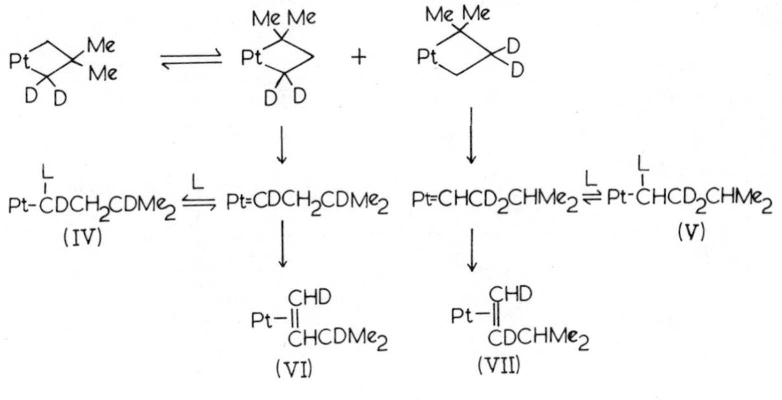

Scheme II

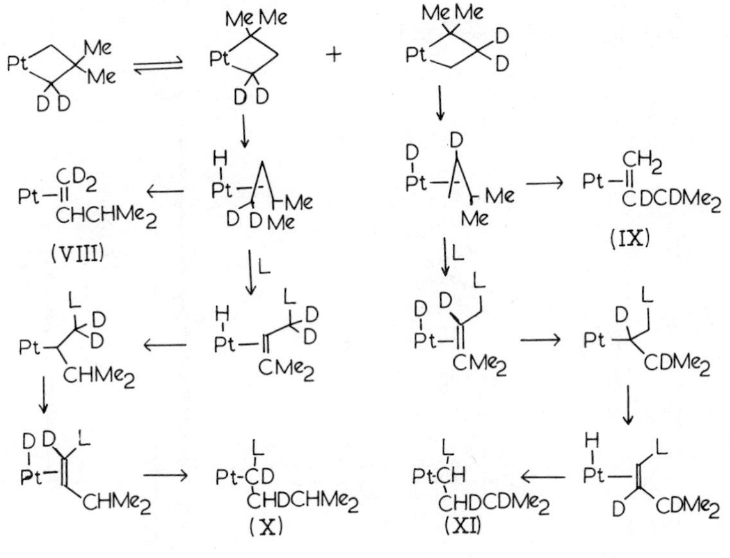

Scheme III

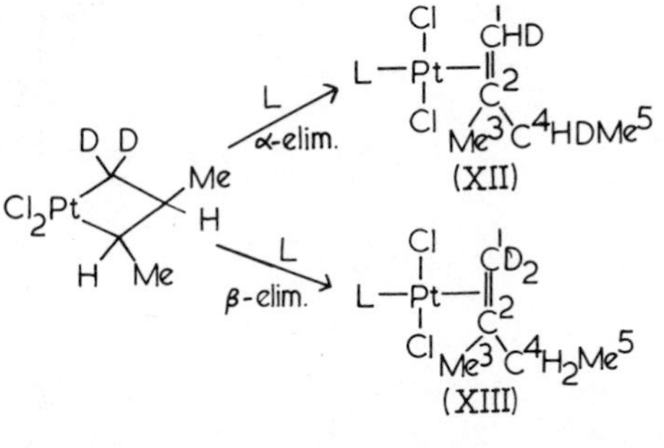

(4)

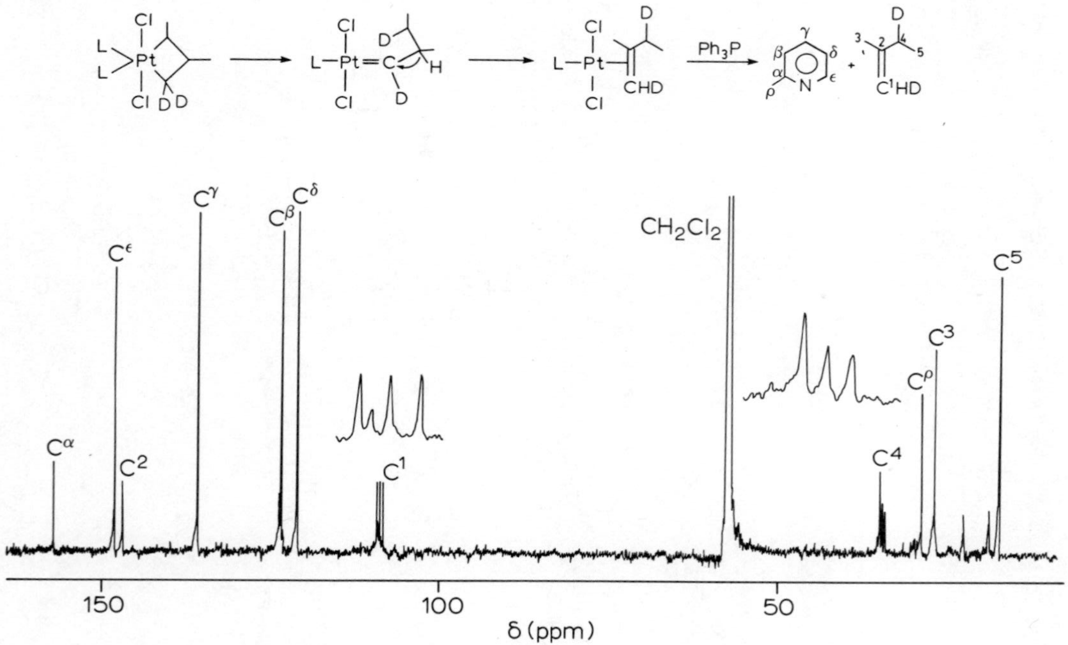

Figure 1. $^{13}C\{^{1}H\}$*-NMR spectrum of products of Equation 4 after treatment with triphenylphosphine.*

(ii). The first hydride shift occurs as a 1,3-hydride shift with transfer of hydride from a CH_2 group to the most substituted carbon atom, to give a platinum-carbene complex. Some examples to illustrate this selectivity are given in equations (2) and (3) and in Scheme II.

(iii). In some cases the carbene complex is trapped as an ylide complex by attack of the ligand L. This occurs if the β-carbon is primary (e.g. for $Pt{=}CHCH_2CHMe_2$ but not for $Pt{=}CHCHMe{-}CH_2Me$), if the ligand is compact (e.g. pyridine or 2-methylpyridine but not usually 2,6-dimethylpyridine) and if the ligand is a strong base (e.g. pyridine but not acetonitrile). The first two conditions are steric in origin while the third is electronic.

(iv). If any of the above conditions is not met, then a subsequent 1,2-hydride shift yields a platinum-alkene complex.

The observed 1,3-shift together with the high selectivity suggest a possible mechanism for the Ziegler-Natta polymerisation of alkenes to add to the mechanisms already put forward (17, 18, 19), as illustrated in equation (5), P = growing polymer.

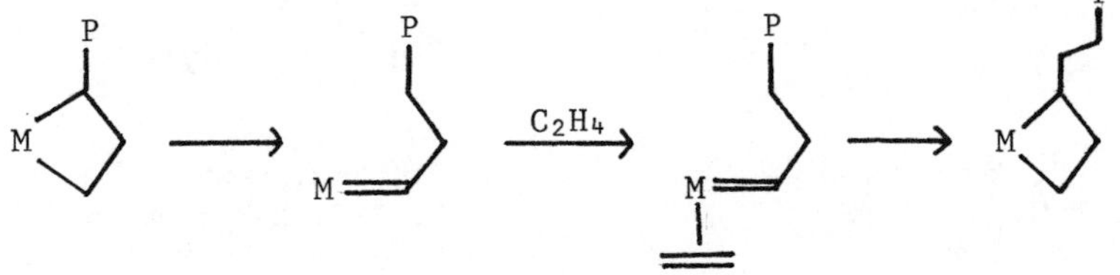

(5)

Amongst other predictions that can be made is that polymerization of $CH_2{=}CD_2$ should give -CHD- units as well as CH_2 and CD_2 units in the linear polymer, but we know of no experimental evidence on this point.

Skeletal Isomerization and Ring Expansion

Since β-elimination reactions are often rapid and reversible, it is surprising that no examples of metallacyclobutane to metallacyclopentane ring expansion reactions according to equation (6) have been found.

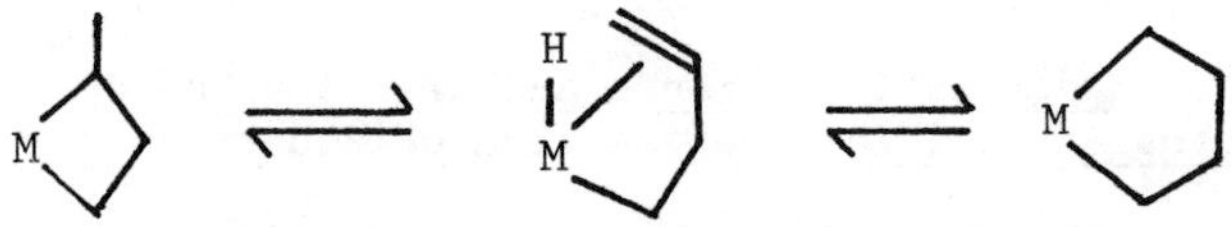

(6)

Schrock has proposed that the reverse reaction occurs during some catalytic alkene dimerisation reactions (20) but, in studies of decomposition of alkyl substituted platinacyclobutanes, no platinacyclopentanes have ever been observed even though they would be thermally stable under the reaction conditions used. A different approach was therefore used, based on the ring expansion commonly observed during solvolysis of cyclopropylmethyl or cyclobutylmethyl esters (equation 7) (21).

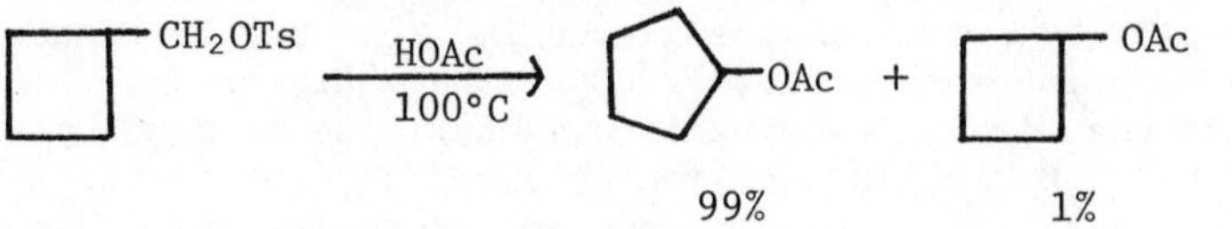
CH2OTs
HOAc
100°C
OAc
+
OAc
99%
1%

(7)

The required platinacyclobutanes were prepared and solvolysed according to equation (8) (OMs = mesylate).

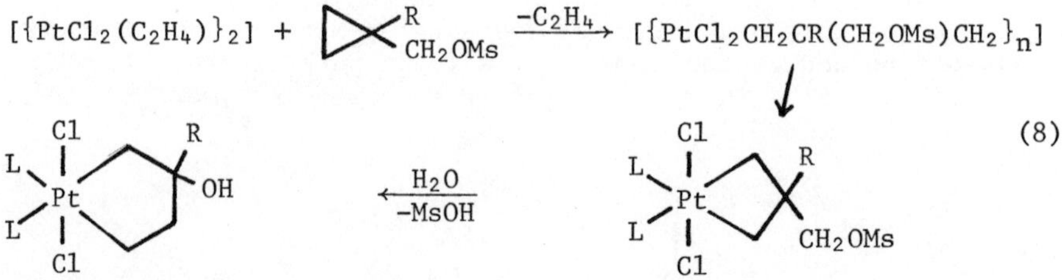
[{PtCl2(C2H4)}2] +
R
CH2OMs
-C2H4
[{PtCl2CH2CR(CH2OMs)CH2}n]
Cl
L
Pt
L
Cl
R
OH
H2O
-MsOH
Cl
L
Pt
L
Cl
R
CH2OMs

(8)

XVII, R=H, L=py
XVIII, R=H, L=½ bipy
XIX, R=Me, L=py

XIV, R=H, L=py
XV, R=H, L=½ bipy
XVI, R=Me, L=py

The solvolyses were conducted in 60% aqueous acetone at 36°C, and products were isolated in good yield and shown to be essentially pure platinacyclopentane derivatives (22). Characterisation of products included elemental analysis and ^{1}H and ^{13}C NMR spectra. Particularly striking are the differences in the coupling constants $^{1}J(PtC)$ which are ∿350 Hz in platinacyclobutanes but 490-550 Hz in platinacyclopentanes, no doubt due to opening up of CPtC angles in the larger ring.

In order to determine the mechanism of this novel rearrangement, some kinetic and labelling studies were carried out. Some results are given in Table I.

TABLE I

First Order Rate Constants and Distribution of Isotopic Label for the Reaction of Equation (9)

			% product	
L	[py]	$k_{obs}(36°C)/s^{-1}$	$1\text{-}D_2$	$3\text{-}D_2$
py	0	2.4×10^{-5}	86	14
py	0.32 M	6.0×10^{-6}	32	68
½ bipy	0	5.9×10^{-6}	27	73

The solvolyses followed simple first order kinetics but the observation that XIV, with pyridine ligands, was solvolysed four times faster than XV, with 2,2'-bipyridine ligands, suggested that ligand dissociation might be involved at some stage. In confirmation the solvolysis of XIV was found to be retarded in the presence of pyridine, the observed first order rate constants being given by the two term rate expression $k_{obs} = 5.90 \times 10^{-6} + 1.81 \times 10^{-5}/\{1 + 275 [py]\}$. The component which is independent of [py] is equal to the rate constant for solvolysis of (XV).

Next, labelling experiments were carried out using the complexes $[PtCl_2L_2CH_2CH(*CH_2OMs)CH_2]$ or $[PtCl_2L_2CH_2CH(CD_2OMs)CH_2]$, where *C = carbon enriched to 7.5% with ^{13}C. In either case, analysis by $^{13}C\{^1H\}$ NMR spectroscopy of the solvolysis product when L = py showed that the labelled carbon largely occupied the 1-position but with some in the 3-position (equation 9, Figure 2).

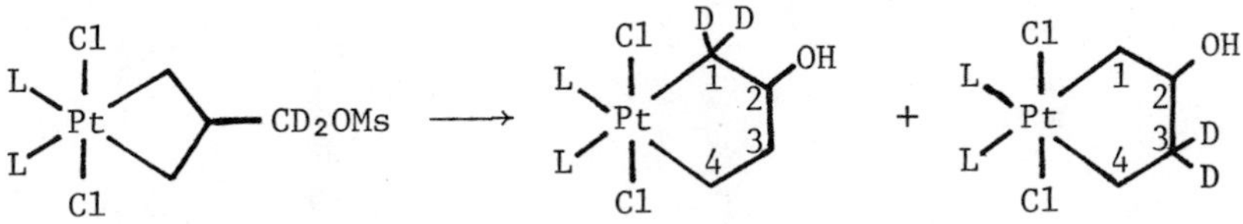

(XX) (9)

However, the ^{13}C NMR method is not quantitative and the products were therefore also analysed by the quantitative $^2H\{^1H\}$ NMR spectra. Typical spectra are shown in Figure 3 and results are summarised in Table I. It can be seen that solvolysis of (XX), L = pyridine, in the absence of added pyridine gives largely the 1-D_2 isomer but that, when sufficient pyridine is added, the selectivity is reversed and the product is largely the 3-D_2 isomer. In no case is there evidence for deuterium at either the 1- or 4-position. Solvolysis of (XX), L_2 = 2,2'-bipyridyl gives largely the 3-D_2 isomer. These results are interpreted in terms of the mechanism shown in Scheme IV.

For (XX), L = py, it is likely that the major reaction path involves initial skeletal isomerization to give (XXI) followed by rapid solvolysis of this isomer. The solvolysis of this isomer is strongly metal-assisted since the intermediate carbonium ion is stabilised by the metal-alkene resonance form as shown in the Scheme. The product is the 1-D_2 isomer. Now, the skeletal isomerization of (XX) is expected to be retarded by free pyridine and cannot occur when L_2 = 2,2'-bipyridyl (7). Hence under these conditions the reaction must occur by solvolysis of (XX) giving largely the 3-D_2 isomer. However, the product formed under these conditions is still about 30% of the 1-D_2 isomer (Table I). Either there is a route to skeletal isomerization which can occur without ligand dissociation or else solvolysis of isomer (XX), Scheme IV, can give 30% of the 1-D_2 isomer.

These solvolysis reactions essentially transform the isobutyl to the n-butyl skeleton and the above results clearly show that

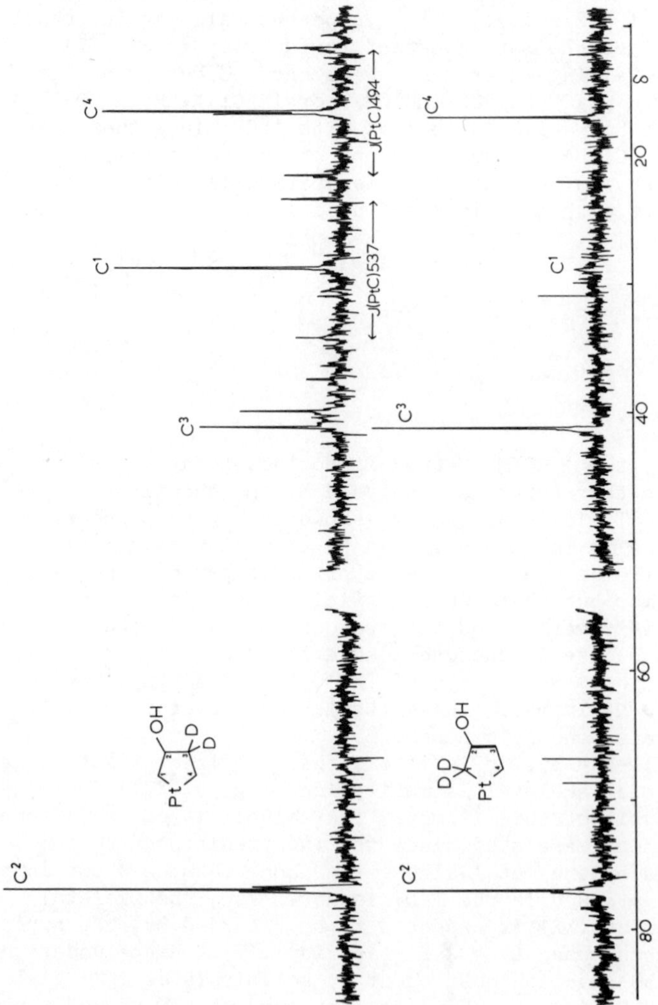

Figure 2. $^{13}C\{^{1}H\}$*-NMR spectra of products of hydrolysis of* $\{PtCl_2L_2[CH_2CH(CD_2OMs)CH_2]\}$. *Key: top, with added pyridine; and bottom, without added pyridine.*

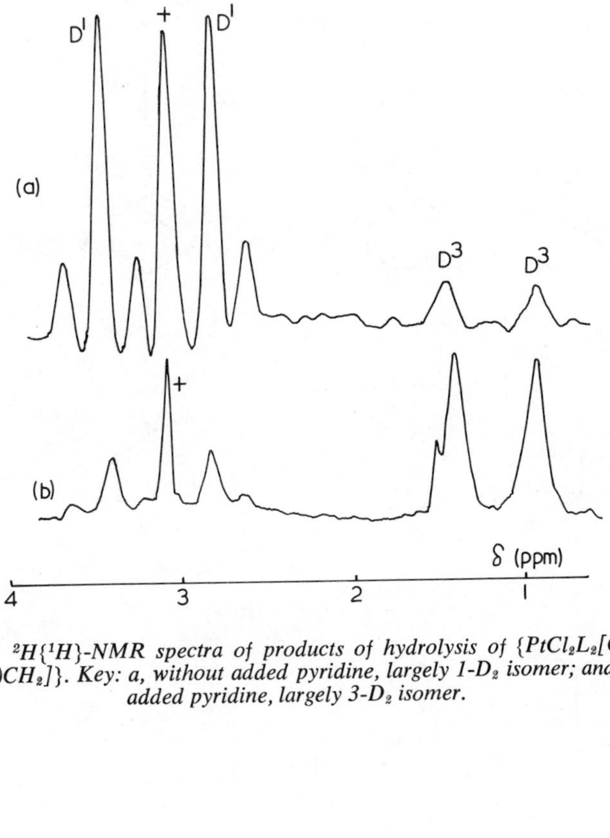

Figure 3. $^2H\{^1H\}$*-NMR spectra of products of hydrolysis of* $\{PtCl_2L_2[CH_2CH(CD_2OMs)CH_2]\}$*. Key: a, without added pyridine, largely 1-*D_2 *isomer; and b, with added pyridine, largely 3-*D_2 *isomer.*

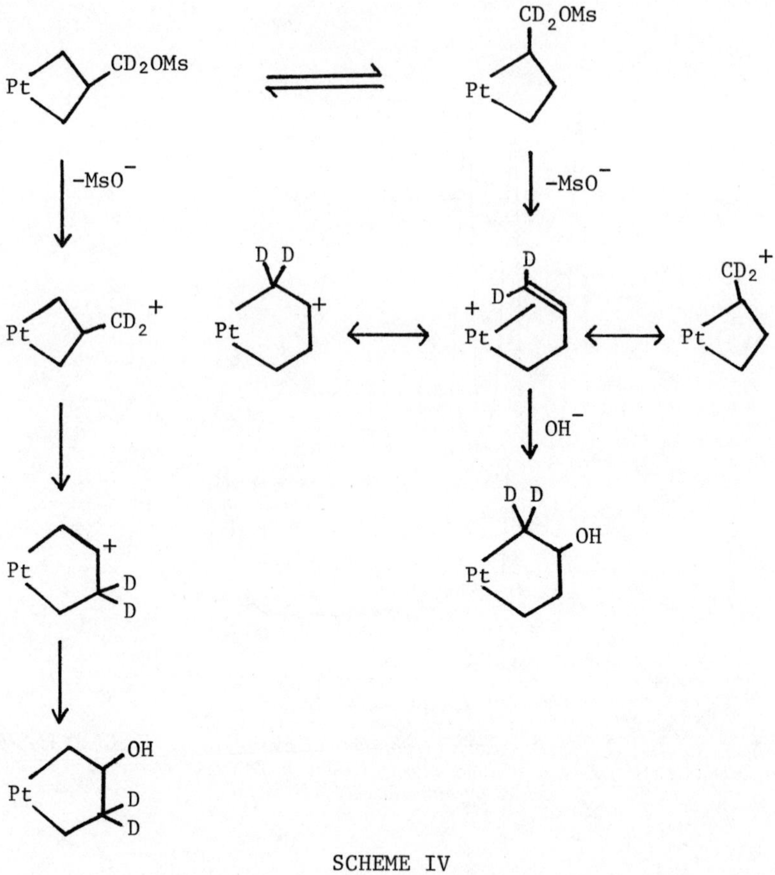

SCHEME IV

this can be accomplished either by skeletal isomerization of the platinacyclobutane or by a normal carbonium ion rearrangement. Taken together with the earlier rearrangements to ylide or alkene complexes, the results establish clearly that the skeletal rearrangement is of fundamental importance in the chemistry of platinacyclobutanes and lend credence to the proposed mechanism of alkane isomerization given in Scheme I.

Acknowledgments

The new results described are from the work of graduate students S.M. Ling (α-elimination) and J.T. Burton (ring expansion). We thank NSERC (Canada) for financial support. Acknowledgment is made to the donors of the Petroleum Research Fund, administered by the American Chemical Society, for partial support. We thank Dr. J.B. Stothers for the ^{13}C and ^{2}H NMR spectra.

Literature Cited

1. Amir-Ebrahami, V.; Garin, F.; Weinsang, F; Gault, F.G. Nouv. J. Chim. 1979, 3, 529. Gault, F.G. Gazz. Chim. Ital. 1979, 109, 255.
2. Parshall, G.W.; Herskovitz, T.; Tebbe, F.N.; English, A.D.; Zeile, J.V. Fundam. Res. Homogen. Catal. 1979, 3, 95. Parshall, G.W., "Homogeneous Catalysis", Wiley, N.Y., 1980.
3. Al-Essa, R.J.; Puddephatt, R.J.; Thompson, P.J.; Tipper, C. F.H. J. Am. Chem. Soc. 1980, 102, 7546. Puddephatt, R.J. Co-ord. Chem. Rev. 1980, 33, 149.
4. Janowicz, A.H.; Bergman, R.G. J. Am. Chem. Soc. 1982, 104, 352.
5. Tulip, T.H.; Thorn, D.L. J. Am. Chem. Soc. 1981, 103, 2448.
6. Foley, P.; Whitesides, G.M. J. Am. Chem. Soc. 1979, 101, 2732.
7. Puddephatt, R.J.; Quyser, M.A.; Tipper, C.F.H. J. Chem. Soc. Chem. Comm. 1976, 626. Al-Essa, R.J.; Puddephatt, R.J.; Quyser, M.A.; Tipper, C.F.H. J. Am. Chem. Soc. 1979, 101, 364.
8. Casey, C.P.; Scheck, D.M.; Shusterman, A.J. J. Am. Chem. Soc. 1979, 101, 4233.
9. Al-Essa, R.J.; Puddephatt, R.J. J. Chem. Soc. Chem. Comm. 1980, 45.
10. Whitesides, G.M.; Gaasch, J.F.; Stedronsky, E.R. J. Am. Chem. Soc. 1972, 94, 5258.
11. Tebbe, F.N.; Parshall, G.W.; Reddy, G.S. J. Am. Chem. Soc. 1978, 100, 3611.
12. Schrock, R.R. Accounts Chem. Res. 1979, 12, 98.
13. Johnson, T.H.; Cheng, S.-S. J. Am. Chem. Soc. 1979, 101, 5277.
14. Cushman, B.M.; Brown, D.B. Inorg. Chem. 1981, 20, 2490.

15. Ling, S.M.; Puddephatt, R.J. J. Chem. Soc. Chem. Comm. 1982, 412.
16. Al-Essa, R.J.; Puddephatt, R.J.; Perkins, D.C.L.; Rendle, M.C.; Tipper, C.F.H. J. Chem. Soc. Dalton Trans. 1981, 1738.
17. Watson, P.L. J. Am. Chem. Soc. 1982, 104, 337 and refs. therein.
18. Turner, H.W.; Schrock, R.R. J. Am. Chem. Soc. 1982, 104, 2331 and refs. therein.
19. McKinney, R.J. J. Chem. Soc. Chem. Comm. 1980, 490.
20. McLain, S.J.; Sancho, J.; Schrock, R.R. J. Am. Chem. Soc. 1979, 101, 5451.
21. Richey, H.G. Jr.; Wiberg, K.B.; Hess, B.A. Jr.; Ashe, A.J. III, in "Carbonium Ions", vol. III, Olah, G.A.; Schleyer, P. von R., eds., Wiley, N.Y., 1972, pp. 1201 and 1295.
22. Burton, J.T.; Puddephatt, R.J. J. Am. Chem. Soc., in the press.

RECEIVED August 19, 1982

Discussion

W.L. Gladfelter, University of Minnesota: At this time, can you make any generalizations which would allow you to predict when you will see α-hydrogen abstraction versus β-hydrogen abstraction?

R. J. Puddephatt: Not really. I think almost all organometallic chemists would have predicted β-elimination from platinacyclobutanes and that, if platinacyclobutanes did α-eliminate, then most other metallacyclobutanes should also. Both predictions would have been wrong and the reasons are still obscure. In a more general observation, although we now know the mechanisms of many of the fundamental organometallic reactions, we know very little about factors which influence selectivity. In the present case, one could argue in terms of the relative stabilities of carbene-hydride vs. η-alkyl-hydride intermediates but, since nothing is known about either class of compound in platinum (IV) chemistry, the arguments are not convincing. More experimental work is needed to establish general patterns of behavior for other metals.

A.J. Carty, Guelph-Waterloo Centre: It might be expected that the facility with which metallocyclobutanes undergo α- or β-eliminations would be rather dependent on the proximity of hydrogen atoms on the α- or β-carbon atoms to the metal. In other words, there might be a structural dependence. Is there any evidence that such structural features contribute substantially to the preferred reaction pathway?

R. J. Puddephatt: The β-elimination mechanism requires considerable puckering of the platinacyclobutane ring to bring the β-hydrogen atom close to platinum. We, in collaboration with J.A.Ibers' group, have shown that some platinacyclobutanes have almost planar PtC_3 rings in the solid state but that the rings are puckered by about 25° in solution, indicating that the activation energy for puckering is small(1). However, it is possible in these Pt((IV) complexes that the more extreme puckering which must precede β-elimination is prevented by steric hindrance of the axial halogen substituents. It would be interesting to find if platinum(II) derivatives, for which this hindrance should be much less, undergo α- or β-elimination.

(1) J. T. Burton, R. J. Puddephatt, N.L. Jones and J. A. Ibers, J. Am. Chem. Soc., submitted for publication.

R.R. Schrock, M.I.T.: Have you or anyone else prepared platinum(IV) metallacyclobutane complexes with alkoxide ligands in place of chlorides? One might expect the alkoxide complexes to behave considerably differently than the chloro complexes, perhaps like early transition metal complexes.

R. J. Puddephatt: No. Nobody has prepared such complexes and the synthesis is not trivial. Substitution of halide ligands in octahedral platinum(IV) derivatives is typically very slow, and a better route (suggested by J. K. Kochi) might involve oxidation of platinum(II) metallacyclobutanes with peroxides. It would certainly be worthwhile to attempt this synthesis in view of the promise of enhanced reactivity.

25

Multiple Metal–Carbon Bonds in Catalysis

RICHARD R. SCHROCK

Massachusetts Institute of Technology, Department of Chemistry, Cambridge, MA 02139

The presence of alkoxide ligands slows down the rate of rearrangement of tantalacyclobutane rings and probably also speeds up the rate of reforming an alkylidene complex and an olefin. However, tantalum and niobium alkylidene complexes are not good olefin metathesis catalysts because either intermediate methylene complexes decompose rapidly, or because intermediate alkylidene ligands rearrange to olefins. Tungsten(VI) oxo and imido alkylidene complexes will metathesize olefins, probably because rearrangement processes involving a β-hydride are even slower as a result of the π-electron donor abilities of the oxo or imido ligand. Disubstituted acetylenes are metathesized by tungsten(VI) alkylidyne complexes containing t-butoxide ligands. When chloride ligands are present instead of t-butoxides, a tungstenacyclobutadiene complex can be isolated. It reacts with additional acetylene to give a cyclopentadienyl complex. Tantalum neopentylidene hydride complexes react with ethylene to form new alkylidene hydride complexes in which many ethylenes have been incorporated into the alkyl chain. The polymer that slowly forms in the presence of excess ethylene is approximately a 1:1 mixture of even and odd carbon olefins in the range C_{50}-C_{100}.

E.O. Fischer's discovery of $(CO)_5W[C(Ph)(OMe)]$ in 1964 marks the beginning of the development of the chemistry of metal-carbon double bonds (1). At about this same time the olefin metathesis reaction was discovered (2), but it was not until about five years later that Chauvin proposed (3) that the catalyst contained an alkylidene ligand and that the mechanism consisted of the random reversible formation of all possible metallacyclobutane rings. Yet low oxidation state Fischer-type carbene complexes were found not to be catalysts for the metathesis of simple olefins. It is now

0097-6156/83/0211-0369$06.00/0

virtually certain that the alkylidene chain transfer mechanism is correct and that the most active catalysts are d^0 complexes (counting the CHR ligand as a dianion). In this article I'd like first to trace the events and findings which led to our concluding that d^0 alkylidene complexes were responsible for the rapid catalytic metathesis of olefins. Then I want to present some recent results concerning the polymerization of ethylene by an alkylidene hydride catalyst, and finally some results concerning the metathesis of acetylenes by tungsten(VI) alkylidyne complexes. We will see that an important feature of much of the chemistry of multiple metal-carbon bonds is the role played by alkoxide, oxo, or other π-bonding ligands. Such observations are congruent with some recent ideas and results Chisholm discusses elsewhere in this volume concerning alkoxide ligands in organometallic chemistry.

Tantalum and Niobium Neopentylidene Complexes (4)

The first neopentylidene complex was prepared by the reaction shown in equation 1 (5). Although the exact details of this reac-

$$Ta(CH_2CMe_3)_3Cl_2 + 2LiCH_2CMe_3 \xrightarrow[-CMe_4]{pentane} Ta(CHCMe_3)(CH_2CMe_3)_3 \quad (1)$$

tion are still unclear (6), it is almost certainly a version of what has come to be called an α-hydrogen atom abstraction reaction. The best studied example of α-hydrogen atom abstraction is the intramolecular decomposition of $Ta(\eta^5\text{-}C_5H_5)(CH_2CMe_3)_2Cl_2$ to $Ta(\eta^5\text{-}C_5H_5)(CHCMe_3)Cl_2$ (7). But the simplest and most general is the α-hydrogen atom abstraction in $M(CH_2CMe_3)_2X_3$ M = Nb or Ta, X = Cl or Br) promoted by oxygen, nitrogen, or phosphorus donor ligands (8). The resulting octahedral molecules of the type $M(CHCMe_3)L_2X_3$ offered an ideal opportunity to study how a neopentylidene complex of Nb or Ta reacts with a simple olefin.

A complex such as $Ta(CHCMe_3)(PMe_3)_2Cl_3$ reacts readily with ethylene, propylene, or styrene to give all of the possible products (up to four) which can be formed by rearrangement of intermediate metallacyclobutane complexes (two for substituted olefins) by a β-hydride elimination process (e.g., equation 2) (9). We saw

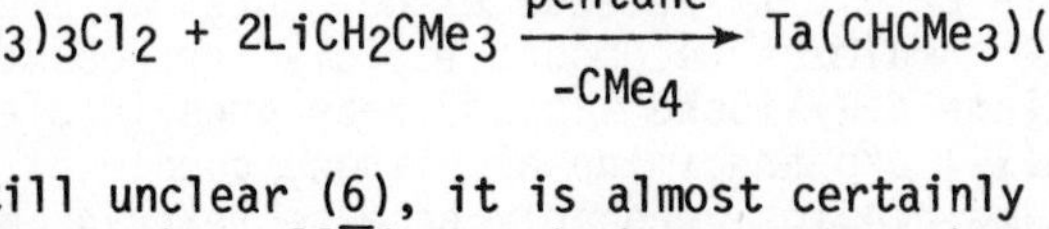
Bu^t
Ta
R
Bu^t
R
+
Bu^t
R

(2)

absolutely no evidence for a metathesis-like reaction until we studied the complexes $M(CHCMe_3)(THF)_2Cl_3$ (M = Nb or Ta). In this case we found low but reproducible yields (5-15%) of 3,3-dimethyl-1-butene upon reacting $M(CHCMe_3)(THF)_2Cl_3$ with ethylene, and in the case of cis-2-pentene, ~6 turnovers to 3-hexenes and 2-butenes. We reasoned that the rate of metathesis of the MC_3 ring was faster relative to the rate of rearrangement of the MC_3 ring when an oxygen donor ligand was present in place of a phosphine ligand. Therefore we prepared t-butoxy complexes such as $Ta(CHCMe_3)$-$(OCMe_3)_2(PMe_3)Cl$ in order to see if selective metathesis of an incipient MC_3 ring would result.

$Ta(CHCMe_3)(OCMe_3)_2(PMe_3)Cl$ reacts with ethylene, styrene, or 1-butene to give largely t-butylethylene (equation 3).

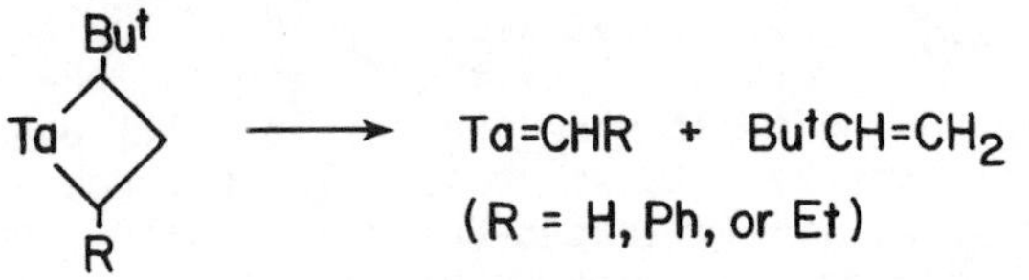

$$Ta \quad \overset{But}{\underset{R}{\diamondsuit}} \quad \longrightarrow \quad Ta=CHR \; + \; Bu^tCH=CH_2 \qquad (3)$$
$$(R = H, Ph, or \; Et)$$

When styrene is the olefin the resulting benzylidene complex can be trapped in the presence of additional PMe_3 to give $Ta(CHPh)$-$(OCMe_3)_2(PMe_3)_2Cl$. Neither the methylene nor the propylidene complex could be observed, but in the case of 1-butene we could trace the fate of intermediate metallacyclobutane and alkylidene complexes. Metathesis of 1-butene was not successful for two reasons. First, an intermediate β-ethylmetallacyclobutane complex rearranges to 2-methyl-1-butene. Second, intermediate methylene complexes decompose by a bimolecular reaction to give ethylene.

In contrast to the failure to metathesize terminal olefins, internal olefins such as cis-2-pentene can be metathesized to the extent of ~50 turnovers. The chain terminating reaction in this case is rearrangement of intermediate ethylidene and propylidene complexes (equation 4). Both rearrangement of intermediate trisub-

$$M=C\overset{H}{\underset{CH_2R}{\Big\langle}} \quad \longrightarrow \quad CH_2=CHR \quad (R = H \; or \; Me) \qquad (4)$$

stituted metallacyclobutane complexes and bimolecular decomposition of monosubstituted alkylidene complexes must be slow enough to allow a significant number of steps in the metathesis reaction to proceed. Rearrangement of intermediate alkylidene complexes then becomes the major termination step.

There had been some evidence that alkoxide ligands slow down reactions which involve elimination of a β-hydride from an alkyl ligand. α-Olefins are dimerized to a mixture of head-to-tail and tail-to-tail dimers by olefin complexes of the type $Ta(\eta^5-C_5Me_5)$-$(CH_2=CHR)Cl_2$ (10). The β,β'- and α,β'-disubstituted tantalacyclopentane complexes are intermediates in this reaction. Their decomposition involves the sequence shown in equation 5. When one

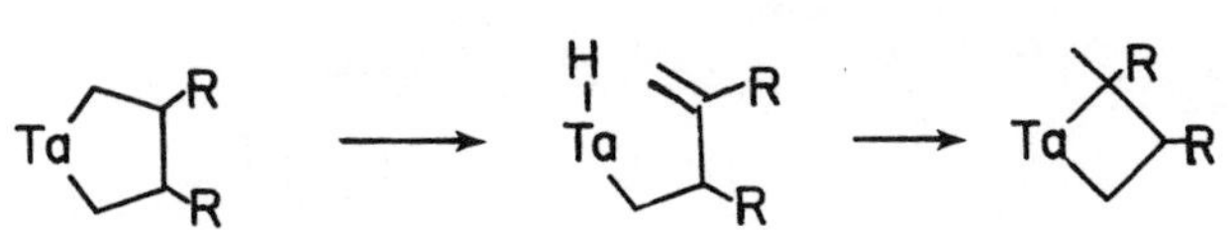

chloride ligand in the catalyst is replaced by a methoxide ligand the rate of olefin dimerization decreases by a factor of approximately 10^2 as a result of the greater stability of the tantalacyclopentane complexes. Although it could not be proven, it was felt that the first step, β-hydride elimination, had been slowed down significantly. Therefore it was felt that the β-elimination process by which tantalacyclobutane complexes rearranged to olefins at least would be slowed down by replacing two chloride ligands in $Ta(CHCMe_3)(PMe_3)_2Cl_3$ with t-butoxide ligands. The question as to whether the rate of metathesis of the TaC_3 ring increases upon replacing chloride by t-butoxide ligands is still open. However, on the basis of some results we present later concerning acetylene metathesis it seems likely that t-butoxide ligands encourage reformation of the metal-carbon double bond.

Olefin Metathesis by Tungsten Oxo and Imido Complexes

In retrospect it is not surprising that the niobium and tantalum alkylidene complexes we prepared are not good metathesis catalysts since these metals are not found in the "classical" olefin metathesis systems (2). Therefore, we set out to prepare some tungsten alkylidene complexes. The first successful reaction is that shown in equation 6 ($L = PMe_3$ or PEt_3) (11). These oxo

$$Ta(CHCMe_3)L_2Cl_3 + W(O)(OCMe_3)_4 \longrightarrow Ta(OCMe_3)_4Cl + W(O)(CHCMe_3)L_2Cl_2 \quad (6)$$

alkylidene complexes are octahedral species in which the oxo and alkylidene ligands are cis to one another and the $W(O)(CHC_\beta)$ atoms all lie in the same plane (12). This type of structure can be rationalized easily on the basis of the fact that the oxo ligand is an excellent π-electron donor (13). Therefore, the oxo ligand uses two of the available three d orbitals of π-type symmetry to bond to W, leaving only one to form the π-bond between W and the alkylidene ligand. Several different types of oxo neopentylidene complexes have been prepared including $W(O)(CHCMe_3)(L)Cl_2$, $[W(O)(CHCMe_3)L_2Cl]^+$, and $[W(O)(CHCMe_3)L_2]^{2+}$. Characterizeable oxo neopentylidene complexes have not yet been prepared directly from oxo neopentyl complexes by α-hydrogen abstraction reactions, although Osborn has presented some evidence that they could be (14). Another potentially important method of preparing oxo alkylidene complexes is by adding water or hydroxide to alkylidyne complexes (see later) as shown in equation 7 (15).

$$[W(CCMe_3)Cl_4]^- + 2PEt_3 + Et_3N + H_2O \longrightarrow W(O)(CHCMe_3)(PEt_3)_2Cl_2 \quad (7)$$

Imido alkylidene complexes were first prepared by a reaction analogous to that shown in equation 6. Recently they have been prepared from imido alkyl complexes by well-behaved α-hydrogen abstraction reactions (16). Imido neopentylidene complexes seem to be more stable than oxo neopentylidene complexes, possibly because the oxo ligand is sterically more accessible to Lewis acids, including another tungsten center.

Oxo alkylidene complexes react with olefins in the presence of a trace of $AlCl_3$ to give new alkylidene complexes (e.g., benzylidene, methylene, ethylidene) (11a). Both terminal and internal olefins can be metathesized slowly in the presence of aluminum chloride. Probably the best catalysts are the ionic species, $[W(O)(CHCMe_3)(PEt_3)_2Cl]^+AlCl_4^-$ and $[W(O)(CHCMe_3)(PEt_3)_2]^{2+}(AlCl_4^-)_2$ in dichloromethane or chlorobenzene (17). Of the order of 10-20 turnovers per hour for a day or more are possible with these cationic catalysts. These studies demonstrate that transalkylidenation is possible with a d^0 tungsten alkylidene complex that will metathesize olefins slowly, but convincingly. There is still considerable doubt concerning the role of the Lewis acid. However, the fact that $W(O)(CHCMe_3)(PEt_3)Cl_2$ (18) metathesizes olefins more rapidly initially than the six-coordinate complexes (in the presence of $AlCl_3$) establishes that a Lewis acid is not required. On the basis of these studies and some calculations by Rappé and Goddard (19) it would seem incontrovertible that the oxo ligand prevents reduction of the metal and perhaps also enhances the rate of reforming an alkylidene complex from a metallacyclobutane complex. The next question was whether other strong π-donor ligands such as alkoxides could take over the oxo's function (11b).

Osborn's discovery (14) that aluminum halides bind to oxo ligands in tungsten oxo neopentyl complexes, and that these complexes decompose to give systems which will efficiently metathesize olefins, raised more questions concerning the role of the Lewis acid. A subsequent communication (20) answered some of the questions; the aluminum halide removes the oxo ligand and replaces it with two halides to yield neopentylidene complexes (equation 8).

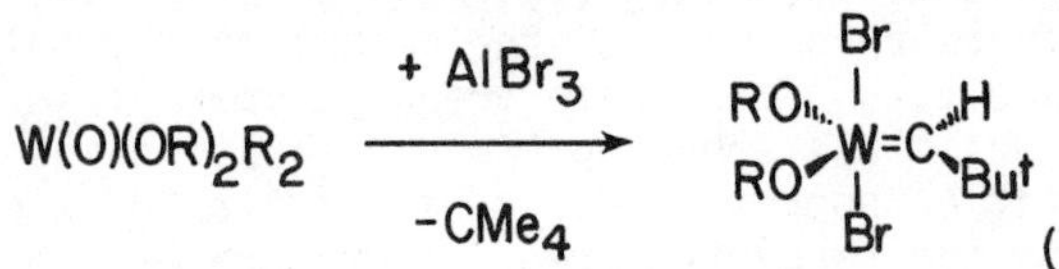

$$W(O)(OR)_2R_2 \xrightarrow[-CMe_4]{+\ AlBr_3} \underset{\underset{Br}{|}}{\overset{\overset{Br}{|}}{RO_{\prime\prime\prime}W=C}}\overset{H}{\underset{Bu^t}{}} \qquad (8)$$

$$(R = CH_2Bu^t)$$

Additional aluminum halide coordinates to an axial halide to give a species which will metathesize olefins extremely efficiently. These studies demonstrate that two alkoxide ligands can take the place of an oxo ligand and that aluminum halides, by coordinating to a halide ligand, can generate an efficient and long-lived catalyst. It is possible that $[W(CHR)(OR)_2(Br)]^+AlBr_4^-$ is responsible for the catalytic activity, but at low concentrations, and in the presence of excess olefin, bimolecular decomposition of the cationic species should be slow.

The Reaction of Tantalum Neopentylidene Hydride Complexes with Ethylene

A few years ago Ivin, Rooney, and Green made a provocative suggestion for which there was only tenuous experimental support (21). They suggested that stereospecific propylene polymerization by Ziegler-Natta catalysts could be explained by a mechanism involving reaction of the olefin with an alkylidene ligand in an alkylidene hydride catalyst. We set out to test this proposal by preparing and studying tantalum alkylidene hydride complexes (22). One of these, $Ta(CHCMe_3)(H)(PMe_3)_3Cl_2$, reacted readily with ethylene to give neither products of rearrangement nor metathesis of an intermediate tantalacyclobutane complex. High boiling products were formed but we could not obtain consistent results. However, results using $Ta(CHCMe_3)(H)(PMe_3)_3I_2$ were consistent and reproducible (23).

$Ta(CHCMe_3)(H)(PMe_3)_3I_2$ is probably a pentagonal bipyramidal complex containing an axial neopentylidene ligand with the phosphines, one iodide, and the hydride in the pentagonal plane (cf. $Ta(CCMe_3)(H)(dmpe)_2(ClAlMe_3)$ (24)). The ^1H NMR signal for the hydride ligand is a characteristic octet at δ 8.29 while the broad alkylidene α-proton signal is found at δ -1.76. On addition of a limited quantity of ethylene virtually identical ^1H NMR patterns appear at δ 7.74 and δ -0.73 consistent with formation of a new complex, $Ta(CHR)(H)(PMe_3)_3I_2$; ~50% of the original $Ta(CHCMe_3)(H)(PMe_3)_3I_2$ remains. The volatiles formed on treatment of this mixture with CF_3CO_2H consist of neopentane (~50%), and the alkanes $Me_3C(CH_2CH_2)_nCH_3$ where n = 1, 2, 3, and 4 (~50% total yield), consistent with hydrolysis of $Ta[CH(CH_2CH_2)_nCMe_3](H)(PMe_3)_3I_2$. When CD_2CD_2 is used the new product is $Ta(CDR)(D)(PMe_3)_3I_2$.

When excess ethylene is added to $Ta(CHCMe_3)(H)(PMe_3)_3I_2$ a pale green polymer slowly forms which weighs approximately four times the original weight of $Ta(CHCMe_3)(H)(PMe_3)_3I_2$ after two days; at this point any further increase of the weight of the polymer is negligible. Hydrolysis of the pale green polymer yielded a white organic polymer which was shown by field desorption mass spectral studies to consist of approximately a 1:1 mixture of even and odd carbon olefins in the range C_{50}-C_{100}. Therefore, the pale green polymer is largely organic. By similarly studying the polymer prepared from $Ta(CDCMe_3)(D)(PMe_3)_3I_2$ we showed that only the odd carbon polymers increased by two mass units. Therefore, most of the even carbon polymers must be polyethylene. The mechanism of chain transfer is at present unknown, but the preceding result suggests that it is not metathesis of metallacyclobutane rings.

There are two ways of viewing the reaction between $Ta(CHCMe_3)(H)(PMe_3)_3I_2$ and ethylene. The one shown in equation 9 (nonessen-

$$\overset{\displaystyle H}{\overset{|}{\mathrm{Ta}}}\mathrm{{=}CHCMe_3} \rightleftharpoons \mathrm{TaCH_2CMe_3} \xrightarrow{\mathrm{C_2H_4}} \mathrm{TaCH_2CH_2CH_2CMe_3} \rightleftharpoons \overset{\displaystyle H}{\overset{|}{\mathrm{Ta}}}\mathrm{{=}CHCH_2CH_2CMe_3} \qquad (9)$$

tial ligands omitted) contains in part the classical Cossee-type step (25) where ethylene "inserts" into the tantalum(III)-neopentyl and subsequent tantalum(III)-alkyl bonds. It cannot yet be ruled out since magnetization transfer experiments show that the alkylidene α-proton and the hydride ligand in $Ta(CHR)(H)(PMe_3)_3I_2$ exchange readily, most likely by forming $Ta(CH_2R)(PMe_3)_3I_2$. The alternative shown in equation 10 is analogous to that proposed by Ivin, Rooney and Green. We do not think it will be easy to distinguish between these two possibilities, if it is possible at all. But since alkylidene ligands in other tantalum(V) complexes react rapidly with olefins, and since there are few examples of isolable transition metal alkyl complexes that react readily with ethylene (26), we feel that the second alternative is more plausible.

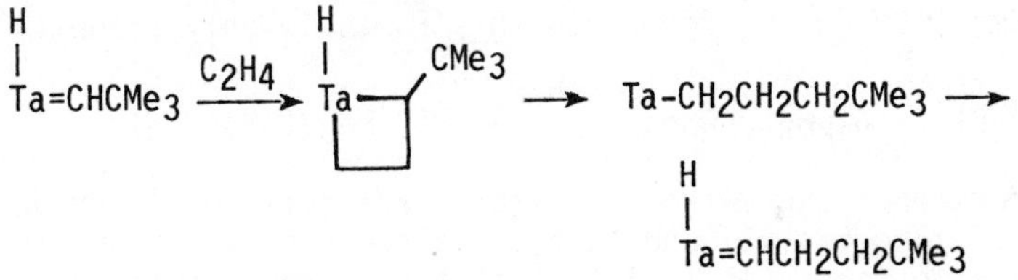

(10)

One of the most interesting aspects of the mechanism shown in equation 10 is the last step, an α-elimination reaction to give the new alkylidene hydride complex. Our results do not imply that β-elimination to give an olefin hydride intermediate is relatively slow. It is possible that although $K_2 > K_1$, $k_1 > k_2$ (equation 11), i.e., β-elimination is still faster. If this is true, it must also

$$\overset{\displaystyle H}{\overset{|}{\mathrm{Ta}}}\text{-}\Vert^{\mathrm{CH_2}}_{\mathrm{CHR}} \underset{k_{-1}}{\overset{k_1}{\rightleftharpoons}} \mathrm{TaCH_2CH_2R} \underset{k_{-2}}{\overset{k_2}{\rightleftharpoons}} \overset{\displaystyle H}{\overset{|}{\mathrm{Ta}}}\mathrm{{=}CHCH_2R} \qquad (11)$$

be true that the olefin hydride complex is relatively stable toward displacement of $CH_2{=}CHR$ by ethylene under the reaction conditions which we employ.

These results at least demonstrate that ethylene can be polymerized by an alkylidene hydride catalyst, probably by forming a metallacyclobutane hydride intermediate. The extent to which this is relevant to the more classical Ziegler-Natta polymerization systems (27) is unknown. Recent results in lutetium chemistry (28), where alkylidene hydride complexes are thought to be unlikely, provide strong evidence for the classical mechanism.

Tungsten(VI) Alkylidyne Complexes and Acetylene Metathesis

On the basis of the fact that tungsten(VI) alkylidene complexes will metathesize olefins one might predict that acetylenes should be metathesized by tungsten(VI) alkylidyne complexes (29). Acetylene metathesis is not unknown, but the catalysts are inefficient and poorly understood (30, 31).

The first tungsten(VI) alkylidyne complex was prepared in low yield (~20%) by reacting WCl_6 with six equivalents of neopentyl lithium (32). Three equivalents of the lithium reagent are used simply to reduce W(VI) to W(III). Therefore the yield is limited and the mechanism by which $W(CCMe_3)(CH_2CMe_3)_3$ forms obscure. A higher yield route to $W(CCMe_3)(CH_2CMe_3)_3$ consists of the reaction shown in equation 12 (33). Reproducible yields of 50-70% can be obtained on a relatively large scale (30 g). The mechanism by which $W(CCMe_3)(CH_2CMe_3)_3$ forms via this route is only slightly

$$W(OMe)_3Cl_3 + 6NpMgCH_2CMe_3 \xrightarrow{\text{ether}} W(CCMe_3)(CH_2CMe_3)_3 \qquad (12)$$

better understood; the methoxide ligands are believed to prevent reduction of tungsten(VI) and so allow a tungsten(VI) neopentylidene complex to form. It is felt that once a neopentylidene complex forms, formation of a neopentylidyne complex would be fast. Since both $W(OMe_3)_3(CH_2CMe_3)_3$ and $W(OMe)_2Np_4$ can be prepared, and shown not to be converted into $W(CCMe_3)(CH_2CMe_3)_3$ under the reaction conditions, the crucial intermediate most likely still contains some halide(s). $W(OMe)_2(CH_2CMe_3)_2Cl_2$ is an interesting possibility since $W(OCH_2CMe_3)_2(CH_2CMe_3)_2Br_2$ is a plausible precursor to $W(CHCMe_3)(OCH_2CMe_3)_2Br_2$ (equation 8).

$W(CCMe_3)Np_3$ reacts with three equivalents of HCl in ether or dichloromethane in the presence of NEt_4Cl to yield blue $[NEt_4]$-$[W(CCMe_3)Cl_4]$ quantitatively (34). If 1,2-dimethoxyethane is present instead of NEt_4Cl, the product is purple $W(CCMe_3)(dme)Cl_3$. Either reacts smoothly with three equivalents of LiX to give $W(CCMe_3)X_3$ (X = $OCMe_3$, $SCMe_3$, NMe_2). All are thermally stable, sublimable, monomeric pale yellow to white, crystalline species.

$W(CCMe_3)(OCMe_3)_3$ reacts rapidly with symmetric acetylenes to give the new alkylidyne complexes shown in equation 13. $W(CPh)(OCMe_3)_3$ is orange and $W(CCH_2CH_2CH_3)(OCMe_3)_3$ is white. Both can be sublimed. The latter is an important species since it

$$W(CCMe_3)(OCMe_3)_3 + RC{\equiv}CR \longrightarrow W(CR)(OCMe_3)_3 + RC{\equiv}CCMe_3 \quad (13)$$

R = Ph or Pr (excess RC≡CR required)

proves that β-hydrogen atoms are tolerated in the alkylidyne ligand, and that the bulk of the substituent in $W(CR)(OCMe_3)_3$ is probably not a factor in determining whether $W(CR)(OCMe_3)_3$ decomposes to $W_2(OCMe_3)_6$ and RC≡CR, or not. (So far we have not observed this reaction.) We have shown that the stoichiometric reaction is first order in tungsten and first order in acetylene over a wide range of concentrations (35).

$W(CCMe_3)(OCMe_3)_3$ reacts rapidly with unsymmetric acetylenes to give the initial metathesis products, $RC{\equiv}CCMe_3$ and/or $R'C{\equiv}CCMe_3$, and the symmetric acetylenes catalytically. The most impressive is the metathesis of 3-heptyne where the value for k (M^{-1} sec^{-1}) is between 1 and 10. Therefore, in neat 3-heptyne (~1 M) at 25° the number of turnovers is of the order of several per second. If we assume that W(VI) or Mo(VI) alkylidyne sites or complexes are responsible for the relatively slow metathesis in the known heterogeneous (30) and homogeneous (31) systems, then it becomes clear that the concentration of active species on the surface or in solution must be extremely small.

$W(CCMe_3)(OCMe_3)_3$ is not the only alkylidyne complex which will metathesize acetylenes. $W(CCMe_3)(NMe_2)_3$ will also, although the data so far have not been quantitated. All others ($W(CCMe_3)Np_3$, $[W(CCMe_3)Cl_4]^-$, $W(CCMe_3)(dme)Cl_3$, and $W(CCMe_3)(SCMe_3)_3$) will not. Of these, we have studied the reaction between the halide complexes and alkyl acetylenes most closely. Addition of excess 2-butyne to $W(CCMe_3)(dme)Cl_3$ yields red, paramagnetic, soluble $W(\eta^5\text{-}C_5Me_4Bu^t)$-$(MeC{\equiv}CMe)Cl_2$, and orange, paramagnetic, $[W(\eta^5\text{-}C_5Me_4Bu^t)Cl_4]_2$, each in ~50% yield (36). The identity of $W(\eta^5\text{-}C_5Me_4Bu^t)(MeC{\equiv}CMe)Cl_2$ was proven by an x-ray structural study which showed it to be similar to $Ta(\eta^5\text{-}C_5Me_5)(PhC{\equiv}CPh)Cl_2$ (37) and related species. The W(V) dimer and W(III) acetylene complex probably form by disproportionation of some intermediate W(IV) complex. What we were most interested in was whether any intermediates not containing a $\eta^5\text{-}C_5Me_4Bu^t$ ligand could be isolated.

$W(CCMe_3)(dme)Cl_3$ reacts with one equivalent of 2-butyne to give a violet complex whose ^{13}C NMR data are consistent with it being a tungstenacyclobutadiene complex (equation 14) (36). In particular, two signals are found at 268 and 263 ppm (cf. 335 for the neopentylidyne α-carbon atom in $W(CCMe_3)(dme)Cl_3$) and a third

$W(CBu^t)(dme)Cl_3$ + MeC≡CMe ⟶ 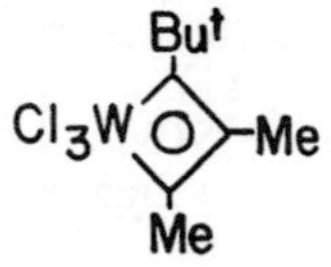 (14)

at 151 ppm. An x-ray structural study showed that it is indeed a tungstenacyclobutadiene complex, a trigonal bipyramidal monomer with axial chloride ligands and a planar WC_3 ring. Perhaps the most surprising feature of this molecule is the fact that W•••C_β (2.12Å) is <u>less</u> than a typical W(VI)-C_{alkyl} bond distance. This is probably the reason why the C_α-C_β-C_α angle is so large (119°), and could be the reason why the α-substituents are bent away from the metal. There is clearly plenty of room for additional 2-butyne to coordinate to tungsten, probably between Cl_{eq} and C_α ($\angle Cl_{eq}$-W-$C_\alpha \approx$ 140°), to produce a WC_5 ring which then collapses to a cyclopentadienyl system (equation 15). As a result of this process the metal

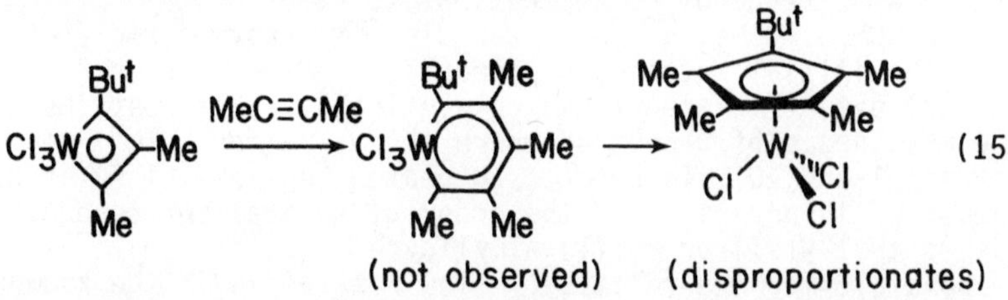

(15)

is reduced from W(VI) to W(IV). Perhaps at least one role of a t-butoxide ligand is to make the metal more difficult to reduce. It probably also destablizes the tungstencyclobutadiene complex relative to the alkylidyne complex so that the actual concentration of a tungstenacyclobutadiene tri-t-butoxide complex is small. Interestingly, one t-butoxide ligand can be added in the equatorial position but if addition of a second is attempted, only tri-t-butoxyalkylidyne complexes result (equation 16).

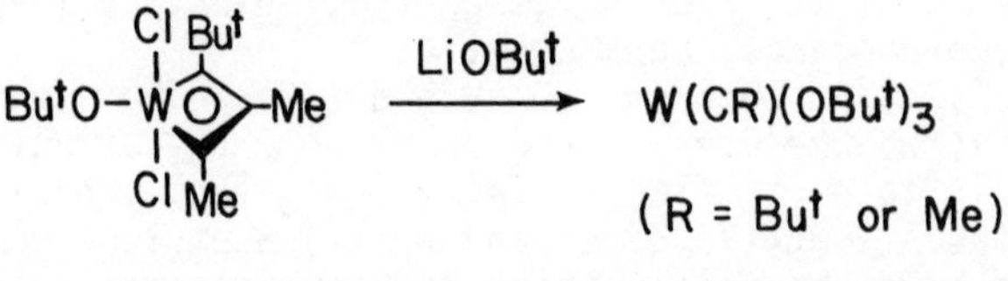

(16)

In view of the sensitivity of olefin metathesis catalysts to functional groups we were somewhat surprised to find that acetylene metathesis catalysts are apparently much more tolerant of functional groups than olefin metathesis catalysts. For example, $W(CCMe_3)(OCMe_3)_3$ will metathesize 3-heptyne in the presence of many equivalents of acetonitrile, ethylacetate, phenol, triethylamine, or internal olefins (<u>35</u>). Consequently, $W(CCMe_3)(OCMe_3)_3$ will metathesize some functionalized acetylenes. Preliminary studies show that it will metathesize $EtC{\equiv}CCH_2NMe_2$, and that it will cross metathesize $Me_3SiOCH_2C{\equiv}CCH_2OSiMe_3$ with 3-hexyne. However, first attempts to metathesize $EtC{\equiv}CCH_2Cl$, $HOCH_2C{\equiv}CCH_2OH$ with 3-hexyne, or $EtC{\equiv}CCO_2Me$ failed. There are several possible reasons why and we are in the process of attempting to find out in detail what they are.

Literature Cited

1. Fischer, E.O.; Maasböl, A. Angew Chem. Int. Ed. Eng. 1964, 3, 580.

2. (a) Grubbs, R.H. Prog. Inorg. Chem. 1978, 24, 1-50; (b) Katz, T.J. Adv. Organomet. Chem. 1977, 16, 283-317; (c) Calderon, N.; Lawrence, J.P.; Ofstead, E.A. Adv. Organomet. Chem. 1979, 17, 449-492; (d) Rooney, J.J.; Stewart, A. Spec. Period. Rep.: Catal. 1977, 1, 277.

3. Hérrison, J.L.; Chauvin, Y. Makromol. Chem. 1970, 141, 161.

4. Schrock, R.R. in "Reactions of Coordinated Ligands"; Braterman, P.S., Ed.; Plenum: New York, in press.

5. Schrock, R.R. J. Am. Chem. Soc. 1974, 96, 6794.

6. Schrock, R.R.; Fellmann, J.D. J. Am. Chem. Soc. 1978, 100, 3359.

7. Wood, C.D.; McLain, S.J.; Schrock, R.R. J. Am. Chem. Soc. 1979, 101, 3210.

8. Rupprecht, G.A.; Messerle, L.W.; Fellmann, J.D.; Schrock, R.R. J. Am. Chem. Soc. 1980, 102, 6236.

9. Rocklage, S.M.; Fellmann, J.D.; Rupprecht, G.A.; Messerle, L.W.; Schrock, R.R. J. Am. Chem. Soc. 1981, 103, 1440.

10 McLain, S.J.; Sancho, J.; Schrock, R.R. J. Am. Chem. Soc. 1980, 102, 5610.

11. (a) Schrock, R.R.; Rocklage, S.; Wengrovius, J.; Rupprecht, G.; Fellmann, J. J. Molec. Catal. 1980, 8, 73; (b) Wengrovius, J.H.; Schrock, R.R. Organometallics 1982, 1, 148.

12. Churchill, M.R.; Rheingold, A.L. Inorg. Chem. 1982, 21, 1357.

13. Griffith, W.P. Coord. Chem. Rev. 1970, 5, 459.

14. Kress, J.; Wesolek, M.; Le Ny, J.; Osborn, J.A. J. Chem. Soc. Chem. Commun. 1981, 1039.

15. Rocklage, S.M.; Schrock, R.R.; Churchill, M.; Wasserman, H.J. Organometallics, in press.

16. Pedersen, S.F.; Schrock, R.R. J. Am. Chem. Soc., in press.

17. Wengrovius, J.H.; Ph.D., Thesis, Massachusetts Institute of Technology, 1981.

18. Wengrovius, J.; Schrock, R.R.; Churchill, M.R.; Missert, J.R.; Youngs, W.J. J. Am. Chem. Soc. 1980, 102, 4515.

19. Rappé, A.K.; Goddard, W.A. III J. Am. Chem. Soc. 1982, 104, 448; 1980, 102, 5114.

20. Kress, J.; Wesolek, M.; Osborn, J.A. J. Chem. Soc. Chem. Commun. 1982, 514.

21. Ivin, K.J.; Rooney, J.J.; Stewart, C.D.; Green, M.L.H.; Mahtab, R. J. Chem. Soc. Chem. Commun. 1978, 604.

22. Fellmann, J.D.; Turner, H.W.; Schrock, R.R. J. Am. Chem. Soc. 1980, 102, 6608.

23. Turner, H.W.; Schrock, R.R. J. Am. Chem. Soc. 1982, 104, 2331.

24. Churchill, M.R.; Wasserman, H.J.; Turner, H.W.; Schrock, R.R. J. Am. Chem. Soc. 1982, 104, 1710.

25. Cossee, P. J. Catal. 1964, 3, 80.

26. An example of a complex that reacts slowly with ethylene in a manner consistent with insertion of ethylene into the metal alkyl bond is $Co(\eta^5\text{-}C_5H_5)(PPh_3)Me_2$: Evitt, E.R.; Bergman, R.G. J. Am. Chem. Soc. 1979, 101, 3973.

27. Boor, J. Jr. "Ziegler-Natta Catalysts and Polymerizations"; Academic: New York, 1979.

28. Watson, P.L. J. Am. Chem. Soc. 1982, 104, 337-339.

29. Katz, T.J. J. Am. Chem. Soc. 1975, 97, 1592-1594.

30. (a) Panella, F.; Banks, R.L.; Bailey, G.C. J. Chem. Soc. Chem. Commun. 1968, 1548-1549;
(b) Moulijn, J.A.; Reitsma, H.J.; Boelhouwer, C. J. Catal. 1972, 25, 434-436.

31. (a) Mortreux, A.; Delgrange, J.C.; Blanchard, M.; Labochinsky, B. J. Molec. Catal. 1977, 2, 73;
(b) Devarajan, S.; Walton, O.R.M.; Leigh, G.J. J. Organomet. Chem. 1979, 181, 99-104.

32. Clark, D.N.; Schrock, R.R. J. Am. Chem. Soc. 1978, 100, 6774.

33. Schrock, R.R.; Clark, D.N.; Sancho, J.; Wengrovius, J.H.; Rocklage, S.M.; Pedersen, S.F. Organometallics, in press.

34. Wengrovius, J.H.; Sancho, J.; Schrock, R.R. J. Am. Chem. Soc. 1981, 103, 3932.

35. Sancho, J.; Schrock, R.R. J. Molec. Catal. 1982, 15, 75-79.

36. Schrock, R.R.; Pedersen, S.F.; Churchill, M.R.; Wasserman, H.J. J. Am. Chem. Soc., in press.

37. Smith, G.; Schrock, R.R.; Churchill, M.R.; Youngs, W.J. Inorg. Chem. 1981, 20, 387.

RECEIVED August 11, 1982

Discussion

W.L. Gladfelter, University of Minnesota: Does the stability of the trichlorotungstenacyclobutadiene complex imply that the "special effect" of the OR ligand is to enhance the rate of decomposition of this intermediate?

R.R. Schrock: We believe so, but the t-butoxide ligand seems to be a unique alkoxide. We have prepared tungstenacyclobutadiene complexes containing other alkoxide ligands which are stable toward decomposition to an alkylidyne complex.

M.H. Chisholm, Indiana University: I should like to make a prediction. In view of the ability of tungsten to form multiple bonds with carbon, as is well illustrated by your work with alkylidene and alkylidyne ligands, I believe we shall soon see the emergence of a new class of tungsten compounds incorporating carbide (C^{4-}) as a ligand. The probable modes of bonding for carbide should compliment that found for oxo and nitrido complexes.

W.L. Gladfelter, University of Minnesota: Can you polymerize propylene with your tantalum catalyst?

R.R. Schrock: No. Propylene reacts to give primarily propane and a complex of the type $Ta(CHCMe_3)(CH_2PMe_2)(PMe_3)_2I_2$.

J.M. Burlitch, Cornell University: Is there any evidence for an exchange of alkylidyne ligand of the sort shown below?

$$L_2'Cl_3W{\equiv}Cr + L_2Cl_3W{\equiv}CR' \rightarrow L_2'Cl_3W{\equiv}CR' + L_2Cl_3W{\equiv}CR$$

R.R. Schrock: No. The required labelling experiment has not been done.

A.J. Carty, Guelph-Waterloo Centre: What are the frequencies of the metal-carbon ($\nu(M{\equiv}C)$) stretching frequencies in the tungsten alkylidyne compounds? One might expect these to be quite high in view of the x-ray data showing short M≡C bonds and evidence of multiple bond reactivity.

R.R. Schrock: There are bands at ca. 1300 cm^- in the IR spectra of several alkylidyne complexes which might be assigned to W≡C stretching modes analogous to those observed by Fischer in complexes of the type $(X)(CO)_5W{\equiv}CR$ (X = halide), but we have not done the appropriate labelling or Raman studies to confirm the assignments.

The Alkylidyne Compound $[(\eta\text{-}C_5H_5)(OC)_2W{\equiv}CC_6H_4Me\text{-}4]$

A Reagent for Synthesizing Complexes with Bonds Between Tungsten and Other Metals

F. GORDON A. STONE

The University, Bristol, Department of Inorganic Chemistry,
Bristol BS8 1TS England

In 1979 we reported (1) that the compound $[(\eta\text{-}C_5H_5)(OC)_2W{\equiv}CR]$ (R = C_6H_4Me-4), discovered by Fischer et al. (2), would displace ethylene from the complexes $[Pt(C_2H_4)(PR_3)_2]$ to give species (1), with platinum-tungsten bonds and bridging tolylidyne ligands. Because of the isolobal relationship existing between the groups CR and $W(CO)_2(\eta\text{-}C_5H_5)[W(d^5), WL_5]$, the compound (1) may be compared with the long known alkyne-platinum complexes (2). It thus became apparent that the molecule $[(\eta\text{-}C_5H_5)(OC)_2W{\equiv}CR]$ would show a reactivity pattern towards other transition metal complexes similar to that of an alkyne, thus affording numerous complexes with metal - metal bonds involving tungsten. This area of study has grown sufficiently to merit a review. A survey is particularly timely for two reasons. Firstly, considerable interest has developed recently in the chemistry of transition metal complexes containing heteronuclear metal - metal bonds. Secondly, in relation to current awareness of the importance of C_1 chemistry, the study of compounds having CR groups bridging metal - metal bonds is important. Finally, in the context of the theme of this Conference, perhaps the work described represents a modest step towards placing the synthesis of compounds with metal - metal bonds on a more logical basis than hitherto.

Dimetal Compounds

Compounds (3) - (11) have been isolated from reactions between $[(\eta\text{-}C_5H_5)(OC)_2W{\equiv}CR]$ and the species $[Co(CO)_2(\eta\text{-}C_5Me_5)]$ (3), $[Rh(CO)L(\eta\text{-}C_9H_7)]$ (L = CO or PMe_3) (4,5), $[Rh(acac)(CO)_2]$ (6), $[M(CO)_2(thf)(\eta\text{-}C_5H_4R')]$ (M = Mn, R′ = Me; M = Re, R′ = H) (4), $[Cr(CO)_2(thf)(\eta\text{-}C_6Me_6)]$ (4), and $[M(CO)_2(\eta\text{-}C_5H_5)_2]$ (M = Ti or Zr) (7), respectively. The iron-tungsten compound (12) has been obtained by treating $[(\eta\text{-}C_5H_5)(OC)_2W{\equiv}CR]$ with $[Fe_2(CO)_9]$ in ether solvents (8). It is especially reactive, readily transforming into diirontungsten or ironditungsten cluster species in the presence of an excess of one or other of the reactants.

0097-6156/83/0211-0383$06.00/0

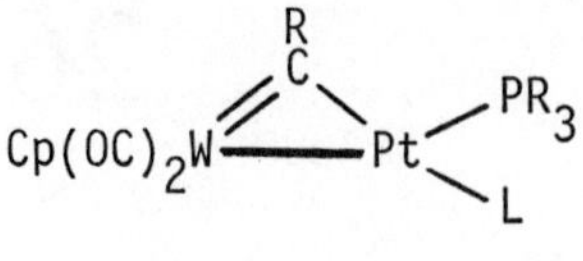

(1)* L = PR_3

(14) L = CO

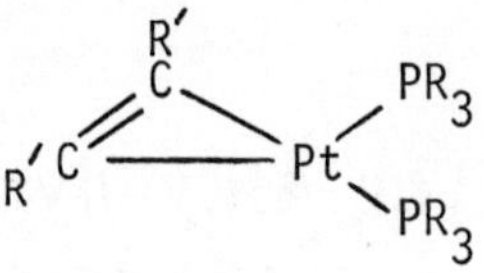

(2)

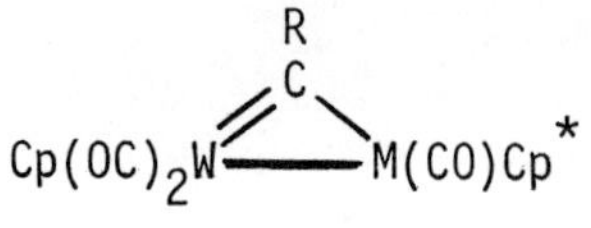

	M
(3)	Co
(13)	Rh

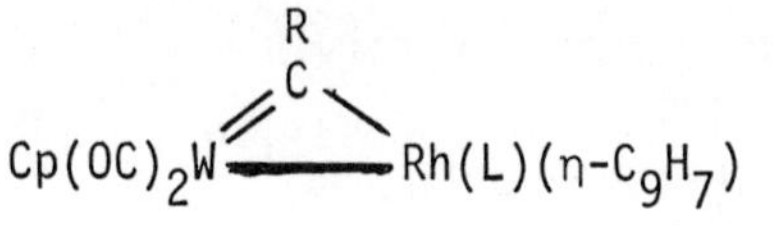

	L
(4)	CO
(5)	PMe_3

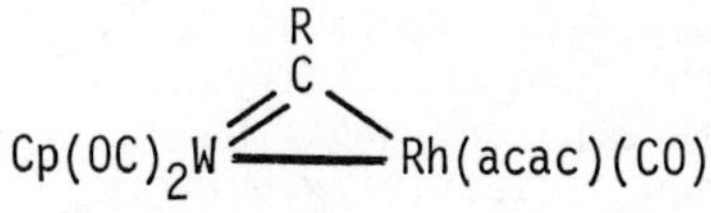

(6)

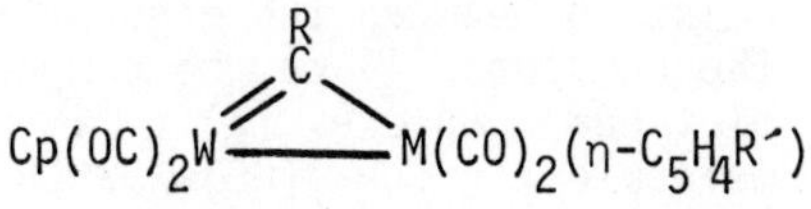

	M	R
(7)	Mn	Me
(8)	Re	H

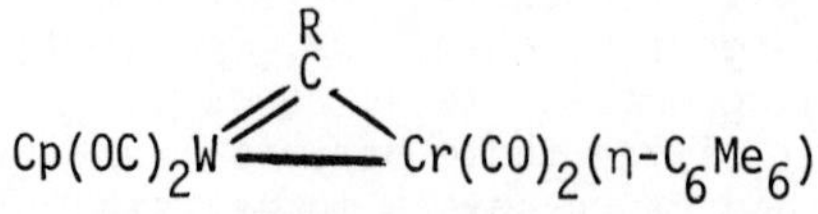

(9)

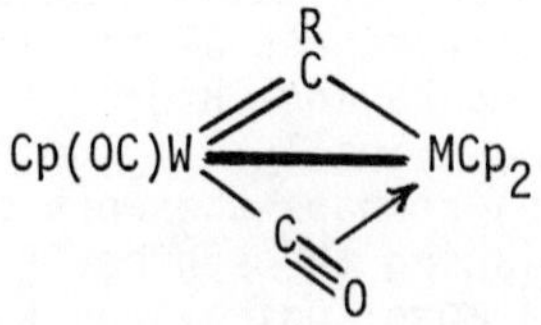

	M
(10)	Ti
(11)	Zr

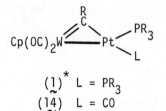

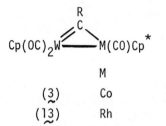

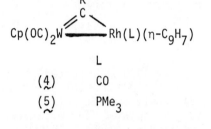

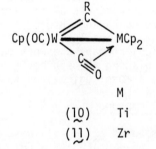

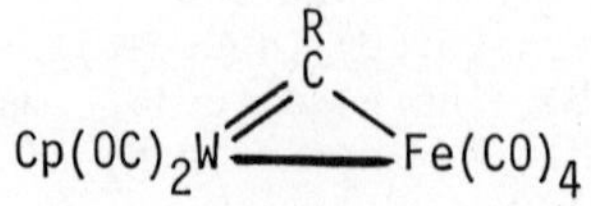

(12)

Throughout this paper, in the structural formulae, CR = CC_6H_4Me-4, Cp = η^5-C_5H_5, Cp = η^5-C_5Me_5, η-C_9H_7 = η^n-indenyl (n = 3 or 5).

Complex (13) is one product (see later) of the reaction of $[Rh_2(\mu\text{-}CO)_2(\eta\text{-}C_5Me_5)_2]$ with $[(\eta\text{-}C_5H_5)(OC)_2W{\equiv}CR]$ (9). The compounds (3) - (13) are formulated with dimetallacyclopropene rings, and in the case of (3), (5), (9) and (10) this is supported by the results of single-crystal X-ray diffraction studies. Several of the species contain a semi-bridging CO ligand for which there is either i.r. (ν_{CO} circa 1 850 cm^{-1}) or X-ray evidence. An interesting feature of the structure of (10) is that the X-ray diffraction results reveal the presence of an $\eta^2\text{-}C{\equiv}O$ bridging group. A similar ligand is probably present in (11) since both (10) and (11) show CO stretching bands in the i.r. at 1 638 and 1 578 cm^{-1}, respectively, as well as terminal CO bands at 1 930 and 1 937 cm^{-1}.

For the dimetal compounds the most useful spectroscopic property of a diagnostic nature is the resonance due to the μ-C ligated carbon in the ^{13}C n.m.r. spectra. In the precursor $[(\eta\text{-}C_5H_5)(OC)_2W{\equiv}CR]$ the alkylidyne carbon atom resonates in the ^{13}C n.m.r. spectrum at δ 300 p.p.m. In the dimetal compounds (3) - (13) the signal for the μ-C nucleus is even more deshielded: typical δ values are 341 (3), 327 (5), 431 (9), and 391 p.p.m. (11).

It is becoming increasingly apparent that a very extensive chemistry is associated with the dimetal species (1), and (3) - (13). Three types of reaction may be identified: (i) addition of other metal ligand fragments, discussed in the next Section, (ii) replacement of peripheral ligands on the metal centres, and (iii) reactions at the bridging carbon atom. One example of (ii) and (iii) will suffice to illustrate the scope and potential for new chemistry.

The PR_3 ligands transoid to the μ-CR group in (1) are readily displaced by CO gas (1 bar), affording initially the dimetal compounds (14), which subsequently in solution yield the trimetal complexes (15). Evidently compounds (14) partially dissociate into $[(\eta\text{-}C_5H_5)(OC)_2W{\equiv}CR]$ and $[Pt(CO)(PR_3)]$, and the latter adds to undissociated (14) to give the complexes (15) (10).

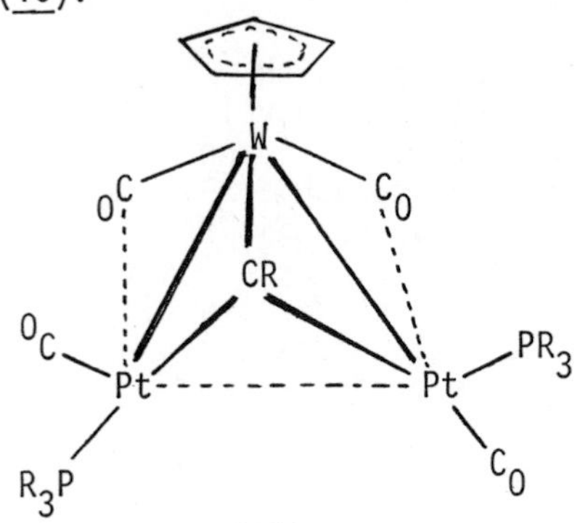

(15)

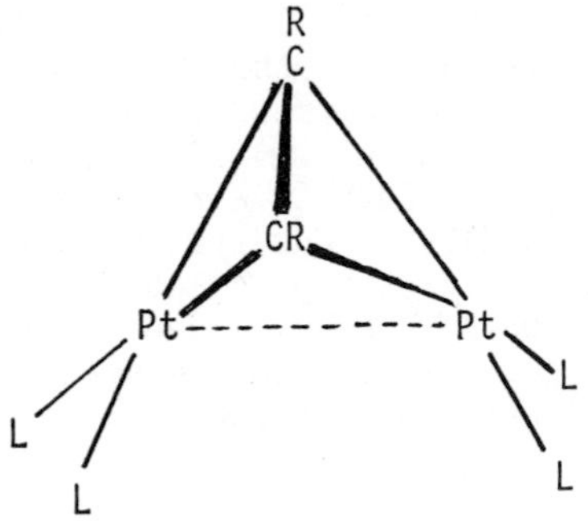

(16)

It is interesting to compare the structure of the species (15) (X-ray diffraction study for PR_3 = $PMePh_2$) with those of the isolobal bridged alkyne complexes (16) (L = PR_3 or L_2 = cyclo-octa-1,5-diene) (11). In both (15) and (16) the Pt····Pt vectors (circa 2.9 Å) are too great for any appreciable metal - metal interactions, and the platinum atoms can be regarded as having 16-electron configurations.

We referred earlier to the significance of reactions at the alkylidyne carbon atoms of the dimetal species. Our studies in this area are in a preliminary stage, but Schemes 1 and 2 summarise some chemistry at the bridged carbon centres for the compounds (1) and (3)(12). It will be noted that protonation of the neutral bridged alkylidyne compounds yields cationic alkylidene species in which one C—C bond of the tolyl group is η^2 co-ordinated to tungsten, a feature revealed by both n.m.r. and X-ray diffraction studies.

Finally, in our survey of dimetal compounds derived from $[(\eta\text{-}C_5H_5)(OC)_2W{\equiv}CR]$, we refer to the reaction of the latter with $[(Ph_3P)_2N][HW(CO)_5]$ which affords (80 %) the novel salt (17)(13). This is apparently the first example of an anionic dimetal species with a bridging alkylidene group, and the structure has been confirmed by X-ray diffraction.

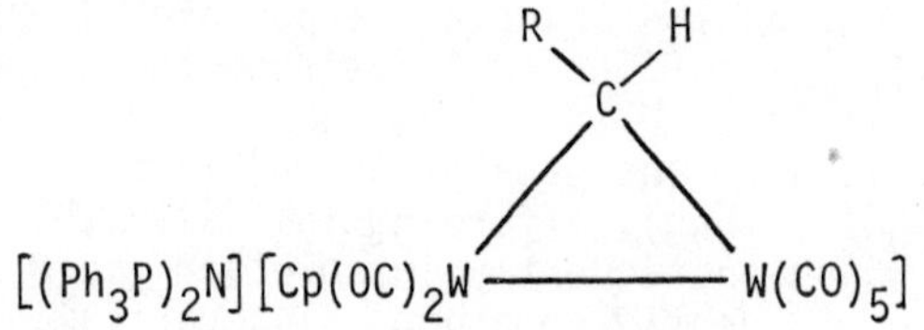

(17)

The salt (17) undergoes an unusual reaction with $[AuCl(PPh_3)]$ in the presence of $TlPF_6$, giving the novel bridged alkylidene compound $[AuW(\mu\text{-}CHR)(CO)_2(PPh_3)(\eta\text{-}C_5H_5)]$, the structure of which has been elucidated by X-ray crystallography. By treating (17) with Ph_3SnCl a related species, which may contain a $W(\mu\text{-}CHR)Sn$ bridge system, has been obtained. In the ^{13}C n.m.r. spectra of the gold and tin compounds resonances for the ligated carbon of the alkylidene ligand occur at δ 229 and 214 p.p.m., respectively. In the spectrum of (17) the bridge carbon resonates at 148 p.p.m.

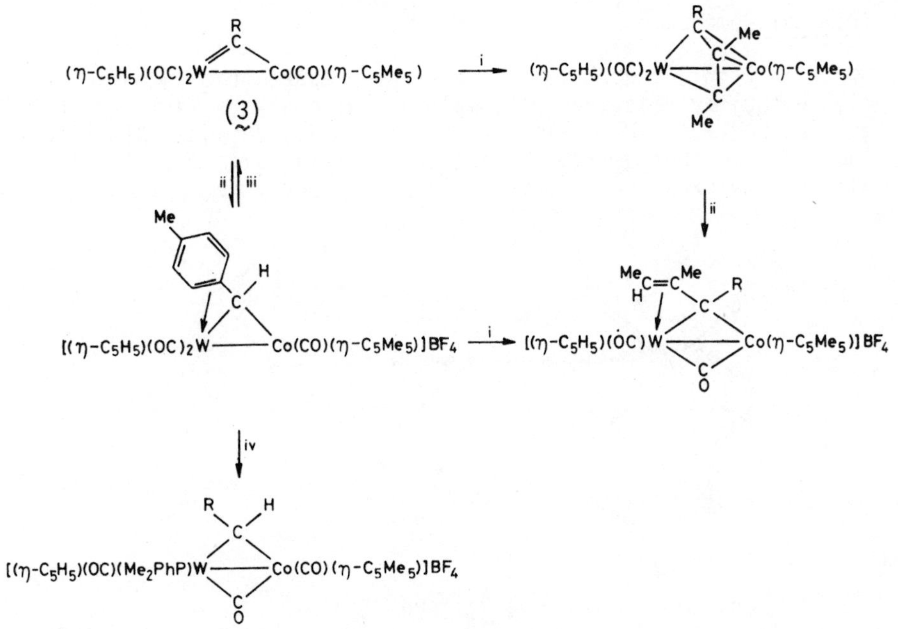

Scheme 1. Chemistry at the bridged carbon centers for Compound 3. Key: R, C_6H_4Me-4; i, MeC_2Me in PhMe; ii, $HBF_4 \cdot Et_2O$; iii, k[BH(CHMeEt)$_3$] in tetrahydrofuran (THF); and iv, PMe_2Ph. (Reprinted with permission from Ref. 12. Copyright 1981, Royal Society of Chemistry.)

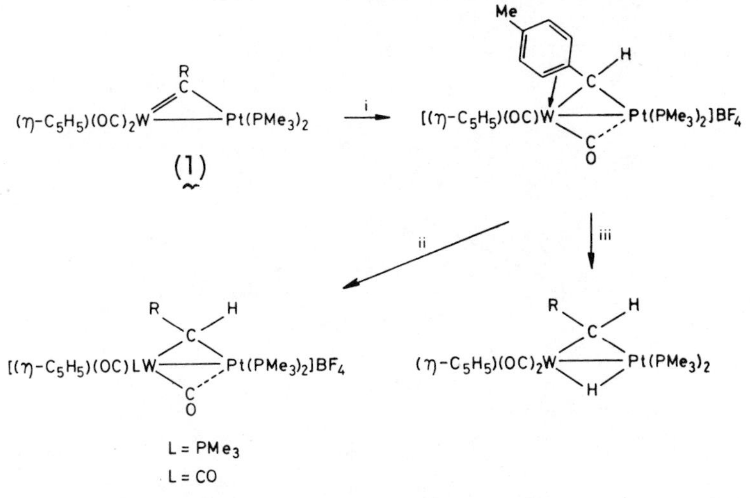

Scheme 2. Chemistry at the bridged carbon centers for Compound 1. Key: R, C_6H_4Me-4; L, PMe_3 or CO; i, $HBF_4 \cdot Et_2O$; ii, CO or PMe_3; and iii, K[BH(CHMe-Et)$_3$] in THF. (Reprinted with permission from Ref. 12. Copyright 1981, Royal Society of Chemistry.)

Trimetal Compounds

Early in our work in this area we prepared the compounds (18) - (20) by reacting the complexes [Ni(cod)$_2$], [Pd(C$_7$H$_{10}$)$_3$] and [Pt(C$_2$H$_4$)$_3$], in turn, with [(η-C$_5$H$_5$)(OC)$_2$W$\equiv$CR] (14). The platinum species (20) is isolobally related to a series of alkyne complexes [Pt(alkyne)$_2$] (15). Recently we have

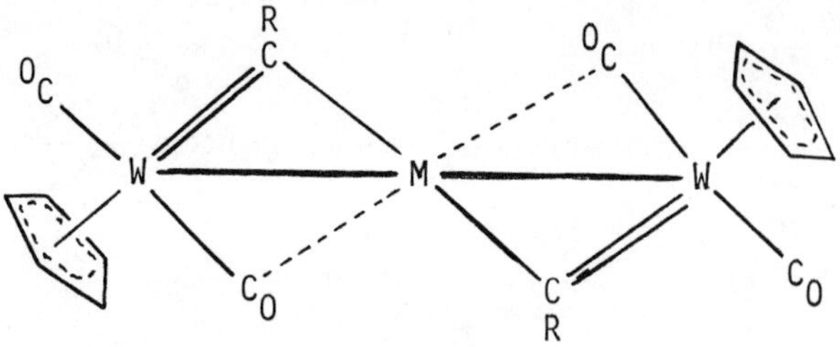

	M
(18)	Ni
(19)	Pd
(20)	Pt

synthesised the molybdenumditungsten compound (21) and its tritungsten analog (22) by reacting the compounds [M(CO)$_3$(NCMe)$_3$] (M = Mo or W) with [(η-C$_5$H$_5$)(OC)$_2$W$\equiv$CR] in refluxing hexane (16). Complexes (20) and (22) are isolobally related:

$$ML_0(d^{10}) \underset{\circ}{\longleftrightarrow} C^{4+} \underset{\circ}{\longleftrightarrow} ML_2(d^6)$$

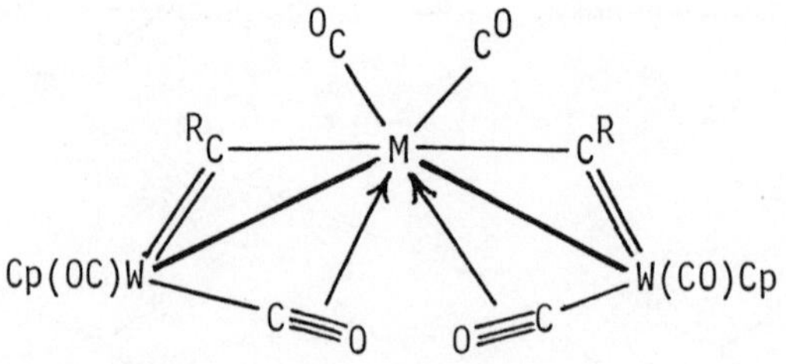

	M
(21)	Mo
(22)	W

In (21) and (22) the central metal atoms acquire 18-electron configurations by η^2-bonding from a CO ligand on each $W(CO)_2(\eta-C_5H_5)$ moiety, a feature established by X-ray diffraction. The ^{13}C n.m.r. spectra of (21) and (22) are also informative showing CO groups in three environments (rel. int. 2:2:2), and resonances for the μ-CR groups at δ 360 (21) and 376 p.p.m. (22).

In an ever increasing number of compounds tolylidyne ligands triply-bridge a triangle of metal atoms, one of which is tungsten and the other two may or may not involve the same element. Thus two types of core structure are possible: $[M_2W(\mu_3-CR)]$ (Class A) or $[M^1M^2W(\mu_3-CR)]$ (Class B). Two methods are available for the synthesis of clusters of Class A. viz

(i) Successive addition of similar metal ligand fragments to $[(\eta-C_5H_5)(OC)_2W\equiv CR]$

Thus (23) is prepared by adding $[Rh(C_2H_4)_2(acac)]$ to (6), the latter being preformed by treating the mononuclear tungsten alkylidyne compound with $[Rh(CO)_2(acac)]$ (6). Similarly, step-wise addition of $[Rh(CO)_2(\eta-C_9H_7)]$ to $[(\eta-C_5H_5)(OC)_2W\equiv CR]$ yields (24)(9).

We referred earlier to the reactivity of (12). If this species is generated in the presence of excess $[Fe(CO)_4(thf)]$ it rapidly affords (25)(8). If, however, (12) is generated in the presence of excess $[(\eta-C_5H_5)(OC)_2W\equiv CR]$ the alkyne bridged ironditungsten cluster (26) is the major product. This species is of interest on two counts. Firstly, it is an example of a cluster in which a coupling of CR fragments of the carbyne has occurred, a reactivity pattern referred to below. Secondly, (26) is an unsaturated metal cluster (46 valence electrons) and the W — W distance [2.747(1) Å] suggests a degree of double bonding.

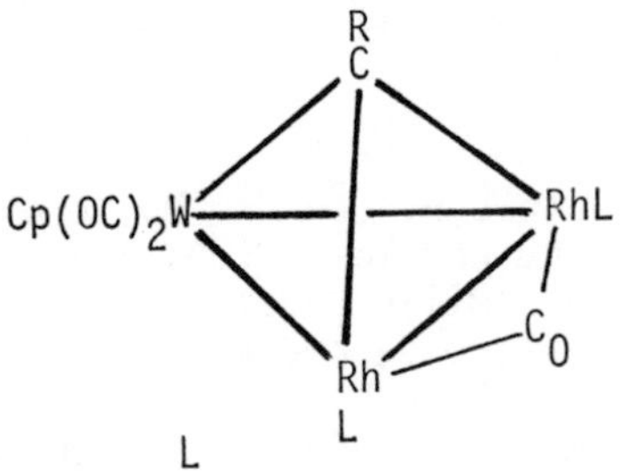

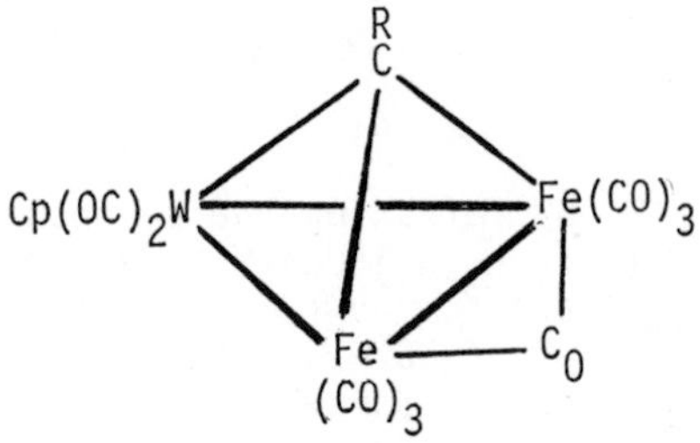

(23) acac

(24) $\eta-C_9H_7$

(33) $\eta-C_5Me_5$

(25)

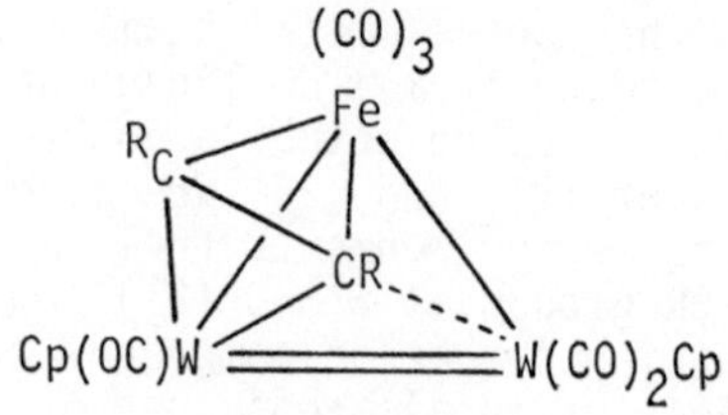

(26)

(ii) Reaction of $[(\eta-C_5H_5)(OC)_2W\equiv CR]$ with dimetal complexes

It has long been known that alkynes react with dimetal compounds, e.g. $[Co_2(CO)_8]$ (17,18), $[Ni_2(\mu-CO)_2(\eta-C_5H_5)_2]$ (19,20), or $[Mo_2(CO)_6(\eta-C_5H_5)_2]$ (21,22), to give species with dimetalla-tetrahedrane core structures. It seemed likely, therefore, that $[(\eta-C_5H_5)(OC)_2W\equiv CR]$ would show a similar reactivity pattern, and in confirmation of this compound (27) is produced in quantitative yield by reacting the alkylidyne-tungsten compound with $[Co_2(CO)_8]$ in pentane at room temperature (6).

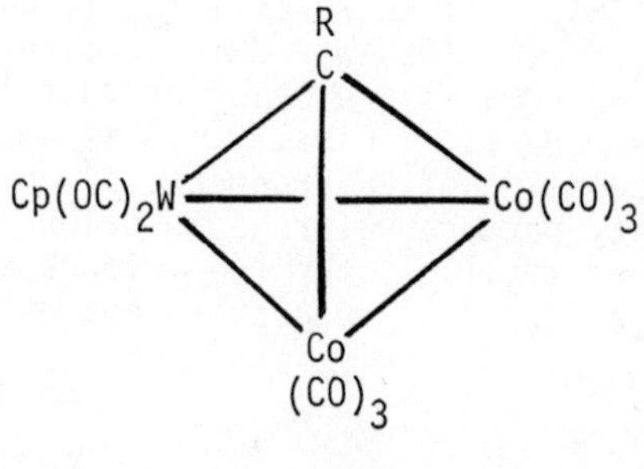

(27)

By refluxing the compounds $[M_2(CO)_4(\eta-C_5H_5)_2]$ (M = Mo or W) with $[(\eta-C_5H_5)(OC)_2W\equiv CR]$ in toluene, the complexes (28) and (29) are formed (23). Interestingly, a similar reaction with $[Cr_2(CO)_4(\eta-C_5H_5)_2]$ affords (30) quantitatively, and no (31) is produced. Somewhat similar behaviour is shown by $[Ni_2(\mu-CO)_2(\eta-C_5H_5)_2]$ which with the alkylidynetungsten compound produces a mixture of (30) and (32). It should be noted that $[(\eta-C_5H_5)(OC)_2W\equiv CR]$ does not dimerise to (30) when heated by itself; indeed such a transformation is unallowed. In the reactions of the dimetal compounds with $[(\eta-C_5H_5)(OC)_2W\equiv CR]$ it seems probable that the former reactants dissociate initially to

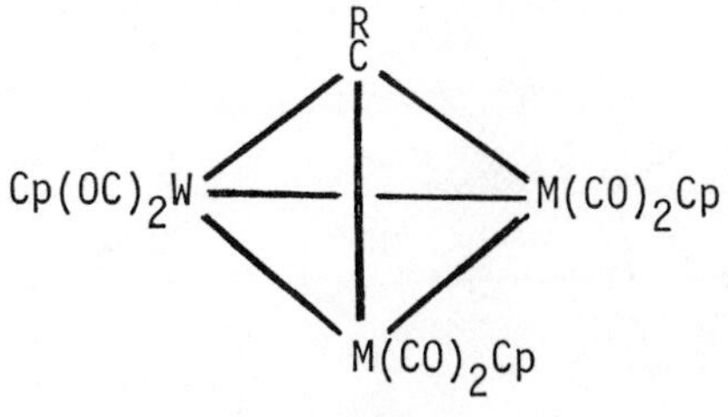

	M
(28)	Mo
(29)	W
(31)	Cr

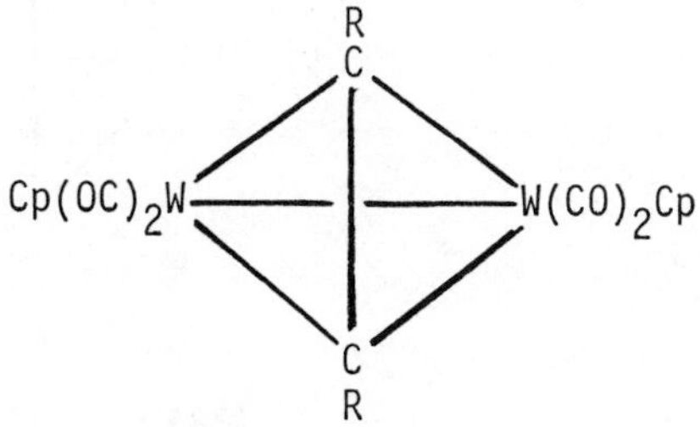

(30)

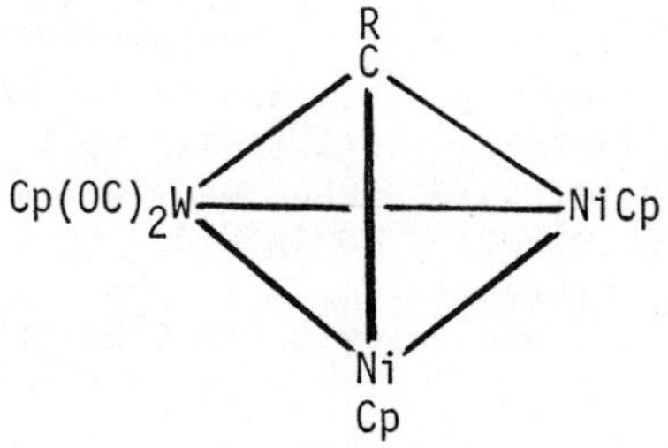

(32)

give mononuclear metal ligand fragments (24), and that the latter are captured by [$(\eta$-$C_5H_5)(OC)_2W\equiv CR$] to give dimetal species which rapidly undergo further reaction, affording the final products (9). In this context it is interesting that the alkylidynetungsten compound reacts with [$Rh_2(\mu$-$CO)_2(\eta$-$C_5Me_5)_2$] to give a mixture of the dimetal species (13) and the trimetal compound [$Rh_2W(\mu_3$-$CR)(\mu$-$CO)(CO)_2(\eta$-$C_5H_5)(\eta$-$C_5Me_5)_2$] (33).

In some instances formation of (30) may be avoided by employing 'alkyne' displacement reactions. Thus (32) is formed in essentially quantitative yield by reacting [$(\eta$-$C_5H_5)(OC)_2W\equiv CR$] with the bridged alkyne complex [$Ni_2(\mu$-$Me_3SiC_2SiMe_3)(\eta$-$C_5H_5)_2$] (9).

Clusters of Class B may be prepared by adding appropriate metal ligand fragments to the dimetal species. Thus treatment of (3) with [$Fe_2(CO)_9$] affords (34), and similarly (4) reacts with [$Fe_2(CO)_9$] to give (35)(9).

The species (1) react readily with [$Fe_2(CO)_9$] in tetrahydrofuran at room temperature. However, a mixture of

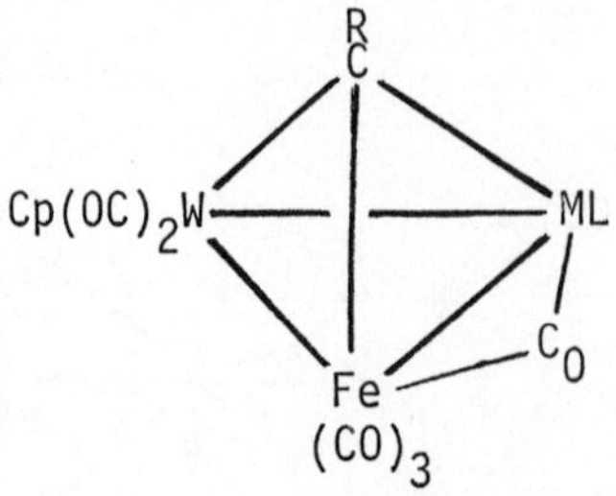

	ML
(34)	CoCp*
(35)	$Rh(\eta-C_9H_7)$

clusters (36) - (38) is produced (25). The trimetal compounds (36) result from addition of an $Fe(CO)_4$ fragment to (1), presumably from [$Fe(CO)_4$(thf)] (26). Loss of a molecule of CO would afford (38). Isolation of the clusters (37) can be understood in the following way. It was mentioned earlier that the dimetal compounds (1) readily react with CO to give (14). Enneacarbonyl diiron in tetrahydrofuran provides a source of CO (from [$Fe(CO)_5$]), thereby leading to the in situ formation of (14). Addition of an $Fe(CO)_4$ fragment to the latter, accompanied by loss of CO, would yield the clusters (37).

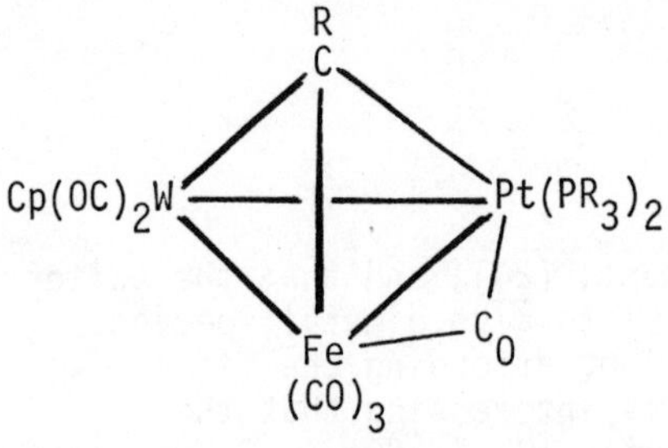

(36)

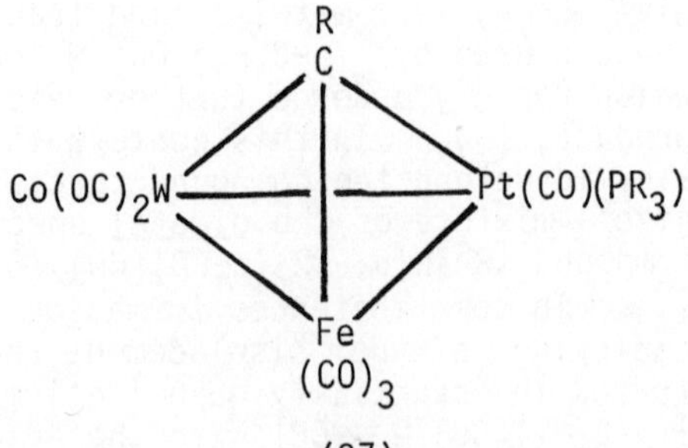

(37)

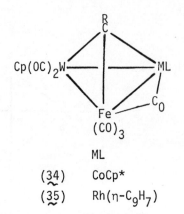

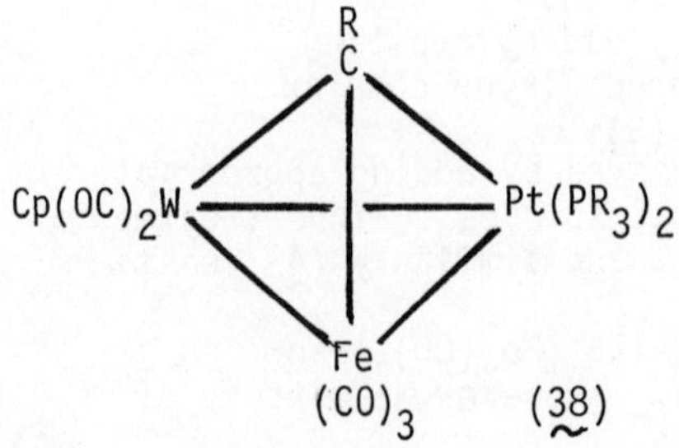

(38)

Interestingly, compounds of Class A or Class B have 50 cluster valence electrons if the three metal atoms are individually to have 18-electron configurations.* Several of the compounds meet this requirement, e.g. (25), (27), (31), (34) or (36). However, compound (23) with 46 cluster valence electrons, and compounds (37) and (38) with 48 cluster valence electrons, possessing four and five skeletal bond electron pairs, respectively, reflect the tendency of rhodium or platinum atoms to adopt 16-electron configurations.

By refluxing [$Ru_3(CO)_{12}$] with [(η-C_5H_5)$(CO)_2W{\equiv}CR$] in toluene the isomeric clusters (39) are produced (8). The osmium analogs (40) form in the reaction between [$Os_3(\mu$-$H)_2(\mu$-$CH_2)(CO)_{10}$] and the alkylidynetungsten compound in thf at 60 °C. In the formation of (39) and (40) alkylidyne groups have coupled to produce trimetal species with co-ordinated RC_2R groups. In the context of isolobal relationships, it is interesting to regard (39) and (40) as dimetallacyclobutadienes stablised by complexation with $M(CO)_3$ or $W(CO)_2$(η-C_5H_5) groups. A further point of interest is their ready interconversion in solution, which we have studied by variable temperature n.m.r. measurements (8). Reaction of [(η-C_5H_5)$(OC)_2W{\equiv}CR$] with polynuclear metal carbonyl species is likely to afford many new heteronuclear cluster compounds containing tungsten. Illustrative of this is the formation of (41) in the reaction with [$Os_3(CO)_{10}$(cyclo-$C_8H_{14})_2$] (8)

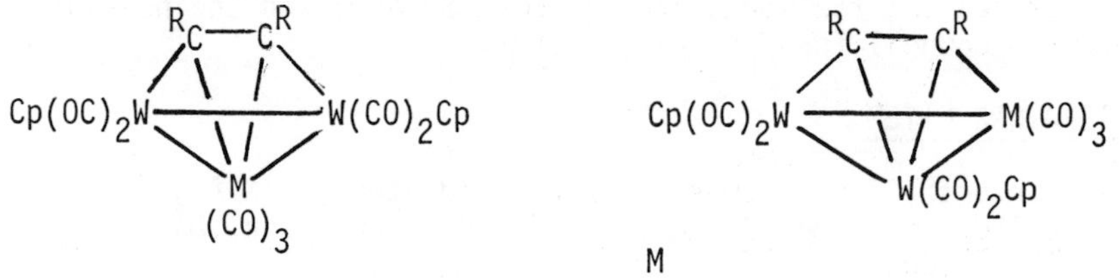

	M
(39)	Ru
(40)	Os

*We have chosen to treat the μ_3-C atom as part of the cluster core, and the RC group (5 valence electrons) as contributing three electrons for cluster bonding. Alternatively, the trimetal compounds, in which each metal atom has an 18-electron configuration, can be regarded as 48 electron species having three-electron μ_3-CR ligands. By adopting this formalism, compound (23) would be a 44-electron cluster.

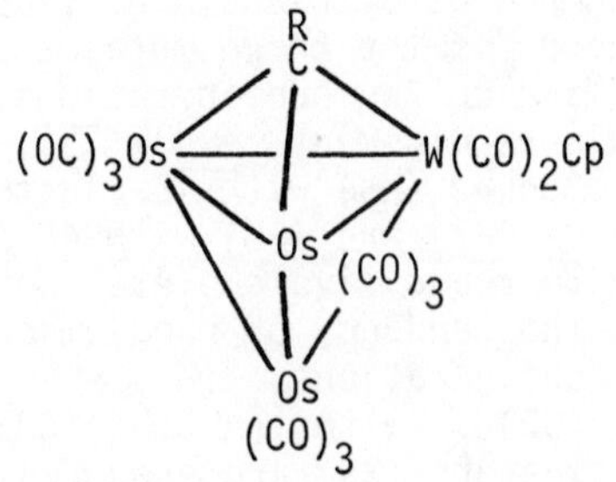

(41)

Conclusion

We seem to have been the first to appreciate that $W(CO)_2(\eta\text{-}C_5H_5)$ (W,d^5) is isolobal with CH (27), and to use this concept in designed syntheses of compounds with bonds between tungsten and other metals. We are hoping to extend the method to other systems containing $L_nM \equiv CR$ groups, involving both early and late transition metals, e.g. $ML_4(d^7)$ { $OsCl(CO)(PPh_3)_2$ (28) } and $ML_6(d^3)$ { $TaCl(PMe_3)_2(\eta\text{-}C_5Me_5)$ (29) }.

Acknowledgments

I thank my very talented co-workers cited in the references who carried out the work, and the U.K. Science and Engineering Research Council for financial support.

LITERATURE CITED

1. Ashworth, T.V., Howard, J.A.K., and Stone, F.G.A., J.Chem.Soc.,Chem.Commun., 1979, 42; J.Chem.Soc.,Dalton Trans., 1980, 1609.
2. Fisher, E.O., Lindner, T.L., Huttner, G., Friedrich, P., Kreissl, F.R., and Besenhard, J.O., Chem.Ber., 1977, 110, 3397.
3. Razay, H., Ph.D. Thesis, Bristol University, 1982.
4. Chetcuti, M.J., Green, M., Jeffery, J.C., Stone, F.G.A., and Wilson, A.A., J.Chem.Soc.,Chem.Commun., 1980, 948.
5. Jeffery, J.C., Schmidt, M., and Stone, F.G.A., unpublished results.
6. Chetcuti, M.J., Chetcuti, P.A.M., Jeffery, J.C., Mills, R.M., Mitrprachachon, P., Pickering, S.J., Stone, F.G.A., and Woodward, P., J.Chem.Soc.,Dalton Trans., 1982, 699.
7. Dawkins, G.M., Green, M., Salaun, J.-Y., and Stone, F.G.A., unpublished results.
8. Busetto, L., Green, M., Howard, J.A.K., Hessner, B., Jeffery, J.C., Mills, R.M., Stone, F.G.A., and Woodward, P., J.Chem.Soc.,Chem.Commun., 1981, 1101.

9. Green, M., Jeffery, J.C., Porter, S.J., Razay, H., and Stone, F.G.A., J.Chem.Soc.,Dalton Trans., in press.
10. Chetcuti, M.J., Marsden, K., Moore, I., Stone, F.G.A., and Woodward, P., J.Chem.Soc.,Dalton Trans., in press.
11. Boag, N.M., Green, M., Howard, J.A.K., Spencer, J.L., Stansfield, R.F.D., Thomas, M.D.O., Stone, F.G.A., and Woodward, P., J.Chem.Soc.,Dalton Trans., 1980, 2182; Boag, N.M., Green, M., Howard, J.A.K., Stone, F.G.A., and Wadepohl, H., ibid., 1981, 862.
12. Jeffery, J.C., Moore, I., Razay, H., and Stone, F.G.A., J.Chem.Soc.,Chem.Commun., 1981, 1255.
13. Hodgson, D., Jeffery, J.C., Marsden, K., Stone, F.G.A., Went, M.J., and Woodward, P., unpublished results.
14. Ashworth, T.V., Chetcuti, M.J., Howard, J.A.K., Stone, F.G.A., Wisbey, S.J., and Woodward, P., J.Chem.Soc.,Dalton Trans., 1981, 763.
15. Boag, N.M., Howard, J.A.K., Green, M., Grove, D.M., Spencer, J.L., and Stone, F.G.A., J.Chem.Soc.,Dalton Trans., 1980, 2170.
16. Carriedo, G.A., Marsden, K., Stone, F.G.A., and Woodward, P., unpublished results.
17. Sternberg, H.W., Greenfield, H., Friedel, R.A., Wotiz, J., Markby, R., and Wender, I., J.Am.Chem.Soc., 1954, 76, 1457.
18. Dickson, R.S., and Fraser, P.J., Adv.Organomet.Chem., 1974, 12, 323.
19. Tilney-Bassett, J.F., J.Chem.Soc., 1961, 577; 1963, 478.
20. Jolly, P.W., and Wilke, G., "The Organic Chemistry of Nickel", Vol. 1, Academic Press, N.Y.C., 1974.
21. Nakamura, A., and Hagahara, N., Nippon Kagaku Kaishi, 1963, 84, 344.
22. Knox, S.A.R., Stansfield, R.F.D., Stone, F.G.A., Winter, M.J., and Woodward, P., J.Chem.Soc.,Dalton Trans., 1982, 173; and references cited therein.
23. Green, M., Porter, S.J., and Stone, F.G.A., to be published.
24. Madach, T., and Vahrenkamp, H., Chem.Ber., 1980, 113, 2675.
25. Chetcuti, M.J., Howard, J.A.K., Mills, R.M., Stone, F.G.A., and Woodward, P., J.Chem.Soc.,Dalton Trans., in press.
26. Cotton, F.A., and Troup, J.M., J.Am.Chem.Soc., 1974, 96, 3438.
27. Hoffmann, R., 'Les Prix Nobel 1981', Almqvist and Wiksell, Stockholm, 1982.
28. Clark, G.R., Marsden, K., Roper, W.R., and Wright, L.J., J.Am.Chem.Soc., 1980, 102, 6570.
29. Fellmann, J.D., Turner, H.W., and Schrock, R.R., J.Am.Chem.Soc., 1980, 102, 6608.

RECEIVED August 24, 1982

Discussion

A.W. Adamson, University of Southern California: You demonstrated the usefulness of the isolobal approach as a correlating one. Can you, however, give some examples of how the approach has led to experimental chemistry that would not have been suggested by other rationales?

F.G.A. Stone: I had thought that the answer to your question had been embodied in the lecture. However, perhaps I failed to stress that the isolobal approach in our work has been predictive rather than a mere correlation of apparently unrelated results. I would highlight two experiments which emphasize the 'predictive' character of the research.

(i) We were led to prepare the trimetal compound (**20**) as a consequence of the earlier preparation of the compounds $[Pt(alkyne)_2]$ (15) from $[Pt(C_2H_4)_3]$ and RC≡CR. The isolobal relationship between $[(\eta\text{-}C_5H_5)(OC)_2W{\equiv}CR]$ and RC≡CR prompted us to investigate the corresponding reaction between the tungsten compound and $[Pt(C_2H_4)_3]$, as discussed elsewhere (14).

(ii) It was the long known observation that alkynes with $[Co_2(CO)_8]$ afford the dicobalt species $[Co_2(\mu\text{-alkyne})(CO)_6]$ which prompted our discovery of the cluster compound (27). It seemed likely (6) that $[(\eta\text{-}C_5H_5)(OC)_2W{\equiv}CR]$ would react with $[Co_2(CO)_8]$ in a similar manner to RC≡CR because of the relationship $CR \xleftrightarrow{0} W(CO)_2(\eta\text{-}C_5H_5)$.

W.L. Gladfelter, University of Minnesota: What is the possibility of forming carbon-carbon bonds using two metal alkylidyne fragments?

F.G.A. Stone: With increasing frequency, as our work in this area develops, we are observing reactions in which the metal alkylidyne fragments link to form C-C bonds. Please refer to compounds (**39**) and (**40**), and the relevant references to these species.

G.A. Ozin, University of Toronto: In view of the current interest in electrically conducting, undoped and doped polyacetylene thin films, do you envisage any possibility of initiating a controlled polymerization of a metal-metal triply bonded organometallic complex to produce a species of the form:

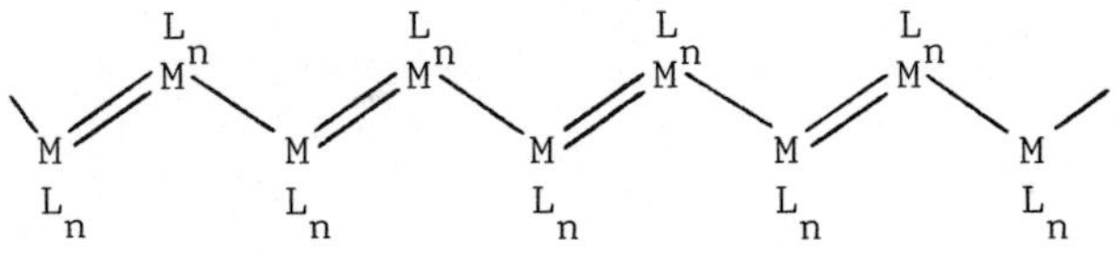

(cis or trans)

analogous to:

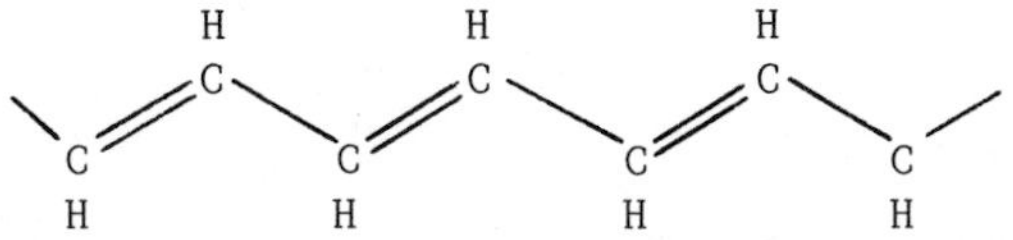

F.G.A. Stone: This is a very interesting thought, but it seems more likely that the model compound employed in our work [(η-C_5H_5)$(OC)_2$W≡CR] will form cyclic oligomers rather than afford linear polymers, when treated with appropriate transition metal salts. You will recall that many low-valent metal complexes react with alkynes to produce arenes catalytically. We have not as yet succeeded in trimerising [(η-C_5H_5)-$(OC)_2$W≡CR], but the latter in the presence of certain complexes affords the bridged-alkyne ditungsten compound (30). I refer you to our papers:

Green, M.; Porter, S.J.; Stone, F.G.A. J. Chem. Soc. Dalton Trans., in press.

Green, M.; Jeffrey, J.C.; Porter, S.J.; Razay, H.; Stone, F.G.A. J. Chem. Soc., Dalton Trans., in press.

L. Lewis, General Electric: Following you isolobal argument, would you expect or do you observe any metal carbon bond scission reactions? Vollhardt *et al.* (1) observe C≡C bond scission to form carbynes using CpCo$(CO)_2$. Professor Schrock just told us how to break C≡C using high valent metals. Does your W≡C show any similar reactivity?

(1) Fritch, J.R.; Vollhardt, K.P.C. Angew. Chem. 1980, 19.

F.G.A. Stone: Yes, I would expect cleavage of the W≡C bond in [(η-C_5H_5)$(OC)_2$W≡CR] to occur in certain reactions. Indeed, we believe we have observed C≡W bond cleavage in a reaction of the tungsten compound with a carbido(carbonyl)iron cluster. Interestingly, [(η-C_5H_5)$(OC)_2$W≡CR] reacts with sulphur to yield [(η-C_5H_5)(OC)•W•S•CR:S] (I. Moore, Bristol University).

27

Structure of $[\{Os_3(\mu\text{-}H)_3(CO)_9C(\mu\text{-}O)\}_3(B_3O_3)]$

SHELDON G. SHORE

The Ohio State University, Department of Chemistry, Columbus, OH 43210

Professor Stone's paper points out that the reactivity of $[(\eta\text{-}C_5H_5)(OC)_2W{\equiv}CR]$ towards transition metal complexes is similar to that of an alkyne. It would be of interest to examine this compound and several of its derivatives which contain C=W double bonds with respect to their reactivity patterns towards the BH_3 group to determine if reactions analogous to the hydroboration reaction of alkynes and olefins would occur (1) or reactions similar to the attempted hydroboration described below would take place.

In our laboratory we have examined the reactivity pattern of $[Os_3(\mu\text{-}H)_2(CO)_{10}]$, an unsaturated cluster which can be represented as possessing an osmium-osmium double bond in its classical valence bond representation. We find (2,3) that this compound undergoes a number of reactions with metal carbonyls which in some cases can be formulated as proceeding through intermediates analogous to metal olefin complexes.

We have also examined the reaction of $THFBH_3$ with $[Os_3(\mu\text{-}H)_2(CO)_{10}]$(4,5). Our purpose was to determine if a reaction analogous to hydroboration would occur to give a compound isoelectronic with the known $[Os_3(\mu\text{-}H)_2(\mu\text{-}CH_2)(CO)_{10}]$(6). The following reaction was observed.

$$3[Os_3(\mu\text{-}H)_2(CO)_{10}] + 3\ C_4H_8OBH_3 \longrightarrow [\{Os_3(\mu\text{-}H)_3(CO)_9C(\mu\text{-}O)\}_3(B_3O_3)] + 3\ C_4H_{10}$$

The x-ray structure of $[\{Os_3(\mu\text{-}H)_3(CO)_9C(\mu\text{-}O)\}_3(B_3O_3)]$ (Figure 1) was determined by Dr. J. C. Huffman (Molecular Structure Center, Indiana University). It contains a boroxine ring (B_3O_3) and to each boron there is attached, through a bridging oxygen between boron and carbon, the cluster unit $Os_3(\mu\text{-}H)_3(CO)_9C$. The proton nmr spectrum (τ28.5) of this compound is consistent with the presence of bridging hydrogens around the Os_3 plane.

0097-6156/83/0211-0399$06.00/0

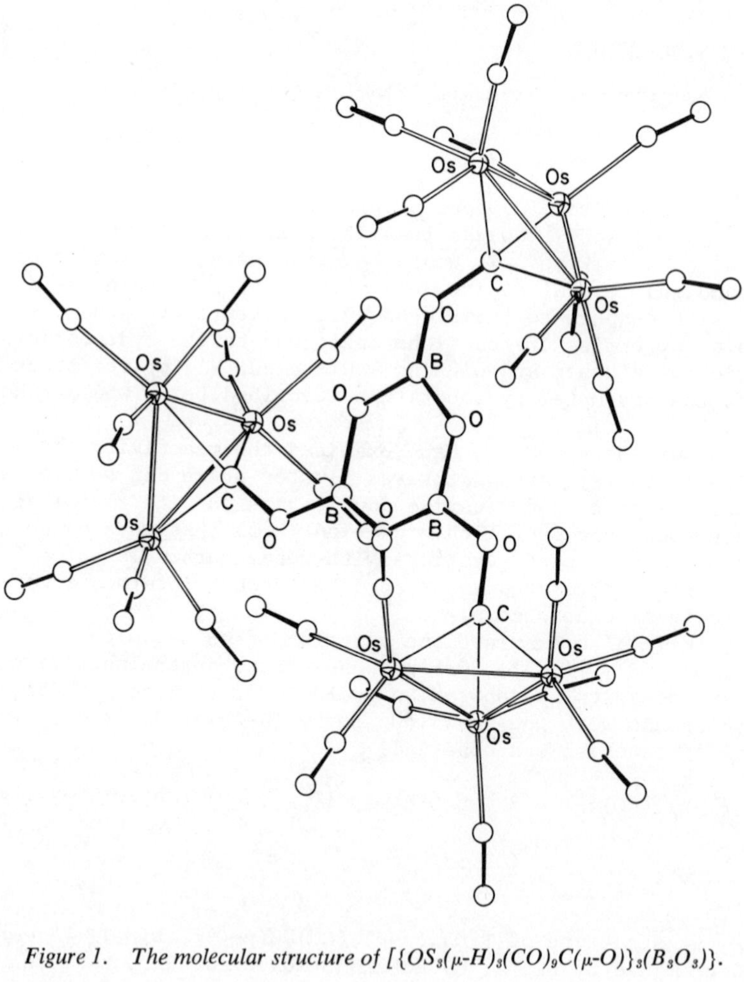

Figure 1. The molecular structure of $[\{OS_3(\mu\text{-}H)_3(CO)_9C(\mu\text{-}O)\}_3(B_3O_3)\}$.

The above reaction is close to being quantitative. Stoichiometry was determined by varying molar ratios of reactants. The C_4H_{10} was identified by mass spectrometry. In this reaction oxygen is abstracted from C_4H_8O (THF) by boron to form the B_3O_3 unit. Two hydrogens from BH_3 add to the C_4H_8 unit to form C_4H_{10} while the third hydrogen adds to the osmium cluster.

Literature Cited

1. Brown, H. C., "Hydroboration" Benjamin, New York, 1962.
2. Plotkin, J. S., Alway, D. G., Weisenberger, C. R., and Shore, S. G. J. Am. Chem. Soc., 1980, 102, 6157.
3. Shore, S. G., Hsu, W. L., Weisenberger, C. R., and Caste, M. L. Organometallics, in press.
4. Kennedy, S., Ph.D. Thesis Ohio State University, 1982.
5. Shore, S. G., Kennedy, S., Hsu, W. L., and Huffman, J. C. In preparation.
6. Calvert, R. B. and Shapley, J. R. J. Am. Chem. Soc., 1977, 99 5525.

RECEIVED December 14, 1982

NEW ANALYTICAL AND SPECTROSCOPIC TECHNIQUES

28

High-Resolution Magic Angle Spinning and Cross-Polarization Magic Angle Spinning Solid-State NMR Spectroscopy

Analytical Chemical Applications

C. A. FYFE, L. BEMI, H. C. CLARK, J. A. DAVIES, G. C. GOBBI, J. S. HARTMAN, P. J. HAYES, and R. E. WASYLISHEN

University of Guelph, The Guelph-Waterloo Centre for Graduate Work in Chemistry, Guelph Campus, Department of Chemistry, Guelph, Ontario N1G 2W1 Canada

The techniques of cross-polarization and 'magic-angle' spinning used to obtain high-resolution solid-state NMR spectra are described and their application to inorganic systems illustrated. In general the experiments are complementary to diffraction techniques, being applicable to amorphous systems (such as surface immobilized species and glasses) where diffraction data cannot be obtained and to crystalline systems where diffraction measurements yield only partial structural data (as in the case of dynamic solid-state structures where the molecular motions are not detected, and zeolites where Si and Al atoms cannot be distinguished). In addition NMR spectroscopy provides a valuable 'bridge' between solid-state diffraction-determined molecular structures and those which exist in solution.

There has been substantial interest in recent years in the chemical applications of high-resolution NMR spectroscopy of solids. The experiments are the result of a series of advances over a number of years and have now reached the routine stage with commercial systems being available. Usually they involve some combination of high-power proton decoupling, cross-polarization and magic-angle spinning techniques. Below we review the experiment and discuss its application in particular to inorganic systems indicating the role of each of the components of the total experiment.

The Experiment

NMR of solids usually give broad featureless absorptions due to the dipolar interactions which, at least in the case of protons, are orders of magnitude larger than the characteristic chemical shifts and spin-spin couplings used for structure elucidation

0097-6156/83/0211-0405$07.25/0

etc. Scheme I illustrates the way in which the high-resolution characteristics of the spectrum may be recovered in high-resolution NMR of solids.

Scheme I

^{1}H: 'Abundant' nucleus

$$H_{Tot} = H_{Zeeman} + H_{H\text{-}H\ dipolar} + H_{Other\ (negligible)} \quad [1]$$

^{13}C: 'Dilute' nucleus

$$H_{Tot} = H_{Zeeman} + H_{H\text{-}C\ dipolar} + H_{C\text{-}C\ dipolar} + H_{Other\ (shift\ anisotropy)} \quad [2]$$

Hydrogen is an example of an 'abundant' nucleus. That is, there is a high concentration of nuclei with a nuclear isotope of high natural abundance (^{1}H, I = ½, 99.8%) in the sample. In this case the dipolar interactions between the nuclei dominate the spectra giving broad featureless absorptions. In the case of ^{13}C, a 'dilute' nucleus (i.e., there are few NMR active nuclei in the system), the situation is considerably simplified. The ^{1}H-^{13}C dipolar interactions are between different nuclei and may be removed by powerful proton decoupling fields and most importantly, the homonuclear ^{13}C-^{13}C dipolar interactions do not exist because of the low concentration of ^{13}C in the system due to the low natural abundance of ^{13}C (I = ½, 1.1%).

The largest remaining interaction is the 'chemical shift anisotropy' which is the three-dimensional magnetic shielding of the nuclei. In the solid state, a single nucleus will give rise to a signal which is dependent on the orientation of the crystal to the field and since all possible orientations exist for a polycrystalline sample, a broad absorption (or shift anisotropy pattern) will result. This is averaged to the isotropic average value by the random motion of the molecules in solution. The same averaging may be achieved for a polycrystalline sample by spinning it rapidly about an axis inclined at an angle of 54°44' to the magnetic field vector (Figure 1), the so-called 'Magic-Angle' (5, 6,7). Various designs are available for rapid sample spinning, most of which are based on the original designs of Andrew (5) and Lowe (8).

In addition, the technique of 'cross polarization' introduced and developed by Pines, Gibby and Waugh (9) is used to increase the signal-to-noise ratio of the spectrum. The proton magnetization is 'spin-locked' along the y' axis with a spin-locking field H_H and the carbons subjected to an RF pulse chosen such that the two fields fulfill the 'Hartmann-Hahn' condition (10), equation [3] (Figure 2).

$$\gamma_H H_H = \gamma_C H_C \quad [3]$$

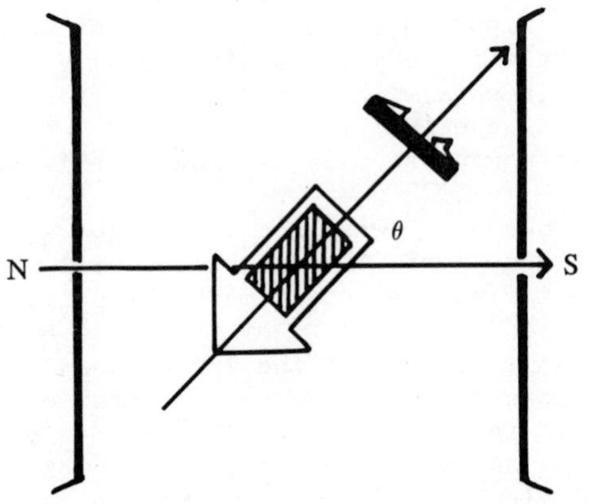

Figure 1. Representation of the geometric arrangement for a sample spinning at the magic angle to the magnetic field vector H_0. (Reproduced with permission from Ref. 40. Copyright 1982, Royal Society of London.)

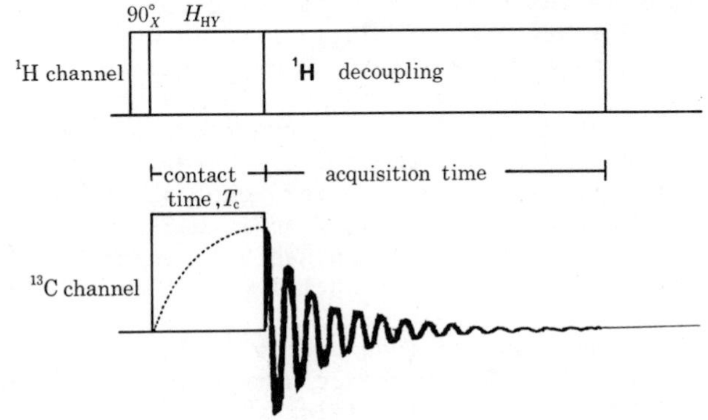

Figure 2. Cross-polarization timing diagram (see text for discussion). (Reproduced with permission from Ref. 40. Copyright 1982, Royal Society of London.)

The net effect is that the carbon nuclei are polarized by the proton magnetization and the S/N is increased both directly by the ratio γ_H/γ_C (4 in the case of ^{13}C) and indirectly because the experiment now depends only on the recovery of the proton magnetization which will usually be much faster than the ^{13}C relaxation.

The combined CP/MAS experiment was first applied by Schaefer and Stejskal (11, 12) and has since found wide application in a number of areas. Figure 3 shows the effect of spinning and the kind of resolution which may be obtained from very crystalline materials (13). Spectra with this degree of resolution obviously contain substantial chemical information.

Several points may usefully be made at this stage concerning the experiment:

Firstly, although ^{13}C was used in the above example and in the practical applications, the experiment will work for any nucleus which satisfies the requirements of being 'dilute'. In some cases, eg. ^{13}C ($I = \frac{1}{2}$, 1.1%) or ^{29}Si ($I = \frac{1}{2}$, 4.7%) this occurs automatically because of the low abundance but will work even for ^{31}P as long as there are not too many of them, too close to each other. In fact, apart from ^{1}H and possibly ^{19}F in 'concentrated' samples any spin ½ nucleus will fulfill the NMR requirements of being 'dilute'. Results reported to date on other nuclei suggest that many are amenable to study in the solid state.

Secondly, in many important inorganic systems, a very considerable simplification exists in that often there are no protons present which are directly incorporated into the lattice (eg. zeolites, glasses and calcogenides) and the only mechanism of line-broadening is the chemical shift anisotropy, which can be removed by MAS. Further, cross-polarization is not possible and the removal of the need for high-power decoupling means that it is possible to use a conventional high-resolution spectrometer and the experiment is thus easily within the reach of most NMR spectroscopists. In many cases, there are substantial advantages of working at high fields in the superior homogeneity and stability of superconducting solenoid magnets, and a MAS probe for narrow bore superconducting magnet systems has been described (14). Examples of spectra of different nuclei obtained in this way are shown in Figure 4.

Thirdly, in the case of quadrupolar nuclei with non-integral spins (eg. ^{27}Al, ^{11}B, ^{17}O) there are very definite advantages in working at the highest possible field. The ($\frac{1}{2} \leftrightarrow -\frac{1}{2}$) transition is not subject to quadrupolar interactions to first order (Figure 5), but is affected by second order quadrupolar interactions, resulting in line-broadening and shifts. This interaction is inversely proportional to the magnetic field strength (equation [4]) where $\omega_{\frac{1}{2}}$ is the linewidth at half-height of the peak, ν_Q is the quadrupolar frequency and ν_L is the Larmor frequency, and is thus minimized at high fields when the chemical shift is maximized. MAS reduces, but does not completely remove, the interaction and substantial improvement is found at high fields (Figure 6). It

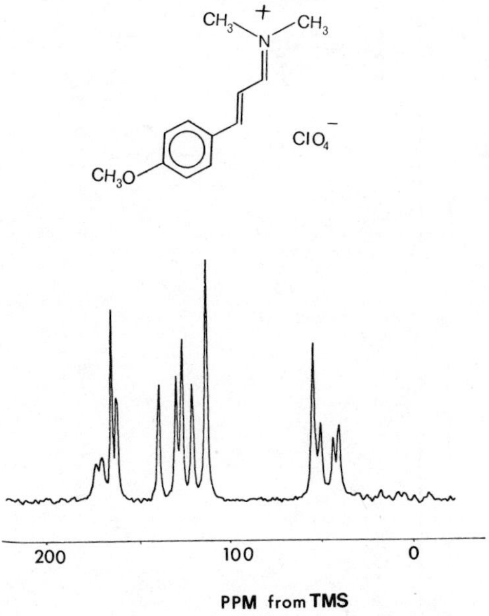

Figure 3. ^{13}C-CP/MAS solid-state spectrum of the cation shown illustrating the resolution obtainable from crystalline samples.

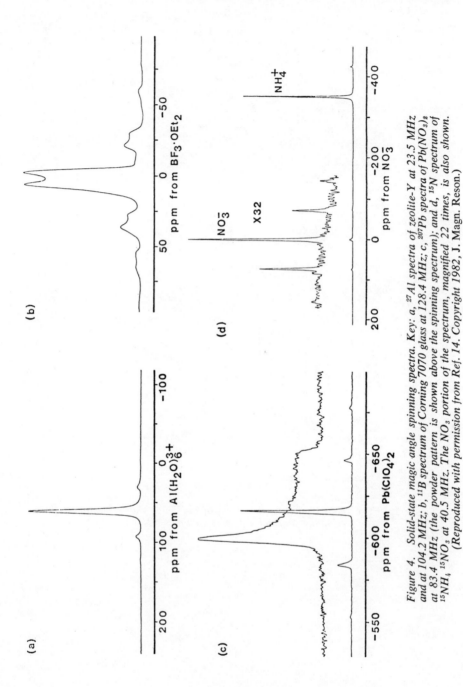

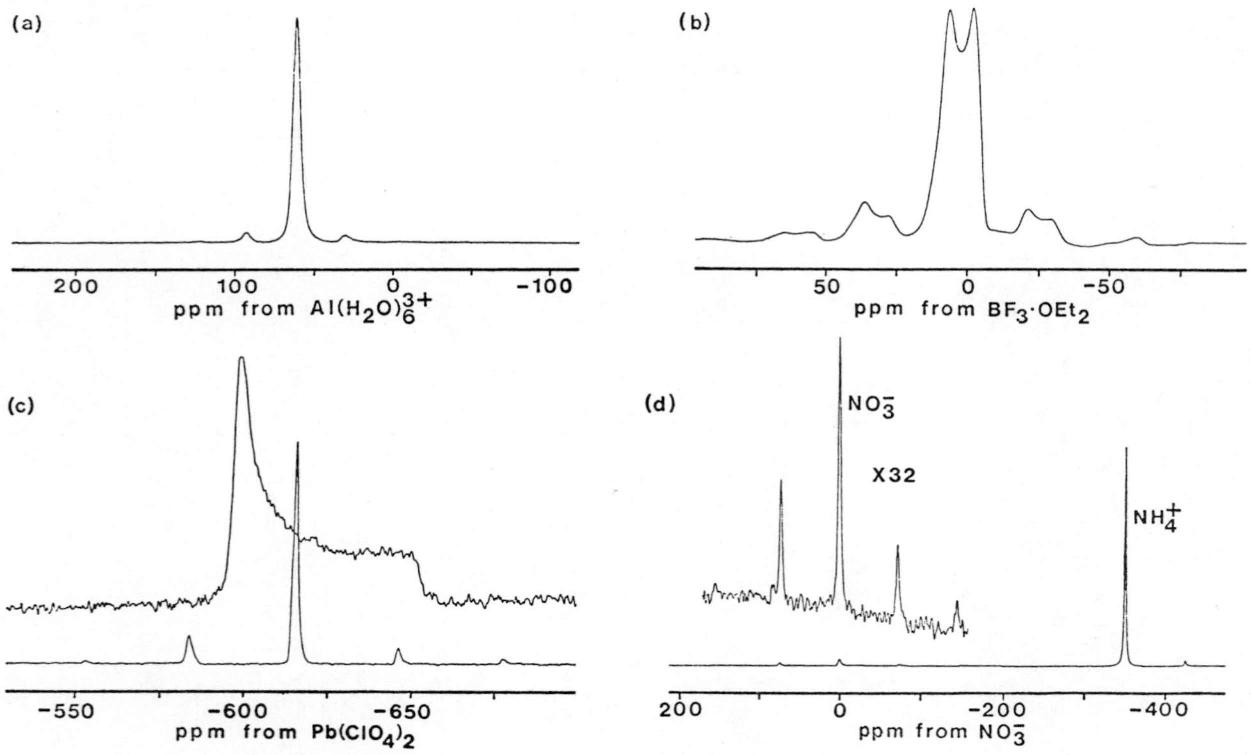

Figure 4. Solid-state magic angle spinning spectra. Key: a, ^{27}Al spectra of zeolite-Y at 23.5 MHz and at 104.2 MHz; b, ^{11}B spectrum of Corning 7070 glass at 128.4 MHz; c, ^{207}Pb spectra of $Pb(NO_3)_2$ at 83.4 MHz (the powder pattern is shown above the spinning spectrum); and d, ^{15}N spectrum of $^{15}NH_4$ $^{15}NO_3$ at 40.5 MHz. The NO_3 portion of the spectrum, magnified 22 times, is also shown. (Reproduced with permission from Ref. 14. Copyright 1982, J. Magn. Reson.*)*

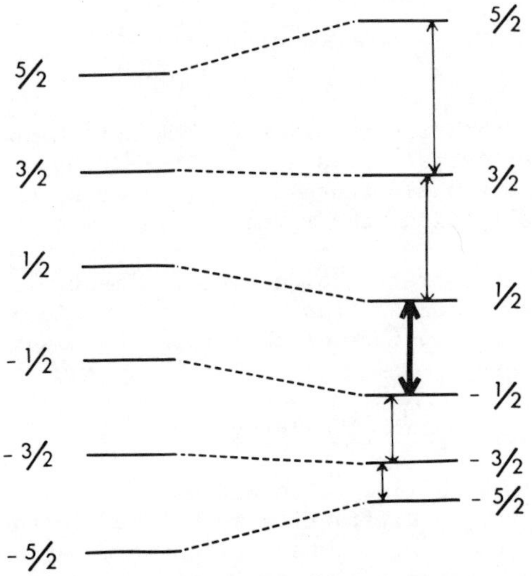

Figure 5. Energy level diagram for a spin 5/2 nucleus showing the effect of the first-order quadrupolar interaction on the Zeeman energy levels. The (m = ½ ⇆ m = –½) *transition (shown in bold) is independent of the quadrupolar interaction to first order.*

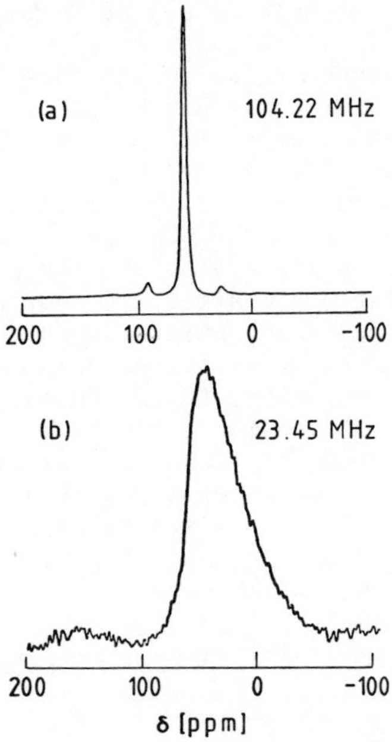

Figure 6. High-resolution micrograph together with corresponding scalar structural drawing, computed image, and appropriate diffraction pattern showing structural resolution of $Bi_2W_2O_9$ with some $Bi_2W_3O_{12}$ intergrowths (24).

$$\omega_{\frac{1}{2}}\ (\text{quadrupolar}) = \frac{25\nu_Q^{\ 2}}{18\nu_L} \qquad [4]$$

should be noted that the majority of NMR active nuclei are quadrupolar with non-integral spins and high-field MAS experiments will be applicable to a whole range of solid systems containing such nuclei, especially since their relaxation times are typically short.

Thus, by the correct choice of experiment, it is possible to obtain high-resolution spectra from a wide variety of nuclei in the solid state and spectra of this type have many applications to inorganic chemistry.

Applications in Inorganic Chemistry

In general, high-resolution NMR of solids is found to be very much complementary to diffraction techniques in the investigation of solid-state structures. The applications may be classified into three very general groupings:

Firstly, for amorphous systems (some of which are of considerable commercial importance) the NMR data provides essential information as diffraction techniques are not applicable at all due to the lack of order in the lattice.

Secondly, for some crystalline systems, the structure obtained by diffraction techniques may be incomplete. For example, in some cases the diffraction data may not reveal dynamic aspects of the solid-state structure (as in the case of fluxional organometallics) and in others it may not be possible to distinguish clearly between different atoms (as for example ^{27}Al and ^{29}Si in zeolites) and a combination of the NMR and x-ray data will yield a more complete and meaningful description of the structure.

Thirdly, in cases where the NMR spectra can be obtained both in solution and in the solid state, the NMR studies will act as a valuable 'bridge' between the structure characterized by diffraction techniques and that which exists in solution. Where gross structural changes do occur, as for example, where exchange equilibria are present in solution, the solid-state spectra will serve as 'benchmark' values for an interpretation of the solution studies. In the following sections we will illustrate applications of the technique within the general groupings described above, using examples from our own work and relating the choice of the experimental conditions to the characteristics of the nuclei being studied as discussed in Part 2 above.

Amorphous Systems (Polymer and Surface Immobilized Catalysts and 'Inorganic' Glasses)

Immobilized Polymers. There has been considerable interest in recent years in the synthesis and use of reagents covalently immobilized on insoluble support materials such as glass and polymers because of their potential advantages in terms of ease of separation, conservation of valuable materials, control of noxious reagents etc. (15,16,17). The immobilization of transition metal catalysts has been extensively studied as a method of combining the most desirable characteristics of homogeneous and heterogeneous catalysts. Systematic research efforts in this field have been severely restricted because of the lack of suitable analytical techniques for the analysis and most importantly, structural characterization of surface immobilized species (diffraction measurements clearly not being applicable). Many of the complexes contain phosphine donor ligands and ^{31}P CP/MAS spectra may be used to probe their structure and geometry.

Immobilization on glass surfaces can be carried out by either routes [5] or [6] in Scheme 2 both yielding the immobilized complex (1), reaction [5] being the procedure most commonly used.

Scheme 2

$$\text{(S)}-OH + (EtO)_3SiCH_2CH_2PPh_2 \rightarrow \text{(S)}-O-\overset{|}{\underset{|}{Si}}-CH_2CH_2PPh_2$$

$$\downarrow [ML_n]$$

$$\text{(S)}-O-\overset{|}{\underset{|}{Si}}-CH_2CH_2-\overset{Ph}{\underset{Ph}{\overset{|}{P}}}-ML_{(n-1)} \qquad [5]$$

(1)

$$[ML_n] + (EtO)_3SiCH_2CH_2PPh_2 \rightarrow (EtO)_3Si-CH_2CH_2-\overset{Ph}{\underset{Ph}{\overset{|}{P}}}-ML_{(n-1)}$$

$$\downarrow \quad \text{(S)}-OH \qquad\qquad [6]$$

$$\text{(S)}-O-\overset{|}{\underset{|}{Si}}-CH_2CH_2-\overset{Ph}{\underset{Ph}{\overset{|}{P}}}-ML_{(n-1)}$$

(1)

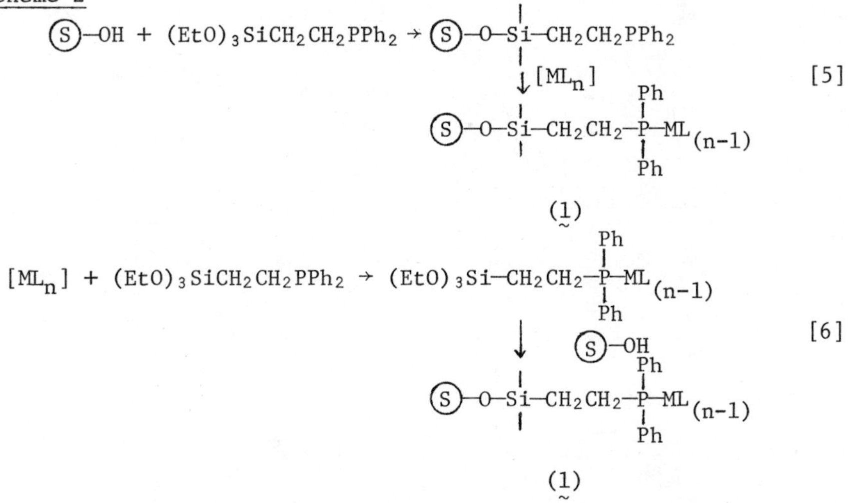

Figure 7 shows the ^{31}P CP/MAS spectrum of cis[PtCl$_2$(PPh$_2$Me)$_2$] (18,19). There are two central resonances indicating that the two phosphine ligands in the complex are not equivalent in the solid, probably due to the freezing in of a fixed conformation (the equivalence in solution resulting from rotation about the Pt—P bonds). The sidebands are from the interaction of the ^{31}P nuclei with the 33% abundant ^{195}Pt isotope (I = ½, 33.4%). Most importantly, the magnitude of the $^1J(^{195}Pt, ^{31}P)$ coupling indicates that the geometry of the complex is cis as indicated in the figure. The middle spectrum is of a very similar complex, cis-[PtCl$_2$(PPh$_2$-CH$_2$CH$_2$SiOEt$_3$)$_2$] (2), where the phosphine ligands are functionalized with siloxy groups for covalent attachment to an activated glass support. In this case, there are three central resonances

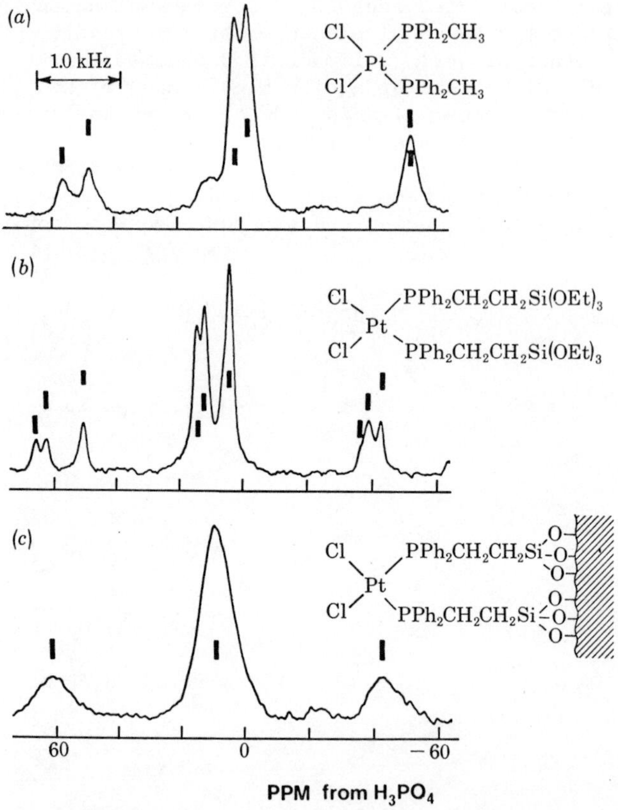

Figure 7. ^{31}P-CP/MAS solid-state spectrum of $[PtCl_2(PPh_2CH_3)_2]$ at 36.442 MHz (a), ^{31}P-CP/MAS solid-state spectrum of $[PtCl_2(PPh_2CH_2CH_2Si(OEt)_3]$ at 36.442 MHz (b), and ^{31}P-CP/MAS solid-state spectrum of $\{PtCl_2[PPh_2CH_2CH_2Si(OEt)_3]\}$ covalently immobilized on glass beads. All spectra referenced to external 85% H_3PO_4 (18). (Reproduced with permission from Ref. 40. Copyright 1982, Royal Society of London.)

but it is not possible to explain this multiplicity without information from other sources. (A single resonance is again observed in the solution ^{31}P spectrum.) The couplings indicate that there has been no change in the geometry of the complex. The above spectra demonstrate a definite sensitivity of the ^{31}P nucleus to local environmental effects and it might be anticipated that relatively broad peaks would be observed for a disordered environment. This is observed in the lowest spectrum in the series which is complex (2) immobilized on an activated glass surface indicating a highly disordered environment with no preferred orientations. The coupling to ^{195}Pt is still present indicating that the coordination to the metal atom is still intact and its magnitude shows that the geometry is still 'cis'.

Thus, a very efficient catalyst immobilization can be carried out by route [6]. However, route [5] is much less successful: Even under the most inert atmosphere conditions, the ^{31}CP/MAS spectrum of the immobilized ligand showed a major signal at δ = 42 ppm (wrt 85% H_3PO_4) characteristic of phosphine oxide rather than phosphine. This could be quantitatively reduced by $HSiCl_3$ and this surface reaction monitored by NMR but the subsequent exchange reaction (equation [5]) generated substantial quantities of phosphine oxide and a number of different isomeric complexes were formed.

A similar situation is seen from the ^{31}P CP/MAS spectra of polymer immobilized catalysts (20,21). Again, two routes are possible as shown in Scheme 3, equations [7] and [8] forming similar immobilized complexes (3), (3)'. Route [8], where the complex is formed with a polymerizable ligand and directly incorporated into the polymer during the polymerization process, quantitatively produces well characterized immobilized catalysts as shown by the ^{31}P CP/MAS spectra in Figure 8. The alternate route [7], which is by far the most widely used approach, again yields phosphine oxide when attaching the ligand and gives a mixture of products during the exchange process. It is possible to probe the structures of these complexes via the ^{31}P CP/MAS spectra of attached phosphine ligands even where they are not the linking group to the surface as indicated in Figure 8c.

Scheme 3

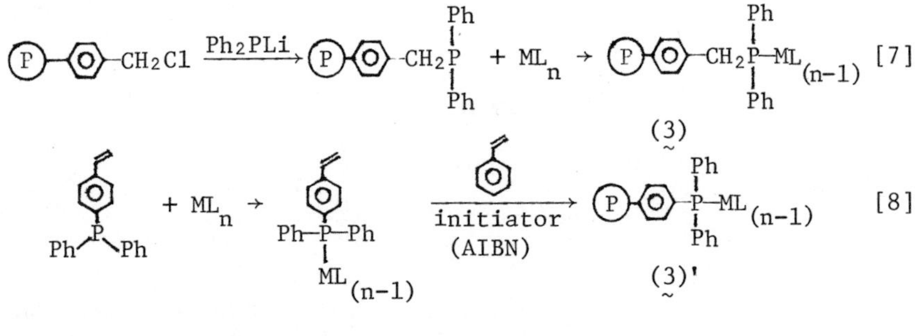

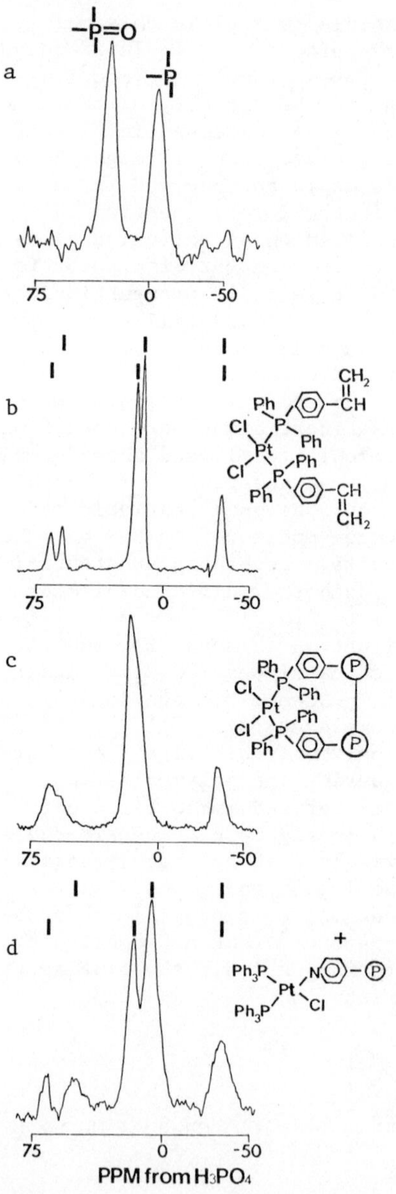

Figure 8. ^{31}*P-CP/MAS solid-state spectrum of polymer-bound (polystyrene cross-linked with 2% divinylbenzene) triphenylphosphine (a),* ^{31}*P-CP/MAS solid-state spectrum of* cis-*[$PtCl_2(PPh_2$-C_6H_4-$CH{=}CH_2)_2$] (referenced to external 85% H_3PO_4) (b),* ^{31}*P-CP/MAS solid-state spectrum of a copolymer of 65% styrene, 31% divinylbenzene and 4%* cis-*[$PtCl_2(PPh_2$-C_6H_4-$CH{=}CH_2)_2$] after soxhlet extraction (c) and* ^{31}*P-CP/MAS solid-state spectrum of* cis-*[$PtCl(PPh_3)_2$, (N P)]$^+ClO_4^-$ after soxhlet extraction (d). All spectra referenced to external 85% H_3PO_4. (Reproduced from Ref. 21. Copyright by American Chemical Society.)*

Thus ^{31}P CP/MAS NMR may be used to characterize solid and immobilized transition metal catalyst systems. The results indicate that high yields of single complex may be formed by the correct choice of preparative route, but that the most commonly used procedures are relatively inefficient and that the measured catalytic activities cannot represent the optimal performance of these systems.

Inorganic Glasses. The important class of inorganic glasses is another situation where the amorphous nature of the system precludes the use of diffraction techniques. Commonly, they are typically composed of Al, B, Na, O and Si and their composition from elemental analysis is expressed in terms of the component oxides (i.e., Al_2O_3, B_2O_3, Na_2O and SiO_2). Thus, the composition of a glass, Pyrex (Corning 7740) for example, is typically given as 81% SiO_2, 2% Al_2O_3, 13% B_2O_3 and 4% Na_2O. In these systems, solid spectra can be obtained for ^{11}B, ^{27}Al, ^{29}Si, possibly ^{23}Na (and even ^{17}O, if enrichment is used) and a complete characterization of the system is possible. There are not protons present in the system and the experiment is reduced to one of simple MAS. Further, two of the most common nuclei, ^{27}Al and ^{11}B, are quadrupolar and the MAS spectra will show least residual distortion when run at the highest possible field strength. Figure 9 shows ^{11}B MAS spectra obtained at 128 MHz corresponding to a proton frequency of 400 MHz on a conventional high-resolution spectrometer (13). The top left spectrum is of a "soda glass" where the boron atoms have tetrahedral coordination and a single sharp absorption is observed as the near spherical symmetry produces a very small quadrupolar interaction. The bottom-right spectrum is of a glass where the boron atoms have trigonal coordination and a large quadrupolar interaction which produces the characteristic 'doublet' structure observed. The intermediate spectra are superpositions of these two extreme cases and can be used to determine the proportions of the two boron coordinations if care is taken to assure the spectra are quantitatively reliable. When taken together with measurements on the other NMR active nuclei, the spectra yield a complete detailed description of these systems.

Incomplete X-Ray Structures

Molecular Motions and Dynamic Structures. Molecular motions are of quite general occurrence in the solid state for molecules of high symmetry (22,23). If the motion does not introduce disorder into the crystal lattice (as, for example, the in-plane reorientation of benzene which occurs by 60° jumps between equivalent sites) it is not detected by diffraction measurements which will find a seemingly static lattice. Such molecular motions may be detected by wide-line proton NMR spectroscopy and quantified by relaxation-time measurements which yield activation barriers for the reorientation process. In addition, in some cases, the molecular reorientation may be coupled with a chemical exchange process as, for example, in the case of many 'fluxional' organometallic molecules.

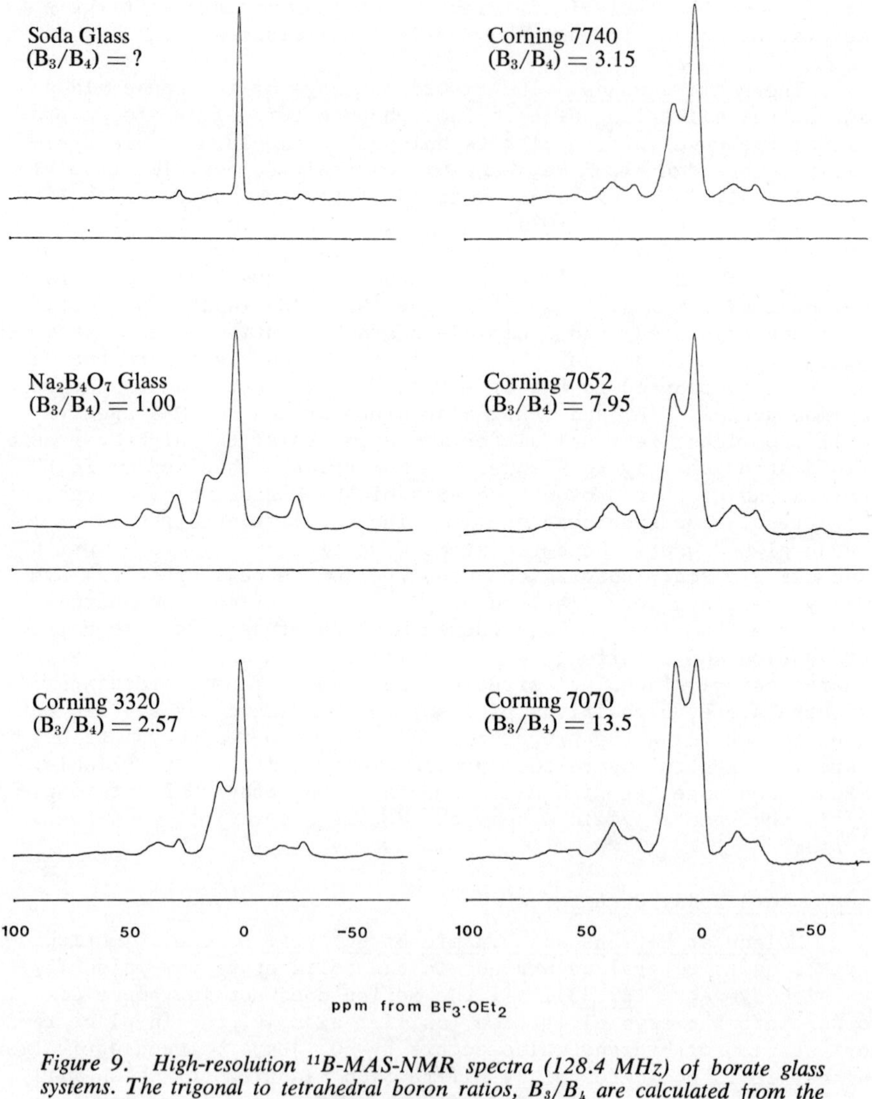

Figure 9. High-resolution ^{11}B-MAS-NMR spectra (128.4 MHz) of borate glass systems. The trigonal to tetrahedral boron ratios, B_3/B_4 are calculated from the oxide formula of the glass.

The proton NMR measurements, however, indicate only that motion is taking place, not that it involves a chemical exchange. This information must come from a comparison with data from other sources which may themselves be suspect if they were obtained assuming the lattice to be rigid. High-resolution CP/MAS NMR provides an opportunity for the direct investigation of the mechanism of the exchange process itself, thus the only mechanism by which chemically inequivalent nuclei will yield a single absorption (apart from artificial degeneracy) is a chemical process, just as in solution.

This is illustrated in Figure 10 which shows the ^{13}C CP/MAS spectrum of the fluxional organometallic cyclooctatetraene-diiron pentacarbonyl (4) (13), whose 'localized' bonding scheme determined by diffraction measurements is shown inset in the figure. All eight carbons show a single sharp absorption due to the exchange which does not change down to 77°K, consistent with the activation barrier of 2.0 ± 0.1 kcal mol^{-1} from proton relaxation data. In some cases, the exchange can be frozen out as in solution. For example, in the ^{13}C CP/MAS spectrum of bis-cyclooctatetraene triruthenium tetracarbonyl (5), at ambient temperatures all eight carbons of the organic moieties contribute to a single exchange averaged absorption whereas at low temperatures this absorption becomes a multiplet when the exchange process is frozen out (24). Spectra at intermediate temperatures show the expected broadening and coalescence of the signals (13) and may be used to determine the mechanism of the exchange, under conditions where the geometry of the ground state is well defined from diffraction measurements. Evidence for chemical exchange in other organometallic systems has been found for η^1-C_5H_5 HgX complexes (6) (where X - Cl, Br, I and η^1-C_5H_5) (13).

Limited Structural Data: Zeolites. Zeolites represent another situation where diffraction data on crystalline compounds is limited, in this case, different atoms (^{29}Si, ^{27}Al) cannot be clearly distinguished because they have very similar scattering factors, and the gross structure may be determined relatively easily but the distribution of the individual atoms is not known. In collaboration with Professor J.M. Thomas and Dr. J. Klinowski of the University of Cambridge, U.K., we have used ^{29}Si, ^{27}Al and ^{17}O solid-state NMR spectroscopy to investigate the detailed microstructures of these materials. Zeolites have very loose frameworks formed from corner or face-sharing SiO_4 and AlO_4 tetrahedra and are versatile in their industrial applications, especially in the petroleum industry becuase of their catalytic and sorptive properties. The way in which the frameworks may be built up is illustrated in Figure 11 which shows a 'truncated octahedron' made from six and four membered rings as a basic building unit where the vertices are either Si or Al atoms joined by bridging oxygens being combined to form the repeat unit of zeolite-A by joining units via the four-membered rings by bridging

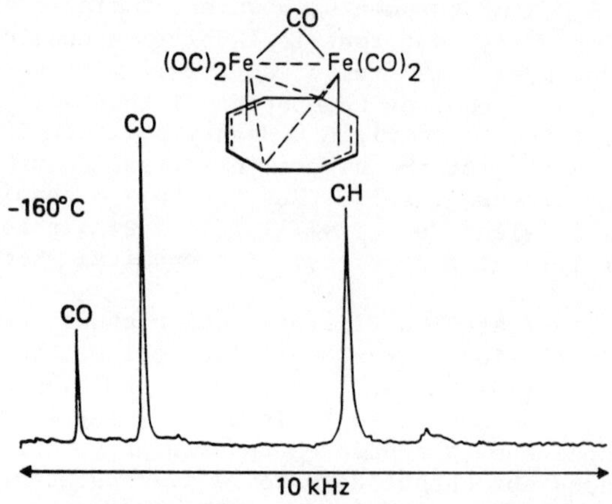

Figure 10. ^{13}*C-CP/MAS spectrum at 22.6 MHz (90 MHz* ^{1}*H) of cyclooctatetraenediiron pentacarbonyl (4) shown along with the localized bonding scheme suggested from x-ray diffraction measurements.*

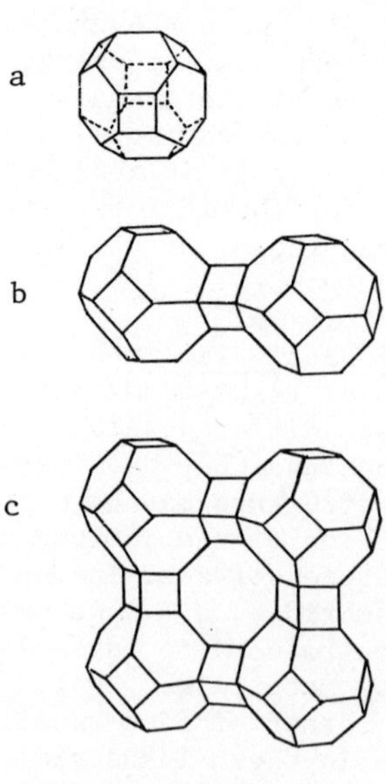

Figure 11. The truncated octahedron building block (also termed "sodalite cage" or "β-cage") (a); tetrahedral atoms (usually Si or Al) are located at the corners of the polygons with oxygen atoms halfway between them. Illustration of the linkage, through double four-membered rings, of two truncated octahedra (b); and the structure of zeolite-A (c).

oxygens. There is a large central cavity accessible through three orthogonal linear channels which gives the system its molecular sieve properties. The same building units may be combined in alternate ways to yield faujasite and sodalite zeolites (Figure 12). The gross overall structures are thus well defined and NMR can provide complementary information regarding the distribution of Si and Al atoms in the lattice. Further, there are no hydrogen atoms covalently incorporated into the lattice and a simple MAS experiment will suffice to give line narrowing and yield quantitatively reliable spectra. The experiments may be advantageously carried out at high field for better resolution in the case of ^{29}Si (Figure 13) and for more reliable spectra in the case of ^{27}Al (Figure 6). It was first shown by Lippmaa et al. (25,26) that the five signals observed in the ^{29}Si spectra corresponded to the five possible second coordinations of Si, i.e., Si[4Al]; Si[3Al,1Si]; Si[2Al,2Si]; Si[1Al,3Si] and Si[4Si], and that the absorptions occurred within characteristic chemical shift ranges (Figure 14) and could thus be used to define the microdistribution of Si within the lattice. Further, the Si/Al ratio may be obtained from the spectrum using equation [9] where n is the number of attached Al atoms for a given peak and I is the peak intensity (27,28,29). It should be noted that the ratio here is

$$\frac{Si}{Al} = \frac{\sum_{n=0}^{4} I_{Si(nAl)}}{\sum_{n=0}^{4} 0.25\, n \cdot I_{Si(nAl)}} \qquad [9]$$

only for Si and Al in the framework. Chemical analysis, however, will include all Si and Al atoms (including contributions from intercalated species) and thus may not provide an accurate composition of the zeolite framework. The ^{27}Al spectrum does not provide direct information regarding the second coordination sphere, but is very sensitive to whether the Al coordination is tetrahedral (δ = 50-70 ppm, wrt $Al(H_2O)_6{}^{3+}$) or octahedral ($\delta \simeq 0$ ppm wrt $Al(H_2O)_6{}^{3+}$) and when combined with the ^{29}Si spectra can be used to investigate the mechanism of dealumination and other chemical modifications of zeolites carried out to modify their catalytic and sorptive properties.

Thus, dealumination may be effected by reaction with $SiCl_4$, the reaction being postulated to occur as in equation [10].

$$Na_x[(AlO_2)_x(SiO_2)_y] + SiCl_4 \longrightarrow Na_{(x-1)}[(AlO_2)_{x-1}(SiO_2)_{y+1}] + AlCl_3 + NaCl \qquad [10]$$

Figure 15 shows the ^{29}Si spectra of the precursor zeolite and the dealuminated material (30). The sharpness of the single peak

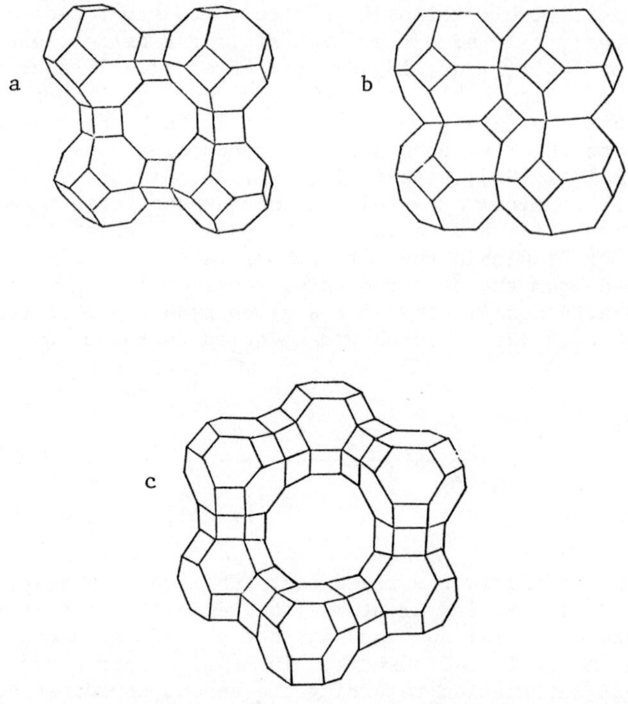

Figure 12. The structure of zeolite-A formed by linking truncated octahedra through double four-membered rings (a), the sodalite structure formed by direct face-sharing of four-membered rings in the neighboring truncated octahedra (b), and the faujasite structure formed by linking the truncated octahedra through double six-membered rings (c).

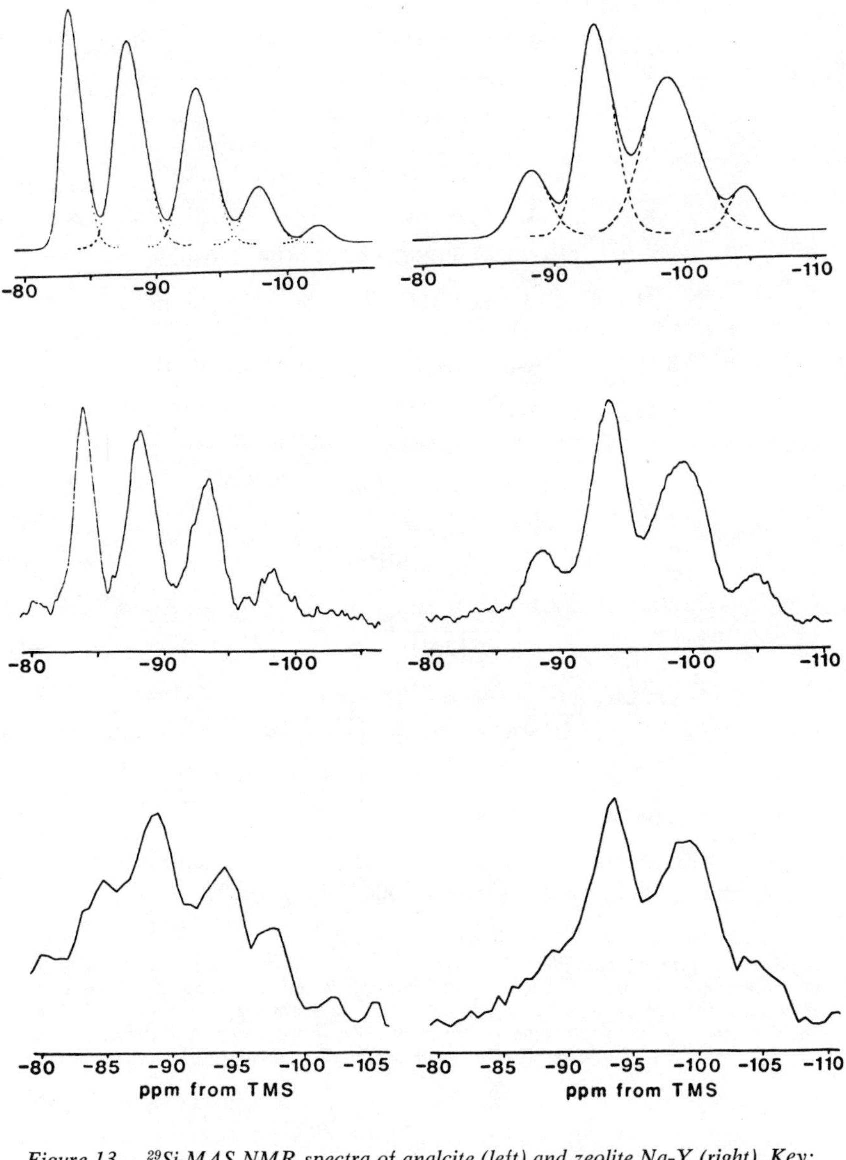

Figure 13. *^{29}Si-MAS NMR spectra of analcite (left) and zeolite Na-Y (right). Key: top, deconvoluted spectra; middle, 79.5 MHz spectra (400 MHz ^{1}H); and bottom, 17.9 MHz spectra (90 MHz ^{1}H). (Reproduced with permission from Ref. 14. Copyright 1982,* J. Magn. Reson.*)*

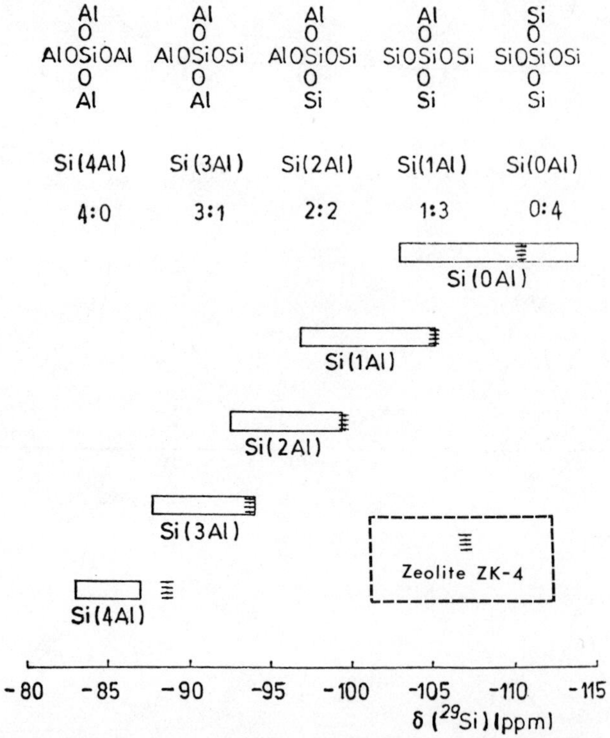

Figure 14. The five possible local environments of a Si atom tetrahedron together with their characteristic ^{29}Si-chemical shift ranges (boxed areas). Si(nAl) represents Si bonded, through oxygen bridges, to nAl atoms where n ranges from 0 to 4. The chemical shifts of the five resonances of zeolite ZK-4 are shown in broken vertical lines. (Figures adapted from References 25 and 35).

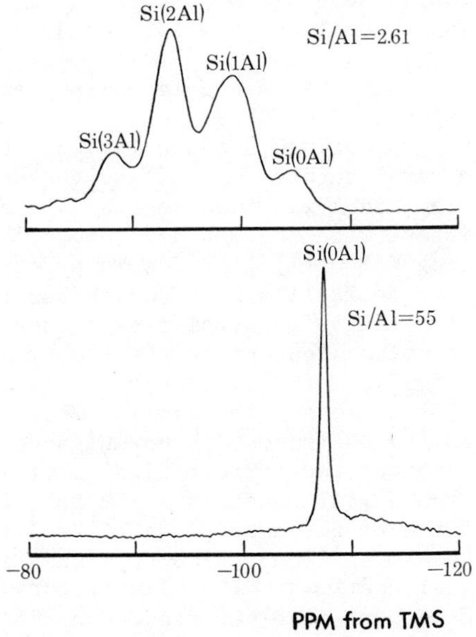

Figure 15. Dealumination of zeolite Na-Y using $SiCl_4$ vapor studied by ^{29}Si-MAS NMR at 79.8 MHz. Key: top, parent Na-Y zeolite; and bottom, after treatment with $SiCl_4$. (Reproduced from Ref. 30. Copyright 1982, American Chemical Society.)

indicates that the material has retained its crystallinity and diffraction measurements confirm that the sample retains the faujasite structure. The ^{27}Al spectrum (Figure 16b) (30), confirms that the aluminum has been lost from the lattice, but shows aluminum in octahedral as well as tetrahedral (lattice) coordinations indicating that some of the aluminum removed from the lattice remains behind as interstitial species formed from $AlCl_4^-$ which may be detected in unwashed samples. This can be removed by extensive washing as shown in the lower figure, but some octahedral aluminum remains behind and bulk chemical analysis would yield Si/Al ratios unrepresentative of the lattice composition. Similar analyses have recently provided important information regarding the relation between the structure of ZSM-5 and Silicalite (two new and important zeolitic materials) and have detailed the lattice composition (31).

In the case of Zeolite-A, a single sharp peak is observed at a chemical shift value of $\delta = -88.9$ ppm (25,32) which corresponds (in the original ranges defined, Figure 14 (26)) to Si[3Al,1Si] coordination and several authors (19-33) suggested that the original structure for Zeolite-A was incorrect and have proposed alternate schemes consistent with the NMR data where 'Loewenstein's Rule' (34) forbidding (Al—O—Al) linkages is systematically broken. Recent work, however, on Zeolite ZK-4 (which has the same basic structure as Zeolite-A but with Si/Al ratios greater than 1) has shown that the original assignment of the spectrum of this zeolite was incorrect (35,36,37), a timely cautionary reminder than any technique has its limitations. The Zeolite ZK-4 spectrum clearly shows five peaks so the assignment is unambiguous (Figure 17) and use of equation [9] gives the correct Si/Al ratio. It can be seen that the shift of the Si[4Al] peak, $\delta = -89$ ppm, is identical to the shift value of the single absorption of Zeolite-A. This is possibly due to the almost linear Si—O—Al linkages in the structure and the original chemical shift ranges, for the effect of the second coordination sphere must be extended as indicated in Figure 14.

The Relation Between Solution and Solid-State Structures

Solution and solid-state structures need not be the same and sometimes quite drastic changes can occur (as exemplified by the early work of Andrew (38,39) on PCl_5 in which the ^{31}P MAS spectrum reveals that PCl_5 crystallizes as $PCl_4^{\oplus}$ $PCl_6^{\ominus}$, Figure 18). In most cases, however, the changes are not nearly as dramatic. The solid-state NMR measurements on compounds of known solid-state structure can be used to confirm that the structure remains unchanged in solution, to put such correlations as ^{31}P chemical shift/bond angle relationships on an absolute basis and to serve as benchmark values for the interpretation of solution data where exchange processes can occur. This area is one of intense current activity and substantial contributions will probably be made,

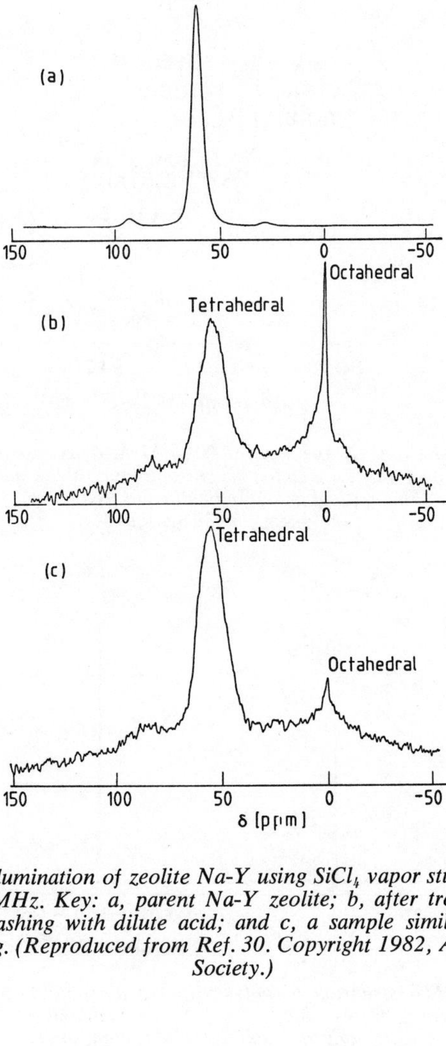

Figure 16. Dealumination of zeolite Na-Y using $SiCl_4$ vapor studied by ^{27}Al-MAS NMR at 104.2 MHz. Key: a, parent Na-Y zeolite; b, after treatment with $SiCl_4$ and moderate washing with dilute acid; and c, a sample similar to b, but after extensive washing. (Reproduced from Ref. 30. Copyright 1982, American Chemical Society.)

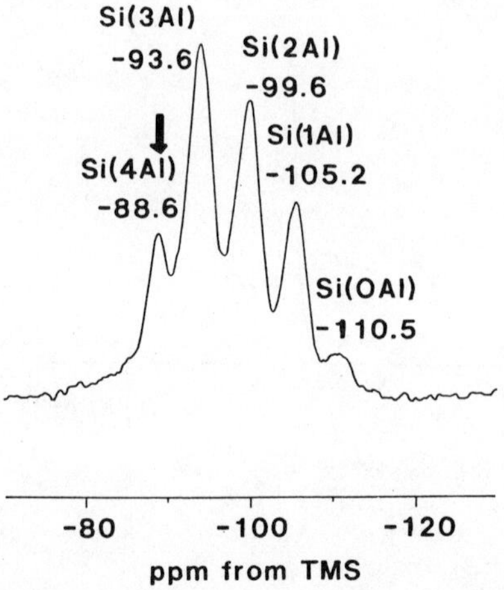

Figure 17. ^{29}*Si-MAS NMR spectrum at 79.8 MHz of zeolite ZK-4 showing five peaks at the chemical shifts indicated, corresponding to the silicon environments shown in the figure. The vertical arrow indicates the frequency of the single peak in the spectrum of zeolite-A.*

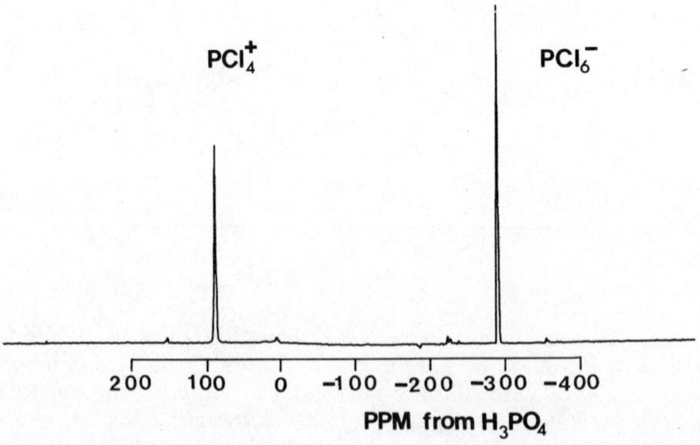

Figure 18. ^{31}*P-NMR spectrum of polycrystalline phosphorus pentachloride with magic angle spinning. The two lines arise from the differently shielded* PCl_4^+ *and* PCl_6^- *ions of which the solid is composed.*

especially in the area of 'other' nuclei including many quadrupolar and metal nuclei.

Acknowledgments

The authors would like to acknowledge the financial support of the Natural Sciences and Engineering Research Council of Canada in the form of Operating Grants (C.A.F., H.C.C., J.S.H. and R.E.W.) and a Graduate Scholarship (G.C.G.). The high-field MAS spectra were obtained at the South-Western Ontario High-Field NMR Facility, Manager, Dr. R.E. Lenkinski. The work on zeolites is a joint project with Professor J.M. Thomas, F.R.S., and Dr. J. Klinowski of the University of Cambridge.

Literature Cited

1. Author to whom all correspondence should be addressed.
2. Present address: Department of Chemistry, University of Toledo, Toledo, Ohio, U.S.A.
3. Present address: Department of Chemistry, Brock University, St. Catharines, Ontario, Canada.
4. Present address: Department of Chemistry, Dalhousie University, Halifax, Nova Scotia, Canada.
5. Andrew, E.R.; Bradbury, A.; Eades, R.G. Nature 1958, 182, 1659.
6. Andrew, E.R. in Prog. in NMR Spectroscopy 1971, 8, 1.
7. Andrew, E.R. in International Review of Science, Physical Chemistry, Series 2 1975, 4 173.
8. Lowe, I.J. Phys. Rev. Lett. 1959, 2, 285.
9. Pines, A.; Gibby, M.G.; Waugh, J.S. J. Chem. Phys. 1973, 59, 569.
10. Hartmann, S.R.; Hahn, E.L. Phys. Rev. 1962, 128, 2042.
11. Schaefer, J.; Stejskal, E.O. J. Amer. Chem. Soc. 1976, 98, 1031.
12. Schaefer, J.; Stejskal, E.O.; Buchdahl, R. Macromolecules 1977, 10, 384.
13. Fyfe, C.A. and co-workers, unpublished results.
14. Fyfe, C.A.; Gobbi, G.C.; Hartman, J.S.; Lenkinski, R.E.; O'Brien, J.H.; Beange, E.R.; Smith, M.A.R. J. Mag. Reson. 1982, 47, 168.
15. Hartley, F.R.; Vesey, P.N. Adv. Organomet. Chem. 1977, 15, 189.
16. Scurrell, M.S. "Catalysis"; London, 1978; 2, p.215.
17. Collman, J.P.; Hegedus, L.S. "Principles and Applications of Organotransition Metal Chemistry"; University Science Books: California, 1980; p.370.
18. Bemi, L.; Clark, H.C.; Davies, J.A.; Drexler, D.; Fyfe, C.A.; Wasylishen, R.E. J. Organomet. Chem. 1982, 224, C5.
19. Bemi, L.; Clark, H.C.; Davies, J.A.; Wasylishen, R.E. J. Amer. Chem. Soc. 1982, 104, 438.

20. Clark, H.C.; Davies, J.A.; Fyfe, C.A.; Hayes, P.J.; Wasylishen, R.E. J. Amer. Chem. Soc., in press.
21. Clark, H.C.; Davies, J.A.; Fyfe, C.A.; Hayes, P.J.; Wasylishen, R.E. Organometallics, in press.
22. Fyfe, C.A. in "Molecular Complexes", Foster, R., Ed.; Crane-Russak: New York, 1973; Vol. 1, Chapter 5.
23. Allen, P.S. M.T.P. Int. Rev. Sci., Phys. Chem. Ser. 1 1972, 4, 41.
24. Campbell, A.J.; Cottrell, C.E.; Fyfe, C.A.; Jeffrey, K.R. J. Inorg. Chem. 1976, 15, 1321.
25. Lippmaa, E.; Mägi, M.; Samoson, A.; Engelhardt, G.; Griminer, A.-R. J. Amer. Chem. Soc. 1980, 102, 4889.
26. Lippmaa, E.; Mägi, M.; Samoson, A.; Tarmak, M.; Engelhardt, G. J. Amer. Chem. Soc. 1981, 103, 4992.
27. Engelhardt, G.; Lohse, V.; Lippmaa, E.; Tarmak, M.; Mägi, M. Z. Anorg. Allg. Chem. 1981, 482, 49.
28. Fyfe, C.A.; Gobbi, G.C.; Hartman, J.S.; Klinowski, J.; Thomas, J.M. J. Phys. Chem. 1982, 86, 1247.
29. Klinowski, J.; Ramdas, S.; Thomas, J.M.; Hartman, J.S. J. Chem. Soc. Farad. Trans. II 1982, in press.
30. Klinowski, J.; Thomas, J.M.; Fyfe, C.A.; Gobbi, G.C.; Hartman, J.S. Inorg. Chem. 1982, in press.
31. Fyfe, C.A.; Gobbi, G.C.; Klinowski, J.; Thomas, J.M.; Ramdas, S. Nature 1982, 296, 530.
32. Thomas, J.M.; Klinowski, J.; Fyfe, C.A.; Hartman, J.S.; Bursill, L.A. J. Chem. Soc. Chem. Comm. 1981, 678.
33. Thomas, J.M.; Bursill, L.A.; Lodge, E.A.; Cheetham, A.K.; Fyfe, C.A. J. Chem. Soc. Chem. Comm. 1981, 276.
34. Loewenstein, W. Am. Mineral. 1954, 39, 92.
35. Thomas, J.M.; Fyfe, C.A.; Ramdas, S.; Klinowski, J.; Gobbi, G.C. J. Phys. Chem. 1982, 86, 3061.
36. Cheetham, A.K.; Fyfe, C.A.; Smith, J.V.; Thomas, J.M. J. Chem. Soc. Chem. Comm. 1982, in press.
37. Melchior, M.T.; Vaughan, D.E.W.; Jarman, R.H.; Jacobson, A.J. Nature 1982, 298, 455.
38. Andrew, E.R.; Bradbury, A.; Eades, R.G.; Jenks, G.J. Nature 1960, 188, 1096.
39. Andrew, E.R. Int. Rev. Phys. Chem. 1981, 1, 195
40. Fyfe, C.A.; Bemi, L.; Childs, R.; Clark, H.C.; Curtin, D.; Davies, J.; Drexler, D.; Dudley, R.; Gobbi, G.; Hartman, S.; Hayes, P.; Klinowski, J.; Lenkinski, R.E.; Lock, C.J.L.; Paul, I.C.; Rudin, A.; Tchir, W.; Thomas, J.M.; Wasylishen, R.E. Phil. Trans., R. Soc. London, Ser. A 1982, 305, 591.

RECEIVED December 15, 1982

29

Polarized x-Ray Absorption Spectroscopy

JAMES E. HAHN and KEITH O. HODGSON

Stanford University, Department of Chemistry, Stanford, CA 94305

The applications of polarized x-ray absorption spectroscopy (PXAS) for structure determination in inorganic and bioinorganic systems are discussed. PXAS studies of oriented samples add angular detail to the information obtained from x-ray absorption edges and from EXAFS. In some cases, PXAS can be used to determine molecular orientation. In other cases, PXAS can be used to infer the details of electronic structure or of chemical bonding. Some of the potential future applications of PXAS are discussed.

During the last decade there has been a dramatic increase in the use of x-ray absorption spectroscopy (XAS) for determining and understanding chemical structures. This increase is in large part attributable to the availability of synchrotron radiation from high-energy electron storage rings (1,2,3). The high intensity of synchrotron radiation light sources allows rapid collection of high quality XAS data.

XAS data comprises both absorption edge structure and extended x-ray absorption fine structure (EXAFS). The application of XAS to systems of chemical interest has been well reviewed (4,5). Briefly, the structure superimposed on the x-ray absorption edge results from the excitation of core-electrons into high-lying vacant orbitals (6,7) and into continuum states (8,9). The shape and intensity of the edge structure can frequently be used to determine information about the symmetry of the absorbing site. For example, the 1s→3d transition in first-row transition metals is dipole forbidden in a centrosymmetric environment. In a non-centrosymmetric environment the admixture of 3d and 4p orbitals can give intensity to this transition. This has been observed, for example, in a study of the iron-sulfur protein rubredoxin, where the iron is tetrahedrally coordinated to four sulfur atoms (6).

0097-6156/83/0211-0431$06.00/0

The EXAFS, which occurs at higher energies above the edge, is due to the interference between the outgoing and the backscattered photoelectron waves (10-14). EXAFS provides information about the local structure of the x-ray absorbing atom. Typically, nearest neighbor bond lengths and coordination numbers can be determined to ±0.02 Å (1%) and one atom in four (25%) (4). The accuracy of these determinations is somewhat worse for outer-shell atoms, for disordered systems, or for systems with asymmetric distributions of atoms within a shell (15,16).

Angular information is notably absent from the list of structural parameters normally obtained from XAS. One approach to obtaining angular detail is to make use of multiple scattering effects (17). Unfortunately, this technique is only useful for outer shells (non-nearest neighbor atoms) where there are atoms intervening between the absorber and the scatterer. This technique suffers from complications if the shells of interest overlap in distance with other shells of atoms.

An alternate method for obtaining angular information is to make use of the plane polarized nature of synchrotron radiation. It has long been known that XAS should exhibit a polarization dependence for anisotropic samples (18); however it is only recently that attempts have been made to exploit this effect. Early attempts to observe anisotropic XAS suffered from the low intensity and incomplete polarization of conventional x-ray sources. This work has been reviewed by Azaroff (19).

In this chapter, we briefly discuss the theoretical background of polarized x-ray absorption spectroscopy (PXAS). Many of the recent applications of synchrotron radiation to polarized absorption edge structure and to EXAFS are discussed, with particular emphasis being given to the study of discrete molecular systems. We present here some indication of the potential applications of PXAS to systems of chemical and biological interest.

Theory

The general expression for a dipole-coupled absorption crosssection σ is

$$\sigma = \Sigma\ |\langle i|\ \hat{e}\cdot\hat{r}\ |f\rangle|^2 \qquad (1)$$

where $\langle i|$ and $|f\rangle$ represent the initial and final state wavefunctions, $\hat{e}$ is a unit vector pointing in the direction of the polarization, $\hat{r}=(\hat{x},\hat{y},\hat{z})$ is a unit vector pointing in the direction of the x, y or z axes, and the sum is taken over all orientations (20). The EXAFS, χ, is proportional to absorption cross-section. Equation 1 can be simplified to

$$\chi = \Sigma\ |\langle i|\ \hat{r}\ |f\rangle|^2 \cos^2\theta \qquad (2)$$

where θ is the angle between $\hat{e}$ and $\hat{r}$. The matrix element $\langle i|\hat{r}|f\rangle$ can be evaluated to give the familiar isotropic EXAFS expression (*c.f.* references 10-14). In the single-electron approximation, this matrix element is only significant if $\hat{r}$ points in the direction of the unoccupied state of interest (edge structure) or the backscatterer of interest (EXAFS). Thus, we can identify θ in Equation 2 with the angle between $\hat{e}$ and an unoccupied orbital (edge structure) or an absorber-backscatterer vector (EXAFS). This is illustrated for EXAFS in Figure 1. For a rigorous development of the theory of PXAS, the reader is referred to, for example, reference 20.

For crystalline samples of cubic or higher symmetry, for polycrystalline samples, or for amorphous materials (*e.g.*, for most XAS experiments), symmetry requires that the observed XAS be spherically symmetric (20,21). In these cases the angular dependence in Equation 2 must be averaged over all orientations, giving a factor of 1/3.

Edges

Although there are notable exceptions, most applications of edge structure for determination of molecular geometry have concentrated on an empirical correlation of absorber site symmetry and chemical environment with edge structure. This is a result of the difficulty in quantitatively interpreting much of the edge structure. When the incident photon has an energy just barely above the absorption threshold, the resulting photoelectron can undergo numerous multiple scattering events, greatly complicating the theoretical analysis (22). Thus, while transitions to bound states are reasonably well understood, a complete theoretical treatment of the edge structure is difficult.

Oriented samples allow the possibility of greatly simplifying the interpretation of the edge structure. The information obtained from polarized edge studies is in many ways analogous to the information obtained from polarized optical spectroscopy. Since edge structure includes transitions into high-lying molecular orbitals (for discrete molecular systems) or high-lying bands (for extended solid-state structures), the polarization dependence of these transitions can, in principle, provide information about the symmetry properties of specific orbitals or bands. The observed symmetry is a product of the symmetries of the initial state, the final state, and the operator coupling these states. K and L(I) absorption edges thus have an important simplification over optical spectroscopy since the initial state (1s or 2s) is spherically symmetric and does not influence the observed orientational dependence of the transition. Unfortunately, the identification of x-ray absorption edge transitions is not always unambiguous, especially for transitions into states that are near the continuum.

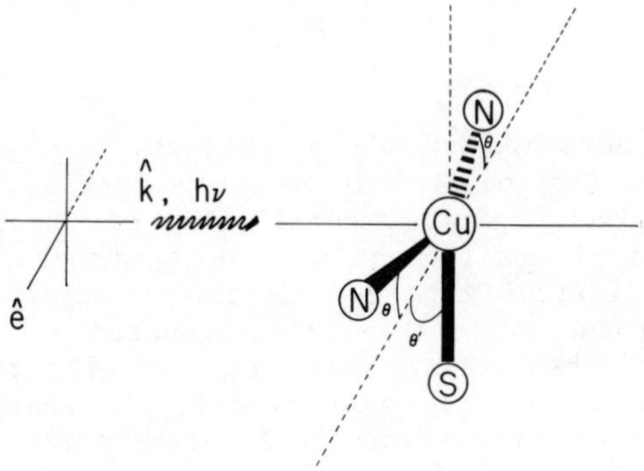

Figure 1. Polarized XAS. The orientations of ê and k and the angle θ (see Equation 2) are shown for a typical PXAS experiment. For this orientation, there can be no Cu–S contribution to the EXAFS (θ′ = 90°). The Cu–N contribution to the EXAFS will be 3cos²θ-times that observed for an isotropic sample.

Electronic Structure Determination. Cox and Beaumont have studied the polarized x-ray absorption edge of a single crystal of ZnF_2 (23), in which the Zn has tetragonal (D_{4h}) site symmetry. The observed anisotropic K-absorption edges were explained in terms of a $1s \rightarrow 4p_z$ and a $1s \rightarrow (4p_x, 4p_y)$ transition.

Templeton and Templeton have observed anisotropic edges for the "oxo" species $[V{=}O]^{2+}$ and $[UO_2]^{2+}$ (24,25). In the vanadium case, a very strong pre-edge transition was observed only when the electric vector of the polarization was oriented parallel to the V=O bond. From this p_z type symmetry and from the strength of the feature (indicating an allowed transition), it was suggested that this peak is due to a transition into a molecular orbital containing vanadium $3d_{z^2}$ and 4s orbitals and a sigma orbital (having p_z symmetry) from the vanadyl oxygen. In the uranyl case, the L(I) (2s initial state) and the L(III) (2p initial state) edges were both studied (25). Because of their different initial-state symmetries, these two edges give mutually exclusive dipole selection rules. In agreement with this, the authors observed not only orientational dependence for both edges, but also a different orientational dependence for the different edges.

Orientation Determination. While polarized edge studies, together with a known sample orientation, can provide information about the electronic structure of the absorber, one can also use polarized edges to probe ordered systems of unknown orientation. This sort of approach was used in a study of Br_2 adsorbed on graphite (26,27). In this case, the orientational dependence of an edge transition was used to calculate the degree of orientational purity of the graphite surface.

Another example of the potential utility of polarized edge spectra for structure determination is found for $[MoO_2S_2]^{2-}$ (28). This molecule has C_{2v} symmetry and the C_2 axes of all of the molecules in the unit cell are collinear. Thus, when the crystal is oriented with the polarization parallel to the S-S interatomic vector, the polarization is perpendicular to the Mo-O bonds and nearly parallel to the Mo-S bonds. Similarly, the crystal can be oriented with the polarization perpendicular to the Mo-S bonds and nearly parallel to the Mo-O bonds. For both orientations, excellent agreement was obtained with SCF-Xα calculations of the edge structure (8).

Moreover, excellent agreement was also observed between the polarized edge spectra of $[MoO_2S_2]^{2-}$ and the corresponding "pure" isotropic spectra of $[MoO_4]^{2-}$ and $[MoS_4]^{2-}$ (*e.g.*, with the polarization parallel to the O-O vector, an $[MoO_4]^{2-}$ spectrum was observed) (8). In general, if a molecule contains ligands which individually give very different edge structures, it may be possible to determine the orientation of these ligands using polarized absorption edges. As shown here, the isotropic spectrum of a "pure" compound, containing only one of the ligands, can be used to determine the ligand-dependent edge structures.

Determination of the Nature of the Coupling Operator. If both the nature of the edge transition and the orientation of the absorber are known, one can used polarized x-ray edge spectroscopy to determine the nature of the operator coupling the initial and final states. The previously discussed studies all involved intense transitions and hence used the reasonable assumption that dipole coupling was involved. However, in the case of the first-row transition metals, a weak transition is observed which is attributed to a 1s→3d transition, which is dipole forbidden for centrosymmetric compounds. This transition is weak, but not absent, even in rigorously centrosymmetric complexes. The intensity of this transition could arise from vibronically induced distortions from centrosymmetry or from a direct coupling to the quadrupolar component of the radiation. Although the quadrupolar coupling is allowed, it is expected to be very weak, with an intensity comparable to that expected for a vibronically allowed transition.

Polarized x-ray edge spectroscopy has been used to distinguish between these two coupling mechanisms in bis-creatinium tetrachlorocuprate (29). In a crystal of this compound, the $[CuCl_4]^{2-}$ unit has D_{2h} symmetry. There are two molecules per unit cell, but they are fortuitously located such that the molecular x, y, and z axes are nearly parallel (the x and y axes are defined to be pointing in the direction of the Cl atoms, as shown in Figure 2a). The amplitude of the 1s→3d transition was observed to have a four-fold periodicity as a function of angle for rotations of a crystal about the molecular z axis, as shown in Figure 2b. This experiment unambiguously demonstrates the presence of direct quadrupolar coupling.

Since the $[CuCl_4]^{2-}$ molecule is only slightly distorted from D_{4h} symmetry, vibronically allowed coupling is expected to be independent of rotation about the z axis. Indeed, an angularly independent component to the crosssection is also observed, with approximately 1/3 the strength of the quadrupole component. Thus, in the rigorously centrosymmetric $[CuCl_4]^{2-}$, the 1s→3d intensity is due approximately 25 % to vibronically allowed dipole coupling and approximately 75% to direct quadrupole coupling. In addition, from the orientation of the maximum in the 1s→3d cross section, it was possible to determine that the vacant orbital had symmetry properties of $d_{x^2-y^2}$, as expected from simple ligand field theory (29). It should be noted that it would be difficult to extract detailed information of this sort from an absorption edge if the transitions were not as well characterized and unambiguously identified as the 1s→3d transition.

Future Applications. These examples illustrate a few of the applications of polarized x-ray absorption edges to the deter-

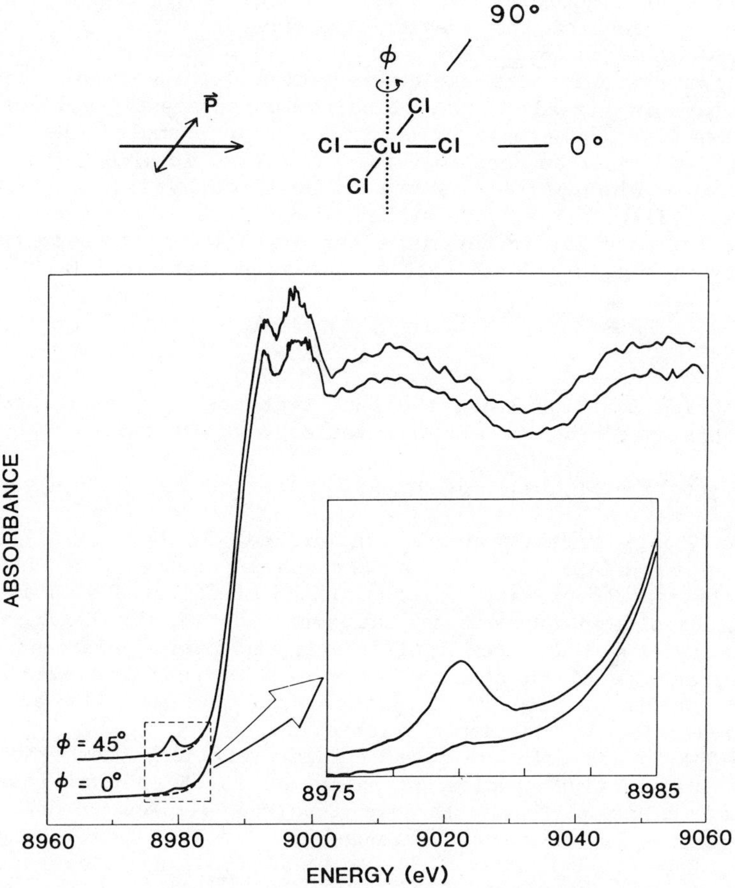

Figure 2. *$[CuCl_4]^{2-}$ absorption edge. Orientation of the $CuCl_4$ unit showing the definition of the angle φ with respect to the polarization direction →**p** (top), and spectra of $CuCl_4^{2-}$ in the orientations that minimize (φ=0°) and maximize (φ=45°) the 1*s*→3*d *transition.*

mination of molecular structure. Other potential applications include determining the symmetry properties of various transitions on an absorption edge. This determination would be one step toward understanding the nature of these transitions. For example, the blue copper protein plastocyanin has been shown to have very anisotropic edge structure (30,31). By determining the symmetry of edge features, polarized edge spectroscopy could be useful in testing the predictions made in various theoretical studies of edges (8,22,32).

For systems where the polarization dependence of the edge structure is already known, polarized measurements could be used to determine the sample orientation. For example, the intense pre-edge transition described for $(acac)_2V{=}O$ is also found in the isotropic absorption spectra of M=O(porphyrin) (M=Ti,V,Cr) systems (33). Strongly polarized transitions like these could be used, for example, to determine the orientation of the porphyrin moiety within an ordered system such as a biological membrane or fiber (24).

EXAFS

Using Equation 2 and the fact that $\cos^2\theta$ gives a factor of 1/3 when averaged over all orientations, we see that

$$\chi(pol) = 3[\chi(iso)\cos^2\theta]$$

where $\chi(pol)$ is the oriented (polarized) EXAFS, $\chi(iso)$ is the isotropic EXAFS and θ is the angle between $\hat{e}$ and the absorber-scatterer vector. Thus, EXAFS of oriented samples gives a three-fold enhancement in the amplitude of the EXAFS from a given scatterer. More significantly, however, polarized EXAFS measurements allow a comparison of the radial distribution function in a number of directions, potentially allowing determination of the sample orientation.

Due to the relative ease of interpretation, there have been more quantitative studies of polarized EXAFS than of polarized edges. This work has largely involved samples in which the orientation is known. The extended lattice systems WSe_2 and TaS_2 (21), Zn, (34,35), ZnF_2 (23), and GeS (36), and the molecular solids $[MoO_2S_2]^{2-}$ (28), $(acac)_2V{=}O$ (24), $[UO_2]^{2+}$ (25), and the blue copper protein plastocyanin (30,31) have all been studied with polarized EXAFS using synchrotron radiation sources. Also, Br_2 adsorbed on graphite has been studied as a function of concentration and temperature (26,27). Anisotropic effects have also been noted using the analogous technique of extended electron energy-loss spectroscopy (37) and have frequently been used in surface EXAFS studies (38-41).

Structure Determination. In general, good agreement was found between the observed EXAFS and the known structure in all of the studies cited above. For example, in the tetrahedral $[MoO_2S_2]^{2-}$, the molecules were oriented within the unit cell such that only Mo-O, only Mo-S, or a combination of Mo-O and Mo-S EXAFS could be selected (28). The observed EXAFS spectra agreed well with the isotropic spectra of $[MoO_4]^{2-}$, $[MoS_4]^{2-}$, and $[MoO_2S_2]^{2-}$, respectively.

In the case of the Br_2 on graphite, polarized EXAFS was used to determine molecular orientation (26,27). The system consisted of a stack of parallel graphite sheets in which the EXAFS could be measured parallel or perpendicular to the plane of the sheets. The authors were able to show that the bromine was bound as Br_2 and to determine the orientation of the Br_2 molecules with respect to the plane of the graphite sheets. Polarized EXAFS proved to be a useful complement to LEED, which had been used to probe the long-range order in this system.

A different sort of information was obtained in a study of the L(II) and L(III) edges of W in WSe_2 (21). Since the initial state in these cases is a 2p orbital, both s and d final states are permitted. From the anisotropy of this EXAFS, the authors determined that the average contribution of the final s state to the total absorption edge was 0.02 of that of the final d states.

Investigations of Chemical Bonding. In addition to anisotropies in bond length and coordination, an analysis of the polarized EXAFS data for Zn revealed anisotropies in the mean square relative displacement of the Zn atoms (34). This illustrates the utility of single crystal EXAFS even for studying systems where the crystal structure is known. Specifically, crystallography determines the mean square motion of each atom about its mean position, while EXAFS is sensitive to the mean square relative displacement of a pair of atoms about their mean separation. Thus the combination of crystallography and single crystal EXAFS can provide a powerful tool for studying chemical bonding.

Another example of using polarized EXAFS to study bonding is found in a study of the blue-copper protein plastocyanin (30). In this case, crystallography had shown that the Cu atom was bonded to two imidazole nitrogens at 2.0 Å and one cysteine sulfur at 2.15 Å (42). In addition, there was a methionine sulfur [S(Met)] located at 2.9 Å from the copper. Previous solution XAS studies had failed to show any evidence of Cu-S(Met) EXAFS. Possible explanations were that Debye-Waller damping was reducing the amplitude of the Cu-S(Met) EXAFS to below the detection limit or that destructive interference was occuring between the Cu-S(Met) EXAFS wave and a Cu-C(imidazole carbon) wave, resulting in the cancellation of the EXAFS.

A single-crystal study of oxidized plastocyanin demonstrated that the absence of Cu-S(Met) EXAFS was due to Debye-Waller damping (30). The analysis benefits from the three-fold enhancement of Cu-S(Met) scattering amplitude expected for oriented EXAFS. However, the decisive factor in this study is the ability to compare an orientation in which the EXAFS should be dominated by Cu-S(Met) EXAFS with an orientation which can contain no S(Met) contribution to the Cu EXAFS (Figure 3). In neither orientation is there a feature above the noise level which can be attributed to an atom at 2.9 Å from the copper (Figure 4). This indicates that the amplitude reduction is due to Debye-Waller damping, a somewhat surprising result since neither the Cu nor the S(Met) has an anomalously large temperature factor in the crystal structure. As in the Zn case above, this highlights the fact that crystallography and EXAFS are sensitive to different sorts of vibrational damping.

These studies indicate the variety of ways in which polarized EXAFS can be used to understand chemical structures. In general, such studies are most useful when used in conjunction with other structural techniques (e.g. LEED or crystallography).

Future Applications. In addition to probing bonding in structurally characterized systems (as described above), future polarized EXAFS studies may also study oriented but structurally uncharacterized systems. This would be similar to the Br_2 adsorbed on graphite study described earlier. Future work might include polarized XAS studies of biological fibers or membranes. For example, optical dichroism might be used to establish the orientation of a heme-containing, membrane-bound protein. Polarized EXAFS could then be used for selective study of the axial ligation of the heme. Even in systems with unknown orientation, polarized EXAFS could conceivably be useful. For example, proteins which can be crystallized, but for which high-resolution diffraction data do not exist, might be likely candidates for study. In these cases, the known space group and unit cell orientation of the protein could be used to select unique orientations for examination with polarized XAS.

Conclusions

Polarized x-ray absorption spectroscopy offers a number of advantages over isotropic XAS. We have presented a brief theoretical description of PXAS and have described a variety of applications of PXAS to the study of inorganic and bioinorganic systems. These applications involve adding angular detail to the electronic information obtained from x-ray absorption edges and the radial distribution information obtained from EXAFS. This angular detail can be used to determine the orientation of a sample within a crystal, membrane, or fiber. Alternately, if the sample orientation is known, the angular detail of PXAS allows detailed probing of the electronic structure (using edges) or of the bonding (using EXAFS). If both the sample geometry and the detailed nature of an edge transition are known, PXAS can be used to study the nature of the operator coupling the initial and final states involved in an edge transition.

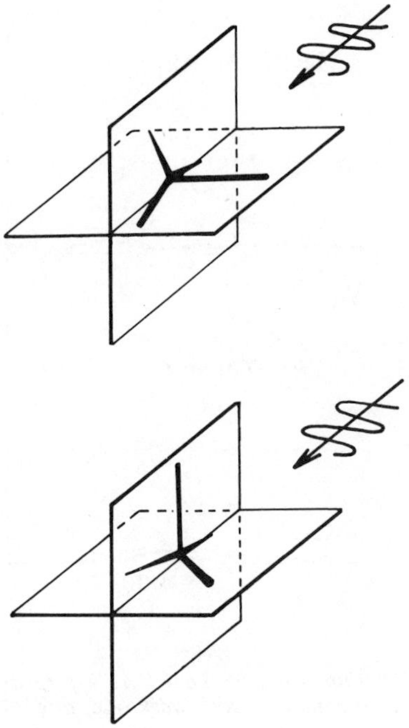

Figure 3. Plastocyanin orientations. The $Cu(N\text{-}Imid)_2(S\text{-}Cys)(S\text{-}Met)$ unit has approximately trigonal pyramidal symmetry, with the S(Met) ligand at the apex of the pyramid. Key: top, orientation giving predominantly Cu–S(Met) EXAFS; and bottom, orientation giving no Cu–S(Met) EXAFS. (Reproduced from Ref. 30. Copyright 1982, American Chemical Society.)

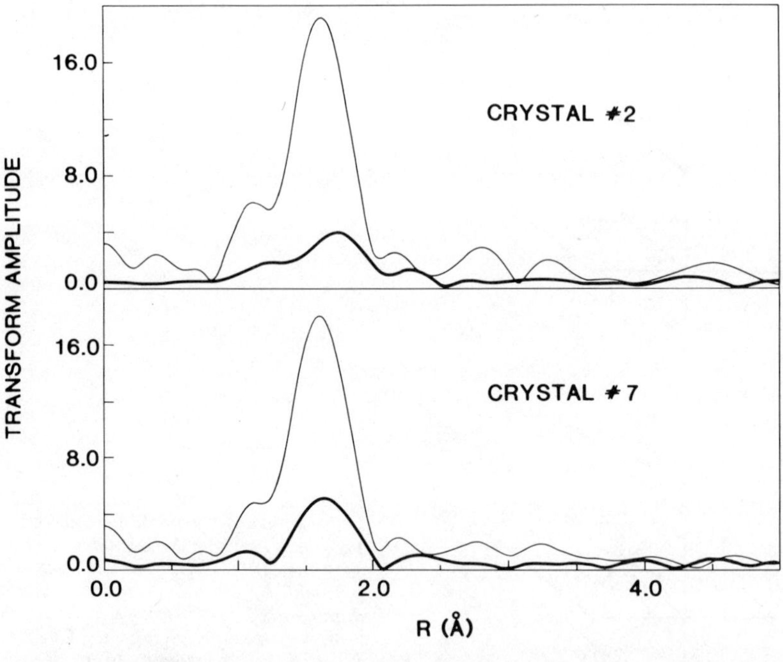

Figure 4. Plastocyanin transforms. Fourier transforms of the EXAFS for two different plastocyanin crystals in two different orientations. Key: dark line, orientation maximizing Cu–S(Met) EXAFS (see *Figure 3a*)*; light line, orientation containing no Cu–S(Met) EXAFS* (see *Figure 3b*)*. In neither orientation is a peak observed at approximately 2.5 Å, corresponding to an atom at 2.9 Å. (Reproduced from Ref. 30. Copyright 1982, American Chemical Society.)*

It should be noted however, that only a small percentage of crystals are suitable candidates for single crystal XAS. Not only must the symmetry of the absorbing site be lower than cubic, but also the different sites within the unit cell must be oriented in such a way that the EXAFS (or edge structure) due to the scatterer of interest can be isolated. Similar constraints apply to biological fibers or membranes, with the additional problem that these systems will have at best uniaxial symmetry. Despite these limitations, there are a large number of systems which are suitable for study using PXAS techniques. We expect that in coming years PXAS will prove to be a useful complement to other structural methods.

Acknowledgments

This work was supported by a grant from the National Science Foundation (PCM 79-04915). JEH is the recipient of a National Science Foundation predoctoral fellowship. The synchrotron radiation used in many of these studies was provided by the Stanford Synchrotron Radiation Laboratory with the financial support of the National Science Foundation (under contract DMR 77-27489) in cooperation with the Department of Energy and by the National Institutes of Health SSRL biotechnology resource (RR-01749).

Literature Cited

1. Winnick, H.; Doniach, S., Eds.; "Synchrotron Radiation Research"; Plenum: New York, 1980.
2. Rowe, E. M. Physics Today, 1980, 28-37.
3. Sparks, C.J. Jr. Physics Today, 1980, 40-49.
4. Cramer, S.P.; Hodgson, K.O. Prog. Inorg. Chem. 1979, 25, 1-39.
5. Teo, B.-K. Accts. Chem. Res. 1980, 13, 412-419.
6. Shulman, R.G.; Yafet, Y.; Eisenberger,P.; Blumberg, W.E. Proc. Nat. Acad. Sci. USA 1976, 73, 1384-1388.
7. Hu, V.W.; Chan, S.I.; Brown, G.S. Proc. Nat. Acad. Sci. USA 1977, 74, 3821-3825.
8. Kutzler, F.W.; Natoli, C.R.; Misemer, D.K.; Doniach, S.; Hodgson, K.O. J. Chem. Phys. 1980, 73, 3274-3288.
9. Kutzler, F.W. Ph.D. Thesis, Stanford University 1981.
10. Sayers, D.E.; Stern, E.A.; Lytle, F.W. Phys. Rev. Lett. 1971, 27, 1204-1207.
11. Stern, E.A. Phys. Rev. B 1974, 10, , 3027-3037.
12. Ashley, C.A.; Doniach, S. Phys. Rev. B 1975, 11, 1279-1288.
13. Lee, P.A.; Pendry, J.B. Phys. Rev. B 1975, 11, 2795-2811.
14. Stern, E.A.; Sayers, D.E.; Lytle, F.W. Phys. Rev. B 1975, 11, 4836-4846.

15. Eisenberger, P.; Brown, G.S. Solid State Commun. 1979, 29, 481-484.
16. Crozier, E.D.; Seary, A.J. Can. J. Phys. 1980, 58, 1388-1399.
17. Teo, B.-K. J. Amer. Chem. Soc. 1981, 103, 3990-4001.
18. Stephenson, J.J. Phys. Rev. 1933, 44, 349-352.
19. Axaroff, L.V.; Pease, D.M. "X-ray Spectroscopy", Azaroff, L.V., Ed.; McGraw Hill: New York, New York, 1975; pp. 284-337.
20. Stern, E. A. Phys. Rev. B 1974, 10, 3027-3037.
21. Heald, S.M.; Stern, E. A. Phys. Rev. B 1977, 16, 5549-5559.
22. Greaves, G.N.; Durham, P.J.; Diakum, G; Quinn, P. Nature 1981, 294, 139.
23. Cox, A.D.; Beaumont, J. H. Philosophical Magazine 1980, 42, 115-126.
24. Templeton, D. H.; Templeton, L.K. Acta Cryst. A 1980, 36, 237-241.
25. Templeton, D. H.; Templeton, L.K. Acta Cryst. A 1982, 38, 62-67.
26. Stern, E. A.; Sayers, D. E.; Dash, J. G.; Shechter, H.; Bunker, B. Phys. Rev. Lett. 1977, 38, 767-770.
27. Heald, S. M.; Stern, E. A. Phys. Rev. B 1978, 17, 4069-4081.
28. Kutzler, F.W.; Scott, R.A.; Berg, J.M.; Hodgson, K.O.; Doniach, S.; Cramer, S.P., and Chang, C.H. J. Amer. Chem. Soc. 1981, 103, 6083-6088.
29. Hahn, J.E.; Scott, R.A.; Hodgson, K.O.; Doniach, S.; Desjardins, S.R.; Solomon, E.I. Chem. Phys. Lett. 1982, 88, 595-598.
30. Scott, R.A.; Hahn, J.E.; Doniach, S.; Freeman, H.C.; Hodgson, K.O. J. Amer. Chem. Soc. 1982, in press.
31. Hahn, J.E.; Scott, R.A., Freeman, H.C.; Doniach, S.; Hodgson, K.O. Unpublished work.
32. Goddard, W.A.; Bair, R.A. Phys. Rev. B 1980, 22, 2767-2776.
33. Hahn, J.E.; Reed, C.A.; Hodgson, K.O., Unpublished work.
34. Beni, G.; Platzman, P.M. Phys. Rev. B 1976, 14, 1514-1518.
35. Brown, G.S.; Eisenberger, P.; Schmidt, P. Solid State Commun. 1977, 24, 201-203.
36. Rabe, P; Tolkiehn, G.; Werner, A. J. Phys. C 1980, 13, 1857-1864.
37. Disko, M.M.; Krivanek, O.L.; Rez, P. Phys. Rev. B 1982, 25, 4252-4255.
38. Stohr, J.; Johansson, L.; Lindau, I.; Pianetta, P. Phys Rev. B 1979, 20, 664-680.
39. Citrin, P.H.; Eisenberger,P.; Hewitt, R.C. Surf. Sci 1979, 89, 28-40.
40. Citrin, P.H.; Eisenberger, P.; Hewitt, R.C. Phys. Rev. Lett. 1980, 45, 1948-1951.
41. Brennan, S.; Stöhr, J.; Jaeger, R. Phys. Rev. B 1981, 24, 4871-4874.
42. Freeman, H.C. "Coordination Chemisty - 21"; Laurent, J.P., Ed.; Pergamon Press: Oxford, 1981; pp. 29-51.

RECEIVED August 18, 1982

30

High-Resolution Electron Microscopy and Electron Energy Loss Spectroscopy

J. M. THOMAS

University of Cambridge, Department of Physical Chemistry,
Lensfield Road, Cambridge CB2 1EP England

The merit of high-resolution electron microscopy (HREM) is that it can yield structural information, in real space and at the sub-nanometric level, about materials that are not amenable to structure determination be X-ray crystallographic and other conventional techniques. Specific, localised, rather than spatially averaged information is gleaned in this way, and new structural types as well as new mechanistic details are brought to light through the application of HREM to hitherto intractable systems. The first part of the article illustrates the validity of these statements by specific reference to: (i) quasi-crystalline solids (zeolites and transition metal sulphide catalysts), and (ii) oxides and silicates in which there are intergrowths at the unit cell level as in the pyroxenoids and bismuth tungstates. Some of the new structural types discovered by HREM include a spatial form of graphite intercalate, multiply-twinned zeolites, and ring-chain intergrowths amonsgt silicates and their germanate analogues. HREM is also especially useful in evaluating mechanisms of structural transfrormations and in clarifying the nature of crystalline imperfections.

Electron energy loss spectroscopy (EELS) is a powerful new microanalytical technique, capable of detecting less than 10^{-20}g of a particular element. The basic principles of this new method of chemical analysis is outlined. Extended, electron energy loss fine structure (EXELFS) and its potential as a new method of determining the structure of ordered or amorphous inorganic materials (applicable unlike EXAGS to materials consisting of all atoms, other than hydrogen, lighter than calcium) is also

0097-6156/83/0211-0445$07.75/0

discussed and illustrated. Future prospects, of using Compton scattering and other procedures are also outlined.

As an analytical tool high resolution electron microscopy (HREM) either alone or in conjunction with closely related techniques (such as selected area electron diffraction (SAED), ultramicro X-ray-emission spectrometry (XRES), ultramicro electron-energy-loss spectroscopy (EELS) (In EELS, because resolution is appreciable (<1eV) we talk of spectroscopy, whereas in XRES the term spectrometry is retained, in view of the poor resolution (with energy dispersive detectors) characteristic of such measurement)) or in association with other measurements that can be effected, *in situ*, inside an electron microscopy (such as luminescence spectroscopy, dynamic mass spectrometry or Compton scattering) is a powerful means of extracting information about inorganic materials. Projected, two dimensional structures may be determined, with near atomic resolution, at the localized scale (covering regions of specimen of typically 100Å diameter); and compositional as well as structural heterogeneity can be straightforwardly identified, also at this level of spatial discrimination.

In this review we shall emphasize some of the unique features of HREM and, in so doing, illustrate how, in association with one or other of the additional techniques mentioned above, or in association with separate studies (e.g. magic-angle-spinning solid-state NMR), HREM has clarified several new, or hitherto enigmatic, features of the chemistry of inorganic solids. In particular we discuss, how:

the nature of quasicrystalline solids, particularly those of interest in heterogeneous catalysis, has been elucidated;

the structure of what may be termed "X-ray intractable" solids has been clarified;

new structural types have been discovered and identified;

the mechanism of certain kinds of solid-state transformation has been established;

various kinds of regular and irregular structural defects may be accommodated within a parent matrix.

EELS, chiefly as a tool for ultramicroanalysis, will also be discussed with the aid of some recently studied examples. Electron energy loss microscopy will also be briefly mentioned.

We shall also outline how bond distances and information pertaining to the distribution of electron density may, in principle, be extracted from measurements carried out using electron microscopy. Finally we touch upon future possible lines of development likely to be of value to the inorganic, surface and analytical chemist.

H.R.E.M.

The Principles. The principles of this technique have been adequately discussed in several recent reviews (1-4). Using microscopes possessing lenses that have the lowest acceptable coefficients of spherical aberration as well as lens pole-pieces which offer adequate scope for generous sample tilting about two orthogonal axes and adequate space for the insertion of detectors for emitted X-rays or secondary electrons, a series of high resolution images is recorded as a function of sample thickness and also as a function of lens defocus. These measurements are facilitated by selecting tapered or wedge-shaped specimens. The trustworthiness of the images so recorded is assessed by comparing observed and calculated intensity distribution in two-dimensional projections of the structure. It follows, therefore, that some prior, rudimentary knowledge of the unknown structure is a prerequisite. Structures can be refined by iterative procedures in which computation of image is carried out after each successive alteration of the atomic coordinates until, ultimately, the observed and calculated images match (5).

The Characterization of Quasi-Crystalline Solids. It has become increasingly apparent that our ignorance of many important phenomena involving inorganic materials stems in part from our inability to elucidate the structure of grossly disordered, quasi-crystalline or amorphous solids. There are numerous facets of heterogeneous catalysis, for example, where the paucity of adequate techniques to cope with structural characterisation of imperfectly ordered solids limits our ability to design and control better catalysts. The work of Chiannelli *et al* (6-8) on hydrodesulphurization using the di- tri- and non stoichiometric sulphides of the transition metals as catalysts nicely illustrates this point. Biological mineralization (9) and amorphous semiconductors (10, 11, 12) are two other important examples.

When the nature of the disorder is spatially invariant EXAFS is a viable technique. Under these circumstances, however, unless the specimens are exceptionally thin (<20Å), HREM is of little value (13, 14), as two-dimensional projected 'images' are difficult to interpret. When, however, there are residual crystalline regions still intact in a 'growing' glassy matrix, or when, conversely, the embryonic stages of crystallization of a glassy precursor have been reached, HREM is particularly powerful (14, 15).

Partially Crystalline Zeolitic Catalysts. Materials that are X-ray amorphous often contain vestiges of crystallinity, but the crystallite size is so small that the line-broadening is such as to yield little or no information by conventional X-ray methods. We have shown (15, 16) that HREM can quickly reveal whether a so-called x-ray amorphous zeolite (still capable of ion exchange) is better classified as either a microcrystalline material with a coherence length of ca 10Å, or, rather, as a heterogeneous mixture of both crystalline and essentially amorphous regions, or whether it approaches the classical random network model. Figure 1 shows a micrograph of a UO_2^{2+} - exchanged 'amorphous' zeolite-Y recorded at a point-to-point resolution of 2.4Å. The dark spots represent projected rows (of about 5 or 6) UO_2^{2+} ions. The continuous diffuse ring in the optical diffractogram (taken on an optical bench using the micrograph negative as an object and a CW laser as a monochromatic source) indicates that the predominant spacings in the image range from 2.3 to 3.4Å, peaking at 2.7Å; and the mean correlation length, derived from the width of the diffuse ring, is ca 9Å.

Zeolites possess the remarkable property of exhibiting shape-selective catalysis even when they are X-ray amorphous. Clearly, even though there is no long range order, there is still a degree of structural organization in the aluminosilicate adequate to exert shape-selectivity in the "non-crystalline" regions of the samples. Thanks to HREM we can now understand how this state of affairs arises (17).

Partially Crystalline Transition Metal Sulphide Catalysts. Chiannelli and coworkers (6, 7, 8) have shown how, by precipitation of metal thio-molybdates from solution and subsequent mild heat-treatment many selective and active hydrodesulphurization catalysts may be produced. We have shown (18) recently that molybdenum sulphide formed in this way is both structurally and compositionally heterogeneous. XRES, which yields directly the variation in Mo/S ratio shows up the compositional non-uniformity of typical preparations; and HREM images coupled to SAED (see Figure 2) exhibit considerable spatial variation, there being amorphous regions at one extreme and highly crystalline (18, 19) MoS_3 at the other.

We shall see later that a special variant of EELS (EXELFS) is likely to be of great value in characterizing quasi-crystalline solids.

Dealing with X-ray Intractable Materials

Wollastonite. From time to time inorganic and geochemists encounter materials which, although showing unmistakable signs of crystallinity, yield most unpromising X-ray diffraction patterns. The naturally occurring silicate of calcium,

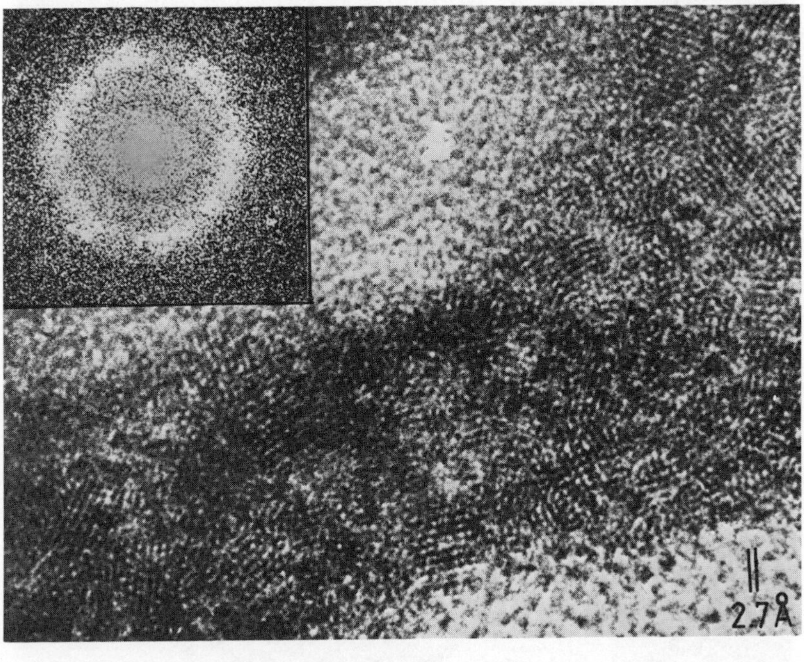

Figure 1. High-resolution electron micrograph and corresponding optical transform (inset) of an x-ray amorphous zeolite-Y specimen that has undergone ion-exchange with a solution containing UO_2^{2+} ions. The microcrystalline regions are rendered visible by the locally ordered UO_2^{2+} ions. (See *text.)*

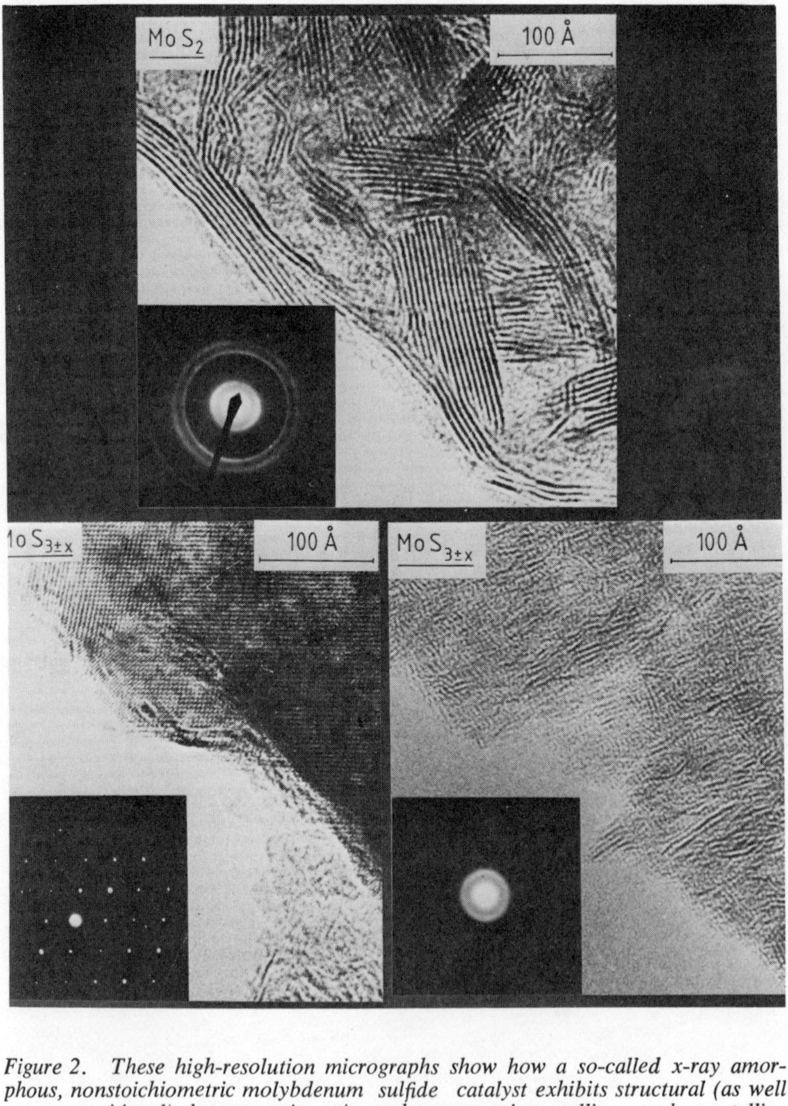

Figure 2. These high-resolution micrographs show how a so-called x-ray amorphous, nonstoichiometric molybdenum sulfide catalyst exhibits structural (as well as compositional) heterogeneity. Amorphous, quasi-crystalline, and crystalline regions coexist at the ultramicro level (18).

wollastonite, $CaSiO_3$, is one such example. Its (hlℓ) reciprocal lattice net, taken by X-ray precession photography (20) is shown in Figure 3. The marked streaking rightaway signifies that there is considerable disorder at the unit cell level. To disentangle the nature of the repeat units along the backbone of the corner-sharing SiO_4 tetrahedra from X-ray photographs of this kind is well nigh impossible. By direct, real-space imaging, however, using HREM (21, 23) great clarification ensures (Figure 4). By constituting one-dimensional (Figure 4 (a) and 4 (b)) and two-dimensional dark-field images (Fig. 4 (d)), i.e. constructed solely from the diffracted beams, bearing in mind that the number of unit cells explored by this electron microscope approach is diminished by a factor of *ca* 10^4 from the number explored in X-ray studies, the ultrastructure of the wollastonite is greatly clarified. We can now understand the reason for the pronounced streaking in the X-ray (hlℓ) (Fig. 3) section, and the streaking of some of the electron diffraction spots (k odd) Figure 4). It should first be recognized that wollastonite is a quasi-layered structure: the triclinic and monoclinic forms differ from one another in that they represent two extreme highly ordered modes of stacking of the (100) layers in the a* direction. In the image shown in Figure 4 (c) (to a lesser degree in Figure 4 (b)) we see that the layers are haphazardly stacked, minute strips - no more than a few unit cell widths - of triclinic and monoclinic structures.

Bismuth Tungstates. A family of structures recently identified (23) to occur in the Bi/W/O system is illustrated schematically in Figure 5. The individual members are made of interleaved Bi_2O_3 and WO_3 layer, the latter consisting of corner-sharing WO_3 octahedra. The resulting homologous series has a formula $Bi_2W_nO_{3n+3}$.

When various compositions of Bi_2O_3 and WO_3 are fused together so as to synthesize individual members of the homologous series, some unexpected difficulties are encountered. Thus, for reasons which are at present obscure, all attempts to isolate the pre n=3 member have failed. Moreover, some complicated intergrowths appear to form under certain circumstances, and the resulting materials are more or less X-ray intractable, in a sense similar to that described above for wollastonite. Combined HREM imaging, computer simulation of images (23, 24) SAED (bright field - i.e. using the 000, or forward scattered, as well as diffracted beams for image construction), leave little doubt as to the nature of these materials (Figure 6). In particular it is seen that some intergrowths of $Bi_2W_3O_{12}$ can be accommodated, possibly stabilized, within a predominantly $Bi_2W_2O_9$ matrix. It appears that the n=3 member survives or exists only when bounded by n=2 or n=1 members. Upon annealing samples that are nominally

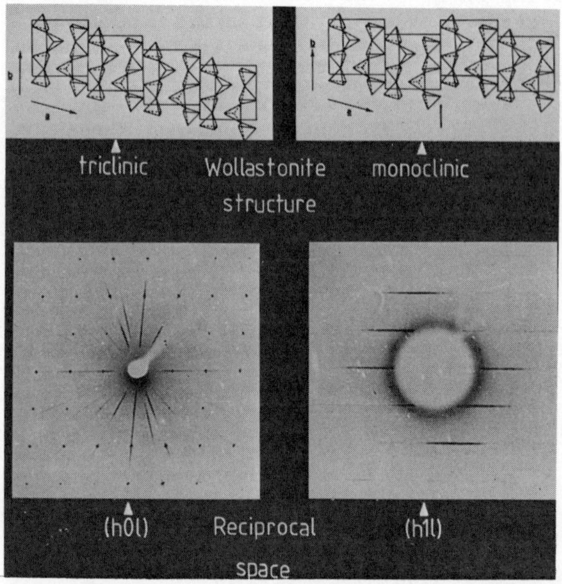

Figure 3. Schematic illustration of the disposition of chains made up of corner-sharing SiO_4 tetrahedra in the triclinic and monoclinic forms of wollastonite. (The Ca^{2+} ions have been omitted for clarity) (top), and (h0l) and (h1l) x-ray diffraction patterns that signify that the specimen in question is disordered and made up of intergrowths of the triclinic and monoclinic forms (bottom).

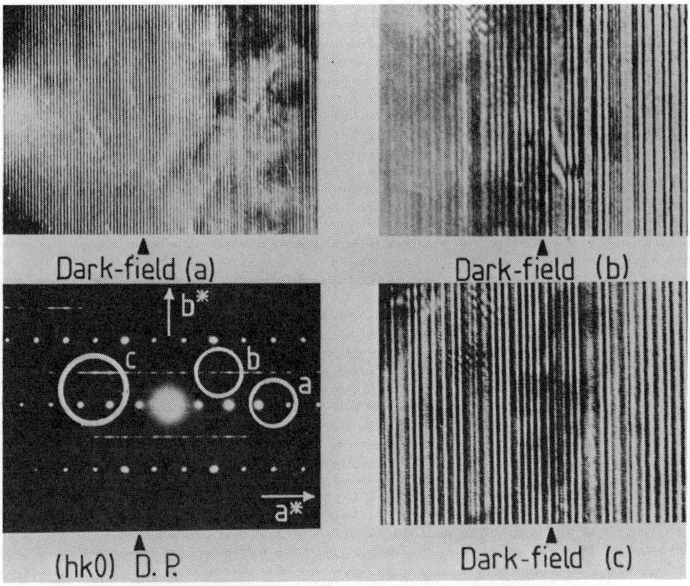

Figure 4. Electron diffraction pattern (bottom left) of a typical 'disordered' wollastonite specimen. Evidence of the disorder comes from the streaked (k odd) diffraction spots. The fringes shown in the dark-field image (a) are ~ 7 Å apart. Dark-field image b, taken from the streaked diffraction spots, shows a one-dimensional image of the disordered wollastonite. Dark-field image c shows a two-dimensional image which shows the haphazard stacking of the triclinic and monoclinic constituents.

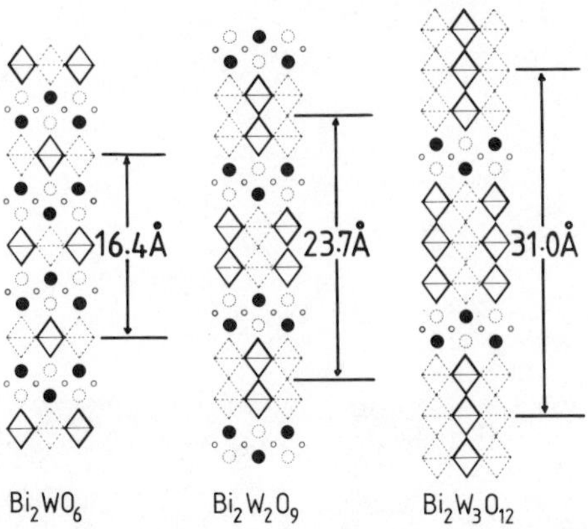

Figure 5. An illustration of the first three members of the structural series $Bi_2W_nO_{3n+3}$, viewed along [100]. The structures consist of 'Bi_2O_2' sheets separated by one, two, three, etc. (n = 1, 2, 3, etc., respectively), corner-sharing sheets of WO_6 octahedra.

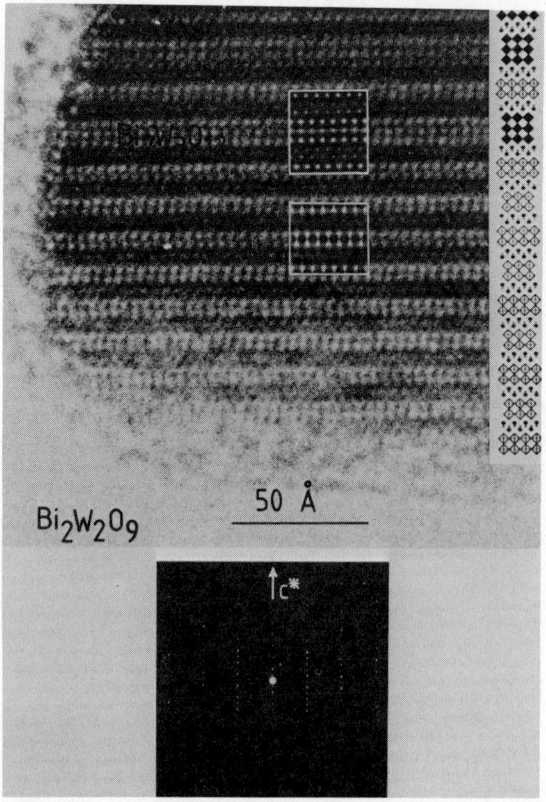

Figure 6. High-resolution micrograph together with corresponding scalar structural drawing, computed image, and appropriate diffraction pattern showing structural resolution of $Bi_2W_2O_9$ with some $Bi_2W_3O_{12}$ intergrowths (24).

$Bi_2W_2O_9$, but which contain very nearly contiguous slabs of $Bi_2W_3O_{12}$ and Bi_2WO_6, it is observed that some kind of mutual annihilation of the n=3 and n=1 members, takes place to yield the n=2 member, as is shown dramatically (24) in Figure 7. The solid-state reaction:

$$Bi_2W_3O_{12} + Bi_2WO_6 \rightarrow 2Bi_2W_2O_9$$

can, therefore, be "seen" directly by HREM at ca 3Å resolution occurring at the unit-cell level.

Pyroxenoid Silicates. Pyroxenoids, of which wollastonite, described above, is but one member, constitute a family of single-chain silicates made up of corner sharing SiO_4 tetrahedra. Their general formula is $MSiO_3$ M=Mg, Ca, Fe, Mn, etc.), and they display a rich diversity of structural types which differ one from the other in the nature of the backbond repeat patterns. The difference between individual members of this family manifests itself in the number of SiO_4 tetrahedra in a backbone repeat. Both naturally occurring pyroxenoids and their synthetic analogues tend to display fine intergrowths so that X-ray patterns are again complicated and tend to be streaked. But HREM enables us to "read off" directly the nature of the repeat patterns (see Figure 8), rather like reading off the constituents of a polypeptide chain (25, 26).

New Structural Types: Their Discovery by HREM. As well as being invaluable in confirming the atomic detail of known structures, HREM has, in the best traditions of a new technique, led to the discovery of hitherto unknown structural types.

Graphite Intercalates. Under optimal imaging conditions (27, 28) the statistical nature of the distribution of stages in a so-called 'second stage' graphite intercalate (i.e. one which, on the basis of X-ray diffractograms has guest species accommodated every other interlamellar space) can be directly viewed (Figure 9 (a)) at a spatial scale ca 10^{-15} times that explored by X-rays. A new type of graphite intercalate structure, consisting of double stacks of guest species ($FeCl_2$) within a given interlamellar space has been discovered (29): see Figure 9 (b).

New Zeolitic Structures. Multiply twinned faujasitic zeolites (typically zeolite-Y) have recently been shown (30, 31) to be capable, by recurrent twinning on {111} planes, to generate a new, hexagonal zeolite in which tunnels replace the interconnected cages of the parent cubic structure.

By exposing certain zeolites to $SiCl_4$ vapour at modest temperatures it is possible to produce (see Fig 10 (a)) crystalline solids that are almost exclusively SiO_2 in

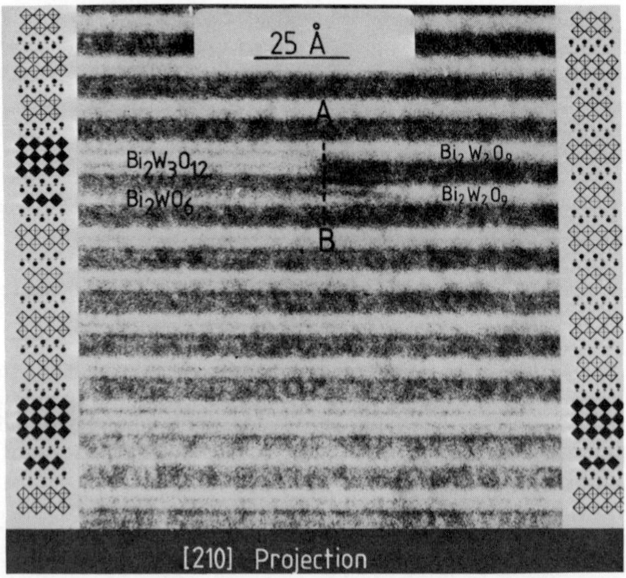

Figure 7. Micrograph showing evidence for the occurrence of the reaction: $Bi_2W_3O_{12} + Bi_2WO_6 \rightarrow 2Bi_2W_2O_9$ at the unit-cell level. To the right of the dashed line AB the composition is $Bi_2W_2O_9$.

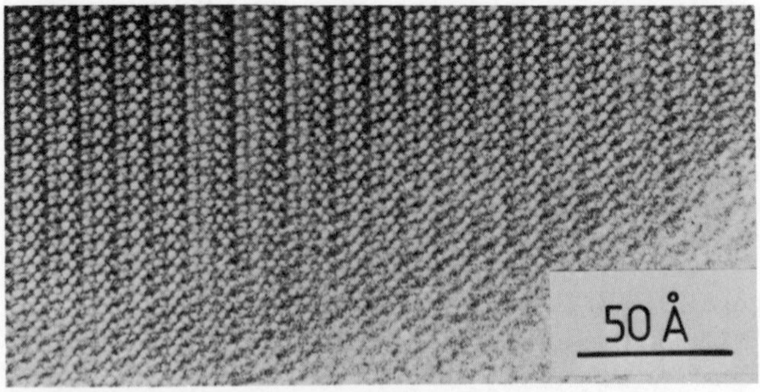

Figure 8. High-resolution image of a disordered synthetic pyroxenoid (of composition $MnFeSi_2O_6$) not amenable to x-ray crystallographic study (see *text). The region shown consists of intergrown, recurrent strips of the pyroxenoid structures known as rhodonite (with 5 linked SiO_4 tetrahedra) and pyroxmangite (with 7 linked SiO_4 tetrahedra).*

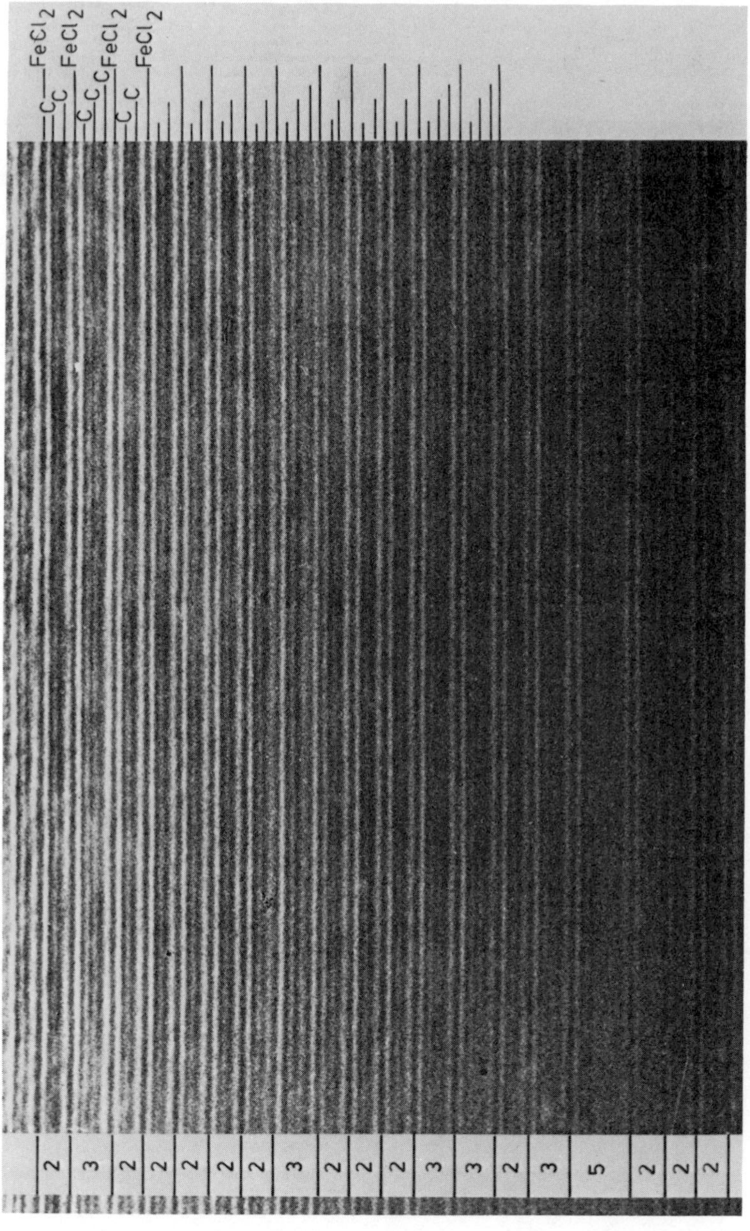

Figure 9a. High-resolution image of the 'second stage' graphite, $FeCl_2$ intercalate viewed along the sheets. There are 'third-stage' and 'fifth-stage' intercalates interspersed amongst the predominant 'second-stage' units shown here. The individual sheets are separated by 3.4 Å.

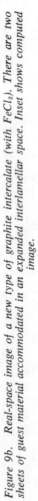

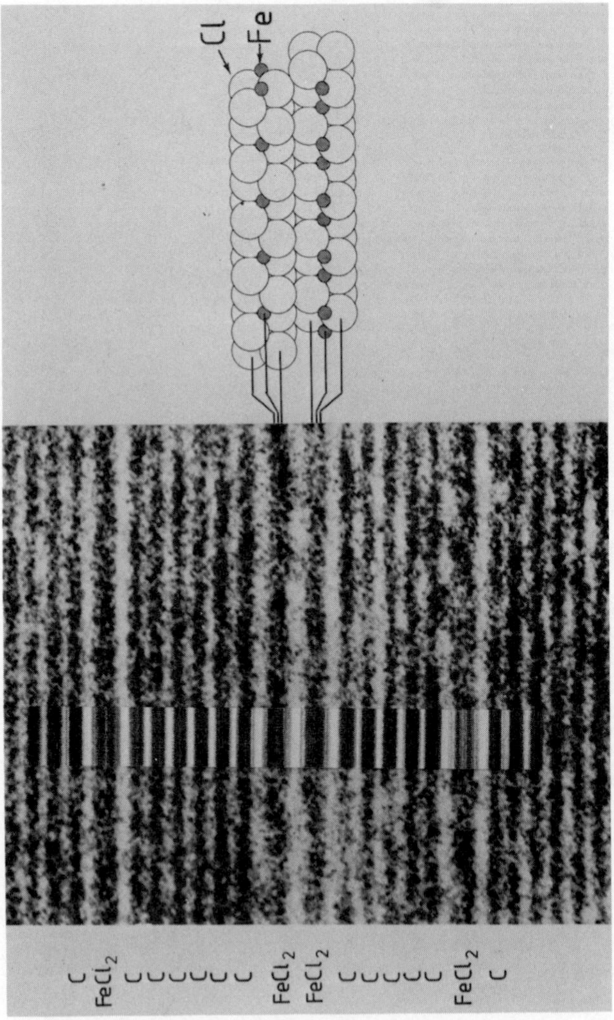

Figure 9b. Real-space image of a new type of graphite intercalate (with $FeCl_2$). There are two sheets of guest material accommodated in an expanded interlamellar space. Inset shows computed image.

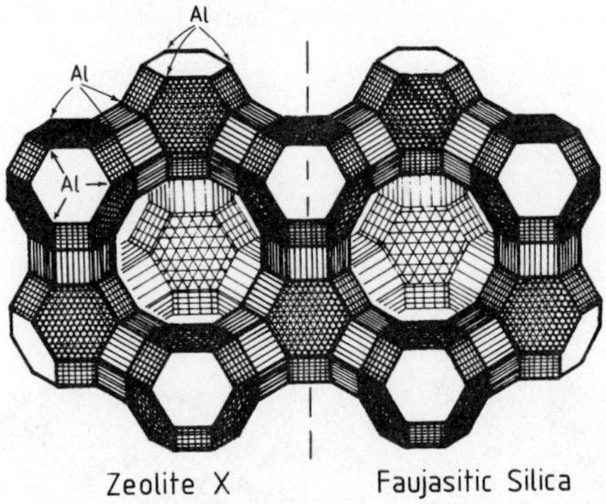

Figure 10a. Scheme of the alumino-silicate framework of a typical faujasitic zeolite Si/Al ratio of 1.18 (arbitrarily chosen to illustrate the ordering among tetrahedral sites) before (left half) and after (right half) exposure to $SiCl_3$, which 'dealuminates' the zeolite (see *Figure 10b).*

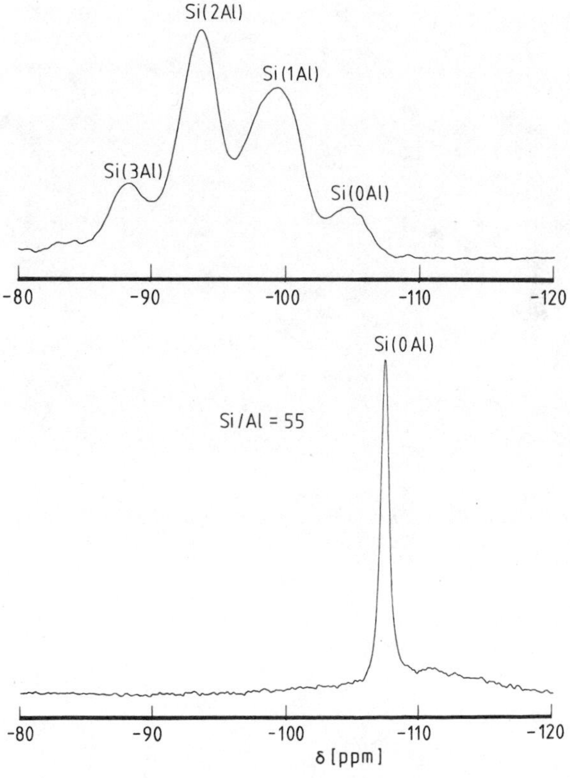

Figure 10b. Magic angle-spinning [29]*Si-NMR spectra of a typical faujasitic zeolite (Si/Al ratio 2.61) before and after dealumination by exposure to $SiCl_4$ vapor* (see *text).*

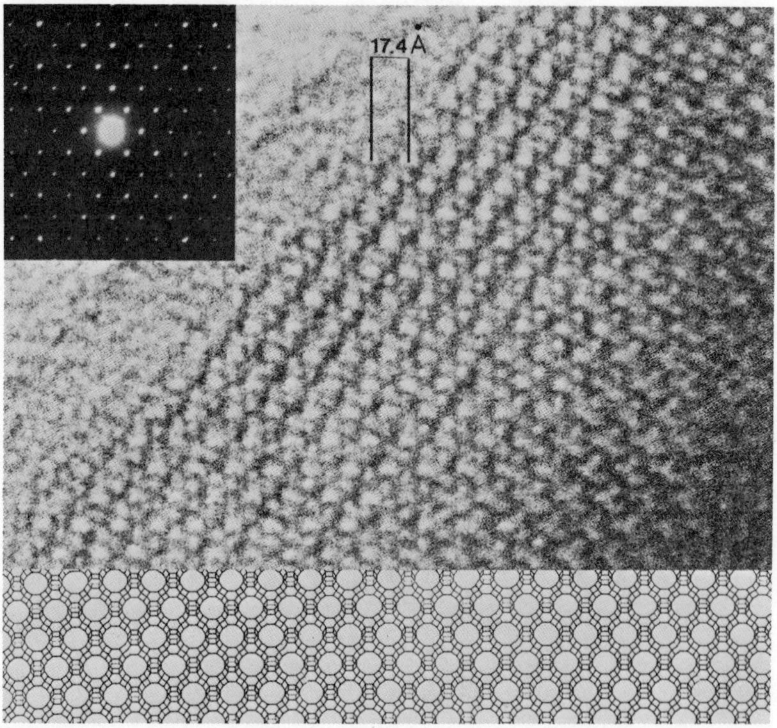

Figure 10c. Typical HREM image (with associated selected area diffraction pattern and schematic illustration of framework structure) after dealumination of a faujasitic zeolite by exposure to $SiCl_4$.

composition but which nevertheless retain the original zeolite structure (32, 32-34, 34). Magic-angle-spinning ^{29}Si and ^{27}Al NMR confirms that Al is removed from the structure, the various proportions of $Si(OAl)_4$, $Si(OAl)_3(OSi)$, $Si(OAl)_2(OSi)_2$ etc., being ultimately replaced by the $Si(OSi)_4$ grouping (see Fig. 10 (b)). HREM and SAED further reveal that the integrity of the precursor zeolitic framework is retained, the <110> projected images being of such quality that the pore-structure is readily apparent (Fig. 10 (c)). The overall reaction here is:

$$Na_x(A\ell O_2)_x(SiO_2)_y + SiC\ell_4 \rightarrow Na_{x-1}(A\ell O_2)_{x-1}(SiO_2)_{y+1} + A\ell C\ell_3 + NaC\ell$$

It is to be noted that the Si/Al ratio, which initially is in the range 1.5 to 2.6, can be increased to over a 100 by this simple gas-solid reaction which converts the original hydrophilic zeolite to a much less hydrophilic microporous crystalline, essentially non-zeolitic structure.

New Ring-Chain Intergrowth Structure. It is well known that among the wide variety of structural types displayed by the silicates the chain structures (which can consist of single chains of linked SiO_4 tetrahedra as in pyroxenes and pyroxenoids, or of double chains as in the amphiboles) and the ring structures (consisting of three, four, six or more tetrahedra) are quite separate. Using HREM and SAED, Wen Shu-Lin *et al*. (34) have recently shown that, over a substantial range of composition, the metagermanates that crystallize out of precursor glasses containing Ca^{2+}, Sr^{2+}, O^{2-} and $GeO_3{}^{2-}$ ions exhibit a microstructure which is composed of pyroxenoid chains (akin to that of wollastonite - see Sections 2.3 (a) and (c) above) and small rings of corner-sharing $GeO_4{}^{4-}$ tetrahedra. There is considerable scope here for combined synthetic and HREM studies to determine the so-called existence ranges of various (unit-cell level) intergrowths of rings and chains in pure or mixed inorganic borates, phosphates, silicates, germanates and vanadates.

Simulating and Predicting New Inorganic Structures. Progress has already been made (26) to adapt the well-known pair potential approach (35, 36), using carefully parameterized atom-atom potentials, to interpret and predict the structures of pyroxenoid silicates, and at the same time to unify the results of HREM studies of this class of crystalline solid. Again the scope for the elaboration of this approach is considerable, and efforts are currently being made (37) to extend these calculations in other areas of inorganic solid-state chemistry. In this connection, it is of interest to note that Parker and Catlow (38) recently predicted the crystal structure of $ThSiO_4$. Their result was at variance with an earlier X-ray based structure, but in agreement with a more refined determination carried out after the results of their calculation were known.

Structural Transformations. Although *in situ* HREM studies are extremely difficult to perform - in view of the demanding experimental conditions - much valuable insight into the mechanism of nucleation and crystallization of new phases from amorphous precursors and into the atomic details of single-crystal → single-crystal transformations. HREM has also shown that some structural transformations which appear on crystallographic arguments to be topotactic (38), solid-state processes take place via the intermediacy of a glassy phase. This has been shown (39) to be the case, in, for example, the conversion of the wollastonite chain structure into pseudowollastonite, a ring structure composed of corner-shared three-membered SiO_4 tetrahedra, at elevated temperatures (*ca* 1250°C). The parent chains break up and simultaneously there is considerable migration of the Ca^{2+} ions leading to the formation of the glassy state. The pseudowollastonite is nucleated and grows from this glass precursor.

A hitherto puzzling process known to occur amongst the graphite intercalates when one stage (40) is rapidly converted into another is seen (28), by HREM, to involve interpenetrating or co-existing stages. By altering the chemical driving force (so as to increase or decrease the 'stage number') one or other of the coexisting stages grows preferentially. In other words, HREM has revealed that present in a given 'stage' of intercalate are nuclei of other (both lower or higher stages) at which growth of a new 'stage number' may proceed in response to changes of chemical potential.

Further examples of how HREM portrays at near atomic scale the mode of solid-state transformation that takes place in solids are the $Bi_2W_3O_{12}$ $Bi_2W_3O_{12}$ + Bi_2WO_6 → $2Bi_2W_2O_9$ conversion discussed in section 2.3 (b) above, and the facile interconversions of the rare-earth oxides and of niobium oxides (41).

Structural Imperfections. In many respects HREM has had a greater impact upon our knowledge of the nature of the atomic reorganization at crystalline imperfections than any other single technique. One of the very first contributions of HREM as a new analytical and structural tool was described in the paper by Iijimia (42) in 1971 on $Ti_2Nb_{10}O_{29}$ viewed down to its *b* - axis. Structural faults, arising from subtle fluctuations in composition, could be clearly seen in the block-structure (based on NbO_6 octahedra) which is a feature of this ternary oxide system. More than a decade later similar materials are yielding to active scrutiny by HREM, and Horiuchi (43), for example, has shown how point defects may be directly viewed (using 1MeV microscopes) in reduced form of nNb_2O_5 mWO_3 and oxidised forms of $Nb_{22}O_{54}$.

In studies of amphiboles (44), isolated strips of triple-chain silicates were discovered embedded in the double-chain parent structure. It was later realized that new types of silicate structures, composed of recurrent triple chains, existed in nature. The part that HREM played in the identification of this new family of triple-chain silicates, which constitute a further step in the progression: pyroxene, amphibole, ... mica, was crucial.

Jefferson's studies of the pyroxenoids has added greatly to our application of the way in which, through the intermediary of planar - or planar and Kinke - faults one structure is converted into another (45). And Audier, Jones and Bowen (46) have revealed how unit cell strips of Fe_3C may be accommodated as extended defects in the Fe_5C_2 structure. Both these carbidic phases can be readily identified by HREM at the interface of iron catalysts used for the disproportionation of CO (to yield $C_{(s)} + CO_2$).

E.E.L.S.

Analytical Electron Microscopy. We have already seen that one way of gaining chemical (compositional) information about the specimens which we image in electron microscopy is to monitor the X-rays that are invariably emitted during the imaging (CXRES). Nowadays this is usually accomplished by employing devices such as Li-drifted Si- detectors coupled to multichannel analyzers, so that the energy dispersed X-rays can be straightforwardly recorded. Another procedure that also permits of analytical electron microscopy entails placing an electron spectrometer in the path of the beam of high-energy transmitted electrons so as to measure the (fractionally) small variations in kinetic energy arising from the various inelastic interactions that have taken place in the specimen (and some which have resulted in the production of X-rays and Auger electrons). The resulting plot of electron intensity as a function of energy loss is termed the electron-energy-loss spectrum. In electron-loss spectroscopy (EELS) we extract information from the inelastically scattered electrons, not from the elastically scattered ones which make HREM possible. Non-crystalline or amorphous solids give little information in HREM (see section 2.2); but the efficiency of EELS is hardly impaired irrespective of whether the specimen is crystalline or not.

Basic Principles. Egerton (47-50) has given admirable summaries of the background theory and the equations required for quantitative application of EELS to the determination of specimen composition: suffice it to note that ionization energies depend upon the type of shell (K, L, etc.) and upon the atomic number (Z) of the atom involved. Their values have been accurately determined for all the elements and are not significantly affected (but see later) by the chemical environment of an atom. Detection of ionization edges in an EEL spectrum, therefore, enables elements present within the specimen to be identified. Moreover EELS, unlike XRES, is very sensitive to light elements (Li, Be, B, C, N, O etc.). Another advantage that EELS possesses - one that has not yet been fully utilized chemically - is that, with EELS, the region of the specimen contributing to the spectrum can be controlled by using an appropriate aperture (since the cone of a inelastically scattered electrons is a rather narrow one around the forward scattered, zero-loss beam). In principle, therefore, EELS is capable of greater (lateral) spatial resolution than, say, XRES; and, indeed, phosphorous distributions in micleosomes have been recently charted (57) with an estimated resolution of 1.4 nm. This technique is so sensitive that it is capable of estimating 48 phosphorous atoms or 2.5 x 10^{-21}g of phosphorous.

Unlike XRES, which yields information as elemental ratios and from comparisons with standard samples of known composition, EELS is capable of yielding absolute results, i.e. the composition is derived without reference to a standard. The concentration, N, of a measured element, in atoms per unit area of the specimen, is given (50, 52) by (see Figure 11.)

$$N \approx \frac{1}{G\sigma_i(\alpha,\Delta)} \left\{ \frac{I_i(\alpha,\Delta)}{I_\ell(\alpha,\Delta)} \right\}$$

Here, G is a factor that takes account of any difference in detector gain between the low-energy and high-energy regions of the spectrum, α is the maximum angle of scattering allowed into the spectrometer, Δ is an energy range of integration within the energy-loss spectrum, and $\sigma_i(\alpha,\Delta)$ is an ionization partial cross section, which can be calculated (47). The quantity $I_\ell(\alpha,\Delta)$ is obtained by integrating the energy-loss intensity over a range Δ, starting at the zero-loss peak. The inner-shell intensity $I_i(\alpha,\Delta)$ (where i indicates the type of shell: K, L, etc.) is obtained by integration under the appropriate ionization edge, making due allowance for the background intensity (Figure 11). Fortunately, this background intensity often follows a power-law dependence on energy loss: AE^{-r}, where A and r can be assumed constant over a limited range of energy loss E. By sampling the background just preceding the ionization edge, the values of A and r can be determined. This above expression (for N) has been found to be sufficiently accurate provided that the thickness of the analyzed region of the specimen is less than the mean free path for valence electron inelastic scattering, which is typically about 1000Å for 100 keV incident electrons. It is to be noted that, even with suitably thin specimens, the ratio of background to characteristic signal is usually higher in EELS than in XRES.

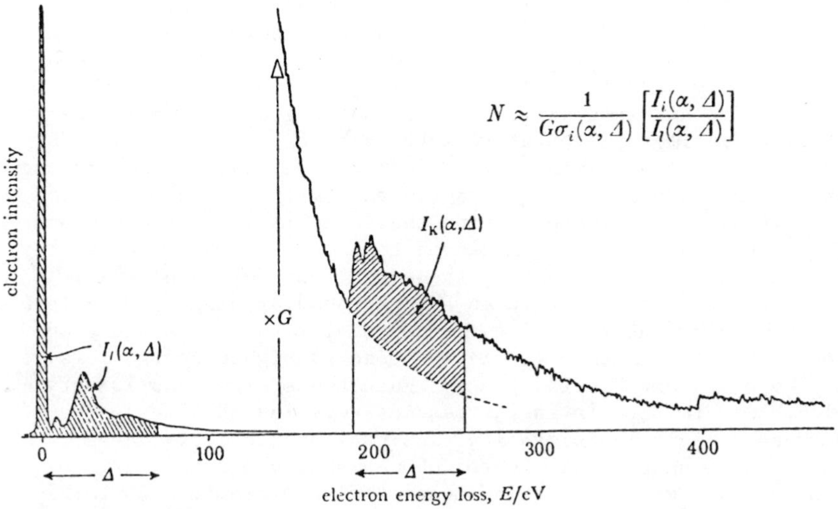

Figure 11. Electron-energy-loss spectrum of crystalline boron nitride, showing the boron K-edge (at 190 eV) and the nitrogen K-edge (at 400 eV). The background intensity, delineated by the dashed curve arises from inelastic scattering by valence electrons. The hatched areas represent the measured values required for the quantitative analysis of boron (see *text) (*50*).*

EELS is, therefore, not good for the detection of elements when their concentration is ca 1 per cent of other constituents in the sample.

Applications of EELS for Chemical Analysis. Apart from the fact that it is an 'absolute' technique requiring no standard for calibration the special advantage possessed by EELS as a means of chemical analysis is that it combines sensitivity with spatial resolution. It is, in a literal sense, an ultramicro method with minimum detectable amounts, in favourable circumstances, of better than 10^{-20}g. Its especial advantage in being able to cope with the light elements makes it a very useful tool for certain applications. Thus, EELS was successfully used by Leapman et al (53) to identify the composition of needle-shaped precipitates in Cr-Mo steel as Cr_2N and that of granular precipitates as $Fe_5Cr_{18}C_6$. In studying the Si-Al-O-N ceramics, EELS has been particularly useful in determining local composition both at crystalline and at contiguous amorphous regions. And in the study (48) of graphite intercalates and graphite fluorides EELS has yielded local compositional information, again illustrating its superiority in coping with the analysis of light elements.

It has to be borne in mind that some ionization edges are more easily detected and measured than others. Below an energy loss E ∿ 100eV, for example, the edges tend to be less visible because they are submerged in a background arising from inelastic scattering by outer-shell (valence) electrons. This background does not vary smoothly (as is the case at higher energy loss) and may even contain several peaks. When the energy loss is very high, the electron intensity becomes very low and the spectrum correspondingly lacks meaningful signal over noise. These considerations suggest 100 to 2000 eV as a preferred range of energy loss, which, in turn, means that the elements Be to Si (Z=4 to 14) can be detected from their K-shell losses, elements Si to Sr (Z=14 to 38) via their L-shell and elements Rb through Os (Z=37 to 76) from their M-losses, with a possibility of detection of even heavier elements by N-shell ionization. It is further worth noting that whereas K- and certain L-edges are rather sharp, other L-edges are somewhat rounded, which renders their detection and measurement more difficult. It is prudent, therefore, to rely more on XRES for the detection of medium and heavy elements.

Structural Information from EELS. Besides yielding chemical composition, EELS is also capable of providing structural information on an atomic scale. It has been known (54) for some time that the fine-structure in the energy-loss spectrum close to an ionization edge reflects the energy dependence of the density of electronic states above the Fermi level.

Pre-Absorption Edge Structure. Egerton (50) and

Ritsko (55) have shown that, for the graphite-ferric chloride intercalate, the additional states due to the intercalant species ($FeCl_3$) show up as an increase in the spectral intensity just ahead of the K-ionization edge of carbon. There is also a small chemical shift in the carbon K-level upon intercalation.

EXELFS-Extended Energy Loss Fine Structure. Modulations in the electron intensity (i.e. in the differential inelastic electron scattering cross-section) extending over *ca* 100eV on the high-energy side of an ionization edge, can be used to provide structural information (bond lengths and coordination numbers as well as Debye-Waller factors) concerning the local environment of the appropriate element. This much was clear from the early work of Leapman and Cosslett (56). It has become (57-69) clear in recent years that, potentially, EXELFS can yield structural information which would be very difficult to extract from its 'photon-analogue' EXAFS. One point to emphasize here is that at, lower energies (characteristic, for example of the K-shell edges of elements with atomic numbers ranging from 3 to 12), X-rays have very little penetrating power. EXAFS is not, therefore, the most appropriate method for ascertaining the atomic environments of typically, B, C, N and O, in solid material. The second point is that EXELFS is especially suitable for the study of inhomogeneous samples (structurally and compositionally heterogeneous in the sense discussed in section 2.2 above) because the primary electron beam can be focussed to a diameter of *ca* 20Å. Other advantages of EXELFS have been discussed elsewhere (60, 61). The limitations of the technique include (i) the need to select an optimal thickness of sample so as to minimize multiple scattering; and (ii) the susceptibility of the samples to suffer radiation damage.

Brown (62) has recently shown that the presently available theory for the interpretation of EXELFS data is inadequate. In his study of the layered BN structure he found that whereas the distances to first nearest neighbours could be satisfactorily extracted from EXELFS data, second nearest-neighbour distances could not. The way ahead (62), here, is for the theory to be extended and tested against several known structures composed of light elements.

Compton Scattering. The electron spectrometers which electron microscopists nowadays use for chemical analyses (section 3.3) or for structural analysis as described in the preceding paragraphs, may also be used (63) to determine the extent of the Doppler broadening of the scattered electrons which, in turn, yield the Compton profile. Using energy and momentum conservation, we can convert the energy scale to one representing the momentum of the electrons in the sample. The Compton profile is a projection of the momentum density on to the direction of the scattering vector. This information then leads (25, 63) to the autocorrelation function of the position space wavefunction. And so, we see that, potentially, this branch of electron energy loss spectroscopy (like EXELFS) is capable of yielding structural information in a spatially resolved manner.

Future Developments

The utilization of higher accelerating voltages (500 to 1000 keV) is already well developed, and the results obtained (64, 65, 66) to date give clear indications that, in the near future, atomic resolutions in the range 1 to 2Å may well become commonplace. Important questions in inorganic chemistry can then be answered by the application of direct, real-space crystallography. To illustrate this point we consider whether, in rendering a perovskite structure (ABO_3) grossly non-stoichiometric, the resulting structure (ABO_{3-x}), is composed either of alternating stacks of BO_6 octahedra, or of uniformly stacked BO_4 pyramids. Bando *et al* (65), using the 1000keV high resolution microscope, answered this question directly and found that in the perovskite related structure ($Ca_4YFe_5O_{13}$) successions of FeO_6 octahedra and FeO_4 tetrahedra are clearly present in the microcrystalline samples, that are not readily amenable to X-ray studies.

As low-temperature operation becomes progressively more feasible, a broader range of specimen types (e.g. organometallic compounds that are stable only below room temperatures) will inevitably be capable of study by HREM. On-line interactive computer facilities, with appropriate means of digitizing images and routines for fast Fourier Transforms, as well as associated video cassette facilities, ought to bring structure analysis by real-space crystallography into greater popularity (5).

HREM and EELS studies are also likely to be of greater importance in the near future, thus enabling structure identification as well as composition and bonding properties to be determined more or less simultaneously and with very high spatial resolution. With the improved vacua attainable in modern microscopes - the current generation of commercial scanning transmission electron microscopes (STEM) operate with vacua of *ca* 10^{-10} Torr at the specimen stage - there is every prospect that many of the measurements employed in studies of catalysis, such as Auger electron spectroscopy, reflection high energy electron diffraction (RHEED) and imaging in the reflection mode (REM) will be possible (67, 68). At the Tokyo Institute of Technology recently, spectacular images of surfaces have been produced (67), using RHEED microscopy with conventional imaging in a ultrahigh vacuum chamber. The (111) face of Si revealed monatomic steps that terminated at emergent dislocations. Yagi and his coworkers (67) have been able to image both (1x1) and (7x7) re-constructed regions at the Si(111) surface, for example.

Developments in electron microscopy also promise to revolutionize other studies of the surfaces of solids in general and/of catalysts in particular. Previously, monatomic steps and other topographical features at the exterior surfaces of solids were best investigated by the powerful but cumbersome, and destructive technique of gold-decoration - see refs 69 and 70 for studies of alkali halide and layered sulphides, respectively. But darkfield conventional transmission electron microscopy can now reveal monatomic steps directly, as the micrograph in Fig. 12 shows (71). Using this kind of approach it should be possible to ascertain quantitatively the extent of the interaction between a catalyst and its underlying support.

Another future development likely to be vigorously pursued is that of electron energy loss microscopy. This functions by combining HREM and EELS in such a way that only electrons that have suffered a well defined loss are "used" to construct the image. The work of Bazelt-Jones and Ottensmeyer (51) is based on this principle. They have already succeeded in resolving structural information of biological specimens to at least the 5Å level while supplying chemical information with the same spatial resolution. Although interest in this novel approach is currently greatest amongst biophysicists (who, for example, can

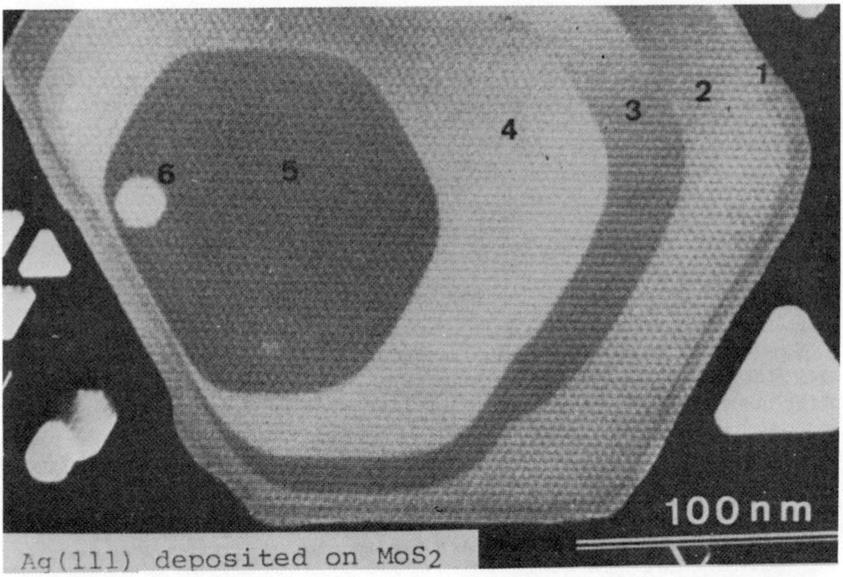

Figure 12. A dark-field transmission electron micrograph of a (111)Ag platelet grown on single-crystal MoS_2. The regions with different contrast differ in thickness by one monatomic step from one another. The larger the number marked on each region, the thicker the crystal (3, 71).

image ferritin molecules either with electrons inelastically scattered from the iron or with those inelastically scattered from the enveloping carbon), the stage is set for the technique to be adopted to inorganic materials.

Acknowledgment

I acknowledge stimulating discussions with my colleagues, Drs. D. A. Jefferson, G. R. Millward, A. Reller and M. A. Uppal and the support of the S.E.R.C.

Literature Cited

1. Thomas, J. M.; Jefferson, D. A. Endeavour New Series 1978, 2, 127.
2. Nobel Symposium No. 47 on "High Resolution Electron Microscopy", Chemica Scripta 1979, 14, 167.
3. Beer, M.; Carpenter, R. E.; Eyring, LeRoy; Lyman, C. E.; Thomas, J. M. Chem. Eng. News 1981, 59, 40.
4. Jefferson, D. A.; Thomas, J. M.; Egerton, R. F. Chemistry in Britain 1981, 17, 514.
5. Smith, A. R.; Eyring, L. Ultramicroscopy 1982, 8, 65.
6. Chianelli, R. R.; Dines, M. B. Inorganic Chemistry 1978, 17, 2758.
7. Chianelli, R. R.; Prestridge, E. B.; Pecoraro, T. A.; DeNeufville, J. P. Science 1979, 203, 1105.
8. Chianelli, R. in Intl. Rev. in Phys. Chem. 1981, in press.
9. Becker, G. L.; Chen. C. H.; Greenawalt, J. W.; Lehninger, A. L. J. Cell Biol. 1974, 61, 316.
10. Gaskell, P. H. Acta Metallurgica 1981, 29, 1203.
11. Elliott, S. R.; Davies, E. A.; J. Non-Cryst. Sol. 1980, 35-36, 849.
12. Bursill, L. A.; Mallinson, L. G.; Elliott, S. R.; Thomas, J. M. J. Phys. Chem. 1981, 85, 3004.
13. Howie, A. ref 2, p 109.
14. Bursill, L. A.; Mallinson, L. G.; Elliott, S. R.; Thomas, J. M. J. Phys. Chem. 1981, 85, 3004.
15. Bursill, L. A.; Thomas, J. M. J. Phys. Chem. 1981, 85, 3007.
16. Thomas, J. M.; Bursill, L. A. Angew. Chemie. 1980, 19, 745.
17. Thomas, J. M.; Millward, G. R.; Ramdas, S.; Audier, M.; Bursill, L. A. Faraday Society Discussions 1981, in press.
18. Reller, A.; Jefferson, D. A.; Thomas, J. M. unpublished work.
19. Hayek, E.; Pallasser, U. Monatsch Chemie 1968, 99, 2126.
20. Jefferson, D. A. Ph.D. Thesis, University of Cambridge, 1973.
21. Jefferson, D. A.; Thomas, J. M. in Proc. of EMAG 1975: Developments in Electron Microscopy (ed. J. A. Venables), Academic Press, 1976, 275.
22. Morales Palomino, J.; Jefferson, D. A.; Hutchison, J. L.; Thomas, J. M. J. Chem. Soc. Dalton Trans. 1977, 1834.
23. Jefferson, D. A. Phil. Trans. Roy. Soc. 1981, in press.
24. Jefferson, D. A.; Uppal, M. A., in preparation.

25. Thomas, J. M. Ultramicroscopy 1982, 8, 13.
26. Catlow, C. R. A.; Thomas, J. M.- Parker, S. C.; Jefferson, D. A. Nature 1982, 295, 658.
27. Thomas, J. M.; Millward, G. R.; Davies, N. C.; Evans, E. L. J. Chem. Soc. Dalton Trans. 1976, 2443,
28. Thomas, J. M.; Millward, G. R.; Schlögl, R.; Boehm, H. P. Materials Research Bulletin 1980, 15, 671.
29. Millward, G. R.; Thomas, J. M., unpublished work.
30. Thomas, J. M.; Audier, M.; Klinowski, J. J. Chem. Soc. Chem. Comm. 1981, 1221.
31. Audier, M.; Thomas, J. M.; Klinowski, J.; Jefferson, D. A.; Bursill, L. A. J. Phys. Chem. 1982, 86, 581.
32. Beyer, H. K.; Belenkaja, I. in "Catalysis by Zeolites" Ed. Imelik, B. Elsevier, 1980, 203.
33. Klinowski, J.; Thomas, J. M.; Audier, M.; Vasudevan, S.; Fyfe, C. A.; Hartman, J. S. J. Chem. Soc. Chem. Comm. 1981, 570.
34. Shu-lin, Wen,; Jefferson, D. A.; Thomas, J. M. J. Chem. Soc. Chem. Comm. 1982, in press.
35. Catlow, C. R. A.; Norgett, M. J. UKAEA Rep. M2936, 1978.
36. Catlow, C. R. A.; James, R. Chem. Soc. Spec. Per. Rep. (Chem. Physics of Solids and Surfaces) 1980, 8, 108.
37. Catlow, C. R. A. private communication.
38. Parker, S. C.; Catlow, C. R. A. private communication. For definition of 'topotactic' and its distinction from 'topochemical' see Thomas, J. M. Phil. Trans. Roy. Soc. 1974, A277, 251.
39. Shu-lin, Wen; Ramdas, S.; Jefferson, D. A. J. Chem. Soc. Chem. Comm. 1981, 662.
40. The stage number n(=1, 2, 3 ...) stands for the number of individual graphite sheets which separate sheets of guest in the intercalate. A higher stage (n high) therefore means a more dilute intercalate.
41. Eyring, L. Ultramicroscopy 1982, 8, 39.
42. Iijima, S. J. Appl. Phys. 1971, 42, 5891.
43. Horiuchi, S. Ultramicroscopy 1982, 8, 27.
44. Ref 2, p 167.
45. Pugh, N. J.; Jefferson, D. A. Ultramicroscopy 1982, 8, 59.
46. Audier, M.; Jones, W.; Bowen, P. J. Cryst. Growth, submitted.
47. Egerton, R. F. Ultramicroscopy 1979, 4, 169.
48. Egerton, R. F. in 37th Annual Proc. Elec. Mic. Sco. Am., Ed. G. W. Bailey, p 128, Baton Range, Claitor's Publishing Div.
49. Egerton, R. F. Ultramicroscopy 5 521.
50. Egerton, R. F. Phil. Trans. Roy. Soc. 1982, in press.
51. Basett-Jones, D. P.; Ottensmeyer, F. P. Science, 1981, 211, 169.
52. Leapman, R. D.; Rez, P.; Mayers, D. F. J. Chem. Phys. 1980, 72, 1232.

53. Leapman, R. D.; Sanderson, S. J.; Whelan, W. J. Metal Sci. 1978, 215.
54. Egerton, R. F.; Whelan, M. J. J. Elect. Spect. Rel. Phenomena 1974, 3, 232.
55. Ritsko, J. J. in 39th Annual Proc. Elect. Mic. Soc. Am. p 174, Baton Range: Claitor's Publishing Div., 1980.
56. Leapman, R. D.; Cosslett, V. E. in J. Phys. D. 1976, 9, 129.
57. Leapman, R. D.; Silcox, J. Phys. Rev. Lett. 1979, 42, 1361.
58. Isaacson, M.; Utlaut, M. Optic 1978, 50, 213.
59. Baston, P. E.; Craven, A. J. Phys. Rev. Lett. 1979, 42, 893.
60. Csillag, S.; Johnson, D. E.; Stern, E. A. in 'EXAFS' Ed. by D. C. Joy (Plenum Press, 1981), p 241.
61. Leapman, R. D.; Grunes, L. A.; Fejes, P. L.; Silcox, J. ref 60, p 217.
62. Brown, L. M., private communication.
63. Williams, B. G.; Parkinson, G. M.; Eckhardt, C. J.; Thomas, J. M.; Sparrow, T. G. Chem. Phys. Lett. 1981, 78, 434.
64. Uyeda, N.; Fujiyoshi, Y.; Kokayashi, T.; Ishizuka, K.; Ishida, Y.; Hanada, Y. Ultramicroscopy 1980, 5, 459.
65. Bando, Y.; Sikikawa, Y.; Yamamura, H.; Matsui, Y. Acta. Cryst. 1981, A37, 723.
66. Smith, D. J.; Cosslett, V. E.; Stobbs, W. M. Interdisciplinary Sci. Rev. 1981, 6, 155.
67. Osakabe, N.; Tanishiro, Y.; Yagi, K.; Honjo, G. Surface Sci. 1981, 102, 424.
68. Ladas, S.; Poppa, H.; Boudart, M. Surface Sci. 1981, 102, 151.
69. Betghe, H.; Phys. Stat. Solid 1962, 2, 3.
70. Bahl, O. P.; Evans, E. L.; Thomas, J. M. Proc. Roy. Soc. 1968, A306, 53.
71. Takayanagi, K. Ultramicroscopy 1982, 8, 145.

RECEIVED September 2, 1982

Discussion

E.A.V. Ebsworth, Edinburgh University: Has Vibrational Electron Energy Loss spectroscopy any relevance in this context?

J.M. Thomas: The energy resolution attainable with the electron spectrometers that have been available up to the present is inadequate to detect the fine structure that may be expected from phonon or local modes. With continued improvements, one may reasonably expect some progress in this direction; but, at present, more information is retrievable from the fine structure, discussed in the text, that arises from causes other than vibrational modes.

31

Lasers, Laser Spectroscopy, and Laser Chemistry

WILLIAM H. WOODRUFF

The University of Texas at Austin, Department of Chemistry, Austin, TX 78712

The historical development and elementary operating principles of lasers are briefly summarized. An overview of the characteristics and capabilities of various lasers is provided. Selected applications of lasers to spectroscopic and dynamical problems in chemistry, as well as the role of lasers as effectors of chemical reactivity, are discussed. Studies from these laboratories concerning time-resolved resonance Raman spectroscopy of electronically excited states of metal polypyridine complexes are presented, exemplifying applications of modern laser techniques to problems in inorganic chemistry.

The advent of the laser has revolutionized many areas of experimental science and technology. The unique properties of laser light - extreme brightness, monochromaticity, spatial and temporal coherence, time resolution, directionality, and polarization - have engendered entirely new fields of study in optics, physics, biology, medicine, engineering, and, of course, chemistry. Many previously existing fields have been transformed by experimental approaches which could not have been contemplated in the pre-laser era. This article will review briefly the chemically relevant characteristics of lasers and laser radiation, and provide a selective survey of current chemical applications of lasers. Concurrently we shall examine present and possible future influences of the laser revolution upon "inorganic chemistry toward the 21st century". A case study is included which illustrates potential applications of lasers to problems of interest to inorganic chemists.

Most chemical applications of lasers have fallen into three categories. Firstly, spectroscopic applications have utilized

0097-6156/83/0211-0473$07.75/0

the extreme spectral purity of laser light for high resolution, and also the high light intensities that lasers can produce to generate saturation or nonlinear spectroscopic effects. Secondly, dynamical applications have utilized the short pulses of light that can be obtained from lasers to observe extremely rapid chemical processes. Finally, several properties of laser light combine to make lasers a unique source of energy for initiating or modifying chemical reactivity, leading to applications which may be termed laser chemistry.

In all of these applications, the emphasis to date has been on the use of lasers to study chemically and physically well characterized systems, that is, simple molecules in the gas phase, or in ordered phases such as molecular crystals, or in cryogenic matrices. There are exceptions to this statement, but the basic fact is that the great strides in chemical applications of lasers have been made by the chemical physics and analytical chemistry communities and largely ignored by inorganic, organic, and biological chemists.

In many respects laser technology is now mature enough so that, in the first place, existing techniques may be profitably applied to complex chemical systems (e.g. solutions, large molecules, "dirty" surfaces, etc.) and, in the second place, at least some laser devices are simple enough to be usable by chemists other than full-time laser technologists. What is not fully developed among many chemists is a knowledge of the availability and capabilities of lasers and related devices, an appreciation of the chemical techniques that these devices make possible, and a realistic view of their present and potential applicability to problems of interest in inorganic chemistry. One purpose of this article is to provide some insight into these areas.

Lasers: Historical Perspective

The word LASER is an acronym for Light Amplification by Stimulated Emission of Radiation. The physical process upon which lasers depend, stimulated emission, was first elucidated by Einstein in 1917 (1). Einstein showed that in quantized systems three processes involving photons must exist: absorption, spontaneous emission, and stimulated emission. These may be represented as follows:

Absorption	$S + h\nu \rightarrow {}^*S$
Spontaneous Emission	${}^*S \rightarrow S + h\nu$
Stimulated Emission	${}^*S + h\nu \rightarrow S + 2h\nu$

where S and *S denote the system in its ground and excited state, respectively. Absorption and stimulated emission are exactly analogous processes for the ground and excited states. The practical consequence of stimulated emission is that for an ensemble of excited-state systems a chain-reaction-like process can take place, resulting in net production of photons (amplification) because one photon directed into an excited system can yield two exiting photons. Furthermore, the exiting photons will have the same energy, direction, phase, polarization; in short, they will be coherent.

The first experimental observation of stimulated emission is credited to C.H. Townes and coworkers who in the early 1950s developed the maser, which was the microwave-frequency precursor of the laser. The laser ("optical maser") was first predicted by Schalow and Townes (2) in 1958. The first observation of stimulated emission at optical wavelengths, and thus the actual invention of a working laser, is due to T.H. Maiman (3) who, in 1960, was able to produce pulsed laser emission from a ruby crystal excited by a photographic flashlamp. Continuous laser emission was first produced by Javan and coworkers (4) who developed the c.w. helium-neon laser in 1961. The ruby and He-Ne lasers were followed closely in development by lasers based upon Nd^{3+}-doped crystals, semiconductors, N_2, CO_2, HF, argon, krypton, and other media, such that by the mid-1960s a number of lasers existed with wavelength capability ranging from the mid-infrared (CO_2, HF) to the near ultraviolet (N_2).

The early lasers either produced a single discrete wavelength or a series of selectable but discrete wavelengths. Thus successful chemical and spectroscopic application of these devices generally required the accidental coincidence of an available laser wavelength with a transition of a chemical system of interest. In fact, for a considerable time a "system of interest" was defined as one which possessed a transition near a laser wavelength. That is, chemical systems were chosen to fit devices rather than _vice versa_. General applicability of lasers to the full range of chemical problems obviously awaited the development of tunable lasers.

The first, and still the most commonly used, of the tunable lasers were those based upon solutions of organic dyes. The first dye laser was developed by Sorokin and Lankard (5), and used a "chloro-aluminum phthalocyanine" (_sic_) solution. Tunable dye lasers operating throughout the visible spectrum were soon produced, using dyes such as coumarins, fluorescein, rhodamines, etc. Each dye will emit laser radiation which is continuously tunable over approximately the fluorescence wavelength range of the dye.

Subsequent to the advent of the dye laser, tunable lasers based upon other lasing media were developed which operate over various wavelength ranges. Nobable among these are the f-center lasers and diode lasers which are tunable in the infrared.

The mid 1970s may be viewed as the period during which a large number of lasers (including broadly tunable ones), having a wide variety of capabilities and satisfactory reliability, designed for users other than dedicated laser experts, became widely available on the market. With this, general application of lasers to chemical problems became feasible. As noted previously, however, chemical applications of lasers have continued to be practiced largely by investigators in chemical physics and analytical chemistry.

Lasers: Elementary Operating Principles

All lasers have the following three fundamental components (see Figure 1). The gain medium, the business end of the laser, is the medium in which excited states are created which will produce stimulated emission. The exciter is the source of energy for production of the excited states in the gain medium. Finally, the optical resonator determines the directionality, wavelength selectivity, optical feedback, polarization, and other characteristics of the stimulated emission from the gain medium.

A wide range of types of materials are suitable laser gain media: noble gases, notably neon, argon and krypton; molecular gases such as nitrogen, hydrogen, CO_2, HF, and CO; excimers such as ArF and XeCl; doped crystals such as ruby and Nd^{3+}-doped yttrium aluminum garnet (Nd:YAG); semiconductors such as GaAs and InSb; fluid solutions of organic dyes such as coumarins and rhodamines; metal vapors such as cadmium and copper; and even free electrons. Each of these media must have at least one pair of energy levels, the transition between which has an acceptable cross-section for stimulated emission. The energy levels which comprise this pair are imaginatively known as the upper laser level and the lower laser level. Some media have many such pairs of levels. In order for a laser to operate (that is, to exhibit net gain or to amplify light at the laser wavelength) it is necessary to produce a population inversion, that is, the upper laser level must be more highly populated than the lower. This is necessary because absorption and stimulated emission are identical processes, therefore if stimulated emission (i.e., gain) is to overcome absorption (i.e., loss) for a given transition, the population of the upper level of the transition must be greater than that of the lower. Thus the second fundamental requirement for a laser gain medium, in addition to the presence of at least one suitable laser transition, is that it must be possible to create an inverted population of the levels involved in the transition.

Because it is difficult to invert the population of an excited state with respect to the ground state of a system, the lower laser level of a gain medium is not usually the ground state. Thus typical gain media are either three-level or four-level systems (see Figure 2) with the lower laser level far enough above the ground state to avoid excessive thermal population. Furthermore, because it is increasingly difficult to invert the population of two levels as the energy difference between them increases, it is increasingly difficult and inefficient to create lasers which operate at increasingly short wave-

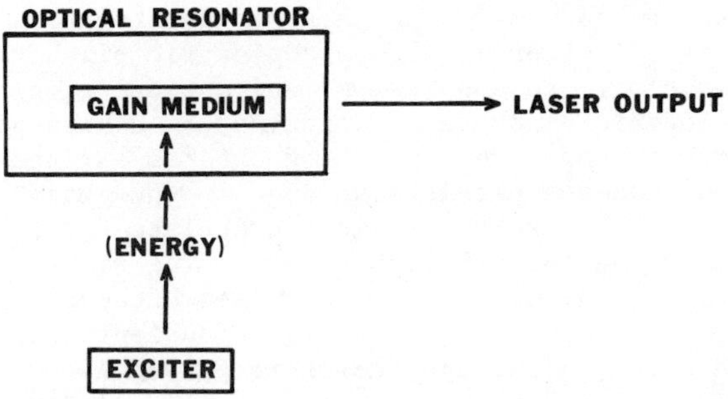

Figure 1. Oversimplified block diagram of the fundamental components of a laser.

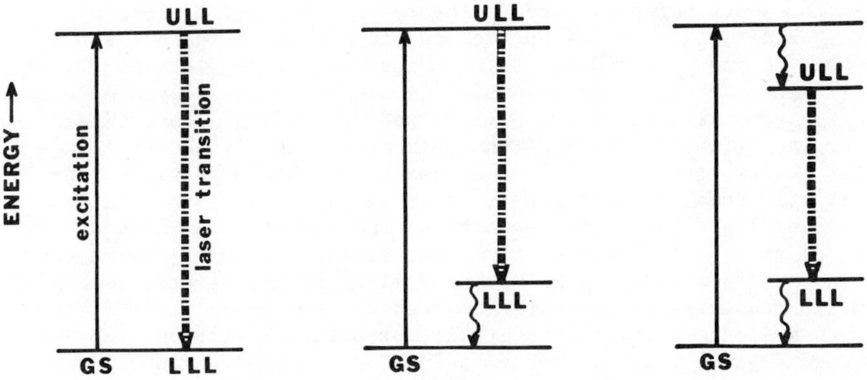

Figure 2. Representations of two- (left), three- (middle), and four-level (right) laser systems. Key: GS, ground state; ULL, upper laser level; and LLL, lower laser level.

Note that operation of a two-level laser system requires inversion of the population of the upper laser level with respect to the ground state—a theoretical impossibility under steady-state conditions if optical excitation is employed.

lengths. It was no accident that the first observation of stimulated emission was in the microwave region, and that the first lasers produced red light.

The creation of a population inversion in the gain medium is the function of the exciter. The characteristics of the exciter are as diverse as those of the gain medium, because a successful method of "pumping" the gain medium into excited states is highly dependent upon the nature of the gain medium itself. The two most common methods of excitation are electric discharge (obviously applicable only to gas lasers) and optical pumping. The exciter for electric discharge pumping may be a high-power D.C. or radio frequency power supply, which creates excited states by inducing a continuous electric discharge in the gaseous gain medium. This type of excitation is appropriate for c.w. gas lasers such as He-Ne, Ar^+, Kr^+, CO_2, etc. The exciter for electric discharge pumping may also be a pulsed high-voltage power supply which will produce a high energy transient electric discharge in the gain medium, and therefore pulsed laser operation. Most gas lasers which can be operated c.w. can also be operated by pulsed excitation. In addition, certain gas laser gain media wherein population inversion is difficult to attain can only be operated by pulsed electric discharge excitation. These are primarily gain media producing ultraviolet laser emission, such as the superradiant N_2 and H_2 lasers and the excimer lasers.

Exciters employing optical pumping are in principle applicable to any gain medium having three or more levels (see Figure 2). In practice optical pumping is almost exclusively used with doped crystal (e.g. ruby, Nd:YAG) and dye lasers. In optical pumping, light is shone on the gain medium and is absorbed by the medium, creating excited states and either directly or indirectly populating the upper laser level. The source of the pumping light may either be a broadband source (e.g. a xenon arc lamp) or a laser. Either broadband or laser pumping may be continuous or pulsed. As is the case with the gas lasers, some gain media are suitable for either pulsed or c.w. operation (Nd:YAG, dye) while some are only suitable for pulsed operation (ruby). Doped crystal lasers, whether pulsed or c.w., are almost always pumped with broadband light, while c.w. dye lasers are invariably pumped by c.w. (typically, Ar^+ or Kr^+) lasers. Pulsed dye lasers may be pumped either by flashlamps or by pulsed laser excitation.

Other methods of excitation are effective or necessary for certain gain media. For example, certain energetic chemical reactions produce molecules in excited states. These excited molecules may then comprise the upper laser level of an inverted-population system. A specific example is the hydrogen fluoride "chemical laser" wherein excitation is provided by the reaction of hydrogen gas with atomic fluorine. Another method of excitation is simply the passage of an electric current through a semiconductor device. This serves as the exciter for diode lasers.

The most common type of optical resonator, in its simplest

form, consists of two parallel mirrors with the laser gain medium placed between them. The mirrors form a Fabry-Perot interferometer, wherein wavelengths of light equal to twice the distance between the mirrors divided by an integer experience constructive interference. This is one of the most important functions of the optical resonator; to provide a resonant cavity for reinforcement of the wavelengths at which the gain medium produces stimulated emission. A second, equally important function of the optical resonator is to provide optical feedback. This means that the optical resonator should reflect one photon having the resonant wavelength back and forth through the gain medium a number of times, in order to stimulate emission of more photons coherent with the first and thus to enhance the gain of the laser. Because only photons emitted parallel to the axis of the interferometer experience enhanced gain due to multiple passes through the medium, the optical resonator is also responsible for the directionality of the laser. Laser emission is usually extracted from the optical resonator by making one of the two cavity mirrors partially transmitting.

The chain-reaction nature of stimulated emission and its coherence dictate that any mode of laser action that has a gain advantage over other possible modes will dominate the output of the laser. Thus the optical resonator can be designed such that the laser output will be extremely selective with respect to wavelength, polarization, phase, timescale, and other characteristics. This selective resonator design may be accomplished by adding optical elements to the basic Fabry-Perot cavity, or by using fundamentally different resonator concepts. In either case design features are incorporated which maximize the quality factor ("Q") of the resonator for the laser output characteristics desired. We shall not consider these resonator hardware features, except as necessary to understand the pragmatic aspects of the laser output.

The availability of a wide range of laser wavelengths produced in a simple manner has been a great concern for chemical applications since the inception of the laser. One solution to this problem, the tunable laser, was discussed above. A second approach, not directly having to do with the fundamental components of the laser itself, involves the use of nonlinear optical devices to produce harmonics or shifted frequencies from a single laser output frequency. Nonlinear optical effects arise when the intensity of light becomes so great that the classical electric field associated with the light exceeds the linear polarizability of the medium. These effects usually require light intensities so great that pulsed lasers are necessary for their observation. Except for a few heroic pre-laser experiments, the pioneering work in nonlinear optics has been done since the advent of the laser, largely by Nicolaas Bloembergen and associates (6).

At present, from the chemist's point of view, the most practical result of nonlinear optical technology is the ability to produce large amounts of energy at the harmonics, sum and difference frequencies, and vibrationally shifted frequencies of pulsed lasers. For example, the Nd:YAG laser emits light at only one wavelength, 1.064 μ. If the laser is pulsed, however, the second (532 nm) and fourth (265 nm) harmonics can be generated simply by passing the 1.064 μ output through appropriately cut crystals of KD_2PO_4 (KD*P). The third (355 nm) and fifth (213 nm) harmonics can be generated by summing the even harmonics with the fundamental in similar crystals. The fundamental or any of the harmonics can be used to generate emission shifted from the harmonic frequencies by the stimulated Raman effect, whereby a substantial fraction of the original energy can be converted into vibrationally shifted frequencies. Utilizing the stimulated Raman shifts (SRS) of H_2 and D_2 gases or a H_2/D_2 mixture, laser light can be produced at discrete wavelengths spaced approximately every 1000 cm^{-1} beginning in the vacuum ultraviolet and continuing into the infrared. Using a minimum number of reliable dyes pumped by the Nd:YAG harmonics along with the H_2/D_2 shifts, continuous tunability throughout the visible and near ultraviolet can be achieved (see Figure 3). A large number of discrete lines are available from the H_2/D_2 shifts between 30,000 and 50,000 cm^{-1}, with energies above 0.1 mJ/pulse. Thus a simple laser system, a Nd:YAG laser with harmonic generators, dye laser, and SRS cell, represents an extraordinarily versatile light source for chemical applications for which pulsed excitation is necessary or tolerable. A similar system based upon an excimer laser rather than Nd:YAG is potentially superior because the excimer laser fundamentals lie in the wavelength range 193-353 nm and therefore may be down-converted into the visible with relatively greater efficiency than the infrared Nd:YAG fundamental may be up-converted into the visible and ultraviolet. Excimer laser technology, however, is at this date far from being as stable and advanced as that of Nd:YAG lasers.

Lasers: Characteristics of the Output

The most obvious characteristics of laser light are its brightness, its spectral purity, and the directionality of the beam. It is not so obvious how extreme these properties are. It has been pointed out (7) that a one-milliwatt He-Ne laser, virtually a toy laser, is 100 times brighter than the surface of the sun in terms of luminous intensity per unit area. Considering only light at the laser wavelength of 632.8 nm, the laser is 2×10^5 times brighter than the sun!

When focussed into a diffraction limited spot (ca. 10 μ in diameter) the 1 mW laser yields a power density of 10^3 watts per square centimeter. Readily available c.w. lasers produce 10 watts of output, whence a focussed power density of 10^7 w/cm^2 can be obtained. Moderately powerful pulsed lasers may produce pulse

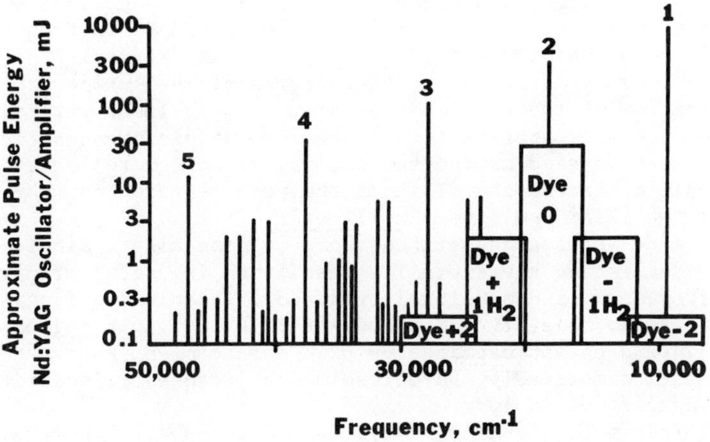

Figure 3. Partial representation of frequencies available from the Nd:YAG fundamental (1) harmonics (2–5), YAG-pumped dye laser fundamental (Dye O), and stimulated Raman H_2 and D_2 shifts thereof. The Dye regions are approximations to the average envelopes of four dyes: Coumarin 500, and Rhodamines 590, 610, and 640.

powers of 100 megawatts which, when focussed, yield power densities of the magnitude of 10^{14} w/cm^2 during the pulse. The electric field associated with such a power density engenders a host of enormously nonlinear optical effects such as breakage of optical components, dielectric breakdown of gases, and the emission of x-rays from metal surfaces. The most powerful pulsed laser yet constructed produces pulse powers of terawatts (1 TW = 10^{12} W) which leads to focussed power densities near 10^{20} w/cm^2. This approaches the power densities thought to occur in the interior of stars, and suggests the planned application of this particular laser in the initiation of fusion reactions.

The spectral resolution of the most monochromatic laser yet devised, expressed as frequency of the laser emission divided by the laser linewidth, is approximately 5 x 10^{13}. The laser in question is a special-purpose He-Ne laser with a nominal wavelength of 632.8 nm (15,308 cm^{-1} or 4.74 x 10^{14} Hz in frequency units) and a linewidth of 7 to 10 Hz. It is difficult to grasp the physical significance of this degree of resolution. One illustration is that, if the spectrum of this laser were displayed on chart paper such that the zero of electromagnetic energy were located at the sun and 15,308 cm^{-1} were located at the orbit of Earth, the width of the peak representing the output of the laser would be 3 millimeters.

The He-Ne laser described above is basically a single-frequency device and therefore its resolution is useful primarily as a frequency or dimensional standard. Tunable dye lasers have been devised, however, which operate in the visible region and have linewidths approaching 100 Hz. More common lasers, readily available commercially, have resolution (frequency/linewidth) in the range of 10^7 to 10^8.

In terms of directionality, an off-the-shelf laser may have a beam divergence of 0.5 milliradians. Such a laser, shone on the moon from the surface of Earth, would project a spot approximately 120 miles in diameter. Very special optics, e.g. collimation of the laser beam by a large optical telescope, can decrease the beam divergence by a factor of ten.

Characteristics of laser light which are less obvious to the unaided eye are its spatial and temporal coherence, time resolution, and polarization. It is common for the output of a laser to be linearly polarized with a polarization ratio of 1000:1 or better. Coherence refers to one's ability to predict the properties of a waveform (amplitude, phase, polarization, etc.) at all positions or times by observation of its properties at a single position or time. A high degree of coherence is inherent in laser light due to the coherent nature of the stimulated emission processes. The actual output of a laser may be more or less coherent, depending upon whether design features are incorporated to maximize this property. The coherence of laser light allows such applications as holography and other types of interferometry, flow visualization, and the creation of coherent superpositions of states (e.g. molecular ensembles vibrating with known phase) in matter.

The temporal resolution available from lasers varies from zero (c.w.) to timescales approaching the shortest of chemical interest. The shortest laser pulses yet produced are approximately 30 femtoseconds (1 fs = 10^{-15} s) in duration. This is a time interval so short that (a) the "thickness" of the pulse, its dimension along its propagation direction, is only 0.01 mm, (b) the pulse duration is short compared to the period of most molecular vibrations, and (c) the Heisenberg principle-limited energy uncertainty (linewidth) of the pulse is approximately 1000 cm^{-1}. The foregoing is a formidable achievement, but lasers are available commercially which, while not trivial to operate, will produce pulses of 10^{-11} s duration. A number of more-or-less idiot-proof lasers are available which generate 10^{-8} s pulses and at the same time yield a great selection of wavelengths (e.g. Q-switched Nd:YAG; see Figure 3 for depiction of wavelength output).

With regard to pulsed lasers, it should be noted that these devices are almost orthogonally divided into two types: the "giant-pulse", low duty cycle lasers (e.g. Nd:YAG, excimer, ruby) and the "quasi-c.w." high repetition rate, small-pulse lasers (e.g. modelocked ion lasers). The giant-pulse lasers typically produce in the neighborhood of one joule of energy in each pulse with peak powers of many megawatts and repetition rates of a few per second. The small-pulse lasers, like the c.w. lasers, may produce the same average power as the giant-pulse ones. However, they do so by repeating pulses of a few nanojoules energy at rates of many megahertz.

There are two practical consequences of interest to chemists in this dichotomy. First, by and large only the giant-pulse lasers effectively induce the nonlinear effects which are the simplest way to generate versatile wavelength selectivity (see Figure 3). Secondly, there is a fundamental chemical difference between giant-pulse illumination on one hand and small-pulse or c.w. illumination on the other. That is that the giant-pulse lasers, when shone on a chemical sample, typically furnish many more photons than there are molecules in the illuminated volume, in a time interval which is short compared to any conceivable transport process (diffusion, flow, etc.) in the sample. The small-pulse or c.w. lasers, on the other hand, typically furnish many fewer photons per pulse than there are molecules in the illuminated volume or (in the c.w. case) the flux of photons is such that molecular transport processes can compete and will furnish previously unilluminated molecules to the illuminated volume. In consequence, sequential single-photon (as well as nonlinear) events are highly probable with giant-pulse illumination and relatively improbable with small-pulse or c.w. illumination.

Laser Spectroscopy

Spectroscopy, the measurement of the interactions of electromagnetic energy with matter, bears upon chemical problems in three important ways. Firstly, spectroscopic measurements may be diagnostic, allowing one to infer the structure, bonding, and other physical and chemical properties of molecules. Secondly, spectroscopy may serve as a detector to determine the presence and abundance of chemical species having known spectroscopic properties. Finally, the interactions of light with matter may be employed as an effector of chemical change (8). Any of these spectroscopic functions may be accomplished with or without temporal resolution. Although an enormous amount of work in this area was accomplished in the pre-laser era, the unique combination of properties available from laser light has revolutionized the entire field. In this section, we shall consider the chemical applications of lasers as diagnostics and detectors. While an understanding of their role as effectors requires considerable spectroscopic knowledge, we shall consider this application in the following sections on laser dynamics and laser chemistry.

Fundamentally, the properties of laser light are concomitants of its coherence, which is in turn a consequence of the nature of stimulated emission. Most of these properties, especially brightness, monochromaticity, directionality, polarization, and coherence itself, are useful (for many applications, indispensible) in a spectroscopic light source. The spectroscopic potential of lasers was recognized even before they were invented. Actual applications remained very specialized until tunable lasers were devised.

An exception to the latter statement is Raman spectroscopy, which in its simplest form does not require a tunable light source but only an extremely bright monochromatic one. In 1928, Landsberg and Mandelstam (9) obtained the first "Raman" spectrum (as opposed to qualitative observation of inelastic light scattering (10)). The apparatus devised by Landsberg and Mandelstam, a mercury arc light source and a photographic spectrograph to detect the spectrum, still represented the state of the art in Raman spectroscopy in 1960 when the laser was invented.

The first spectrum of any sort recorded using a laser light source was a ruby laser-excited Raman spectrum reported by Porto and Wood in 1962 (11). This may also be regarded as the first reported chemical application of a laser. Aside from the light source, the apparatus used by Porto and Wood was essentially the same as that used by Landsberg and Mandelstam. Shortly thereafter, Porto and coworkers developed the double monochromator with cooled photoelectric detection for Raman spectroscopy (12,13), and first used the c.w. gas laser (He-Ne) as a Raman excitation source (14). These developments led to the fabled laser resurgence in Raman spectroscopy. Spontaneous Raman spectroscopy using c.w. laser excitation remains the most common chemical application of lasers, if not indeed the most common application of lasers in general. Despite extensive refinements, the apparatus used for this type of spectroscopy has not changed in any fundamental manner since the landmark innovations by Sergio Porto and his associates.

Raman spectroscopy is primarily useful as a diagnostic, inasmuch as the vibrational Raman spectrum is directly related to molecular structure and bonding. The major development since 1965 in spontaneous, c.w. Raman spectroscopy has been the observation and exploitation by chemists of the resonance Raman effect. This advance, pioneered in chemical applications by Long and Loehr (15a) and by Spiro and Strekas (15b), overcomes the inherently feeble nature of normal (nonresonant) Raman scattering and allows observation of Raman spectra of dilute chemical systems. Because the observation of the resonance effect requires selection of a laser wavelength at or near an electronic transition of the sample, developments in resonance Raman spectroscopy have closely paralleled the increasing availability of widely tunable and line-selectable lasers.

The first laser Raman spectra were inherently time-resolved (although no dynamical processes were actually studied) by virtue of the pulsed excitation source (ruby laser) and the simultaneous detection of all Raman frequencies by photographic spectroscopy. The advent of the scanning double monochromator, while a great advance for c.w. spectroscopy, spelled the temporary end of time resolution in Raman spectroscopy. The time-resolved techniques began to be revitalized in 1968 when Bridoux and Delhaye (16) adapted television detectors (analogous to, but faster, more convenient, and more sensitive than, photographic film) to Raman spectroscopy. The advent of the resonance Raman effect provided the sensitivity required to detect the Raman spectra of intrinsically dilute, short-lived chemical species. The development of time-resolved resonance Raman (TR^3) techniques (17) in our laboratories and by others (18) has led to the routine TR^3 observation of nanosecond-lived transients (19) and isolated observations of picosecond-timescale events by TR^3 (20-22). A specific example of a TR^3 study will be discussed in a later section.

The foregoing paragraphs refer to spontaneous Raman scattering, which is actually a two-photon process involving the incident and scattered photons. Spontaneous Raman is linear in the incident light intensity. Nonlinear or higher order Raman processes have also proliferated with laser development (23). Some of these, notably stimulated Raman shifts (SRS), were among the first nonlinear optical effects to be observed using lasers (6). Other effects involving multiwave mixing require two or more lasers, generally tunable, for their observation, and therefore are more recent techniques. The most prominent of these are coherent antiStokes Raman spectroscopy (CARS) and stimulated Raman gain/loss (SRG/SRL). These and other higher order Raman techniques are difficult and highly specialized, but they have two great advantages. They are coherent processes, and their output is in the form of a laser-like beam which can be spatially separated from potentially interfering light. The coherence means that these spectroscopies can be used to perform the optical analogues of coherent magnetic resonance (e.g., spin echo) experiments and also to study coherent dynamical phenomena (e.g., quantum beats). The spatial property means that CARS and SRG/SRL can be successful with strongly emitting samples such as flames, explosions, and luminescent species in solution. Additionally, the frequency resolution of each of these techniques is limited only by the linewidths of the lasers involved, making them attractive probes for high-resolution vibrational or rotational studies.

Spectroscopic methods based upon one-photon absorption or luminescence obviously existed long before lasers. However, these methods have been transformed by the resolution of laser light sources and their brightness. The latter leads to new schemes for the detection of absorption and to enhanced sensitivity. It is clear that tunable lasers had to be developed before absorption spectrometric methods using laser light could succeed, therefore the field is barely ten years old. In general application of these methods only makes sense in situations where high spectral resolution is required, because of the expense of the tunable lasers and the limited tuning range for a given dye, diode, etc. For this reason laser absorption/laser induced fluorescence (LA/LIF) methods have commonly been limited to isolated molecule conditions (molecular beams, matrices) and ordered systems at cryogenic temperatures (molecular crystals). There have already been cases, however, wherein the need for extreme sensitivity or spatial resolution dominated all other considerations and led to the application of LA/LIF to less well-defined systems. Examples are the detection of femtograms of potent carcinogens (aflatoxins) in vegtable oils (24) and the detection of the luminescence spectra of single, living cells in flow cytometry (25). It is certain that condensed-phase chemists will find more such applications for LA/LIF when new, more reasonably priced tunable laser light sources are developed.

A related technique of potential interest to solid-state inorganic chemists and others able to study their systems in solid or frozen media is laser-induced matrix luminescence (26). It is possible to obtain vibronically resolved luminescence spectra by laser excitation of inhomogeneously broadened absorptions of guest molecules in solid matrices (e.g., frozen solutions, glasses, catalyst supports, etc.) if the exchange of energy among the inhomogeneous sites of the guest molecule in the matrix is slow compared to the emission lifetime. The laser then excites only the molecules in a selected type of inhomogeneous site, and the emission comes only from that site. Information to be gained from this resolved emission includes (a) the structure of the guest moelcule and (b) the number and structures of the matrix sites occupied by the guest. In favorable cases this technique need not require cryogenic conditions, as long as the timescale criterion stated above is met.

One of the earliest spectroscopic applications of lasers was two-photon absorbance. It was possible (although difficult) to practice this technique before the advent of tunable lasers because the two photons involved could be from a fixed-frequency pulsed laser (e.g., ruby) and a high-intensity white light source such as a xenon flashlamp. The development of tunable lasers made the technique much simpler. One interesting aspect of two-photon absorption spectroscopy is that the absorption intensity is governed by quadrapole selection rules rather than dipole selection rules as is the case with single-photon absorption. In addition, in the gas phase, two-photon absorption can be performed with counterpropagating laser beams and thereby doppler-free absorption linewidths can be observed. From most inorganic chemists' point of view, the selection rule aspect of two-photon absorption should be most interesting. This rule means that one-photon and two-photon absorption spectra are complementary in the same sense as infrared and Raman spectra, because odd-parity ($g \rightarrow u$) transitions are allowed in the former and even-parity ($g \rightarrow g$) transitions are allowed in the latter. This characteristic has been of great value in elucidating the electronic structure of aromatic molecules and polyenes, and its potential applicability to the ligand field spectroscopy of transition metal complexes is obvious.

A technique which is not a laser method but which is most useful when combined with laser spectroscopy (LA/LIF) is that of supersonic molecular beams (27). If a molecule can be coaxed into the gas phase, it can be expanded through a supersonic nozzle at fairly high flux into a supersonic beam. The apparatus for this is fairly simple, in molecular beam terms. The result of the supersonic expansion is to cool the molecules rotationally to a few degrees Kelvin and vibrationally to a few tens of degrees, eliminating almost all thermal population of vibrational and rotational states and enormously simplifying the LA/LIF spectra that are observed. It is then possible, even for complex molecules, to make reliable vibronic assignments and infer structural parameters of the unperturbed molecule therefrom. Molecules as complex as metal phthalocyanines have been examined by this technique.

A number of other laser spectroscopic techniques are of interest but space does not permit their discussion. A few specialized methods of detecting laser absorption worthy of mention include: multiphoton ionization/mass spectrometry (28), which is extremely sensitive as well as mass selective for gas-phase systems; optically detected magnetic resonance (29); laser intracavity absorption, which can be extremely sensitive and is applicable to gases or solutions (30); thermal blooming, which is also applicable to very weak absorbances in gases or liquids (31); and photoacoustic detection, which is applicable to any phase of matter and especially to optically poor materials (32).

Laser Dynamics

From the point of view of the study of dynamics, the laser has three enormously important characteristics. Firstly, because of its potentially great time resolution, it can act as both the effector and the detector for dynamical processes on timescales as short as 10^{-14} s. Secondly, due to its spectral resolution and brightness, the laser can be used to prepare large amounts of a selected quantum state of a molecule so that the chemical reactivity or other dynamical properties of that state may be studied. Finally, because of its coherence as a light source the laser may be used to create in an ensemble of molecules a coherent superposition of states wherein the phase relationships of the molecular and electronic motions are specified. The dynamics of the dephasing of the molecular ensemble may subsequently be determined.

The first capability, simply that of a time-resolved light source, has been used to a considerable extent by inorganic photochemists. Although there are conspicuous exceptions it has been largely overlooked by inorganic kineticists interested in thermal reactions, despite the evident fact that a large number of thermal processes can be initiated and probed by laser pulses. This low level of active investigation of thermal reaction dynamics by inorganic chemists between the fastest relaxation method (ca. 10^{-8} s) and the fastest laser method (ca. 10^{-13} s) is difficult to understand in view of the successive lessons on the importance of "immeasurably fast" reactions provided by the development of flow methods and relaxation methods. On the other hand, the chemical reactivity and intrastate dynamics of excited states are being intensely studied by inorganic photochemists. Obvious problems in thermal chemistry which might be studied on nanosecond and shorter timescales include nondiffusional processes such as isomerization, intramolecular electron transfer, geminate recomination in solution, solvent motions, and possible spin restrictions on substitution reactions.

The second dynamical capability of lasers, the ability to excite a large fraction of ground-state molecules to a selected quantum state, has hardly been applied to inorganic systems. To be sure, saturation of excited states by laser pulses is commonplace, but excitation to a selected vibronic level is rare. There are several good reasons for this. The molecular system chosen for such a study must generally be under isolated-molecule conditions (gas phase or inert gas matrices) so that the states of interest are spectroscopically resolved. Furthermore, the spectroscopy of the system must be well enough understood to make the dynamics observed interpretable. These requirements tend to dictate study of extremely simple systems at this time;

few inorganic systems more complex than SF_6 have been studied. No doubt this will change soon as fundamental understanding extends to more complex molecules. For example, the supersonic beam techniques described earlier show promise in elucidating the vibronic states of relatively complex molecules, if they have adequate volatility.

The last mentioned dynamical capability of lasers is in its infancy. It is possible in principle to make optical measurements which are analogous to coherent NMR measurements, and thereby to observe homogeneous linewidths in inhomogeneously broadened systems, to measure optical or vibrational T_1 and T_2 relaxation times directly, and to observe quantum recurrences. In practice these experiments are very difficult and expensive, and have typically been applied to systems such as liquid benzene (33). On the encouraging side, it should be noted that these techniques are indeed applicable to condensed-phase systems and are extremely informative concerning fundamental condensed-phase dynamics.

Laser Chemistry

One of tbe central properties of lasers is the ability to furnish large numbers of photons at very specific energies. This ability has caused many investigators to hope that laser chemistry might be possible, that is that the energy from the laser might be deposited in molecules in very specific ways in order to initiate very selective, interesting, or remunerative chemistry.

The practical questions concerning laser chemistry may be tersely stated. (1) Will it work? If so, (2) is it interesting as opposed to being a trivial extension of known, non-laser photochemistry? (3) Is it a practical tool for real chemists as opposed to full-time laser technologists? And, (4) are the goods worth the price charged - is it economically worth the effort? To make a long story short, the answers are: (1) yes, in many but not all cases; (2) yes; (3) almost; and (4) yes, in some cases, but the product had better be valuable.

A number of different ways in which laser chemistry might work can be imagined.

Bond- or Mode-selective Laser Chemistry.

Suppose we wish to break a specific bond in a molecule or cause a molecule to rearrange in a specific way, and the desired transformation is not the one which will occur if the molecule is simply heated (i.e., it is not the weakest coordinate in the molecule). Can we, by selectively exciting with a laser the bond or motion in question, cause the desired transformation to occur in greater than thermal yield?

This problem is generally thought of in terms of vibrational excitation, and resolves itself into two questions. First, the ground-state vibrations of the molecule are a collection of vibrational modes, each of which are some linear combination of the elementary molecular motions (bond stretches, bends, torsions, etc.). Does there exist a vibrational mode which resembles the desired transformation, so that laser energy can be absorbed specifically into the desired molecular coordinate? Secondly, assuming that the answer to the previous question is affirmative, does energy absorbed into the selected motion remain there long enough to promote the desired transformation, or does the energy rapidly redistribute itself among the molecular motions in a thermal manner before specific chemistry can occur? Because the photon energies used to promote this chemistry are vibrational, and therefore multiphoton absorption is necessary to break bonds, a related question is the following: can a sufficient number of photons be absorbed in a sufficiently short time to beat the rate of intramolecular energy thermalization?

In many cases molecules have vibrational modes that are sufficiently local to satisfy the first criterion above. In many other cases of molecules having strongly coupled oscillators, they do not. For example, isolated C-H stretches are good candidates to be "local", while the stretch of a bond in an aromatic ring is not. The second criterion is more difficult. Intramolecular thermalization of vibrational energy can be extremely rapid, especially under collisional conditions, and generally becomes more rapid for higher vibrational excitations. As a result, there is probably no case observed to date wherein nonthermal bond dissociation occurs as a result of infrared excitation. In a number of cases, however, nonthermal isomerizations may occur as a result of single-state vibrational excitation (34, 35). All in all, the outlook for mode-specific laser chemistry is not bright at present. However, since many theorists have decided that it will not work, the time is probably ripe for an experimental breakthrough.

Molecule-specific Laser Chemistry. It may be desirable to initiate chemical transformations selectively in a subgroup of a collection of similar molecules. Photochemical isotope separation is an obvious case in point. The quantum states of chemically identical molecules will be slightly different depending upon which isotopes of the constituent elements are present. It is possible in principle and also in fact to use the resolution available in laser light to excite specifically molecules containing a chosen isotope of one element. It is also possible that these isotope-selectively excited molecules will undergo reactivity that the ground state molecules, containing the undesired isotope, will not. Subsequently, a chemical step can be carried out which will separate the isotope-enriched, reacted molecules from the remaining unreacted ones. The only requirements for such a scheme to succeed are for there to be sufficient isotope splitting to allow selective excitation, and that the excitation and subsequent reactivity be rapid compared to the intermolecular redistribution of the excitation energy. These requirements are relatively mild compared to those in (a) above. Indeed laser isotope separation has been shown to work in a number of cases (36,37), yielding enrichment of isotopes such as ^{2}D, ^{13}C, ^{15}N, etc. There is obviously much interest in separating much heavier isotopes such as ^{235}U, and much of isotope separation technique is probably subject to the "iceberg effect" due to classification. Nevertheless, it is clear that laser separation of certain light isotopes may be close to being economically competitive with current non-laser separation methods.

Nonspecific Laser Chemistry. If laser-initiated chemistry is nonspecific, why bother with the expensive and troublesome laser? The fact is that unique, or at least uniquely energetic, chemical reactants can be generated by laser irradiation even if no mode- or molecule-selectivity exists. For example, it has been shown (38) that excitation of HCl, HBr and H_2S into their dissociative continua with excimer laser pulses at 193 nm produces hydrogen atoms with average translational energies of 61, 45 and 55 kcal/mol, respectively. Such average energies are absolutely unattainable thermally because, for example, 61 kcal/mol thermal translational energy requires a temperature of approximately 30,000°K! The potential reactivity of these hot atoms must be quite rich, and is virtually uninvestigated. It may be expected that other photodissociation reactions will produce hot atoms of other elements, and that these species will find some utility in gas-phase synthesis.

Effect of the Radiation Field. The great power density which can be generated by lasers makes it appropriate in some circumstances to view the radiation as a classical electric field. The presence of such a field can obviously modify the energy of molecules possessing electric dipoles. As a consequence, chemical reactivity may be modified because the energies of the potential surface of the reaction may be changed. This effect remains the realm of the theorist at present. Power densities of gigawatts per square centimeter or more are calculated to be required to produce an observable effect on model reactions (39). Such high power densities, while easily attaintable, generally lead to other problems such as dielectric breakdown.

"Classical" Photochemistry with Lasers. Any photochemistry which can be done with incoherent light sources can also be done with a laser, assuming that an appropriate laser wavelength is available. Generally the spectral brightness of the laser will be many orders of magnitude greater than that of the incoherent source. Even in the case of very nonspecific photochemistry which can be initiated with broadband incoherent irradiation, it is usually the case that laser irradiation can result in faster production of photoproducts simply because the initiating photon flux is greater. This may be important enough in preparative photochemistry to justify the increased cost of the laser. The narrower the wavelength band which is effective in producing the desired photoproduct, the greater the advantage of the laser becomes. In inorganic systems, which may exhibit different photochemistry depending upon whether ligand-field, charge-transfer, or $\pi \rightarrow \pi^*$ chromophores of a given molecule are excited, the selectivity and spectral brightness of the laser may be important.

The relative ease with which lasers can produce high concentrations of excited states can be important in initiating multimolecular photochemistry. It is trivial to produce 0.1 M or greater photon "concentration" in a 1 μℓ volume over a 10 ns period of time. Subsequent multimolecular reactions of excited states or labile photofragments are limited principally by the unimolecular lifetimes involved.

The ability to supply more photons to initiate photochemistry with lasers compared to incoherent light sources increases the importance of, and the ability to observe, low-yield photochemistry. Aside from the intellectual challenge of understanding the subtleties of condensed-phase inorganic photochemistry, there is potential synthetic utility in this area. Many photofragments are extremely reactive, and if produced in concentrations low enough so that they do not annihilate one another, may function as chain carriers in catalytic processes. In terms of economically practical laser chemistry, this is a possibility that must be investigated because the cost of a laser photon is so high that quantum yields much greater than unity are highly desirable.

The Bottom Line: What do the Photons Cost? We will consider four representative cases: an ArF excimer laser (193 nm) producing 0.2 J/pulse at 10 pulses per second; a c.w. Ar^+ laser producing 5 W at 514.5 nm; a Nd:YAG laser (1064 nm) producing 0.5 J/pulse at 10 pulses per second; and a c.w. CO_2 laser producing 100 W at 10,600 nm. Each of these lasers presently cost approximately $40,000. It is assumed that the useful lifetime of each laser is 5 years. In terms of operation, it is assumed that the lasers can be run 12 hours per day, 300 days per year. Electrical costs are figured at the residential rate in Austin, Texas, $0.08 per kwh. Water expenses are not included, but it should be noted that the Ar^+ laser uses $1000 of water per year at Austin rates ($0.10 per 100 gal). Annual operating expenses exclusive of energy and water (plasma tubes, YAG rods, cavities, gases, etc.) are estimated to average $3300 annually for the ArF and Nd:YAG lasers, $15,000 for the Ar^+ laser, and $1700 for the CO_2 laser. The results of the calculations are given in Table I.

The "bottom line" cost figures are given in the last two columns of Table I. These are the approximate cost of Avogadro's

Table I.
Laser Operating Parameters and Expenses

Laser (λ, nm)	Operation	Photon Flux moles/yr*	Energy Output J/yr*	Annual Operating Expenses Hardware/Energy	Cost of Photons per mole	Cost of Photons per 100 kcal
ArF (193 nm)	0.2 J/pulse 10 pps	42	2.6×10^7	\$14,000 3,000	\$400	\$261
Ar^+ (514.5 nm)	5 W c.w.	280	6.5×10^7	\$26,000 9,000	\$125	\$215
Nd:YAG (1064 nm)	0.5 J/pulse 10 pps	590	6.5×10^7	\$14,000 2,000	\$ 27	\$ 98
CO_2 (10,600 nm)	100 W c.w.	1.2×10^5	1.3×10^9	\$12,000 3,000	\$0.13	\$ 5

* One year = 3600 operating hours (see text).

number of photons (cost per mole) and the approximate cost of the number of photons energetically equivalent to a 100 kcal/mol chemical bond. The figures appear overly optimistic, if that is the correct word, because for the "cost per mole" to transfer directly into chemical terms a reaction would have to be found which used exactly the wavelength of the laser and had a quantum yield of unity. Furthermore, the reduced "cost per 100 kcal bond" compared to "cost per mole" of the excimer laser is spurious unless a way to break more than one bond per photon were devised. Moreover, no account is taken of the energetic penalty assessed or hardware costs incurred when frequency shifts, harmonics, or tunable dye emission is generated. Clearly, the cost of photons mounts as versatility increases.

A number of points are clear. First, in all cases the major expense of laser photons is the hardware, not the energy (even at Austin prices). Secondly, the intrinsically greater efficiency of the lower-energy lasers, especially the economic attractiveness of the CO_2 laser, is evident. One can easily understand why laser chemistry schemes based upon multiphoton infrared absorption attract so much effort. Thirdly, on a per-unit-time basis the ion laser is more than twice as expensive to operate than even the rather exotic excimer laser. This is because of the inherent energetic inefficiency of the rare-gas plasma as a gain medium and because of the extrinsic, and hideous, expense of ion laser plasma tubes (and their poor reliability).

No attempt has been made to assess relative reliability in these figures. Undoubtedly, the Nd:YAG and CO_2 lasers are the most reliable of the group. When the criterion of flexibility and usability for a number of chemical purposes is included, the Nd:YAG system is the winner because it can be used (albeit at lower power) to produce light at the CO_2 wavelengths as well as the excimer wavelengths, while the CO_2 output cannot be up-converted with any acceptable efficiency. One may argue that the only circumstance that keeps ion lasers in existence is the fact that, in the wavelength range covered by the ion lasers, <u>some experiments simply cannot be done with pulsed lasers.</u>

To summarize the state of technology for the chemist wishing to practice laser chemistry: the laser devices exist with the capability one would like, but they are expensive. We may expect that cheaper pulsed laser systems based upon excimer, Nd:YAG, N_2, alexandrite, etc. may be in the offing in the near future. This has already begun to happen with a new generation of N_2-pumped dye lasers from two manufacturers. No such prospects presently exist for c.w. lasers in the visible and ultraviolet, but one may hope that the ion laser will be radically improved or supplanted soon. For chemical applications which can use infrared excitation, satisfactory devices presently exist and the price is right.

Case Study: Time-Resolved Resonance Raman Studies of the Excited States of Tris(Bipyridine)Ruthenium(II).

The following case study contains examples of several topics discussed in previous sections, including some aspects of laser technology, laser spectroscopy and laser chemistry. A variety of lasers and laser techniques are applied in a straightforward manner to the problem of ascertaining structural and dynamical information on an excited electronic state of wide chemical interest. This information is obtained rather simply, illustrating the potential of laser techniques in the resolution of problems in solution chemistry.

We have reported the first direct observation of the vibrational spectrum of an electronically excited state of a metal complex in solution (40). The excited state observed was the emissive and photochemically active metal-to-ligand charge transfer (MLCT) state of $Ru(bpy)_3^{2+}$, the vibrational spectrum of which was acquired by time-resolved resonance Raman (TR^3) spectroscopy. This study and others (19,41,42) demonstrates the enormous, virtually unique utility of TR^3 in structural elucidation of electronically excited states in solution.

The photochemical and photophysical properties of $Ru(bpy)_3^{2+}$ and related d^6 polypyridine complexes have been the subject of intense recent interest (43-49). This is due to their potential in photochemical energy conversion, their intrinsically significant excited state behavior, and their attractive chemical properties. The structures of the excited states of these complexes are clearly of great interest.

At least two possibilities for the structure of the MLCT state exist. It may be formulated as $Ru(III)(bpy)_2(bpy^{\cdot-})^{2+}$, which has maximum symmetry of C_2, or the heretofore commonly presumed $Ru(III)(bpy^{-1/3})_3^{2+}$, which may have D_3 symmetry. We shall refer to the former structure as the "localized" model of the excited state, and the latter as the "delocalized" model. The experimental details of this study are presented elsewhere (19).

The electronic absorption and emission spectra of the ground state and MLCT excited state of $Ru(bpy)_3^{2+}$ are shown in Figure 4. It can be seen that the ground state absorption at the wavelength of the Nd:YAG third harmonic, 354.5 nm, is considerable. Therefore, laser pulses at this wavelength are suitable for generating good yields of the MLCT excited state. It is also evident that the resonance condition for enhancement of Raman scattering in the MLCT excited state should be very favorable if 354.5 nm excitation is used, because this wavelength is virtually on top of the excited state absorption maximum at ca. 360 nm. Finally, the onset of the emission of the MLCT state occurs at long enough wavelengths so that Raman scattering produced by 354.5 nm excitation encounters no interference from the intrinsic luminescence of the complex.

The excited state TR^3 results are obtained by illuminating the sample at an appropriate wavelength with a laser pulse of 5-7

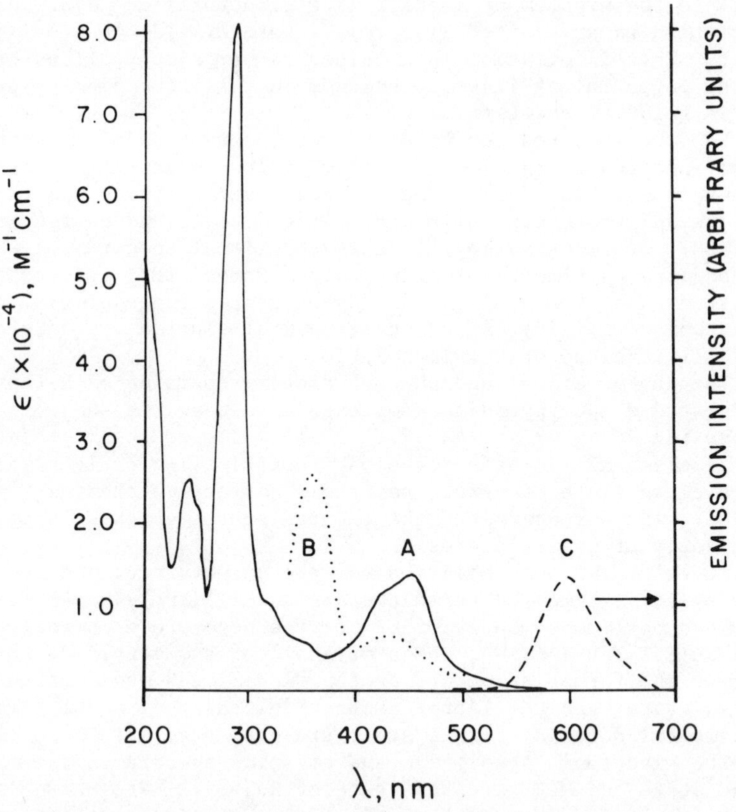

Figure 4. Absorption and emission spectra of $Ru(bpy)_3^{2+}$ and the excited state transient absorption spectrum of $Ru(bpy)_3^{2+}$. Key: A, absorption spectrum of $Ru(bpy)_3^{2+}$; B, absorption spectrum of $*Ru(bpy)_3^{2+}$; and C, emission spectrum. Conditions: room temperature, and aqueous solution. (Reproduced from Ref. 19c. Copyright 1981, American Chemical Society.)

ns duration. The laser pulse contains greater than tenfold excess of photons over molecules in the illuminated volume of solution (a cylinder approximately 1 mm in diameter and 1 mm in depth). The excited state lifetime, ca. 600 ns, is much longer than the laser pulse width. Therefore, the absorbed photons from the laser pulse produce essentially complete saturation of the excited state. Other photons from the same laser pulse are Raman scattered by the sample, both by the ground state and excited state complexes. The dominant spectrum which is observed is that of the excited state, because of the extreme saturation condition. Complementary ground state spectra were obtained using c.w. ultraviolet illumination (350.7 or 356.4 nm) from a krypton ion laser.

The c.w. and TR^3 spectra of $Ru(bpy)_3^{2+}$ in the frequency region of the C-C and C-N stretches (900-1700 cm^{-1}), obtained using near-UV excitation, are compared in Figure 5. The peak at 984 cm^{-1} in Figure 5 is the internal intensity reference, the symmetric stretching vibration of $SO_4^=$ (0.50 M Na_2SO_4 added). The c.w. and TR^3 spectra were obtained on the identical sample using a Shriver rotating cell (50) and 135° backscattering illumination geometry. All of the Raman peaks having sufficient intensity to allow measurement of depolarization ratios are polarized ($\rho \sim$ 0.3-0.5) in both the c.w. and TR^3 spectra.

The ground state spectrum in Figure 5 exhibits the typical features of the Raman spectrum of a bipyridine complex (40,51,52). Seven relatively intense peaks dominate the spectrum. These may be approximately described as the seven symmetric C-C and C-N stretches expected of bipyridine in any point group wherein the two pyridine rings are related by a symmetry element.

The TR^3 spectrum of the MLCT state comprising the predominant fraction of excited $Ru(bpy)_3^{2+}$ species averaged over the timescale of our laser excitation pulse (7 ns) is shown in Figure 5 (lower trace). This predominant MLCT species has been variously denoted "triplet charge-transfer" (^{3}CT) (45), "$d\pi$*" (43), or simply $*Ru(bpy)_3^{2+}$ (44). We adopt the latter nomenclature. It is the emissive and photochemically active state which has a lifetime of ca. 600 ns (room temperature, aqueous solution).

The TR^3 spectra of $*Ru(bpy)_3^{2+}$ exhibit increased complexity compared to the ground state RR spectra (Figure 5, upper trace). If the excited state were not saturated on the average over the duration of the laser pulse, this complexity might be thought to be due simply to the superposition of the bpy modes of $*Ru(bpy)_3^{2+}$ upon the seven bpy modes observed for ground state $Ru(bpy)_3^{2+}$. However, luminescence intensity measurements as a function of laser pulse energy show that the $*Ru(bpy)_3^{2+}$ state is greater than 90% saturated under the TR^3 illumination conditions in Figure 5. Therefore, no peaks observed in Figure 5 are due to the ground state. This can be seen by comparing the intensities of the 1609 cm^{-1} "ground state" peak between c.w. and TR^3 spectra in Figure 5. Despite 90% depletion of the ground state in the TR^3

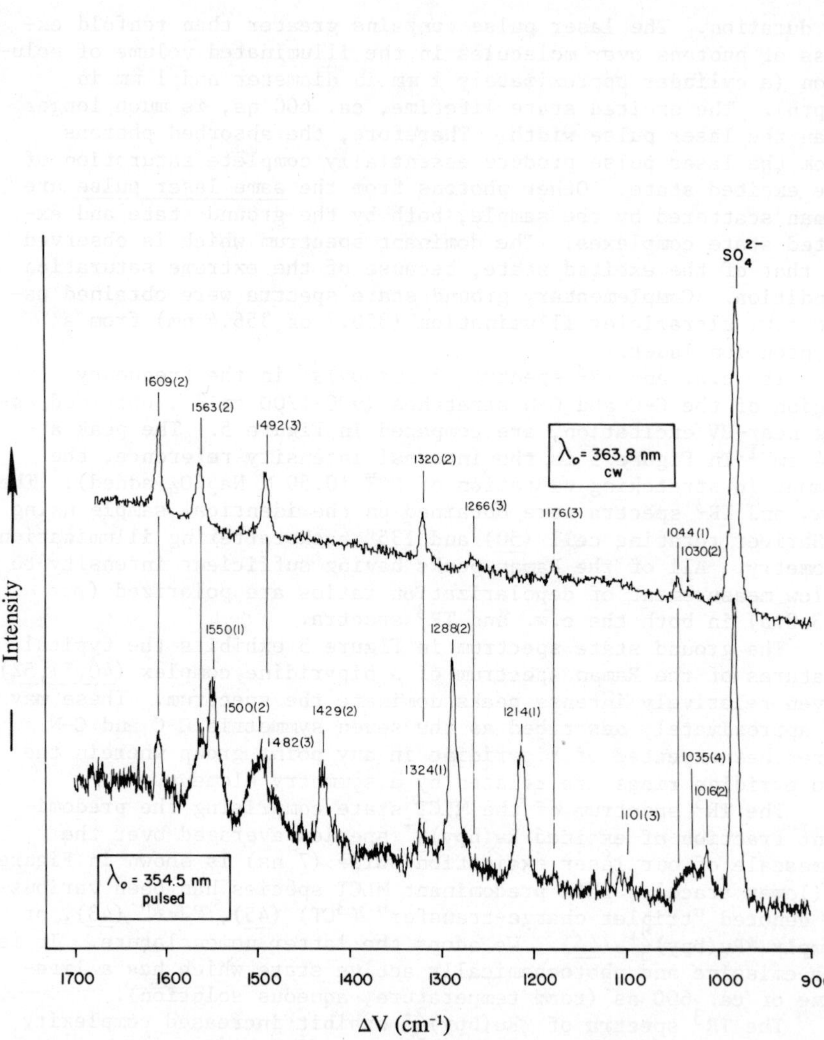

Figure 5. Comparison of the c.w. and TR^3 spectra of $Ru(bpy)_3^{2+}$ (top) and $^*Ru(bpy)_3^{2+}$ (bottom). (Reproduced from Ref. 19c. Copyright 1981, American Chemical Society.)

spectrum, the 1609 cm^{-1} peak retains approximately two thirds of its intensity. Apparently two sets of bpy frequencies are being observed in $*Ru(bpy)_3^{2+}$, one set at approximately the ground state frequencies and another set at considerably lower frequencies. This observation alone allows rejection of the possibility of D_3 symmetry for $*Ru(bpy)_3^{2+}$.

If the "localized" formulation of the structure of $*Ru(bpy)_3^{2+}$ as $Ru(III)(bpy)_2(bpy^{\bar{\cdot}})^{2+}$ is realistic, the resonance Raman spectrum of $*Ru(bpy)_3^{2+}$ can be predicted. A set of seven prominent symmetric modes should be observed at approximately the frequencies seen in $Ru(III)(bpy)_3^{3+}$, with approximately two thirds of the intensity of the ground state bpy modes. The intensity of the isolated 1609 cm^{-1} peak fits this prediction, as do the other "unshifted" peaks. A second set of seven prominent Raman modes at frequencies approximating those of $bpy^{\bar{\cdot}}$ should also be observed. Figure 6 shows that this prediction is correct. The seven $*Ru(bpy)_3^{2+}$ peaks which show substantial (average ~ 60 cm^{-1}) shifts from the ground state frequencies may be correlated one-for-one with peaks of $Li^+(bpy^{\bar{\cdot}})$ with an average deviation of 10 cm^{-1}. In addition, the weak 1370 cm^{-1} mode in $*Ru(bpy)_3^{2+}$ is correlated with a $bpy^{\bar{\cdot}}$ mode at 1351 cm^{-1}. It is somewhat uncertain whether the 1486 cm^{-1} $bpy^{\bar{\cdot}}$ mode should be correlated with the $*Ru(bpy)_3^{2+}$ mode at 1500 cm^{-1} or 1482 cm^{-1}. It appears clear that the proper formulation of $*Ru(bpy)_3^{2+}$ is $Ru(III)(bpy)_2(bpy^{\bar{\cdot}})^{2+}$. This conclusion requires reinterpretation of a large volume of photophysical data (43,45,51 and references therein).

These TR^3 results demonstrate that the localized model of $*Ru(bpy)_3^{2+}$ is valid on the timescales of electronic motions and molecular vibrations. It is virtually certain that delocalization (via, for example, intramolecular electron transfer or dynamic Jahn-Teller effects) occurs on some longer timescale. The present experiments are mute as to the timescale on which delocalization may occur. EPR results on $Ru(bpy)_3^{+}$ demonstrate localization of the $bpy^{\bar{\cdot}}$ electron density in this $Ru(II)(bpy)_2 \cdot (bpy^{\bar{\cdot}})^+$ species on the EPR timescale, but suggest that delocalization may occur on a timescale only slightly longer. It is possible that either time-resolved EPR or temperature dependent fluorescence depolarization experiments may establish the timescale of localization in $*Ru(bpy)_3^{2+}$.

In this system, we have been able to observe the first resonance Raman excitation profile of an electronically excited state. The availability of a number of stimulated Raman shifted (SRS) frequencies between 340 nm and 460 nm, and also tunable sum frequencies adding the Nd:YAG fundamental to the Rhodamine 6G dye laser fundamental, allows the acquisition of the excitation profile of $*Ru(bpy)_3^{2+}$. The SRS frequencies were generated by focussing the laser beam inside a cell containing ca. 100 psi of H_2 or D_2 gas. The laser and SRS frequencies are given in Table II. Saturation is confirmed by comparing spectra at a given wavelength using extremely different pulse energies (e.g., 369 nm,

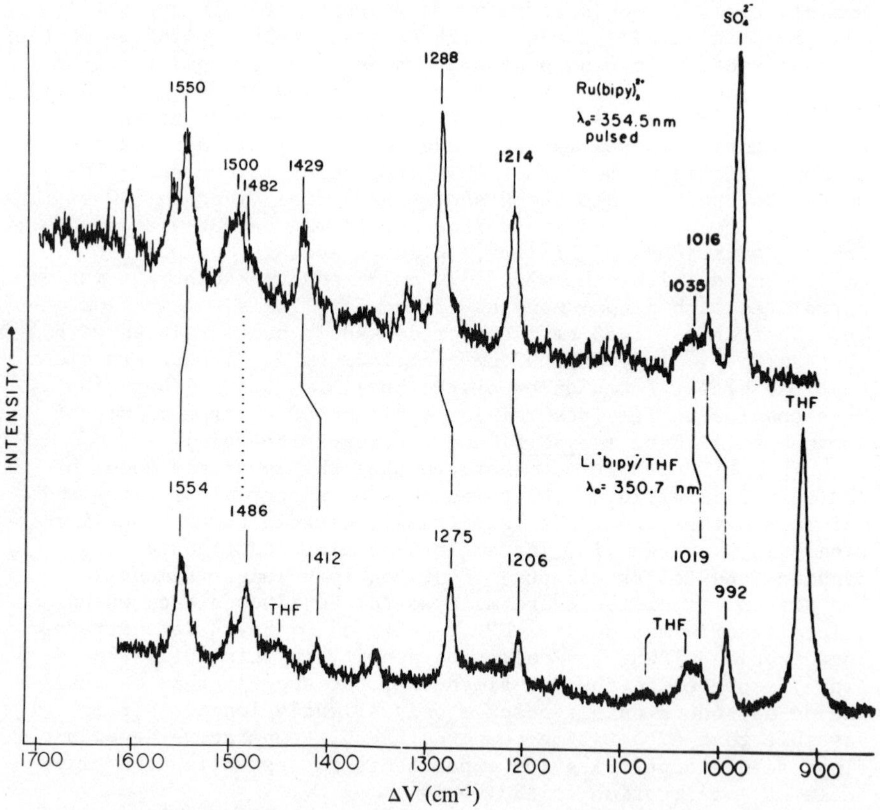

*Figure 6. Comparison of the TR[3] spectrum of *Ru(bpy)$_3^{2+}$ (top) to the c.w. resonance Raman spectrum of bipyridine radical anion (lithium reduction) (bottom). (Reproduced from Ref. 19c. Copyright 1981, American Chemical Society.)*

Table II.
Stimulated Raman Shifts and Nonlinear Sum Frequencies
Available from Nd:YAG/Dye Laser System

λ, nm	ν, cm^{-1}	SRS Shift	Per Pulse Energy
459	21776	$2\nu + 1D_2$	0.3 mJ
450	22209	$3\nu - 2D_2$	3. mJ
436	22946	$2\nu + 1H_2$	0.5 mJ
416	24030	$3\nu - 1H_2$	>4. mJ
396	25198	$3\nu - 1D_2$	>7. mJ
369	27100	$2\nu + 2H_2$	$\leq$0.03 mJ
355	28289	3ν	>7. mJ
342	29253	$4\nu - 2H_2$	2. mJ
		Tunable UV from Nonlinear Mixing	
360-375	ca.27000	---	1Nd:YAG + Dye(R6G)
364	27473	---	$\geq$5. mJ
369	27100	---	$\geq$5. mJ
374	26738	---	$\geq$1. mJ

Nd:YAG Frequencies			SRS Frequencies	
Harmonic	λ, nm	ν, cm^{-1}	Molecule,	Shift, cm^{-1}
2ν	532	18800	H_2	4155
3ν	355	28289	D_2	2940
4ν	266	37563		

Table II). Representative TR^3 spectra at three different wavelengths are shown in Figure 7. It is seen that all of the "$bpy^{\overline{\cdot}}$" Raman peaks show greatly increased intensity with short excitation wavelength, while the "neutral bpy" peaks increase in relative intensity at the long excitation wavelength. Three of the "$bpy^{\overline{\cdot}}$" modes in the excited state also show long-wavelength enhancement, and these are analogous to three modes of chemically reduced bipyridine radical anion which are also enhanced with long-wavelength excitation (53).

The resonance Raman excitation profiles are shown in Figures 8, 9 and 10. The radical modes clearly peak with the electronic transition of $*Ru(bpy)_3^{2+}$ at ca. 360 nm (Figure 8), demonstrating a $bpy^{\overline{\cdot}}$ $\pi^* \rightarrow \pi^*$ assignment for this absorption. The neutral bpy modes peak with the weaker transition at ca. 430 nm (Figure 9), suggesting that this absorbance has charge-transfer character, undoubtably analogous to the LMCT transition of $Ru(III)(bpy)_3^{3+}$ at 420 nm. In addition, the three aforementioned "$bpy^{\overline{\cdot}}$" modes peak near the same wavelength (Figure 10), suggesting that this absorption is actually the superposition of two $*Ru(bpy)_3^{2+}$ transitions, one being the Ru(III) LMCT mentioned above and the other a $bpy^{\overline{\cdot}}$ transition analogous to the long-wavelength absorptions of the chemically produced radical (53).

This study and others in this series (19, 40-42) as well as results from other laboratories clearly establish TR^3 as the most versatile and generally applicable probe presently available which provides direct information on the structures of electronically excited states in fluid media, the relevant conditions for essentially all photobiology as well as much photochemistry and photophysics. Such structural information is crucial (a) to test (as is the present case) conclusions on excited state parameters drawn from less structure-specific experimental probes or from theoretical approaches, (b) to establish excited state potential surfaces experimentally under chemically relevant conditions, and (c) in general to understand the mechanisms whereby light is converted into chemical energy. Our efforts to extend both the temporal range and the chemical range of applicability of TR^3 and related laser spectroscopies are continuing.

Acknowledgments The author is grateful for support of this work by National Science Foundation Grant CHE8109541 and Robert A. Welch Foundation Grant F-733.

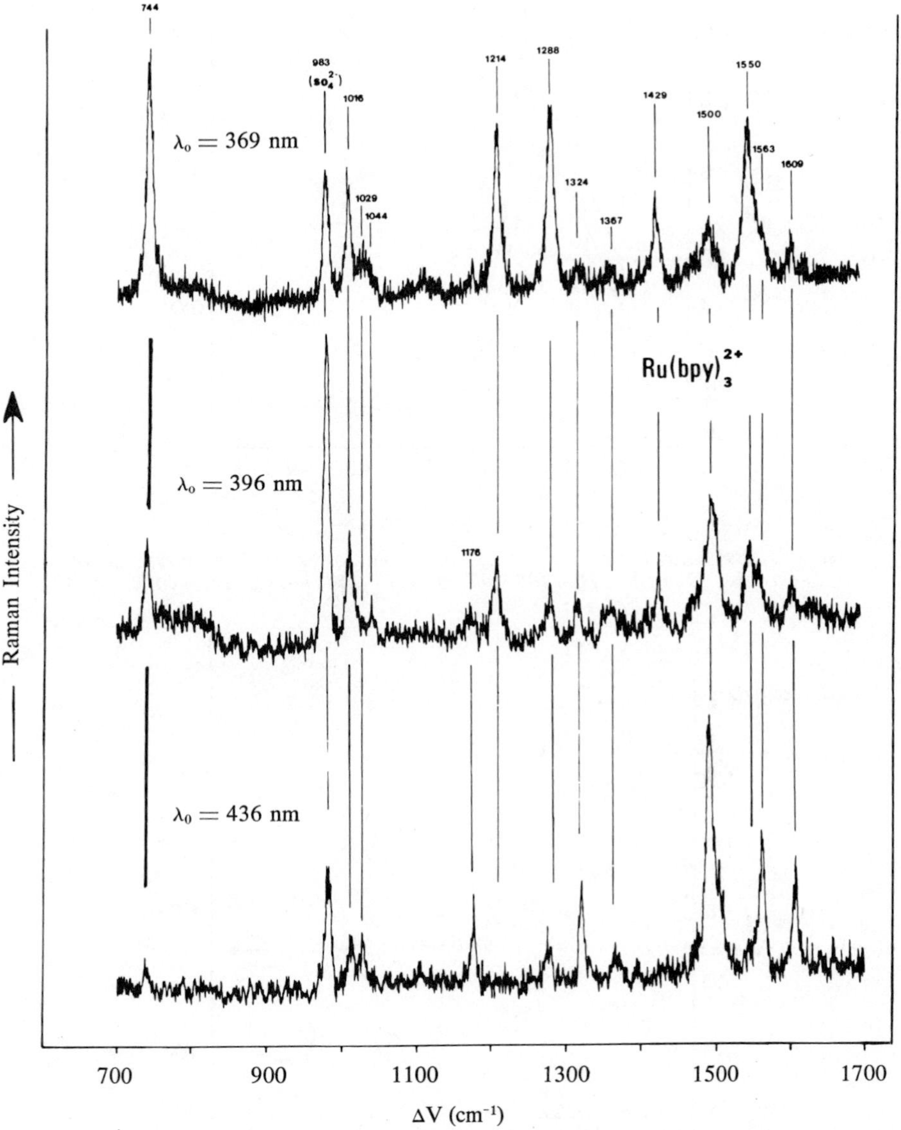

Figure 7. Comparison of TR^3 spectra of $^*Ru(bpy)_3^{2+}$ acquired by using laser excitation at 436 nm, 396 nm, and 369 nm.

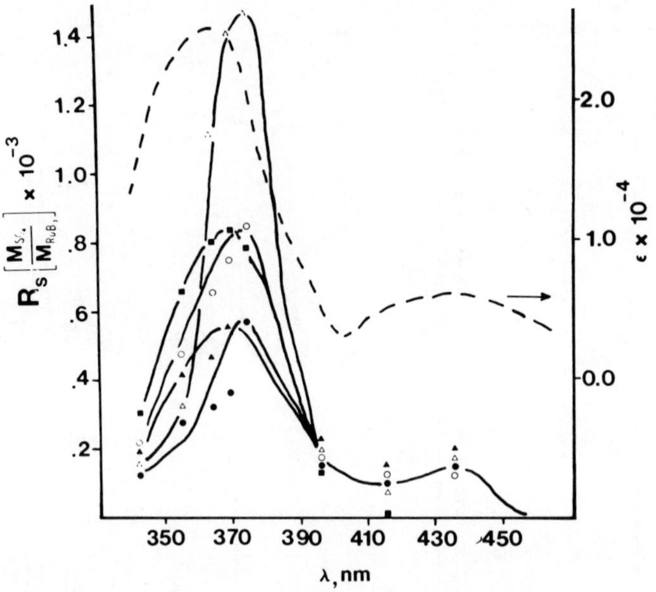

Figure 8. *RR excitation profiles of the "bpy⁻ modes of *Ru(bpy)$_3^{2+}$. Key:* △, *744 cm^{-1};* ○, *1214 cm^{-1};* ■, *1288 cm^{-1};* ●, *1429 cm^{-1}; and* ▲, *1550 cm^{-1}.*

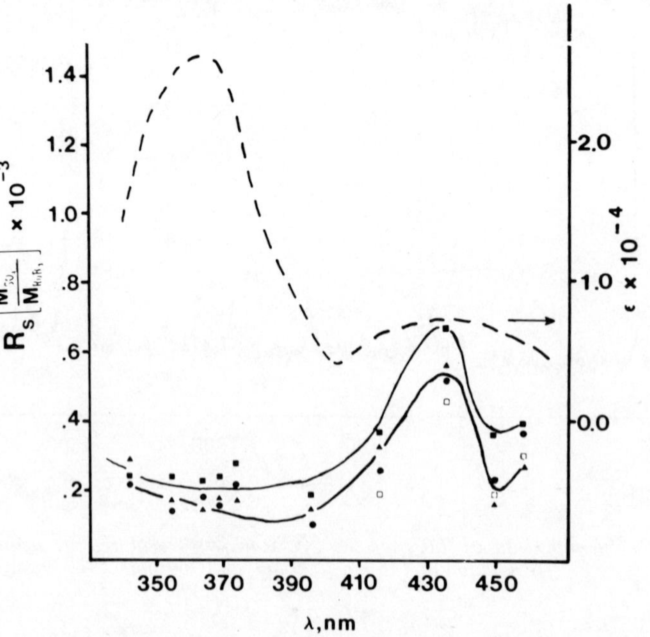

Figure 9. *RR excitation profiles of the "neutral bpy" modes of *Ru(bpy)$_3^{2+}$. Key:* □, *1176 cm^{-1};* ●, *1320 cm^{-1};* ■, *1566 cm^{-1}; and* ▲, *1609 cm^{-1}.*

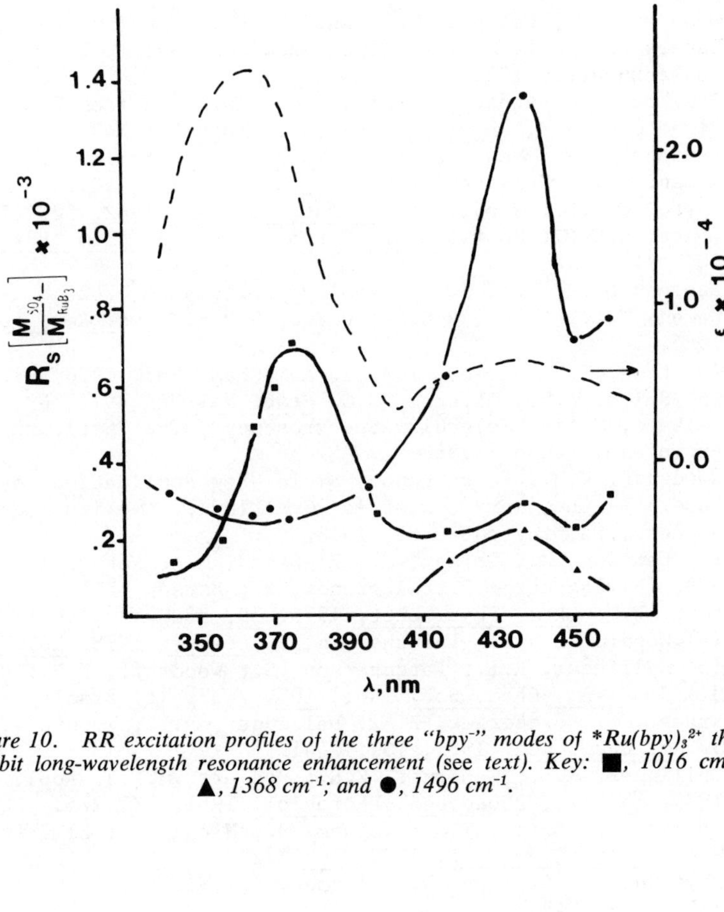

*Figure 10. RR excitation profiles of the three "bpy⁻" modes of *Ru(bpy)$_3^{2+}$ that exhibit long-wavelength resonance enhancement* (see *text). Key:* ■, *1016 cm^{-1};* ▲, *1368 cm^{-1}; and* ●, *1496 cm^{-1}.*

Literature Cited

1. Einstein, A. Phys. Z. 1917, 18, 121.
2. Schalow, A.L.; Townes, C.H. Phys. Rev. 1958, 112, 1940.
3. Maiman, T.H. Nature 1960, 187, 493.
4. Javan, A.; Bennett, W.R.; Herriot, D.R. Phys. Rev. Lett. 1961, 6, 106.
5. Sorokin, P.P.; Lankard, J.R. IBM J. Res. Dev. 1966, 10, 162.
6. Bloembergen, N. "Nonlinear Optics", Academic Press, New York, 1976.
7. O'Shea, D.C.; Callen, W.R.; Rhodes, W.T. "An Introduction to Lasers and Their Applications", Addison-Wesley, Reading, Massachusetts, 1977.
8. The "detector-effector- terminology is taken from Zare, R.N.; Bernstein, R.B. Physics Today 1980, November, 43.
9. Landsberg, G.; Mandelstam, L. Naturwiss.1928, 16, 557.
10. Raman, C.V.; Krishnan, K.S. Nature 1928, 121, 501.
11. Porto, S.P.S.; Wood, D.L. J. Opt. Soc. Am. 1962, 52, 251.
12. Leite, R.C.C.; Moore, R.S.; Porto, S.P.S. J. Chem. Phys. 1964, 40, 3741.
13. Landon, D.; Porto, S.P.S. Appl. Opt. 1965, 4, 762.
14. Damen, T.C.; Leite, R.C.C.; Porto, S.P.S. Phys. Rev. Lett. 1965, 14, 9.
15. (a) Long, T.V.; Loehr, T.M. J. Am. Chem. Soc. 1970, 92, 6384; (b) Spiro, T.G.; Strekas, T.C. Proc. NAS-USA 1972, 69, 2622.
16. Delhaye, M. in "Molecular Spectroscopy", The Institute of Petroleum, London, 1968.
17. Woodruff, W.H.; Farquharson, S. in "New Applications of Lasers in Chemistry", Hieftje, G.M. (ed.), American Chemical Society, Washington, D.C., 1978.
18. (a) Campion, A.; Terner, J.; El-Sayed, M.A. Nature 1977, 265, 659; (b) Pagsberg, P.; Wilbrandt, R.; Hansen, K.B.; Weissberg, K.V. Chem. Phys. Lett. 1976, 39, 538.
19. (a) Woodruff, W.H.; Farquharson, S. Science 1978, 201, 831; (b) Dallinger, R.F.; Farquharson, S.; Woodruff, W.H.; Rodgers, M.A.J. J. Am. Chem. Soc. 1981, 103, 7433; (c) Bradley, P.G.; Kress, N.; Hornberger, B.A.; Dallinger, R.F.; Woodruff, W.H. J. Am. Chem. Soc. 1981, 103, 7441.
20. Dallinger, R.F.; Woodruff, W.H.; Rodgers, M.A.J. Appl. Spect. 1979, 33, 522; Photochem. Photobiol. 1981, 33, 275.
21. Terner, J.; Spiro, T.G.; Nagumo, M.; Nicol, M.; El-Sayed, M.A. J. Am. Chem. Soc. 1980, 102, 3238.
22. Hayward, G.; Carlsen, W.; Siegman, A.; Stryer, L. Science 1981, 211, 942.
23. Eesly, G.L. "Coherent Raman Spectroscopy", Pergammon Press, New York, 1981.
24. Zare, R.N. in "New Applications of Lasers in Chemistry", Hieftje, G.M. (ed.), American Chemical Society, Washington, D.C., 1978.

25. Wade, C.G.; Rhyne, R.H. Jr.; Woodruff, W.H.; Bloch, D.P.; Bartholomew, J.C. J. Histochem. Cytochem. 1979, 27, 1049.
26. Wehry, E.L. in "New Applications of Lasers in Chemistry", Hieftje, G.M. (ed.), American Chemical Society, Washington, D.C., 1978.
27. Smalley, R.; Wharton, L.; Levy, D. Acc. Chem. Res. 1977, 10, 139.
28. Johnson, P.M. Acc. Chem. Res. 1980, 13, 20.
29. El-Sayed, M.A. Acc. Chem. Res. 1971, 4, 23.
30. Reddy, K.V.; Berry, M.J. Chem. Phys. Lett. 1979, 66, 223.
31. Albrecht, A.C. in "Advances in Laser Chemistry", Zewail, A.H. (ed.), Springer, New York, 1978.
32. Gerlach, R.; Amer, N.M. Appl. Phys. 1980, 23, 309.
33. Laubereau, A.; Kaiser, W. Rev. Mod. Phys. 1978, 50, 607.
34. Reddy, K.V.; Berry, Chem. Phys. Lett. 1979, 66, 223.
35. Hall, R.; Kaldor, A. J. Chem. Phys. 1979, 70, 4029.
36. Yeung, E.S.; Moore, C.B. Appl. Phys. Lett. 1972, 21, 109.
37. Letokhov, V.S.; Moore, C.B. in "Chemical and Biochemical Applications of Lasers", Moore, C.B. (ed.), Academic Press, New York, 1977.
38. Quick, C.R. Jr.; Weston, R.E. Jr.; Flynn, C.G. Jr. Chem. Phys. Lett. 1981, 83, 15.
39. George, T.F. (ed.) "Theoretical Aspects of Laser Radiation and its Interactions with Atomic and Molecular Systems", University of Rochester Press, Rochester, New York, 1978.
40. Dallinger, R.F.; Woodruff, W.H. J. Am. Chem. Soc. 1979, 101, 1355.
41. Dallinger, R.F.; Guanci, J.J. Jr.; Woodruff, W.H.; Rodgers, M.A.J. J. Am. Chem. Soc. 1979, 101, 1355.
42. Dallinger, R.F.; Miskowski, V.M.; Gray, H.B.; Woodruff, W.H. J. Am. Chem. Soc. 1981, 103, 1595.
43. (a) Hager, G.D.; Crosby, G.A. J. Am. Chem. Soc. 1975, 97, 7031; (b) Hager, C.G.; Watts, R.J.; Crosby, G.A. J. Am. Chem. Soc. 1975, 97, 7037; (c) Hipps, K.W.; Crosby, G.A. J. Am. Chem. Soc. 1975, 97, 7042; (d) Crosby, G.A.; Elfring, W.H. Jr. J. Phys. Chem. 1976, 80, 2206.
44. Meyer, T.J. Acc. Chem. Res. 1978, 11, 94.
45. Balzani, F.; Bolletta, F.; Gandolfi, M.T.; Maestri, M. Topics Curr. Chem. 1978, 75, 1.
46. DeArmond, M.K. Acc. Chem. Res. 1974, 7, 309.
47. Sutin, N.; Creutz, C. in "Advances in Chemistry Series, No. 138", American Chemical Society, Washington, D.C., 1978, pp. 1-27.
48. Hipps, K.W. Inorg. Chem. 1980, 19, 1390.
49. Felix, F.; Ferguson, J.; Gudel, H.U.; Lüdi, A. J. Am. Chem. Soc. 1980, 102, 4096.
50. Shriver, D.F.; Dunn, J.B.R. Appl. Spect. 1974, 28, 319.
51. Paskuch, B.J.; Lacky, D.E.; Crosby, G.A. J. Phys. Chem. 1980, 84, 2061.

Motten, A.G.; DeArmond, M.K.; Hauck, K.W. Chem. Phys. Lett. 1981, 79, 541.
Hornberger, B.A. Masters of Science Thesis, The University of Texas at Austin, 1980; Woodruff, W.H.; Hornberger, B.A. manuscript in preparation.

RECEIVED August 10, 1982

32

Determining the Geometries of Molecules and Ions in Excited States by Using Resonance Raman Spectroscopy

ROBIN J. H. CLARK

University College London, Christopher Ingold Laboratories,
20 Gordon Street, London WC1H 0AJ England

The magnitudes of geometric changes in molecules on electronic excitation can be determined from the excitation profiles of resonance-enhanced Raman bands, most accurately where both the resonant absorption band and the profiles show vibronic structure.

We are all well aware of the tremendous advances over the past decade in both the quality and number of X-ray structural studies on molecules and ions, aimed at the accurate determination of the ground-state geometries. However, our knowledge of excited-state geometries has not developed at a comparable rate, and few techniques can be brought to bear upon this problem. This situation is unfortunate, since it could be argued that chemists have a greater interest in obtaining a knowledge of excited-state than ground-state geometries of species of chemical interest, since such species will be the more reactive. However, it has recently been shown that resonance Raman spectroscopy can be applied with effect, not only to permit the indication of the nature of the geometric change on excitation to an excited state but also, where the resonant transition is vibronically structured, the magnitude of that change.

At resonance with an electric dipole allowed transition, the Stokes resonance Raman scattering, $I(\pi/2)$, associated with a single totally symmetric mode and its overtones is proportional to

$$I(\pi/2) \propto (\tilde{\nu}_0-\tilde{\nu}_{n0})^4 |[\mu_\rho]^0_{ge}|^4 \left\{ \sum_v \frac{\langle n|v\rangle^2 \langle v|0\rangle^2}{\varepsilon_v^2 + \Gamma_v^2} + \sum_{v \neq v'}\sum \frac{\langle n|v\rangle\langle v|0\rangle\langle n|v'\rangle\langle v'|0\rangle[\varepsilon_v\varepsilon_{v'} + \Gamma_v\Gamma_{v'}]}{[\varepsilon_v^2 + \Gamma_v^2][\varepsilon_{v'}^2 + \Gamma_{v'}^2]} \right\}$$

where ν_0 and ν_{n0} are the wavenumbers of the exciting line and n^{th}

0097-6156/83/0211-0509$06.00/0

harmonic, respectively, $[\mu_\rho]^0_{ge}$ is the pure electronic transition moment of polarization ρ from the ground (g) to the resonant excited (e) electronic states, $\langle n|$, $\langle v|$, $\langle v'|$ and $\langle 0|$ are vibrational wavefunctions, Γ_v and $\Gamma_{v'}$ are complex damping factors related to the lifetime of states v and v', and $\varepsilon_v = \tilde{\nu}_{ev,g0} - \tilde{\nu}_0$. The first summation gives rise to a series of peaks, each at a vibronic transition wavenumber; the Franck-Condon factors responsible for these peaks are the same as those which give rise to vibronic structure in the resonant absorption band. The second summation, however, depends on pairs of excited-state vibrational levels, and may give rise to either constructive or destructive interference effects. The net effect of the two terms is to produce an excitation profile for the a_{1g} Raman band which does not necessarily follow the contour of the resonant absorption band, but may differ from it in critically important ways.

The Franck-Condon integrals depend on (a) the ground-state and excited-state vibrational harmonic wavenumbers (ω_1) of the a_1 mode of the scattering species and (b) the displacement of the excited-state potential minimum with respect to that of the ground state along the a_1 coordinate. This displacement is related to the bond length change (δ) undergone by the scattering species on excitation to the resonant excited state. Thus experimentally obtained excitation profiles, if vibronically structured, can be simulated with the appropriate choice of δ and Γ values. The desired δ values are derived with greater accuracy by this method than they are directly from the vibronic structure in the resonant absorption band since, by the Raman method, only one mode is being monitored, whereas several modes may contribute to overlapping progressions in the resonant absorption band, a situation which tends to confuse the analysis.

Some results are given in Figure 1 for $[MnO_4]^-$, and related ones have been obtained for $[ReS_4]^-$, $[MoS_4]^{2-}$ and $[WS_4]^{2-}$ (1-7), for each of which the lowest electric-dipole-allowed transition ($^1T_2 \leftarrow {}^1A_1$) is clearly vibronically structured (a situation which is brought about by the fact that $\omega_1 > \Gamma$). The best fit δ value derived for each ion (0.05-0.09 Å) indicates that, in the $^1T_2$ state, the change undergone is rather larger than that typical of a one-electron reduction of the ion e.g. 0.03 Å for $[MnO_4]^-$ (8).

Unfortunately, the above analysis can never be widely applicable to the determination of excited-state geometries since so few molecules and ions exhibit vibronically structured absorption bands and excitation profiles, even at low temperatures. Moreover, some questions arise as to the possible breakdown of the Condon approximation. Other types of molecule for which similar analyses have been carried out include β-carotene, carotenoids (9) and certain carotenoproteins such as ovorubin (10). In these cases the excitation profiles of three skeletal a_1 bands are monitored, and estimates for the change in C-C and C=C bonds lengths (∿0.02 Å) have been made.

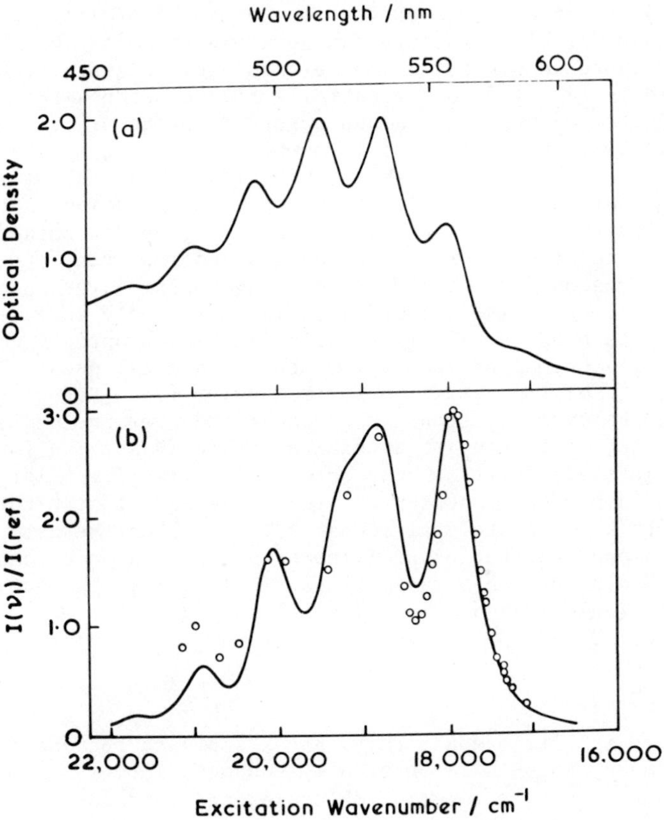

Figure 1. Absorption spectrum of $K[MnO_4]/K[ClO_4]$ mixed crystal at room temperature (a), and experimental (○) and best-fit calculated (—) excitation profile of the ν_1 (a_1) band of $[MnO_4]^-$ in $K[MnO_4]/K[ClO_4]$ mixed crystal for $\Gamma/hc = 300\ cm^{-1}$ and $\delta = 0.092$ Å (b).

Another type of complex for which it is obvious that large structural changes occur on excitation (probably 0.2-0.3 Å) along particular coordinates are the infinite-chain mixed-valence complexes of platinum and palladium (class II, or localized mixed-valence complexes) (11). These Pt^{II}/Pt^{IV}, Pd^{II}/Pt^{IV} or Pd^{II}/Pd^{IV} complexes contain .. M^{II}....X-M^{IV}-X.. chains (X = Cl, Br or I), with amines (NH_3, 1,2-diamino-ethane, -propane, -butane, -pentane, -cyclopentane, -cyclohexane, and 1,3-diamino-propane) in the equatorial positions. On excitation to the M^{III}/M^{III} intervalence state, considerable structure change along the M^{IV}-X bond length would be expected since, in the relaxed excited state, the halide ion would be expected to be centrally placed between the M ions. The very long Raman progressions observed in the ν_1, ν(X-M^{IV}-X) symmetric stretching mode (to 17 members in some cases, Figure 2), at resonance with the $M^{IV} \leftarrow M^{II}$ intervalence band, clearly indicate that the MX bond-length change on excitation is large. The magnitude of the change cannot, however, be determined exactly, since the intervalence band is unstructured, even at 4 K. The reason for the structure change being large and localized along a single coordinate is undoubtedly because the intervalence band is axially polarized; the appropriate coordinate is virtually uncoupled from any of the equatorial ones. This is the opposite of the situation for ruthenium red, $[Ru_3O_2(NH_3)_{14}]^{6+}$, where the valence electrons are delocalized over the whole complex ion; excitation to an excited state of such an ion results in small shifts in the potential minima for a large number of different coordinates, leading to small structural changes to all the skeletal bonds (12,13). Thus the nature of the Raman band excitation profiles of a mixed-valence complex may allow the distinction between class II (localized valence) and class III (delocalized valence) complexes, a matter of considerable contemporary interest.

Conclusion

There is little doubt that, as we approach the end of the 20th century, there will be greatly enhanced interest in determining the structures of molecules and ions in excited states and, in consequence, in techniques which bear upon this problem.

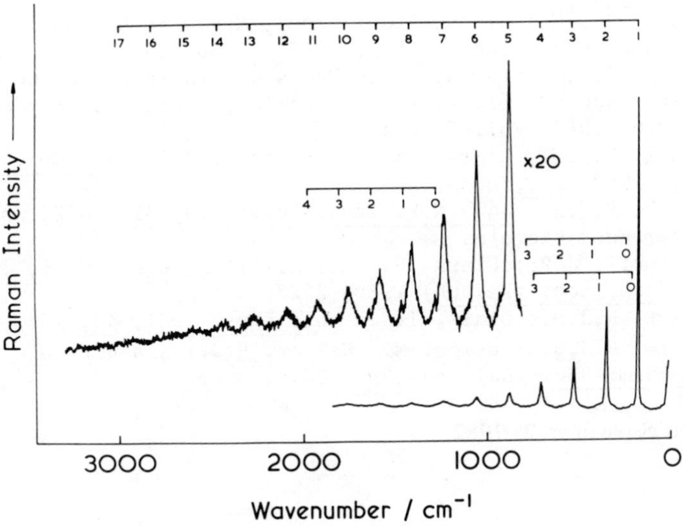

Figure 2. Resonance Raman spectrum of $[Pt(pn)_2][Pt(pn)_2Br_2][Cu_3Br_5]_2$ in a KBr disc at ~ 80 K (λ_0 = 568.2 nm). (Reproduced with permission from Ref.14. Copyright 1980, Royal Society of Chemistry.)

Literature Cited

1. Mingardi, M.; Siebrand, W.; Van Labeke, D.; Jacon, M. Chem. Phys. Lett. 1975, 31, 208.
2. Clark, R.J.H.; Cobbold, D.G.; Stewart, B. Chem. Phys. Lett. 1980, 69, 208.
3. Clark, R.J.H.; Stewart, B. J. Amer. Chem. Soc. 1981, 103, 6593.
4. Samoc, M.; Siebrand, W.; Williams, D.F.; Woolgar, E.G.; Zgierski, M.Z. J. Raman Spectrosc. 1981, 11, 369.
5. Clark, R.J.H.; Dines, T.J.; Wolf, M.L. J. Chem. Soc. Faraday Trans. 2 1982, 78, 679.
6. Clark, R.J.H.; Dines, T.J.; J. Chem. Soc. Faraday Trans. 2 1982, 78, 723.
7. Clark, R.J.H.; Dines, T.J.; Proud, G.P., to be published.
8. Palenik, G.J. Inorg. Chem. 1967, 6, 503, 507.
9. Carey, P.R.; Salares, V.R. "Advances in Infrared and Raman Spectroscopy", Clark, R.J.H.; Hester, R.E., Ed., Heyden: London, 1980: Vol. 7, p.1.
10. Clark, R.J.H.; D'Urso, N.R.; Zagalsky, P.F. J. Amer. Chem. Soc. 1980, 102, 6693.
11. Clark, R.J.H. Ann. N.Y. Acad. Sci. 1978, 313, 672, and references therein.
12. Campbell, J.R.; Clark, R.J.H.; Griffith, W.P.; Hall, J.P. J. Chem. Soc. (Dalton) 1980, 2228.
13. Clark, R.J.H.; Dines, T.J. Mol. Phys. 1981, 42, 193.
14. Clark, R.J.H.; Kurmoo, M.; Keller, H.J.; Traeger, U. J. Chem. Soc. (Dalton) 1980, 2498.

RECEIVED September 2, 1982

33

Laser Studies of Radiationless Decay Mechanisms in $Os^{2+/3+}$ Polypyridine Complexes

THOMAS L. NETZEL and MICHAEL A. BERGKAMP

Brookhaven National Laboratory, Chemistry Department, Upton, NY 11973

The lowest energy excited states in Os(II) polypyridine complexes are of a metal-to-ligand charge transfer (MLCT) type and live for 10-40 μs at 4.2 K.[1] The long wavelength absorptions in the visible region of the spectrum in Os(III) polypyridine complexes arise from ligand-to-metal charge transfer (LMCT) transitions and do not produce detectable luminescence. This suggests that these LMCT states are very short lived. We report here the results of picosecond absorption studies on the lifetimes the LMCT states in OsL_3^{3+} complexes [L = 2,2'-bipyridine(bpy) or 1,10-phenanthroline(phen)] as functions of temperature and isotopic substitution. The LMCT lifetimes at low temperature are then contrasted with the low temperature lifetimes of the MLCT states of OsL_3^{2+} complexes and both are examined from the perspective of a coarse-grained radiationless decay theory developed by Englman and Jortner.[2] In particular we seek to understand which molecular factors are responsible for the experimentally observed lifetimes.

Results

For picosecond kinetic measurements of change-in-absorbance (ΔA) spectra, the samples were degassed in 2 mm path length cells and held at constant temperature in a flowing-helium cryostat. The samples were excited at 527 nm with 6 ps laser pulses and the ΔA spectra of the excited states were measured with 8 ps white probe pulses. The laser system has been described elsewhere.[3] The observed ΔA signals are consistent with the known MLCT spectra of ground state OsL_3^{2+} complexes and support our assignment of the observed optical transients in OsL_3^{3+} complexes to the production of LMCT states.

Table I lists the excited state lifetimes for $Os(phen)_3^{3+}$ at three temperatures. The striking result is the lack of any significant change in the LMCT state's lifetime on going from 295 to 10 K. Table I also presents data on the effects of deuteration on the charge transfer state lifetimes of $Os(bpy)_3^{2+/3+}$ complexes. The lifetimes are lengthened, but only by factors of 2 and 2.5 respectively, for $Os(bpy)_3^{3+}$ and $Os(bpy)_3^{2+}$.

0097-6156/83/0211-0515$06.00/0

Table 1. Charge Transfer Excited State Lifetimes[a]

Compound	Temperature (K)	Lifetime (ps)
$Os(phen)_3^{3+}$,[b]	295	$\leq$ 9
	80	20 ± 3
	10	19 ± 2
$Os(bpy)_3^{3+}$,[b]	5	62 ± 4[c]
$Os(d_8\text{-}bpy)_3^{3+}$,[d]	10	120 ± 10
$Os(phen)_3^{2+}$	4	32 μs[e]
$Os(bpy)_3^{2+}$,[d]	10	1.05 ± 0.04 μs[f]
$Os(d_8\text{-}bpy)_3^{2+}$,[d]	10	2.5 ± 0.2 μs[f]

[a] The following abbreviations are used in this table: phen = 1,10-phenanthroline and bpy = 2,2'-bipyridine.
[b] In H_2O with 9 M H_2SO_4.
[c] The lifetime in H_2O with 9 M D_2SO_4 is 64 ± 4 ps.
[d] In D_2O with 9 M D_2SO_4.
[e] From ref. 1.
[f] Determined in this work by measurement of emission decay.

While the lifetimes of the OsL_3^{2+} and OsL_3^{3+} charge transfer states differ by a factor of 10^5–10^6, the energy gaps between their ground and charge transfer states are similar, 14.5 x 10^3 and 16.0 x 10^3 cm^{-1}, respectively. However, OsL_3^{2+} complexes have IR bands that OsL_3^{2+} complexes don't have. Our observation of these absorptions agrees with a recent report by Kober and Meyer of IR absorptions in $[Os(bpy)_3](PF_6)_3$ at 5090 and 4580 cm^{-1}.[4] These transitions arise because spin-orbit coupling interactions in the trigonal field split the t_{2g} levels of O_h symmetry.

Discussion

We expect the charge transfer excited states of $OsL_3^{2+/3+}$ complexes to have much the same equilibrium nuclear configurations as their corresponding ground states and thus to fall into the weak electron-vibration coupling limit of Englman and Jortner's radiationless decay theory.[2] The small temperature dependence of the lifetime of the LMCT state in OsL_3^{3+} complexes suggests that its nonradiative decay rate (k_{nr}) has little activated component even at room temperature. Thus low frequency ($\leq$ 500 cm^{-1}) molecular modes are not critical to its decay mechanism. Similarly the small increase in charge transfer state lifetime upon deuteration implies that high frequency C-H stretching modes are not critical to the radiationless decay process in $OsL_3^{2+/3+}$ complexes. The above considerations imply that mid-frequency

(1000-2000 cm^{-1}) skeletal stretching modes are likely to be the key energy accepting channels in these molecules. Consistent with this is the 1300 cm^{-1} vibrational progression observed in the emission spectrum of OsL_3^{2+} complexes.[1]

Equation 1 describes the radiationless decay rate for a single-frequency model with weak electron-vibration coupling in the low temperature limit as derived by Englman and Jortner.[2]

$$k_{nr} = \kappa_{el} \cdot \nu_M \cdot F_M \tag{1}$$

where κ_{el} is a dimensionless electronic coupling factor which should be near unity for the charge transfer states of $OsL_3^{2+/3+}$ complexes; ν_M is the frequency ($\sim 4 \times 10^{13}$ s^{-1}) of the critical vibration governing the decay process; and F_M is a Frank-Condon factor describing the vibrational overlap of the initial and final states.

$$F_M = \exp\,(-\gamma \cdot \Delta E/h\nu_M) \tag{2}$$

where

$$\gamma = \log_e\left[(2 \cdot \Delta E)/(d \cdot h\nu_M \cdot \Delta_M^2)\right] - 1 \tag{3}$$

In the above equations, ΔE is the electronic energy gap; d is the number of degenerate (or nearly degenerate) modes of frequency ν_M whose reduced displacement Δ_M is non-zero. (d = 13 and 16, respectively, for $Os(bpy)_3^{2+/3+}$ and $Os(phen)_3^{2+/3+}$.)

As a preliminary test of equations 1-3, we calculated the values of Δ_M required to explain the observed nonradiative deactivation rates for $Os(phen)_3^{2+}$ and $Os(bpy)_3^{2+}$ at 4.2 K.[1] The resulting values, respectively, 0.29 and 0.33 are in good agreement with the value of 0.29 calculated by Byrne <u>et al</u>.[5] for skeletal stretching modes in large aromatic molecules.

A more stringent test of the formalism would be to explain the much shorter lifetimes of the LMCT states of $Os(phen)_3^{3+}$ and $Os(bpy)_3^{3+}$. Since OsL_3^{3+} complexes do not emit in the visible, the energy of the 0-0 level of the lowest LMCT state is not known. If one takes ΔE from the onset of absorption in the visible and uses $h\nu_M$ = 1300 cm^{-1} and Δ_M and d as specified above, the calculated values of k_{nr} are seven orders of magnitude too small relative to the observed nonradiative decay rates. In fact the observation of IR transitions for $Os(bpy)_3^{3+}$ shows that ΔE for this complex can be no larger than 11×10^3 cm^{-1}. With this energy gap, the calculated value of k_{nr} can be brought into exact agreement with the observed decay rate if both the accepting mode frequency and displacement are increased slightly to 1600 cm^{-1} and 0.35, respectively.

Conclusions

The above agreement between experiment and theory suggests the following: 1) Englman and Jortner's theory of radiationless decay is useful for inorganic as well as organic systems, 2) mid-frequency (1300-1600 cm^{-1}) vibrations are the important energy accepting modes for radiationless decay of the charge transfer excited states of $OsL_3{}^{2+/3+}$ complexes, and 3) the 10^5-10^6 difference in lifetimes between the MLCT states of $OsL_3{}^{2+}$ complexes and the LMCT states of $OsL_3{}^{3+}$ complexes is largely due to the difference in their energy gaps.

Acknowledgment

This research was performed in collaboration with Drs. Philipp Gütlich and Norman Sutin at Brookhaven National Laboratory under contract with the U. S. Department of Energy and supported by its Office of Basic Energy Sciences. A full account of these and related studies will be forthcoming.

Literature Cited

1. Lacky, D. E.; Pankuch, B. J.; and Crosby, G. A. J. Phys. Chem. 1980, 84, 2068; ibid. 1980, 84, 2061.
2. Englman, R. and Jortner, J. Molec. Phys. 1970, 18, 145.
3. Creutz, C.; Chou, M.; Netzel, T. L.; Okumura, M.; and Sutin, N. J. Am. Chem. Soc. 1980, 102, 1309.
4. Kober, E. and Meyer, T. J. (private communication).
5. Byrne, J. P.; McCoy, E. F.; and Ross, I. G. Aust. J. Chem. 1965, 18, 1589.

RECEIVED October 7, 1982

ABSTRACTS OF POSTER PRESENTATIONS

SYNTHESIS AND SPECTROSCOPIC INDENTIFICATION OF TIN IV COORDINATION COMPOUNDS. PART IV.

R.H. ABU-SAMN. Chemical Engineering Department, Faculty of Engineering P.O. Box 800, King Saud University, Riyadh, Saudi Arabia.

By the interaction of tin tetrachloride and/or diorgano tin dihalide with disubstituted urea derivatives different tin IV coordination compounds were synthesized. This paper describes the syntheses of different Tin IV coordination compounds obtained by reaction of SnX_4 or R_2SnX_2 and disubstituted urea or thiourea, amines, amides or carboxylic acid amides derivatives. The prepared complexes are of the general formulae R_2SnX_2L where L is a mono or bidentate ligand. The coordination number four of tin is increased in the octahedral configuration complexes to six. The tin atom is bonded in these complexes either to the oxygen, sulpher or nitrogen atoms of the ligands. The properties of the prepared coordination compounds were studied and were identified spectroscopically by physical methods such as i.r., ^{1}H.n.m.r. and mass spectroscopy.

$SnX_4 + L \longrightarrow SnX_4L$ or $SnX_4.2L$

$R_2SnX_2 + 2L \longrightarrow R_2SnX_2.L_2$ or $R_2SnX_2.L$

SYNTHESIS AND CHARACTERIZATION OF THE COMPLEX $Mo_2Br_2(=CHSiMe_3)_2(PMe_3)_4$ (Mo≡Mo); A COMPOUND CONTAINING TERMINAL CARBENE LIGANDS ON ADJACENT METAL CENTERS.

K.J. Ahmed, M.H. Chisholm, and J.C. Huffman. Department of Chemistry and Molecular Structure Center, Indiana University, Bloomington, Indiana 47405.

This novel carbene compound has been prepared by the reaction of PMe_3 with 1,2-Mo_2-$Br_2(CH_2SiMe_3)_4$ (Mo≡Mo) according to the equation:

$$1,2\text{-}Mo_2Br_2(CH_2SiMe_3)_4 + 4PMe_3 \rightarrow Mo_2Br_2(=CHSiMe_3)_2(PMe_3)_4 + 2(Me)_4Si$$

GC/MS, multinuclear nmr (^{1}H, ^{2}H, ^{13}C, ^{31}P) and single crystal X-ray diffraction studies have been used to characterize the products. The new dimolybdenum compound has the skeletal geometry shown in 1 below. The molecule has rigorous C_2 symmetry and the following important structural parameters (distances in Å, angles in degrees): Mo-Mo = 2.276(1), Mo-C = 1.949(5), Mo-P = 2.534(3) (averaged), Mo-Br = 2.636(1), Mo-Mo-Br = 116.5(1), Mo-Mo-P = 101.5(1), 98.0(1); Mo-Mo-C = 109.1(2), Mo-C-Si = 129.8(3), Mo-C-H = 120(3). Interesting synthetic and structural analogies are seen here between mononuclear and dinuclear chemistry. The Me_3SiCH=Mo group is not of the "grossly distorted" type commonly observed in mononuclear early transition metal alkylidene complexes.

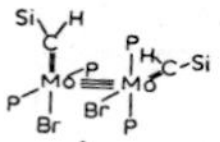

1

REDUCTION OF ORGANONITRILES TO AMINES BY SODIUM BOROHYDRIDE IN THE PRESENCE OF GROUP VI B HEXACARBONYLS.

S. Akhavan, L. K. Chesnut and B. N. Storhoff. Department of Chemistry, Ball State University, Muncie, IN 47306.

The reaction of $Ph_2PCH_2CH(R)CN$ (R = H,CH_3) and $M(CO)_6$ (M = Cr,Mo,W) with an excess of $NaBH_4$ in ethanol provides excellent yields of cis-coordinated $M(CO)_4[Ph_2PCH_2CH(R)CH_2NH_2]$ complexes. Phosphorus nmr spectra of the reaction mixtures indicate that the products arise from phosphine-containing precursors. These precursors have been isolated and identified as the phosphine-imidate complexes, $M(CO)_4[Ph_2PCH_2CH(R)C(OC_2H_5)NH]$. The

corresponding phosphine-amine complexes are obtained when these complexes are subsequently reduced with $NaBH_4$. This suggests that the nitrile groups have been activated toward reduction by nucleophilic attack by ethanol on the CN carbons.

FLUORINE ON THE BENCH TOP: OXIDATION AND FLUORINATION WITH THE FLUOROXYSULFATE ION, SO_4F^-. Evan H. Appelman, Chemistry Division, Argonne National Laboratory, 9700 S. Cass Ave., Argonne, IL 60439, and Richard C. Thompson, Department of Chemistry, University of Missouri, Columbia, MO 65211.

The fluoroxysulfate ion, SO_4F^-, is the only known ionic hypofluorite. As such, it forms stable salts with the heavy alkali metals. These salts are convenient bench-top reagents with which to take advantage of this ion's formidable oxidizing and fluorinating power, which approaches that of F_2 itself. The standard electrode potential of aqueous fluoroxysulfate is approximately 2.5 V., making it thermodynamically one of the most powerful of all known oxidants. Nevertheless, the ion shows in its reactions a remarkable selectivity that bears little or no relation to thermodynamic driving force. The reaction with Ag^+ is particularly rapid, and Ag^+ is therefore a good catalyst for a variety of oxidations by SO_4F^-. Fluoroxysulfate reactions often appear to proceed by transfer of a fluorine atom or a single electron, with concomitant formation of the $SO_4^-\cdot$ radical anion. However, ionic and oxygen-transfer mechanisms are also observed. Examples of oxidations and fluorinations of both inorganic and organic substrates will be presented. The alkali fluoroxysulfates are soluble in acetonitrile, and reactions with organic compounds can be carried out in this medium as well as heterogeneously in methylene chloride. Work supported in part by the Office of Basic Energy Sciences, Division of Chemical Sciences, U. S. Department of Energy.

DIOSMIUM(IV) COMPLEXES CONTAINING BOTH OXIDE AND CARBOXYLATE BRIDGES. J.E. Armstrong, W. R. Robinson and R. A. Walton. Department of Chemistry, Purdue University, West Lafayette, Indiana 47907.

Reactions of trans-$OsO_2X_2(PR_3)_2$(X = Cl or Br, PR_3 = triphenylphosphine or diethylphenylphosphine) with refluxing carboxylic acid/carboxylic anhydride mixtures produce a new class of osmium dimers of the type $Os_2(\mu\text{-}O)(\mu\text{-}O_2CR')_2X_4(PR_3)_2$ (R'=methyl, ethyl, iso-propyl). The x-ray crystal structure of the diethyl ether solvate of $Os_2(\mu\text{-}O)(\mu\text{-}O_2CCH_3)_2Cl_4(PPh_3)_2$ shows the bridging oxygen atom to be bonded in a bent fashion (140°) to the two osmium atoms. This is the only diosmium complex known to contain a single bent bridging oxide ligand. The long osmium-osmium distance of 3.440Å precludes any significant metal-metal bonding interaction. The electrochemical redox properties of these complexes have been explored using cyclic voltammetry. All complexes have a reversible one-electron reduction with the potentials depending on the nature of the halide, carboxylate and phosphine ligands. Reacting $Os_2(\mu\text{-}O)(\mu\text{-}O_2CCH_3)_2Cl_4(PPh_3)_2$ with Na metal in refluxing tetrahydrofuran produces the reduced species of stoichiometry $Na[Os_2(\mu\text{-}O)(\mu\text{-}O_2CCH_3)_2Cl_4(PPh_3)_2]$.

PHOTOCHEMICAL SYNTHESIS OF $FeH(C_6H_4PPhCH_2CH_2PPh_2)$(DPPE). ITS REACTIONS AND USE IN A HYDROGEN STORAGE SYSTEM. Hormoz Azizian and Robert H. Morris. Scarborough College and the Chemistry Department, Lash Miller Chemical Laboratories, University of Toronto, Toronto, Ontario M5S 1A1.

The irradiation of solutions of cis-$FeH_2(DPPE)_2$ (1), DPPE = 1,2-bis(diphenylphosphino)-ethane, with ultraviolet or visible light is an efficient method for preparing FeH $(C_6H_4PPhCH_2CH_2PPh_2)$(DPPE) (2). The 366 nm quantum yield for loss of hydrogen from 1 is 0.4±0.1 and the chemical yield of crystalline 2 is 90%. Hydrogen can be stored by its reaction with 2 at 50°, 1 atm, to give 1 quantitatively and can be released irreversibly at 25° by the photoconversion of 1 to 2 and this cycle can be repeated without degeneration of the storage system. Protonation of 2 with formic acid gives 1 and CO_2 and with fluoroboric acid under N_2 gives trans-$[FeH(N_2)(DPPE)_2]BF_4$. The decarbonylation of acetaldehyde by 2 provides a direct synthesis of Fe(CO)$(DPPE)_2$.

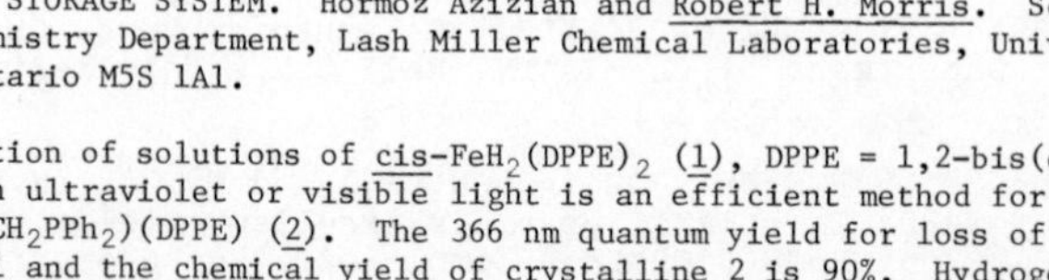

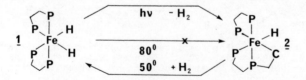

STRUCTURAL STUDIES OF THE HEXAAQUA IONS OF Ru(II) AND Ru(III) AND OF $[(NH_3)_5Ru$ PYRAZINE $Ru(NH_3)_5]^{n+}$, n = 4,5,6.
P. Bernhard[1], H.B. Bürgi[2], U. Fürholz[1], J. Hauser[2], H. Lehmann[1], and A. Ludi[1]. [1]Institut für Anorganische Chemie, [2]Laboratorium für Kristallographie, Universität Bern, CH-3000 Bern 9, Switzerland.

Crystals of $Ru(H_2O)_6(tos)_2$ (II) (tos = p-toluene sulfonate) and $Ru(H_2O)_6(tos)_3 \cdot 3H_2O$ (III) were grown from aqueous solutions. The crystal structure of triclinic II was refined to R = 3.3% for 2302 reflections, the structure of monoclinic III to R = 5.2% for 2989 reflections. The average metal to oxygen distance is 2.122(16) Å for $Ru^{II}-OH_2$ and 2.029(7) Å for $Ru^{III}-OH_2$. The effect of this difference in bond length on the rate of the $Ru(H_2O)_6^{3+/2+}$ self exchange is discussed.

In the course of a re-investigation of the structural properties of the Creutz-Taube complex the crystal structures of the following complexes were determined (pyr = pyrazine): $[(NH_3)_5Ru$ pyr $Ru(NH_3)_5][ZnCl_4]_2$, $[(NH_3)_5Ru$ pyr $Ru(NH_3)_5]Cl_5 \cdot 5H_2O$, $[(NH_3)_5Ru$ pyr $Ru(NH_3)_5](tos)_5 \cdot 5H_2O$, $[(NH_3)_5Ru$ pyr $Ru(NH_3)_5]Cl_6 \cdot 2H_2O$. Important average bond lengths (Å) are

Compound	R (%)	Ru-N (NH_3)	Ru-N (pyrazine)
Ru(II)-Ru(II), n = 4	3.6	2.13	2.02
Ru(II)-Ru(III), n = 5	5.1	2.13	2.01
Ru(III)-Ru(III), n = 6	1.9	2.09	2.12

BIDENTATE PHOSPHORUS COMPLEXES OF PLATINUM(II).
D.E.Berry, K.Beveridge, J.Browning, G.W.Bushnell, K.R.Dixon and A.Pidcock. University of Victoria, Victoria, British Columbia, V8W 2Y2.

N.M.R. spectroscopy (^{31}P and ^{195}Pt) and X-ray crystallography have been used to elucidate the structure of various platinum(II)-phosphorus containing species bearing more than one possible site for binding. One class of compounds has been derived from the β-diketone analogues $Ph_2P(X)CH_2P(Y)Ph_2$, (X,Y=O or S) where the products have all been found to be mononuclear. Other complexes have been produced by the hydrolysis of chlorodiethylphosphite coordinated to platinum(II). Dinuclear species containing P-O-P bridges have been characterised in several instances. The P-O-P link has yet to be shown as being capable of chelation.

KINETIC STUDIES OF METAL INCORPORATION BY APOAZURIN.
Judith A. Blaszak, David R. McMillin, Department of Chemistry, Purdue Universtiy, West Lafayette, IN 47907.

The continuing tale of the kinetics of metal uptake by apoazurin from *Pseudomonas aeruginosa* are reported. The method of apoprotein preparation has been investigated and is now well-defined kinetically. The study of Cu(II) uptake by this apoprotein under a variety of different conditions will be discussed, including pH, temperature, and buffer changes. A possible relationship between metal ion incorporation kinetics and metal selectivity in copper proteins will be discussed. Activation parameters will also be discussed and related to studies of other blue copper proteins, and their relationship to physiological conditions.

SYNTHESIS AND STRUCTURAL CHARACTERIZATION OF ORGANOMETALLIC COMPLEXES OF SAMARIUM.
I. Bloom, W.J. Evans, J.L. Atwood and W.E. Hunter, Department of Chemistry, University of Chicago, Chicago, Illinois 60637 and University of Alabama, University, Alabama 33486.

Investigations of organolanthanide chemistry frequently focus on the late lanthanides, Er, Yb and Lu, since it is easier to obtain stable complexes of these smaller metals by steric saturation of the metal coordination sphere. We have recently been developing the chemistry of a larger, more abundant member of the series, samarium, and report here two synthetic routes to new Sm-C σ bonded complexes. The reaction of $(CH_3C_5H_4)_2SmCl(THF)$ with $LiC{\equiv}CC(CH_3)_3$ in THF generates the alkynide bridged dimer $[(CH_3C_5H_4)_2SmC{\equiv}CC(CH_3)_3]_2$, I, which has been characterized by X-ray diffraction. The study provides the first X-ray data on Sm-C σ bonds and the second example of an unusual, unsymmetrical, electron deficient, alkynide bridge. The Sm-C distance is 2.55(1) Å and the external Sm-C≡C angles in the alkynide bridge are 151.1° and 112.3°. Samarium carbon σ bonded complexes can also be synthesized from the soluble, divalent complex $(C_5Me_5)_2Sm(THF)_2$, II (recently characterized by X-ray diffraction), and its analog, $(C_5Me_4Et)_2Sm(THF)_2$, III. II and III are obtained on a preparative, 1-2g, scale by deposition of samarium vapor into a hexane solution of C_5Me_5H or C_5Me_4EtH, respectively. Reaction of II and III with $Hg(C_6H_5)_2$ provides new samarium phenyl complexes, IV and V, and metallic mercury. Full characterization of I through V will be presented.

Reactions of Edge-Double-Bridged Complexes $[Ru_3(\mu\text{-H},\mu\text{-X})(CO)_{10}]$ [X = Cl, Br, I, OCNu (Nu = Me, Et, Ph, NMe_2)] with π-Acids; a Facile Route to Complexes of the Tri-ruthenium Fragment. N. M. Boag, H.D. Kaesz and C.E. Kampe. Department of Chemistry and Biochemistry, University of California, Los Angeles, CA 90024.

We have found that the title complexes react at ambient temperatures with a variety of π-acids with retention of the trimetallic fragment. With diphenylacetylene or with but-2-yne, complexes such as $[Ru_3\{\mu\text{-O=CNMe}_2,\mu\text{-C(Ph)=CHPh}\}(CO)_9]$ or $[Ru_3\{\mu\text{-O=CNMe}_2\}\{\eta^3\text{-CH}_2\text{-CH-CHMe}\}(CO)_9]$ are obtained in yields of 75% or 45%, respectively. These reactions contrast those of $[Ru_3(CO)_{12}]$ with similar substrates which require more elevated temperatures. At these temperatures the reactions are dominated by pathways in which the cluster dissociates to mononuclear fragments which recombine to give final products frequently leading to complex mixtures. The title complexes may thus serve as an entry into tri-ruthenium chemistry similar to the role of $[Os_3\{\mu\text{-H}\}_2(CO)_{10}]$ in triosmium cluster chemistry.

"Application of Novel Borides as Hydrodesulfurization and Solvent Refined Coal Liquefaction Catalysts", Alan Bonny*, Ruth W. Brewster, and C. Ann Welborn, Department of Chemistry, Furman University, Greenville, SC 29613, U.S.A.

We have been involved in investigating the application of new types of metal catalysts for fossil energy conversions. However, relatively few compounds survive reaction conditions typically required to liquefy coal and upgrade heavy oils; 1000-3000 psig, H_2, 350-450°C, with high heteroatom levels. There were predictions in the literature that metal borides might exhibit the high stabilities required. Further, recent advances with hydride-derived metal boride catalysts have indicated considerable potential in hydrogenation reactions, with turnovers similar to active noble metals reported in some cases.

This paper details the synthesis, characterization and reactions of a series of new class of metal borides prepared at elevated temperatures and under hydrogen pressure from polyborane anions and cobalt or nickel salts. These borides have unusually high boron levels (12-17%), exhibit high sulfidation resistances, and are highly active catalysts for the hydrodesulfurization of thiophene. The most active catalysts are prepared from the anions NaB_5H_8 and $NaB_{10}H_{13}$. Using these catalysts, 88-99% reaction of thiophene is observed. With less active catalyst systems or at lower temperatures, butene and tetrahydrothiophene are produced in addition to butane and H_2S. Under similar reaction conditions, the nickel-pentaborane derived catalyst promotes 93% hydrogenolysis of furan. The boride catalysts do not promote hydrogenation of non-heterocyclic aromatics such as toluene and ethylbenzene.

*address (from June, 1982) The Standard Oil Company of Ohio, Cleveland, Ohio

THE REACTION OF TRIETHYLSILANE WITH $MoCl_4(Me_2S)_2$: SYNTHESIS AND CHARACTERIZATION OF ISOMERIC MOLYBDENUM(III) CONFACIAL BIOCTAHEDRAL COMPLEXES WITH THE GENERAL FORMULA $Mo_2Cl_6(Me_2S)_3$. By P. Michael Boorman, Kelly J. Moynihan and Richard T. Oakley. Department of Chemistry, University of Calgary, Calgary, Alberta T2N 1N4 CANADA.

The metathetical reaction between Et_3SiH and $MoCl_4(Me_2S)_2$ in CH_2Cl_2 produces a mixture of novel dinuclear molybdenum(III) complexes with the general formula $Mo_2Cl_6(Me_2S)_3$. The first isomer, $(Me_2S)Cl_2Mo(\mu\text{-Cl})_3MoCl(SMe_2)_2$ (1), exists as a mixture of

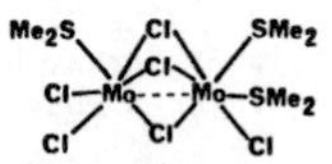

1

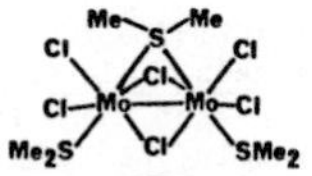

2

stereoisomers (a *d*, *l* pair and a *meso* derivative), all of which can be interconverted by a a rotation of one set of terminal ligands. One of the terminal thioether groups in 1 is labile and is easily displaced by chloride ion to give $[(Me_2S)Cl_2Mo(\mu\text{-}Cl)_3MoCl_2(SMe_2)]^-$. The X-ray crystal structure of $[Ph_4P][(Me_2S)Cl_2Mo(\mu\text{-}Cl)_3MoCl_2(SMe_2)]$ indicates that the anion possesses a slightly distorted confacial bioctahedral framework with a weak metal-metal interaction (d(Mo-Mo)=2.746(9)Å). Variable temperature ^1H nmr studies of 1 and its chloroanionic derivative have shown that the molybdenum atoms in these complexes are anti-ferromagnetically coupled. In contrast, the second isomer that can be isolated from the original reaction mixture, $(Me_2S)Cl_2Mo(\mu\text{-}Me_2S)(\mu\text{-}Cl)_2MoCl_2(SMe_2)$ (2), is diamagnetic. The X-ray crystal structure of 2 reveals that this compound possesses a confacial bioctahedral geometry that is distorted by a short molybdenum-molybdenum vector of length 2.462(2)Å. This suggests that a formal Mo-Mo triple bond exists between the metal centres in 2. Above +60°C, the isomers 1 (*d* & *l* and *meso*) and 2 can be easily interconverted. Kinetic and thermodynamic parameters related to this isomerization process will be presented.

STEREOSELECTIVITY IN LANTHANIDE COMPLEXES OF CHIRAL β-DIKETONES.

Harry G. Brittain, Department of Chemistry, Seton Hall University, South Orange, N.J. 07079

Optical activity has been observed to result in Tb(III) and Eu(III) β-diketone chelated derived from d-camphor upon adduct formation with a variety of substrates. While no optical activity (as detected via circularly polarized luminescence spectroscopy) can be measured in non-coordinating solvents, addition of substrates capable of binding to the lanthanide chelates often yielded strong optical activity even though none of the added substrates were themselves chiral. Results have been obtained where the substrates have been amines, sulfoxides, sulfones, phosphate esters, or phosphine oxides. In the non-coordinating solvents, the tris-chelates exist as a mixture of labile diastereomers (Λ-cis, Λ-trans, Δ-cis, and Δ-trans) and the lack of measurable optical activity in $CHCl_3$ or CCl_4 indicates that the Λ and Δ isomers are present in approximately equal amounts. Adduct formation increases the coordination number of the lanthanide ions and causes severe crowding of the bulky camphorato ligands. Exactly which diastereomer becomes enriched as a result of stereoselective rearrangements is a function of the steric nature of the substrate molecules.

THE PHOTOCHEMICAL DECOMPOSITION OF SULPHURIC ACID CATALYSED BY $[HPt(PEt_3)_3]^+$,

Duncan W. Bruce, Rodney F. Jones and David J. Cole-Hamilton,
Department of Inorganic, Physical and Industrial Chemistry, Donnan Laboratories, University of Liverpool, P.O. Box 147, Liverpool L69 3BX, England.

$Pt(PEt_3)_3$ dissolves in water producing $[HPt(PEt_3)_3]^+$. Near u.v. irradiation (λ = 356 nm) of such solutions acidified with sulphuric acid produces hydrogen and $[Pt(PEt_3)_3(SO_4H)]^+$. Utilising far u.v. radiation (λ = 254 nm), hydrogen is still produced but little or no $[Pt(PEt_3)_3(SO_4H)]^+$ is formed. A catalytic cycle is therefore proposed for the photochemical decomposition of sulphuric acid in which the products are hydrogen and peroxydisulphuric acid, according to:

$$2H_2SO_4 \xrightarrow{2h\nu} H_2 + H_2S_2O_8$$

The mechanism and potential of this system are discussed.

Hydrogen is also produced on visible irradiation of solutions of the iso-electronic rhodium species $[HRh(P^iPr_3)_3]$, although no catalytic cycle has yet been demonstrated. The potential of this rhodium system is also discussed.

C-H BOND ACTIVATION BY ACTINIDE ORGANOMETALLICS: FORMATION AND REACTIVITY OF ACTINACYCLOBUTANES. Joseph W. Bruno and Tobin J. Marks, Department of Chemistry, Northwestern University, Evanston, IL 60201 and Victor W. Day, Department of Chemistry, University of Nebraska, Lincoln, NE 68500

In an effort to further elucidate the chemistry of the actinide-carbon sigma bond, the thermal reactivities of several coordinatively unsaturated compounds $(C_5Me_5)_2ThR_2$ (A) have been investigated. For compounds with β-branched alkyls (R = CH_2SiMe_3, CH_2CMe_3), novel intramolecular activation of a γ-C-H bond yields the highly reactive metallacycles, B and C. These new compounds have been characterized by standard tech-

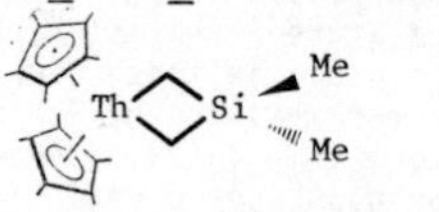

B

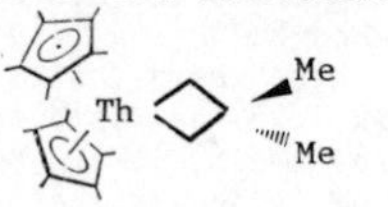

C

niques and in one case (B) by single crystal X-ray diffraction. Mechanistic information on their formation follows from deuterium labeling, kinetic analysis, and activation energy analysis. Furthermore, C is observed to induce catalytic and stoichiometric processes such as alkene polymerization and C-H activation (hydrocarbon metathesis). In contrast to the above, compounds containing β-C-H bonds exhibit an equilibrium involving β-H elimination and alkene reinsertion. Thus, attempted synthesis of A with R = iso-propyl yields only the n-propyl analog, presumably via this sequence. The insertion is favored to the extent that there is no apparent pathway to permanent alkene elimination products.

NEW IRON(III) SPIN CROSSOVER SYSTEMS WITH HEXADENTATE LIGANDS. R. J. Butcher, M. Pouian, R. J. Aviles, Ekk Sinn and T. Thanyasiri, Departments of Chemistry, Howard University, Washington D.C. 20059, and the University of Virginia, Charlottesville, Virginia, 22901.

Complexes of type $(FeL)^+X^-$, where L is shown, and X = PF_6^-, BF_4^-, NO_3^-, ClO_4^-, Cl^- and Br^-, have been synthesised and investigated magnetically and crystallographically to determine the effect of changing the alkyl chain lengths(C,D) and ligand substituents(A,B), as well as the effect of changing anions and occluded solvents in the lattice. Surpris-

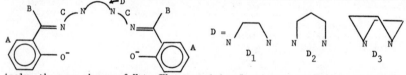

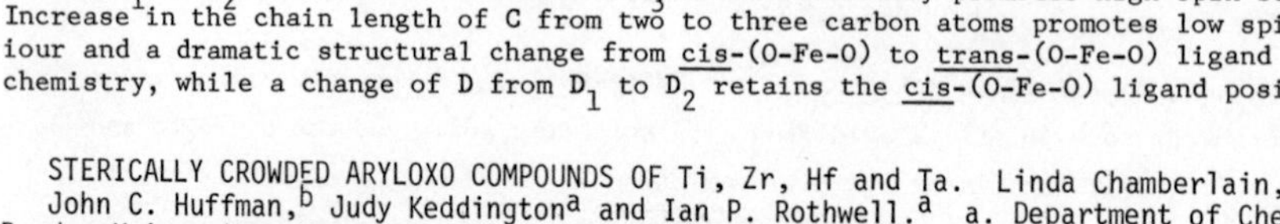

ingly, the mere change of H to CH_3 at position B can cause a complete change in spin state with no structural differences. Similar dramatic spin changes result from changing D from D_1 or D_2 to the more constrained D_3 which invariably promotes high spin behaviour. Increase in the chain length of C from two to three carbon atoms promotes low spin behaviour and a dramatic structural change from cis-(O-Fe-O) to trans-(O-Fe-O) ligand stereochemistry, while a change of D from D_1 to D_2 retains the cis-(O-Fe-O) ligand positioning.

STERICALLY CROWDED ARYLOXO COMPOUNDS OF Ti, Zr, Hf and Ta. Linda Chamberlain,[a] John C. Huffman,[b] Judy Keddington[a] and Ian P. Rothwell.[a] a. Department of Chemistry, Purdue University, West Lafayette, IN 47907. b. Molecular Structure Center, Indiana University, Bloomington, IN 47405.

The coordination chemistry of the 2,6-di-tert-butylphenoxo (OAr´) ligand with the metals Ti, Zr, Hf and Ta has been explored. With MCl_4 (M=Zr, Hf), LiOAr´ reacts to give the monomeric compound $MCl(OAr')_3$ which undergoes further substitution with $LiCH_3$ to give $M(CH_3)(OAr')_3$, but is substitutionally inert to other, bulkier alkyls such as $LiCH_2SiMe_3$. With $TiCl_4$, only the disubstituted product $[TiCl_2(OAr')_2]_n$ can be obtained, but with $[TiCl_3(NMe_3)_2]$ the blue, monomeric, three coordinate $[Ti(OAr')_3]$ is given. A crystal structure determination on $Hf\,Cl(OAr')_3$ shows a very congested metal center with considerable distortions of the aryloxo ligand taking place. $Ta\,Cl_5$ reacts with an excess of LiOAr to give $Ta\,Cl_3(OAr\,)_2$. Treatment of this with $LiCH_3$ or LiPh leads to $Ta(OAr')_2(CH_3)_3$ and $Ta(OAr')_2(Ph)_3$, but with the alkyls $LiCH_2XMe_3$ (X=C or Si) rapid evolution of alkane occurs with the formation of an alkylidene compound.

CHEMISTRY OF OCTAHEDRAL RHENIUM CLUSTERS, S. Chen, B. J. Morris and W. R. Robinson, Department of Chemistry, Purdue University, West Lafayette, Indiana 47907.

The reaction of rhenium metal or ReS_2 in Na_2CO_3 with gaseous H_2S at 900°C yields transparent red $Na_4[Re_6S_8]S_2(S_2)$, a Re(III) compound containing Re_6 octahedra in a cube of sulfide ions. Additional sulfide as well as disulfide ions bridge octahedra. It is interesting to note that this compound possesses structural features similar to those of the isoelectronic Mo(II) octahedral cluster compounds such as $[Mo_6Cl_8]Cl_4$, in which halides bridge the clusters along their 4-fold axes, and the various Chevrel phases, $M_xMo_6S_8$, in which intercluster bridging is accomplished by the positioning of clusters so that cluster sulfides occupy the exo positions of neighboring clusters.

$Na_4[Re_6S_8]S_2(S_2)$ reacts with acid to give semiconducting black $Na_{1.5}[Re_6S_8]S_2(S_2)$ with only slight changes in cell dimensions. The structure of the black compound has been refined sufficiently to show that the Re_6S_8 unit, sulfide, and disulfide bridges remain intact. Sodium sites are partially occupied. The oxidation of the 24 electron Re_6 cluster is accompanied by an expansion of the Re_6 cluster and shifts in the far IR spectra which are consistent with weakening of the Re-Re bonds. The black compound can be reduced by butyl lithium and although the exact effect upon the structure has not been determined, far IR studies suggest that the Re_6S_8 unit is reduced back to a 24 electron cluster.

This research was supported by NSF Grant No. 80-09143 and by NSF Grant No. DMR 80-20249 through the Materials Research Laboratory at Purdue University.

SYNTHESIS AND STRUCTURES OF NEW METAL ARYL COMPLEXES. M.J. Chetcuti, M.H. Chisholm, and J.C. Huffman. Department of Chemistry and Molecular Structure Center, Indiana University, Bloomington, Indiana 47405.

A series of compounds of formula $1,2\text{-}M_2Ar_2(NMe_2)_4$ (M≡M) (M = Mo, W; R = Ph, $C_6H_4Me\text{-}2$, $C_6H_4Me\text{-}4$) has been synthesized. X-ray diffraction studies on the ortho- and para-tolyl-Mo compounds, together with variable temperature ^1H nmr studies, reveal substantial nitrogen-to-metal π-bonding, as evidenced by short M-N bond lengths, restricted rotation about M-N bonds, and unusual conformational effects. The high thermal stability and relative inertness of these $d^3\text{-}d^3$ Mo^{III} dimers towards donor ligands may also be attributed to strong N(pπ) → M(dπ) bonding, tying up otherwise free orbitals.

SYNTHESIS, CHARACTERIZATION AND EQUILIBRIUM OF THE COMPLEXES $M_2(OR)_6L_2$ (M≡M) [M = Mo, W; R = i-Pr, CH_2-t-Bu(Ne); L = PMe_3, PEt_3 or L_2 = $Me_2PC_2H_4PMe_2$(DMPE), $Me_2PC_2H_4NMe_2$(TMAPE)]. M.J. Chetcuti, M.H. Chisholm, J.C. Huffman and J.L. Stewart. Department of Chemistry and Molecular Structure Center, Indiana University, Bloomington, Indiana 47405.

The titled compounds have been synthesized by reactions involving the alkoxide in hexane with the appropriate phosphine. Crystallization at low temperature, which drives equilibrium (1) to the right, yields the adducts. The complexes have been characterized

$$M_2(OR)_6 + L_2 \underset{\text{heat}}{\overset{\text{cool}}{\rightleftharpoons}} M_2(OR)_6L_2 \qquad (1)$$

by ^1H and ^{31}P nmr, together with analytical and infrared data. X-ray diffraction studies have been carried out for the complexes $Mo_2(ONe)_6$(TMAPE) and $W_2(ONe)_6(PMe_3)_2$. The equilibria have been studied by variable temperature ^{31}P nmr and values for K, ΔG, ΔH and ΔS, for various complexes, have been determined. The equilibrium constant at a particular temperature is dependent on the steric bulk of the ligand and the alkoxide group, and a pronounced "chelate-effect" is evidenced. These results also show that Mo complexes are appreciably more labile towards phosphine loss than corresponding W compounds and furthermore that tungsten binds tertiary phosphines more strongly than amines.

SYNTHESIS AND CHARACTERIZATION OF $M_2(OR)_6L_2$ (M≡M) COMPOUNDS. M.H. Chisholm, J.C. Huffman, J. Leonelli and A.L. Ratermann. Department of Chemistry and Molecular Structure Center, Indiana University, Bloomington, Indiana 47405.

There are two synthetic routes to $M_2(OR)_6L_2$ compounds, M = Mo or W. $Mo_2(OR)_6$ (R =

i-Pr, Ne) react with N donor ligands L (L = NC_5H_5, NC_5H_4-4-Me) to give red crystalline compounds as shown below in equation (1).

(1) $$Mo_2(OR)_6 + 2L \rightarrow Mo_2(OR)_6L_2$$

R = Pr, Ne and L = NC_5H_5, NC_5H_4-4-Me

$W_2(NMe_2)_6$ reacts with alcohols ROH (R = i-Pr, Ne) in the presence of L (L = NC_5H_5, NC_5H_4-4-Me, $HNMe_2$) to give black or yellow crystalline compounds according to equation (2) shown below.

(2) $$W_2(NMe_2)_6 + ROH + 2L \rightarrow W_2(OR)_6L_2$$

R = Pr, Ne and L = $HNMe_2$, NC_5H_5, NC_5H_4-4-Me

These $M_2(OR)_6L_2$ compounds, in hydrocarbon solvents, undergo rapid reversible dissociation at room temperature and the position of equilibrium is largely determined by steric factors associated with R and L. X-ray studies of several $M_2(OR)_6L_2$ compounds reveal two four coordinated metal atoms united by a metal-metal triple bond. The four ligands coordinated to each metal atom lie roughly in a square plane.

SYNTHESIS AND CHARACTERIZATION OF DIAMINE ADDUCTS OF HEXAKIS(NEOPENTOXY)DIMOLYBDENUM. M.H. Chisholm, J.C. Huffman and J.W. Pasterczyk. Department of Chemistry and Molecular Structure Center, Indiana University, Bloomington, Indiana 47405.

Hydrocarbon solutions of $Mo_2(ONe)_6$ (M≡M), where Ne = neopentyl, react with diamines to form adducts of empirical formula $Mo_2(ONe)_6$(diamine). For the ethylene diamines $Me_2NCH_2CH_2NMe_2$, $Me(H)NCH_2CH_2N(H)Me$ and $Me(H)NCH_2CH_2NH_2$, the adducts are discrete dinuclear compounds in which the diamines span the Mo≡Mo bond. The molecular structure of $Mo_2(ONe)_6(Me(H)NCH_2CH_2N(H)Me)$, deduced from a single crystal X-ray study, reveals a molecule having virtual C_2 symmetry and low temperature ^1H nmr studies are consistent with the preservation of this structure in solution. At higher temperatures in solution, reversible dissociation of the diamine can be monitored by nmr spectroscopy. The equilibrium reaction $Mo_2(OR)_6 + L\text{-}L \rightleftharpoons Mo_2(OR)_6(L\text{-}L)$ is very dependent on the steric bulk of the alkoxide (R = t-Bu, i-Pr and Ne), the number of methyl substituents on the diamine, and temperature. The adducts formed from diamines having 3 or 6 carbon chains between the nitrogen atoms show very different physico-chemical properties and are believed to be polymeric. The longer chains are less suitable for bridges across the Mo≡Mo bond and thus favor intermolecular association.

SECONDARY ION MASS SPECTROMETRY IN THE ANALYSIS OF SUPPORTED INORGANIC COMPLEXES. R. G. Cooks, J. L. Pierce and R. A. Walton, Department of Chemistry, Purdue University, West Lafayette, Indiana 47907.

Secondary ion mass spectrometry (SIMS) has been used for the direct analysis of complex inorganic molecules on support materials of catalytic significance. This technique employs argon ions to probe solid surfaces. The resulting secondary ions, formed by several processes including electron transfer, cationization and direct sputtering of precharged species, are subsequently collected and mass analyzed. In general, solid samples may be burnished directly onto a metal foil or deposited from solution onto a support prior to SIMS analysis. Nickel phosphine and phosphite complexes such as $CpNi(PPh_3)Cl$ and $Ni[P(OMe)_3]_4$ have been studied on a variety of supports including graphite, silica and alumina in addition to the silver support commonly used in SIMS experiments. The resulting spectra provide clear evidence for the presence of the intact complex on the surface of the support illustrating the versatility of this highly sensitive surface technique for the characterization of complex inorganic molecules.

ANALYSIS OF THE DISTORTIONS OF TRANS BIPYRIDINE LIGANDS. A.W. Cordes, B. Durham, W. Pennington, and P. Swepston, Chemistry Department, University of Arkansas, Fayetteville, Arkansas 72701.

The crystal structures of octahedral and square-planar complexes containing two 2,2' bipyridine (bpy) ligands in the trans configuration experience a steric interaction between opposing alpha hydrogen atoms of the bpy ligands. This interaction causes a significant deviation from planarity of the trans $M(bpy)_2$ system. The crystal structures of trans $Ru(bpy)_2(PPh_3)_2{}^{+2}(PF_6^-)_2$ and $Ru(4,4'\text{-dimethyl bpy})_2py_2{}^{+2}$ $(PF_6^-)_2$, which will be reported, along with the previously reported structures of $Ru(bpy)_2(H_2O)(OH)^{+2}$ and four square-planar di-bpy complexes, provide sufficient data for an analysis of the distortion modes which accomodate the steric strain. Three of the complexes have bpy ligands which are in a "twist" configuration in which the two alpha hydrogen atoms of each bpy are found on opposite sides of the idealized MN_4 plane. The other complexes have bpy ligands which are "bowed" such that the alpha hydrogen atoms of each bpy lie on the same side of MN_4 plane. These distortions are discussed in terms of rotation and pyramidalization parameters within the bpy ligands.

AN APPROACH TO THE METAL-CARBONYL SUBSTITUTION REACTION USING $[(\eta^5\text{-}C_5H_5)Fe(CO)_2]_2$ AS CATALYST. N.J. Coville and M.O. Albers. Chemistry Department, University of the Witwatersrand, Johannesburg, Republic of South Africa.

A catalyst cycle typically comprises a series of steps such as ligand substitution, oxidative addition and insertion and any of these steps may be rate determining in the cycle. Theoretically, the energy barrier of the rate determining step could be lowered by the action of a catalyst and we have commenced an investigation, using model reactions, to explore this possibility. Thisposter presents data on the effect of catalysts on the ligand substitution reaction, specifically for the reaction $M\text{-}CO + L \xrightarrow{cat} M\text{-}L + CO$ ($L = PR_3$,RNC; cat = $[(\eta^5\text{-}C_5H_5)Fe(CO)_2]_2$). This reaction allows for the facile synthesis of a wide range of complexes, *e.g.* $\eta^5\text{-}C_5H_5Fe(CNR)_2I$, $Fe(CO)_{5-n}(CNR)_n$ ($n > 0$), $Fe(CO)_4PR_3$, etc., many of which are not normally accessible *via* direct thermal substitution reactions. A mechanism for the reaction will be proposed.

LASER RAMAN AND CARBON-13 NMR STUDIES OF SIMPLE CARBON-CONTAINING SPECIES IN AQUEOUS SOLUTION. B.J. Cromarty, C.S. Cundy, R.J. Terrell and J.R. Wood. ICI Mond Division Technical Department, The Heath, Runcorn, Cheshire, WA7 4QE, England.

Laser Raman and Carbon-13 Nuclear Magnetic Resonance spectroscopy have been used to identify, and distinguish between, simple carbon-containing species (carbonate, bicarbonate, carbamate) in aqueous solution. Characteristic carbonyl stretching vibrations are seen at ca. 1065 cm^{-1} (CO_3^{2-}), 1030 cm^{-1} (HCO_3^-) and 1040 cm^{-1} ($CO_2NH_2^-$). The $CO_3^{2-}/HCO_3^-/H_2O$ system, in which protons are exchanging rapidly on the NMR time-scale, displays a single, pH-dependent ^{13}C resonance between 162 ppm (100% HCO_3^-) and 170 ppm (100% CO_3^{2-}), which is almost independent of temperature. [Chemical shifts are downfield from TMS (0.0 ppm)]. Solutions containing the carbamate ion (stable only at high pH) show a ^{13}C signal around 167 ppm. When carbonate, bicarbonate and carbamate species are present in solution, they interact through a series of equilibria. Two ^{13}C NMR signals result: one (carbamate) whose chemical shift is almost constant with temperature but whose intensity varies with the temperature-dependent hydrolysis of $CO_2NH_2^-$; the other due to carbonate-bicarbonate whose chemical shift and intensity both vary with temperature as the CO_3^{2-}/HCO_3^- ratio and the total $[CO_3^{2-} + HCO_3^-]$ concentration are perturbed by hydrolysis of the carbamate ion.

THE ABSORPTION AND EMISSION SPECTRA OF THE PHOTOCATALYST TRIS(BIPYRAZINE)RUTHENIUM (II) IN ACIDIC MEDIA. R.J. Crutchley, N. Kress and A.B.P. Lever, Department of Chemistry, York University, 4700 Keele Street, Downsview, Ontario, Canada, M3J 1P3.

The tris(bipyrazine)ruthenium(II) cation I, is an analog of the more well known tris(bipyridine)ruthenium(II) II, [R.J. Crutchley and A.B.P. Lever, J.Am.Chem Soc. 102, (1980) 7128]. The additional nitrogen atoms on the periphery of I, provide acid-base properties permitting I to dissolve in acid media. A detailed study of the spectroscopic (absorption, excitation and emission) characteristics of the free ligand bipyrazine and the ruthenium complex I in acid solutions are reported as a function of pH and the Hammett function -Ho. A series of spectroscopic changes occurs in the visible (MLCT) and ultraviolet (π-π*) regions, and in the emission spectra. In absorption, as the solution changes from 0 - 96% sulphuric acid, six different sets of isosbestic points are associated with six successive protonation steps of this complex. In the case of the free ligand, bipyrazine, three of the expected four steps can be seen up to 96% sulphuric acid. Analysis of the data provides some ground state pKa values. All six of the protonated species are luminescent with lifetimes in the range 27-520ns. The luminescent behaviour of these species as a function of acidity, provides excited state pKa values. For example, in the ground state the first protonation step has $pKa_1 = -2.2$, while in the excited state, $pKa_1^* = 2.0$. In the case of the first three protonation steps, the excited state is a stronger base than the ground state, while the reverse is true of the fourth to sixth protonation steps. Full details will be presented.

ALKENE ADDITION AND ALKANE SUBSTITUTION REACTIONS MEDIATED BY TRANSITION METAL COMPLEXES R. Davis, J.L.A. Durrant, I.F. Groves and C.C. Rowland, School of Chemical and Physical Sciences, Kingston Polytechnic, Kingston upon Thames, United Kingdom.

Mechanistic studies have been performed on the reaction :
$CCl_4 + RCH = CH_2 \rightarrow RCHCl.CH_2CCl_3$ in the presence of $[Mo(CO)_3(\eta\text{-cp})]_2$ (I) and $[Cr(CO)_3(\eta^6 - C_{10}H_8)]$ (II). In both cases, it had previously been suggested that the reactions proceed via an oxidative addition - migratory insertion sequence. Our studies show that (I) acts as a redox catalyst in the early stages of the reaction, but the complex undergoes total decomposition in the early stages of the reaction, which subsequently continues by a radical chain mechanism. (II) acts solely as an initiator of a radical chain reaction. In neither case is there evidence for formation of metal-trichloro alkyl or metal-alkene intermediates.

The reaction:
$CCl_4 + C_6H_{12} \rightarrow CHCl_3 + C_6H_{11}Cl$ proceeds in moderate yield in the presence of a number of transition metal complexes and in good yield (ca. 70%) in the presence of $[Re_2(CO)_{10}]$. Mechanistic possibilities will be presented for this reaction.

SINGLE RING CHARACTER IN $[Ru(bpy)_3]^{2+}$ AND RELATED COMPOUNDS, M. Keith DeArmond and Clifford M. Carlin, Department of Chemistry, North Carolina State University, Raleigh, North Carolina 27650.

The variable temperature studies of Crosby and coworkers led to a phenomenological description of the lowest (emitting) manifold of states for $[Ru(bpy)_3]^{2+}$ as deriving from a combination of large exciton and large spin-orbit coupling. The meticulous single crystal absorption spectra obtained by Ferguson and coworkers for the series of complexes, $[M(bpy)_3]^{2+}$ where M = Fe^{2+}, Ru^{2+}, Os^{2+} indicate that the lowest energy bands can to a good approximation be assigned as spin forbidden transitions to the $d\Pi^*$ charge transfer state, thus S, to a reasonable approximation is a good quantum number. Both models presume an effective D_3 symmetry, however the photoselection data reported by Fujita and Kobayashi and verified by us and the Ferguson group indicate that a symmetry less than D_3 must be utilized to rationalize a P of 0.23 at 21,300 cm^{-1}. Ferguson and coworkers suggest that the high P value results from a solvent heterogeneity. Our present high resolution study examines the $[Ru(bpy)_3]^{2+}$, $[Ru(phen)_3]^{2+}$, $[Os(bpy)_3]^{2+}$ and $[Ru(bpy)py_4]^{2+}$ complexes and includes data into the overlap (absorption-emission) region. On the basis of these data, a calculation of P from the single crystal absorption data and additional experiments, the conclusions are:

1) that the P values greater than 0.14 (planar emission oscillator) indicate an effective symmetry less than D_3 for the tris complexes, and
2) that the symmetry reduction mechanism is not a static heterogeneous interaction but must be a dynamic intramolecular interaction, perhaps a dynamic Jahn-Teller effect.

THE PHOTOPHYSICS OF Cu(I) COMPLEXES. Alan A. Del Paggio, Roland E. Gamache, Jr., Jon R. Kirchhoff, Russell K. Lengel, and David R. McMillin, Department of Chemistry, Purdue University, West Lafayette, IN 47907.

The photophysics of several $Cu(biL)_2^+$ systems will be discussed (biL = 2,2'-bipyridine, 1,10-phenanthroline or substituted derivatives thereof). The emission from methylene chloride solutions of these complexes becomes more intense with increasing temperature. Detailed studies of the effect of temperature upon emission, absorption and lifetime phenomena will be discussed. A generalized energy level scheme which assumes that two excited states participate in the emission process will be discussed to explain this unusual behavior. Replacing specific ring hydrogens of the biL moiety with carbon-containing substituents has a dramatic effect on the lifetimes and emission yields, and possible explanations are also considered.

THE CHEMISTRY OF SINGLY AND DOUBLY BONDED DITUNGSTEN ALKOXIDES OF THE TYPE $W_2Cl_4(OR)_6$ AND $W_2Cl_4(OR)_4(ROH)_2$. D. A. DeMarco and R. A. Walton. Department of Chemistry, Purdue University, West Lafayette, Indiana 47907.

The doubly bonded alkoxides of the type $W_2Cl_4(OR)_4(ROH)_2$ have been prepared by reaction of the quadruply bonded ditungsten(II) complex $W_2(mhp)_4$, where mhp represents the anion of 2-hydroxy-6-methylpyridine, with gaseous hydrogen chloride in alcoholic solvents. These molecules undergo alcohol exchange reactions without disruption of the metal-metal bond. The degree of exchange is dependent upon the steric constraints engendered by the alcohol and leads to two groups of doubly bonded tungsten alkoxides of the types $W_2Cl_4(OR)_4(ROH)_2$ and $W_2Cl_4(OR)_2(OR')_2(R'OH)_2$, where R = methyl, ethyl, n-propyl, n-butyl, n-pentyl, and n-octyl, R' = i-propyl, s-butyl, and s-pentyl. Each set of complexes has different solution dynamics as demonstrated by ^1H NMR spectroscopy and similar redox properties as shown by cyclic voltammetry measurements. Both sets of doubly bonded ditungsten(IV) species are readily oxidized to the singly bonded derivatives $W_2Cl_4(OR)_6$ and $W_2Cl_4(OR)_2(OR')_4$ by O_2 and other oxidants.

HETEROLIGAND PEROXO COMPLEXES OF VANADIUM. C. Djordjevic, P. L. Wilkins, L. G. Gonshor, N. H. Guenther and K. J. Roche. The College of William and Mary, Williamsburg, Virginia 23185

Vanadium peroxides are of special interest because the metal forms a variety of simple V(V) peroxo and oxoperoxo compounds, only a few well characterized heteroligand derivatives are known, and redox potential V(V)/V(IV) falls in the range where highly pH dependent electron transfers on peroxo group occur. In addition, vanadium is an essential trace metal for mammals, but its function remains unknown. We are investigating vanadium peroxo complexes with carboxylato and aminocarboxylato ligands in solid state and aqueous solutions. Proton and ^{13}C NMR of the heteroligand are used as an indirect evidence to characterize the peroxo group environment in water. Metal-peroxo charge transfer bands have been found to show a distinctive pH dependence, which is determined by heteroligand. Electrochemical properties of aqueous solutions of these complexes are especially revealing, since the course and the potentials of peroxide oxidation by strong oxidants varies as dependent upon heteroligands. Potential differences up to 300 mV have been observed. Experiments show that vanadium peroxo complexes are not analogous to corresponding derivatives of heavier metals in the group, niobium and tantalum.

INVESTIGATION OF THE CHEMILUMINESCENT OXIDATION OF AMINES BY $Ru(bpy)_3^{3+}$. Bill Durham, Department of Chemistry, University of Arkansas, Fayetteville, Arkansas 72701.

Chemiluminescent oxidation reactions involving $Ru(bpy)_3^{3+}$ have been known for several years. Recently, a number of related reactions have been incorporated into solar energy conversion schemes. Although one such reaction, that of OH^- in aqueous solution, has been studied extensively the details of the mechanism are poorly understood. The thermodynamics of the process is particularly puzzling. The oxidation of triethylamine in acetonitrile appears to proceed cleanly to a small number of products and thus offers a convenient system for an investigation of the mechanism. Reactions carried out in CD_3CN with 0.06M $Ru(bpy)_3^{3+}$ and a stoichiometric amount of amine produce only a single product when observed by NMR, $HN(C_2H_5)_3^+$. Excess base (2X) or the addition of 10% D_2O gives similar results but with minor amounts of what are tentatively assigned as CH_3CHO and $H_2N(C_2H_5)_2^+$. Gas chromatography of the reaction mixture or of the components left after removal of the solvent and extraction with ethylacetate also indicated only minor amounts of other components (5% or less). Succinonitrile may be among them. Preliminary observations suggest that reaction rate is a function of base strength (pKb). The oxidation proceeds rapidly with the base triethylenediamine (DAPCO) and attempts to trap the transient radical cation have been carried out.

ELECTRONIC INTERACTIONS IN BIMOLECULAR ELECTRON-TRANSFER AND ENERGY-TRANSFER REACTIONS.
J. F. Endicott, G. R. Brubaker, T. Ramasami, R. Tamilarasan, D. C. Gaswick, M. J. Heeg, S. C. Pyke. Wayne State University, Department of Chemistry, Detroit, Michigan 48202

The dependence of electron transfer rates on purely electronic properties of donor and acceptor (e.g., orbital symmetry, orbital overlap, spin orthogonality, etc.) has been a matter of speculation and controversy, largely owing to uncertainties in the dominant Franck-Condon contributions to reactivity. Bimolecular energy transfer reactions have smaller Franck-Condon factors and their rate patterns can be used as probes of the general form of donor acceptor electronic interactions. Studies of the bimolecular quenching of $(^2E)Cr(phen)_3{}^{3+}$ by Co(III) complexes has demonstrated that: (a) kq exp-$(11\bar{r}_{DA})$ where $\bar{r}$ is the donor-acceptor contact separation is nm; and (b) kq increases with the oxidizability of X^- is $Co(NH_3)_5X^{2+}$ (or in $Co^{III}(N_4)XY$) complexes. The distance dependence is of the form and magnitude expected for weak donor-acceptor orbital overlap, and the ligand dependence suggests that the donor-acceptor overlap is enhanced by charge-transfer interactions. This concept has been applied to electron transfer reactions, and some reaction rates (e.g., those involving Co(III)/Co(II) couples) are susceptible to perturbation by weak (ion pair) charge-transfer interactions. For example for the presumably "non-adiabatic" $Co(phen)_3{}^{3+}/Co(sep)^{2+}$ reaction, the $\{Co(phen)_3{}^{3+}$, ascorbate$\}$ ion pair is ∿30 times more reactive than the $\{Co(phen)_3{}^{3+}$, $TFMS^-\}$. In contrast, the nature of the ion pair has very little effect on the more "adiabatic" $Co(phen)_3{}^{3+}/Ru(NH_3)_6{}^{2+}$ reaction. Extentions are being made to determine the ratio of Franck-Condon to electronic factors.

Metal Pentadienyl Compounds.
R.D. Ernst, D.R. Wilson, J.-Z. Liu, L. Stahl, and T.H. Cymbaluk. Department of Chemistry, University of Utah, Salt Lake City, UT 84112

An examination of the general properties of the pentadienyl group has led to the expectation that metal-pentadienyl complexes should be capable of possessing both stability and chemical reactivity. A series of metal pentadienyl complexes has now been prepared and characterized for titanium, vanadium, chromium, manganese, iron, cobalt, and ruthenium. Many of the above have been isolated as bis(pentadienyl)metal complexes, which may be considered "open metallocenes". These compounds are being studied in part to allow an understanding of the relationship of metal-pentadienyl complexes to the better known metal-allyl and metal-cyclopentadienyl systems. In addition, a number of these complexes have been found to possess chemical and catalytic reactivities, one example being in "bare metal" transformations. X-ray structural and other physical studies have played a crucial role in our efforts to date, and these results as well as those from various chemical studies will be presented.

HYDROCARBATION: THE ADDITION OF THE C-H BOND OF A CATIONIC BRIDGING IRON METHYLIDYNE COMPLEX TO ALKENES.
Paul J. Fagan and Charles P. Casey. Chemistry Department, University of Wisconsin, Madison, WI 53706.

The insertion of alkenes into element-to-hydrogen bonds is a fundamental step in many catalytic and stoichiometric transformations. We have found that the bridging iron methylidyne complex [*cis*-$Cp_2Fe_2(CO)_2(\mu$-$CO)(\mu$-$CH)]^+PF_6{}^-$ (1) reacts with a variety of alkenes to add the methylidyne carbon-hydrogen bond across the alkene carbon-carbon double bond. Thus, 1 reacts with ethylene at 25°C in CH_2Cl_2 to yield the propylidyne complex [*cis*-Cp_2-$Fe_2(CO)_2(\mu$-$CO)(\mu$-$CCH_2CH_3)]^+PF_6{}^-$ in good yield. With propene and isobutylene, the reaction with 1 is regiospecific forming exclusively the alkylidyne products [*cis*-$Cp_2Fe_2(CO)_2(\mu$-$CO)$-$(\mu$-$CCH_2CH_2CH_3)]^+PF_6{}^-$ and [*cis*-$Cp_2Fe_2(CO)_2(\mu$-$CO)(\mu$-$CCH_2C(CH_3)_2H)]^+PF_6{}^-$ respectively. Studies of the reaction of 1 with C_2D_4 and 2,3-dimethyl-2-butene, which forms [*cis*-Cp_2Fe_2-$(CO)_2(\mu$-$CO)(\mu$-$CCD_2CD_2H)]^+PF_6{}^-$ and [*cis*-$Cp_2Fe_2(CO)_2(\mu$-$CO)(\mu$-$C(CH_3)_2C(CH_3)_2H)]^+PF_6{}^-$ respectively, show that 1,2-addition of the methylidyne C-H bond of 1 across the alkene double bond has occurred. The reaction of 1 with 1-methylcyclohexene takes a different course and produces the fluxional μ-vinyl cationic complex [*cis*-$Cp_2Fe_2(CO)_2(\mu$-$CO)(\mu$-*trans*-$CH{=}CHC(CH_3)CH_2CH_2CH_2CH_2)]^+PF_6{}^-$. The mechanistic aspects of these reactions will be discussed, and many parallels to hydroboration chemistry are noted. This hydrocarbation reaction promises to be useful synthetically, as well as being an intiguing mode of reactivity for a methylidyne ligand.

CATIONIC η[6]-BENZENE COMPLEXES OF RUTHENIUM(II) WHICH CONTAIN NITROGEN HETEROCYCLES.
F. M. Farah and D. G. Hendricker, Clippinger Laboratories, Department of Chemistry, Ohio University, Athens, Ohio 45701

Cleavage of the halogen bridged dimer $[Ru(C_6H_6)Cl_2]_2$ with the nitrogen heterocycles, N-N, 1,8-naphthyridine, its 2-methyl and 2,7-dimethyl derivative, or one of several mono or disubstituted 1,10-phenanthrolines, followed by treatment with NH_4PF_6 afforded compounds of the type $[Ru(C_6H_6)Cl(N\text{-}N)]PF_6$. These octahedral complexes were characterized by elemental analysis, conductivity studies, ^1H and ^{13}C NMR and IR spectroscopy. The naphthyridines, which form 4-membered chelate rings, are susceptible to displacement by donor solvents, whereas, the phenanthrolines, which form 5-membered chelate rings, are stable. The ^{13}C NMR spectra of the complexes exhibit resonance positions of the coordinated benzene which can be correlated with both the basicity of the heterocycle, as measured by pKa, and with the size of the chelate ring. A similar relationship was also observed for both the ^1H and ^{13}C resonances of the nitrogen heterocycles. A complex which contains two heterocycles, both acting in a monodentate fashion, $[Ru(C_6H_6)Cl(1,8\text{-naphthyridine})_2]PF_6$ has also been characterized.

PHOSPHINE BRIDGED BINUCLEAR COMPLEXES, J.P. Farr, M.M. Olmstead, F. Wood, A. Maisonnat and A.L. Balch, Department of Chemistry, University of California, Davis, CA 95616

Chemical reactivity of hetero-binuclear complexes with bridging 2(diphenylphosphino)-pyridine ligands will be summarized. Examples of substitution, head-to-head/head-to-tail isomerization, and single-site reactivity (in the form of oxidative additions which effect only one of the two metal atoms) will be given. These proceed with little of no fragmentation of th ε binuclear units. Some observations on the mechanism of formation of heterobinuclear complexes ($RhPt(\mu\text{-}Ph_2Ppy)_2(CO)Cl_3$, $PtPd(\mu\text{-}Ph_2Ppy)_2Cl_2$) will be presented.

ELECTRON UPTAKE AND DELIVERY SITES ON COPPER PROTEINS. O. Farver, Y. Blatt, Y. Shahak and I. Pecht. The Weizmann Institute of Science, Rehovot 76100, Israel.

Plastocyanin and azurin are labeled upon reduction with Cr(II) ions yielding substitution inert Cr(III) adducts (Farver and Pecht, Proc. Natl. Acad. Sci. USA (1981) 78,4190; Farver and Pecht, Israel J. Chem. (1981) 21, 13). The effect of this modification on the activities of these proteins with their presumed physiological partners were investigated:
a) Whereas the rates of photoreduction of native and Cr(III)-labeled plastocyanin were indistinguishable, the rates of photooxidation of the modified protein were markedly attennuated relative to those of the native protein. Thus, there are most probably two physiologically significant electron transfer sites on plastocyanin; the site involved in electron transfer to P_{700} in the region of Tyr-83 and the second site at the region of His-87 (Farver, Shahak, and Pecht, Biochemistry, in press).
b) The cytochrome oxidase catalyzed oxidation of reduced native and Cr(III)-labeled azurin by O_2 was found to be unaffected by the modification. The electron exchange reaction between native or chromium-labeled azurin and cytochrome $\underline{c}_{551}$ showed a significant difference in reactivity. The kinetic and thermodynamic data in conjunction with the present knowledge of the azurin structure suggest the involvement of His-35 in the electron exchange with cytochrome $\underline{c}_{551}$. A second active site proximal to His-117 may be involved in the reduction of cytochrome oxidase (Farver, Blatt, and Pecht, Biochemistry, in press).

OXYGENATION KINETICS OF MANGANESE(II) SCHIFF BASE COMPLEXES, Fred C. Frederick and Larry T. Taylor, Virginia Polytechnic Institute and State University, Department of Chemistry, Blacksburg, VA 24061.

The reaction of molecular oxygen and manganese(II) complexes of the type shown has previously been studied both in solution and in the solid state. It was postulated that the reaction proceeded via a two-step overall mechanism passing through an isolable $Mn^{III}\text{-}O^{2-}\text{-}Mn^{III}$ intermediate on the way to the fully oxygenated $Mn^{IV}<(O^{2-})_2>Mn^{IV}$ product. We have now extended this work to include a quantitative kinetic study. Data were generated by following changes in the UV-visible spectrum of the complex as oxidation occurred in either DMSO or pyridine. Rate constants and activation energies have been calculated from the data. Preliminary results indcate that E_a for the process is greater in pyridine than in DMSO (eg. n=2, E_a (DMSO)=11.3 kcal/mole; E_a (PY)=20.5 kcal/mole). Rates depend both on the number of methylene carbons joining azomethine donors and the aromatic substituents. Comparisons will be made with manganese(II) complexes with analogous tridentate ligands dervied from salicylaldehyde and 2-aminomethylpyridine.

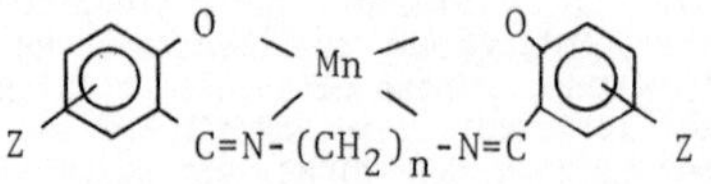

NEW ROUTES TO ORGANOBORANES. Donald F. Gaines, Joseph A. Heppert, and Gail A. Steehler, Department of Chemistry, University of Wisconsin-Madison, Madison, Wisconsin 53706

New synthetic routes for the production of aryl and alkyl derivatives of pentaborane have been studied. Lewis acid catalyzed electrophilic attack by 2-ClB_5H_8 on various alkyl benzene derivatives has been found to produce 2-arylpentaboranes. This represents the first synthesis of simple aryl substituted borane clusters. The site of ring substitution by the B_5H_8 moiety appears to be sterically controlled. Copper lithium reagents of the type R_2CuLi have been shown to react with 2-halopentaboranes to yield 2-alkyl and 2-aryl pentaboranes. While only monosubstitution is observed in reactions of these copper lithium reagents with organic halides, disubstitution to yield 2,3 dialkylpentaboranes has been observed here.

CONTROL OF SPIN STATE IN PORPHYRIN COMPLEXES. D. K. Geiger and W. R. Scheidt. Department of Chemistry, University of Notre Dame, Notre Dame, Indiana 46556.

The nature and number of axial ligands are the primary determinants of metalloporphyrin spin state. For porphinato iron(III) complexes, three different spin states are possible (low, intermediate, and high). All three spin states are found in different crystalline forms of $[Fe(OEP)(3\text{-ClPy})_2]ClO_4$ (OEP = octaethylporphyrin; 3-ClPy = 3-chloropyridine). A triclinic form is a thermal spin-equilibrium of low-(S = 1/2) and high-spin (S = 5/2) states. A monoclinic form has a quantum-admixed intermediate-spin state. Both crystalline forms have been characterized crystallographically and by Mossbauer spectroscopy. Specific axial ligand orientations in the solid state appear to be responsible for these effects. Changes in the porphinato ligand can also lead to different spin states. In solution, $[Fe(OEP)(3\text{-ClPy})_2]ClO_4$ is a thermal spin-equilibrium system. The analogous tetraphenylporphyrin, TPP, complex is low spin. In the OEP and TPP complexes of Fe(P)(NCS)(Py), the OEP complex is high spin while the TPP complex is low spin even though the axial ligands are identical. The structural differences will be described. The effects of porphyrin basicity on spin state are being investigated in a series of bis(3-ClPy)iron(III) porphyrinates.

ACTIVATION OF H_2 BY CU(I) ALKOXIDES: A NEW ROUTE TO SOLUBLE COPPER HYDRIDES. G. V. Goeden, J. C. Huffman, and K. G. Caulton, Department of Chemistry and Molecular Structure Center, Indiana University, Bloomington, Indiana 47405.

New soluble copper hydride complexes of the type $[HCuP]_6$ (P = tertiary phosphine or tertiary phosphite) and $[HCu(Triphos)]_2$ (Triphos = $CH_3C(CH_2PPh_2)_3$) have been prepared by the interaction of molecular hydrogen with $[CuO^tBu]_4$ in the presence of tertiary phosphines. This reaction is remarkably facile at 25°C and 1 atm H_2 pressure. NMR evidence (^1H, ^{31}P) from a pre-reaction solution indicates the presence of an equilibrium mixture of $[Bu^tOCuP]_2$, $PCu(Bu^tO)_2CuP_2$, and Bu^tOCuP_2; the crystal structure of one of these components, $[Bu^tOCu(PPh_3)]_2$, has been determined. The crystal structures of the copper hydride complexes, $[HCuP(O^iPr)_3]_6$ and $[HCu(Triphos)]_2$, were determined at low temperature. The structure of $[HCuP(O^iPr)_3]_6$ displays an octahedral arrangement of copper atoms with 6 "long" and 6 "short" Cu-Cu distances. The structure of the first dimeric copper hydride complex $[HCu(Triphos)]_2$ exhibits a number of unusual and interesting features which include the location of two bridging hydrides, the presence of a "dangling" ligand arm, and an exceedingly short Cu-Cu distance.

CHARACTERIZATION OF SUPEROXIDE-METALLOPORPHYRIN REACTION PRODUCTS: EFFECTIVE USE OF DEUTERIUM NMR SPECTROSCOPY, Harold M. Goff and Ataollah Shirazi, Department of Chemistry, University of Iowa, Iowa City, Iowa 52242.

Nuclear magnetic resonance spectroscopy has been utilized to define the solution electronic structures of complexes generated from combination of superoxide ion with iron(II) and manganese(II) porphyrins. Very large proton NMR linewidths for the iron porphyrin product dictated the use of deuterium NMR for selectively deuterated synthetic porphyrins. Based on the pyrrole deuterium signal at 60 PPM (DMSO solvent, 28°C, TMS reference) and a solution magnetic moment of ≥5.6 B.M., a peroxo-iron(III) porphyrin structure consistent with an earlier formulation is favored. A corresponding manganese porphyrin signal at 32 PPM and a solution magnetic moment of 5.0 B.M. are indicative of a superoxo-manganese(II) porphyrin configuration with spin coupling between the superoxide ligand and the manganese center. This species provides an excellent model for what has been described as superoxide coupling to a low-spin iron(III) center in the oxyhemoglobin molecule. The peroxo-iron porphyrin complex is likely coordinated in the second axial position by a solvent molecule. The axial ligand combination leaves the iron center in a high-spin state, but the peroxo ligand serves to reduce the zero-field-splitting value and indirectly induce large NMR linewidths. Variable temperature NMR Curie law behavior is consistent with monomeric structures. Absence of splitting in the pyrrole signals suggests no attack of porphyrin nitrogen atoms by peroxide or superoxide species.

NICKEL (I) CATALYSIS OF THE EPIMERIZATION OF A NICKEL (II)-CARBON CENTER, J. B. Grutzner and C. R. Roberts, Purdue University, West Lafayette, IN 47907.

We have been studying the equilibrium between 1 and 2. K = .59 ± .05 at 75°C in d_6 ϕH as determined by ^1H NMR. The epimerization is not mechanistically simple: The kinetics are zero order in time and concentration dependent. Addition of dimethyl acetylene dicarboxylate causes inhibition of the epimerization. On the other hand addition of ligands (PDPh$_3$, NEt$_3$, Bipy) increases the rate of epimerization. CpNiBipy (E. K. Barefield et al. J.A.C.S. 103, 6219 (1981)) has been identified in the reaction mixture by its characteristic ESR spectrum. Finally addition of authentic CpNiBipy to 1 at 25°C results in epimerization. Thus it appears that CpNiL$_2$· or species in equilibrium with CpNiL$_2$· (CpNiL, CpNi) are the active catalysts in the epimerization of 1 to 2. Further experiments to determine the details of the catalysis are currently underway.

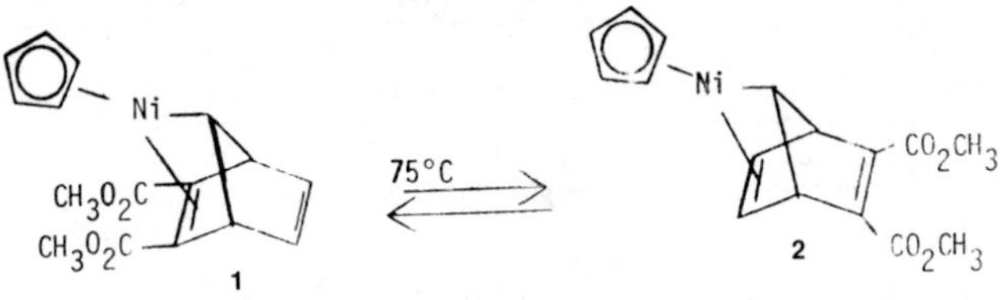

COORDINATION COMPLEXES WITH ORGANOTELLURIUM LIGANDS. H. J. Gysling. Research Laboratories, Eastman Kodak Company, Rochester NY 14650.

Although the first organotellurium ligand (TeEt$_2$) was prepared by Wöhler in 1840 and the first coordination complex with an organotellurium ligand (cis-PtCl$_2$(Te(CH$_2$Ph)$_2$)$_2$) was reported by Fritzmann in 1915, the ligand chemistry of tellurium has been a relatively unexplored research area until very recently.

Some general synthetic routes to organotellurium ligands will be outlined and the synthesis and characterization of some typical coordination complexes with Pd(II) and Pt(II) will be described. A variety of spectroscopic methods (infrared, Raman, ^{13}C and ^{125}Te NMR, ESCA and ^{125}Te Mössbauer spectroscopy) have been used to characterize such complexes.

SYNTHESIS AND REACTIONS OF MAIN GROUP POLYATOMIC ANIONS: FORMATION OF THIN FILMS AND SUPPORTED CRYSTALS VIA TOPOCHEMICAL OXIDATIVE DECOMPOSITION ONTO ORGANIC, INORGANIC, AND METALLIC SUBSTRATES.* Robert C. Haushalter, Raymond G. Teller, and Larry J. Krause, Chemistry Division, Argonne National Laboratory, Argonne, Illinois 60439.

Several types of main group polyanions, such as Sn_9^{4-}, Sb_7^{3-}, and $Tl_2Te_2^{2-}$, have been isolated and structurally characterized by Corbett and coworkers with the aid of cryptate ligands. We will present two areas of main group polyanion chemistry: the isolation of these anions without the use of cryptates and secondly the topochemical oxidative decomposition of these polyanions. We have isolated ammonium salts of group IVa polyanions, e.g., $(Me_4N)_4Sn_9$, as well as several ammonium polychalcogenides. Several of the polychalcogenides have been structurally characterized: $(Bu_4N)_2M_x$ (M=Te, x=5; M=Se, x=6; M=S, x=6). The extremely reactive polyanions have been found to undergo a topochemical reaction with a very large number of solids. The reaction can be explained in terms of a reduction/intercalation/deposition sequence, e.g., reaction of MoS_2 with a solution of K_4Sn_9 gives a reduced, potassium-intercalated MoS_2 lattice with elemental Sn deposited on the surface. This reaction of main group anions with organic polymers (e.g., polyimides), with inorganics (e.g., graphite, transition metal oxides, and chalcogenides), and with metals (e.g., Pt, Au) will be discussed.

*Work performed under the auspices of the Office of Basic Energy Sciences, Division of Chemical Sciences, U. S. Department of Energy, under Contract W-31-109-Eng-38.

PHOTODECOMPOSITION OF TWO FORMS OF THE Cu^{III}-TRI-α-AMINOISOBUTYRIC ACID AMIDE COMPLEX. Joanna P. Hinton, Arlene W. Hamburg, and Dale W. Margerum. Department of Chemistry, Purdue University, West Lafayette, Indiana 47907.

The tripeptide amide of the amino acid α-aminoisobutyric acid, Aib, forms thermally stable copper(III) complexes over the pH range 1-14; however, they show varied photochemical sensitivity. Three different stable species are present in this pH range:

$$Cu^{III}(H_{-2}Aib_3a)^+ \underset{}{\overset{pK_a=.25}{\rightleftharpoons}} Cu^{III}(H_{-3}Aib_3a) \overset{pK_a=12.5}{\rightleftharpoons} Cu^{III}(H_{-4}Aib_3a)^-$$

H_{-n} indicates the number of deprotonated nitrogens coordinated to copper(III). The uv-vis spectra of these complexes are characterized by ligand to metal charge transfer bands. The quantum yield, Φ, for the disappearance of the copper(III) complex at pH 5 exhibits a steplike wavelength dependence similar to that for the tripeptide complex $Cu^{III}(H_{-2}Aib_3)$. The Φ for the H_{-3} species varies from 0.35 mol einstein^{-1} at 250 nm to 0.006 mol einstein^{-1} at 546 nm. The photoreactivity of the $Cu^{III}(H_{-4}Aib_3a)$ complex in 1 M OH^- is greatly reduced with Φ varying from 0.08 to <0.004 mol einstein^{-1} for the same wavelength range. The yield and identity of the peptide oxidation fragments for both complexes indicate a photodecomposition mechanism similar to that of the Cu^{III} tripeptide complexes, $Cu^{III}(H_{-2}Aib_3)$ and $Cu^{III}(H_{-2}AlaAib_2)$.

METHYLIODINE DIFLUORIDE AS A FLUORINATING AGENT. A. F. Janzen and K. Alam. Department of Chemistry, University of Manitoba, Winnipeg, Manitoba, Canada R3T 2N2.

Methyliodine difluoride, $MeIF_2$, prepared from iodomethane and xenon difluoride, has been used as a fluorinating agent in several laboratories. We have investigated its mechanism of reaction and found that $MeIF_2$ slowly liberates iodine monofluoride, which can be trapped and identified

$$CH_3IF_2 \longrightarrow CH_3F + IF$$

by its reaction with cyclohexene to give $C_6H_{10}FI$. Reaction with $Ph_2C{=}CH_2$ gave Ph_2CFCH_2I, but the latter product slowly rearranges and reacts further to give $PhCF_2CH_2Ph$. In reactions of this kind, therefore, the effective fluorinating agent is IF. Also investigated by proton and fluorine NMR was the mechanism of fluorine exchange in $MeIF_2$.

SYNTHESIS AND REACTION CHEMISTRY OF σ-CYCLOPROPYL COMPLEXES OF IRIDIUM AND RHODIUM[1] Nancy L. Jones, Department of Chemistry, Northwestern University, Evanston, IL 60201

Transition metal σ-cyclopropyl complexes have been proposed as intermediates in transition-metal assisted rearrangements of cyclopropanes. Reaction of $[IrCl(C_8H_{14})_2]_2$ and 4 equivalents of $PMePh_2$ with C_3H_5COCl first affords an equilibrium mixture of cis-$IrCl_2(COC_3H_5)(PMePh_2)_2$ and cis-$IrCl_2(C_3H_5)(CO)(PMePh_2)_2$ followed by complete conversion to trans-$IrCl_2(C_3H_5)(CO)(PMePh_2)_2$. Reaction of $[IrCl(C_8H_{14})_2]_2$ and excess $PMePh_2$ with C_3H_5COCl affords mer-$IrCl_2(COC_3H_5)(PMePh_3)_3$ in high yield. Abstraction of the Cl^- ligands creates vacant coordination sites at the Ir(III) center to which the alkyl group can migrate. The complex $RhCl_2(COC_3H_5)$-$(PMePh_2)_2$ is isolated from the reaction of $[RhCl(C_2H_4)_2]_2$ and 4 equivalents of $PMePh_2$ with C_3H_5COCl. Upon prolonged heating of toluene solutions of the Rh-acyl complex alkyl migration occurs.

[1]This work was supported by the U.S. National Science Foundation (CHE80-09671).

A COPPER MONOOXYGENASE MODEL SYSTEM: ACTIVATION OF DIOXYGEN BY A DINUCLEAR THREE-CO-ORDINATE CU(I) COMPLEX. Kenneth D. Karlin, Yilma Gultneh, John P. Hutchinson, Jon C. Hayes, Richard W. Cruse and Jon Zubieta, Department of Chemistry, State University of New York at Albany, Albany, NY 12222.

The binding and/or activation of dioxygen using dinuclear copper moities is of interest because of the relevance to the active site chemistry of the reversible oxygen carrier hemocyanin and the copper monooxygenase tyrosinase, and their potential utility in synthetic systems. We have designed and developed new flexible binucleating ligands where tridentate donor groups are connected by a -xylene bridge. Reaction of a meta substituted ligand with $Cu(CH_3CN)_4PF_6$ produces a dinuclear three-coordinate copper(I) complex(A) which has been characterized by x-ray diffraction. It contains two distorted trigonal-planar Cu(I) groups which are 8.9Å apart in the solid state. Reaction with O_2 ($Cu{:}O_2{=}2{:}1$) results in high yield regiospecific hydroxylation of the aromatic ring to give a phenol bridged binuclear Cu(II) complex (B) suggesting incorporation of dioxygen. Compound (A) is an excellent model for the proposed active sites of deoxyhemocyanin and/or deoxytyrosinase. Structural data on (A) and (B) and mechanistic implications will be discussed.

(A)

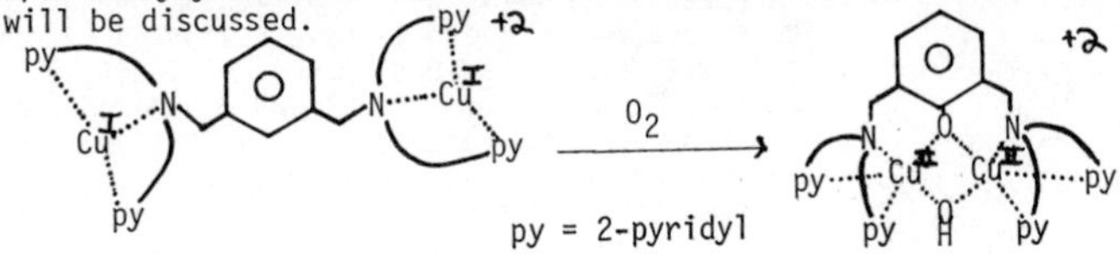

(B)

CARBON MONOXIDE HOMOLOGATION REACTIONS OF ORGANOACTINIDES. ON THE MECHANISM OF DIHAPTOFORMYL COUPLING TO YIELD ENEDIOLATES. Dean A. Katahira and Tobin J. Marks, Department of Chemistry, Northwestern University, Evanston, IL 60201

We have recently shown that bis(pentamethylcyclopentadienyl) thorium alkoxyhydrides (A) react rapidly and reversibly with carbon monoxide to yield highly reactive, carbene-like dihaptoacyls (B). Over a longer time period, quantitative, regiospecific

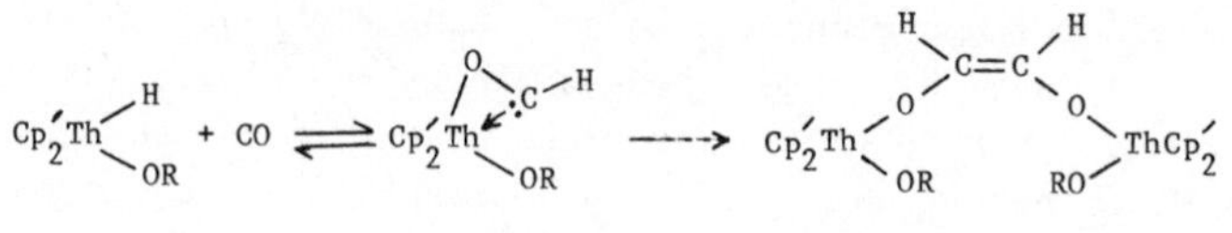

C=C fusion occurs to yield cis-enediolates (C). The present paper reports a detailed kinetic analysis of this latter transformation. On the basis of these data and several model reactions, we implicate a binuclear thoroxyalkyl insertion product ($ThCH_2OTh$) as a key intermediate in the C=C fusion process.

A MODEL FOR THE LOCALIZED EXCITED STATES OF $Ru[bpy]_3^{2+}$ and $Os[bpy]_3^{2+}$.
E. M. Kober and T. J. Meyer, Department of Chemistry, University of North Carolina, Chapel Hill, NC 27514

Evidence from various sources strongly suggests that the metal-to-ligand charge transfer [MLCT] excited states of $Ru[bpy]_3^{2+}$ and $Os[bpy]_3^{2+}$ [bpy = 2,2'-bipyridine] can be considered as having the promoted electron localized in the π^* orbital of a single bpy [versus being delocalized over all three bpy's]. An electronic structural model, which includes the effects of spin-orbit coupling, is developed for such a localized excited state. Four low-lying excited states are predicted for both complexes. Consideration of the symmetry and the amount of singlet character allows a prediction of the relative radiative and nonradiative lifetimes for each of the four states. This leads to an explicit assignment of the observed states which is consistent with the observed polarization data. Differences in the lifetimes of the analogous states for the Ru and Os complexes can be quantitatively accounted for by considering the difference in the energies of the excited states, the amount of singlet character in each state, and the extent of metal[$d\pi$]-bpy[π^*] orbital mixing.

IRON(III) PHENOXIDE COMPLEXES: MODELS FOR IRON(III)-TYROSINE COORDINATION.
Stephen A. Koch, Department of Chemistry, SUNY Stony Brook, Stony Brook, N.Y. 11794 and Michelle Millar, Department of Chemistry, New York University, New York, N.Y. 10003.

The first complexes of Fe(III) with monodentate phenoxide ligands have been synthesized and structurally characterized by x-ray diffraction. The Fe(III) tetraphenoxide complexes, (Et_4N)(Fe(2,3,5,6-tetramethylphenolate)$_4$) (1) and (Ph_4P)(Fe(2,4,6-trichlorophenolate)$_4$) (2), are prepared by the reaction of $FeCl_3$ with the lithium phenolate in ethanol. The structures of 1 and 2 have distorted tetrahedral FeO_4 cores with average Fe-O distances of 1.847(13) Å in 1 and 1.866(6) Å in 2. Compounds 1 and 2 undergo one a one electon reduction at -1.32v and -0.45v (vs SCE), respectively. The two low energy bands in their electronic spectra have been assigned to the ligand to metal charge transfer transitions, $t_1(nb)\rightarrow e$ and $t_1(nb)\rightarrow t_2$. The relationship of these model compounds to the iron-tyrosine proteins will be discussed.

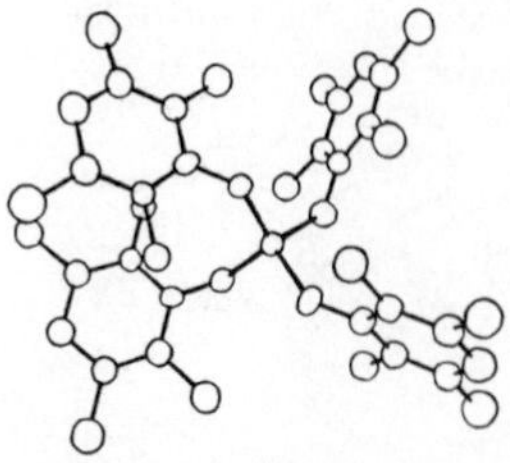

MOLECULAR FRAGMENTS OF SOLID-STATE STRUCTURES: THE STRUCTURES OF TWO MO(V) TETRAMERS
Stephen A. Koch and Sr. Sandra Lincoln S.H.C.J. Department of Chemistry, SUNY Stony Brook, Stony Brook, N.Y. 11794.

Mo_4(Hydrotris(1-pyrazolyl)borate)$_2$ $(O)_4(\mu\text{-}O)_4(\mu\text{-}OMe)_2(HOMe)_2$ (1) has been synthesized by the methanolysis of $(HBpz_3)_2Mo_2O_4$ and characterized by x-ray crystallography.

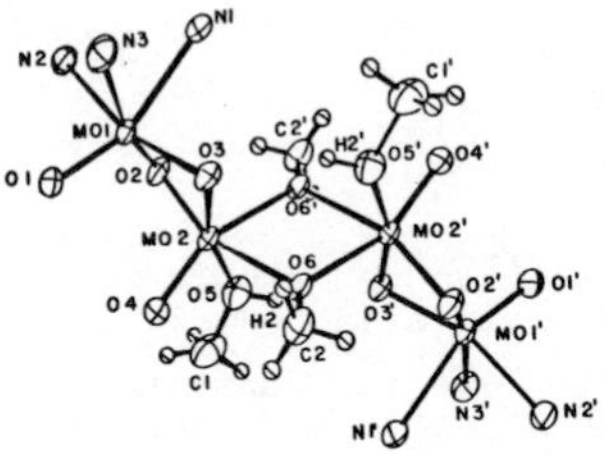

The structure consists of four distorted octahedra linked together by edge sharing to give a fragment of a zig-zag chain. The Mo(V) tetramer is a dimer of dimers with two $Mo(V)_2O_4^{2+}$ dimers joined together by two bridging methoxides. An analysis of the original crystal structure parameters of $Mo_4Cl_4O_6(O\text{-}n\text{-}Pr)_6$ (J. Chem. Soc. Dalton, 1973, 1376-1380) leads to the conclusion that it has been misformulated. Reformulated as $Mo_4Cl_4O_6(O\text{-}n\text{-}Pr)_4(n\text{-}HOPr)_2$ this $Ti_4(OR)_{16}$ type cluster is a Mo(V) tetramer rather than a mixed valence (2Mo(V), 2Mo(VI)) cluster.

STRUCTURE AND BONDING OF MELAMINIUM BETA-OCTAMOLYBDATE. William J. Kroenke, John P. Fackler, Jr., and Anthony M. Mazany. BFGoodrich Research and Development Laboratory, Brecksville, Ohio 44141 and the Chemistry Department, Case Western Reserve University, Cleveland, Ohio 44106.

No crystal structures of organoammonium beta-octamolybdates have been published. In fact, the only $\beta\text{-}Mo_8O_{26}^{4-}$ structure reported is from the hydrated ammonium beta-octamolybdate. (Lindquist, Arkiv Kemi, 2, 349, 1950. L. O. Atovmyan and N. Krasochka, Zh. Strucht. Khim., 13, 342, 1972. H. Vivier, J. Bernard, and H. Djomaa, Rev. Chim. Mine., 14, 584, 1977.) Melaminium beta-octamolybdate is the first melaminium isopolymolybdate reported in the open literature. It crystallizes in the triclinic system with space group $P\bar{1}$ and Z=1. The unit cell dimensions are a = 10.286, b = 10.706, c = 10.209 Å, α = 115.75, β = 85.58, and γ = 93.11°. The three-dimensional x-ray structure was determined by Patterson and Fourier methods with least squares refinement to a conventional R factor of about 0.03. The $\beta\text{-}Mo_8O_{26}^{4-}$ anions in the melaminium and ammonium molybdates are similar, but show significant differences in certain bond lengths and angles. The structure contains ring-protonated melamine rings, similar to the ring-protonated cyclotriphosphazene rings in a cyclophasphazinium hexamolybdate. (H. R. Allcock, E. C. Bissell, and E. T. Shawl, Inorg. Chem., 12, 2963, 1973.) The protonated ring nitrogens are located close to molybdate oxygens (~2.75 Å). Since the protons are displaced toward these close-approach oxygens, hydrogen bonding is indicated. Such hydrogen bonding interactions appear to stabilize the structure, and explain its high thermal stability.

ELECTRONIC CONSTRAINTS ON INNER-SPHERE ELECTRON-TRANSFER REACTIONS. K. Kumar, F. P. Rotzinger, and J. F. Endicott. Wayne State University, Department of Chemistry, Detroit, Michigan 48202

The factors contributing to the reactivity patterns of inner-sphere electron-transfer reactions are not yet well defined. In order to unearth some of these factors we have examined a variety of reactions involving $Co^{III,II}(N_4)$, $Ni^{III,II}(N_4)$, $Cu^{III,II}(N_4)$ complexes (N_4 = a tetraaza macrocyclic ligand). The intrinsic kinetic advantage of the inner-sphere pathway may be defined in terms of the ratio of rate constants for the inner-sphere and the equivalent outer-sphere pathways, extrapolated to zero free energy change: $(k^{IS}/k^{OS})_0$. Such quantities have been inferred from the necessary rate and equilibrium constant measurements for $M^{III,II}(N_4)Cl$ couples and for the ratio of $Co(OH_2)_5Cl^{2+}$ to $Co(OH_2)_6^{3+}$ oxidations of $M^{II}(N_4)$ complexes. After corrections for variations in equilibrium constants and ion pair effects we find $(k^{IS}/k^{OS})_0 \sim 10^{6\pm1}$, $10^{4.5\pm1}$, $10^{2\pm1}$, and $10^{1\pm1}$ respectively for $Co^{III,II}(N_4)$, $Co^{III}/Ni^{II}(N_4)$ (or $Ni^{III}(N_4)/Co^{II}(N_4)$), $Ni^{III,II}(N_4)$, and $Co^{III}/Cu^{II}(N_4)$ couples. These observations indicate that the kinetic advantage of the coupled transfer of an electron and bridging ligand (inner-sphere pathway) can be correlated with the number of electrons along the 3-center transition state bridging axis (considered as d_{z^2}, p_{z^m} d'_{z^2}) and the effectiveness of the 3-center transition state bonding interaction in mixing the donor and acceptor electron transfer orbitals.

SYNTHESIS OF A TRANSITION METAL-SUBSTITUTED PHOSPHORANE: 2-PENTACARBONYLMANGANESE-2,2'-SPIROBI(1,3,2-BENZODIOXAPHOSPHOLE). M. Lattman, B. N. Anand, D. R. Garrett, and M. A. Whitener. Southern Methodist University, Chemistry Department, Dallas, Texas 75275.

The title compound, 2, has been synthesized by the reaction of

2-chloro-2,2'-spirobi(1,3,2-benzodioxaphosphole), $\underset{\sim}{1}$, with $Na^+[Mn(CO)_5]$
according to

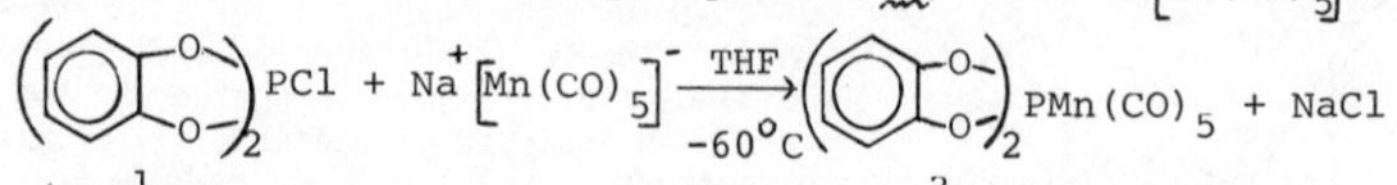

$\underset{\sim}{1}$

$\underset{\sim}{2}$ is the first example of a transition metal-substituted phosphorane, R_4PML_n, where only the phosphorus of the R_4P- group is attached directly to the metal. IR, NMR, and mass spectrometric data are all consistent with a phosphorus(V)-metal bond. The exact geometry about the phosphorus is most likely a square pyramid with the Mn atom at the apex, in accord with previous structural data on related compounds. The synthesis may prove to be general for a hitherto unknown class of molecules.

HETEROGENEOUS KINETICS OF HEXAAMINE Co(III) CAGE COMPLEXES. P. A. Lay and A. M. Sargeson. Research School of Chemistry, Australian National University, P. O. Box 4, Canberra, A. C. T., 2600, Australia.

The heterogeneous kinetics of electron transfer of Co(III) complexes of substituted sar (sar $=$ sarcophagine $= 3,6,10,13,16,19$-hexaazabicyclo[6.6.6]icosane) and absar (absar $=$ absarcophagine $= 3,6,10,13,15,18$-hexaazabicyclo[6.6.5]nonadecane) ligands have been studied at various electrodes and in a variety of different media. The results have been examined in terms of the Marcus-Hush theories of adiabatic electron transfer and compared with those obtained from homogeneous reactions. Problems associated with the outer-sphere reorganizational energy terms have been delineated and will be presented.

OXIDATION OF THE $[Ir(dppe)_2CO]^+$ CATION.[1]   M. A. Lilga.   Department of Chemistry, Northwestern University, Evanston, Illinois 60201.

The $[Ir(dppe)_2CO]^+$ cation (dppe $= 1,2$-bis(diphenylphosphino)ethane) reacts readily with halogens and acid to form the kinetic cis Ir(III) isomers without loss of carbon monoxide.

$$\left[\begin{array}{c} Ph_2P \diagup \text{PPh}_2 \\ Ph_2P \diagdown \overset{|}{Ir} - CO \\ \diagdown \text{PPh}_2 \end{array} \right] [BF_4] \ + \ XY \ \longrightarrow \ \left[\begin{array}{c} Ph_2P \diagup \text{PPh}_2 \\ Ph_2P \diagdown \overset{|}{\underset{|}{Ir}} \diagup CO \\ \diagdown \text{PPh}_2 \diagdown X \end{array} \right] [BF_4][Y]$$

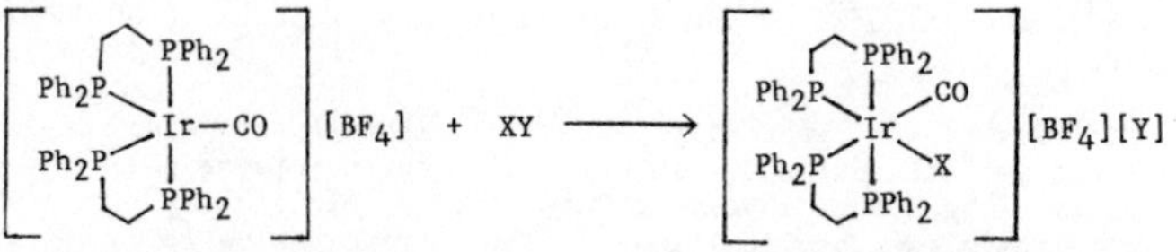

$XY = Cl_2, Br_2, HSO_3CF_3$

For X = halogen, the carbon monoxide ligand is very susceptible to nucleophilic attack. Thus water reacts rapidly with cis-$[Ir(dppe)_2(CO)Cl]Cl_2$ to form trans-$[Ir(dppe)_2(H)Cl]Cl$ with liberation of CO_2 and H^+. Other preparations of both the cis and trans isomers of these Ir(III) dicationic carbonyl complexes, including electrochemical oxidation, and their reactivity with nucleophiles will be discussed.

[1]Research supported by the NSF (CHE80-09671).

SYNERGISM AND PHASE TRANSFER EFFECTS IN HYDROGENATIONS OF MODEL COAL CONSTITUENTS USING IRON CARBONYL CATALYSTS. Thomas J. Lynch, Mahmoud Banah and Herbert D. Kaesz. Dept. of Chemistry, Univ. of California, Los Angeles, CA 90024 and Clifford R. Porter Pentanyl Technologies, 11728 Hwy 93, Boulder, Colorado 80303

The title reactions have been carried out under water gas shift conditions using a variety of iron carbonyl complexes. With CO, H_2O and base, $Fe(CO)_5$ leads to quantitative hydrogenation of the nitrogen containing ring of nitrogen heterocycles. N-formylation of isoquinoline is also observed. Turnover numbers of 22-35 are obtained using excess of nitrogen heterocycle. Anthracene and other aromatic hydrocarbons are more difficult to reduce under the above conditions. The presence of nitrogen heterocycles in the $Fe(CO)_5$/anthracene reaction mixture, however, enhances the reduction of anthracene to 9,10-dihydroanthracene. Nitrogen bases such as dipyridyl and 1,10-phenanthroline are more effective for promoting anthracene reduction than quinoline or isoquinoline. Phase transfer agents also lead to improved yields of reduction products. The nitrogen heterocycles and phase transfer catalysts can be combined to generate a system superior to that using each of these alone with $Fe(CO)_5$.

ELECTRON TRANSFER REACTIONS OF NICKEL(IV) AND NICKEL(III) COMPLEXES. D.H. Macartney and A. McAuley. Department of Chemistry, University of Victoria, Victoria, B.C. V8W 2Y2 Canada.

The stabilisation of higher oxidation states of nickel may be accomplished using either (a) tetraazamacrocyclic (N_4) ligands or (b) polydentate (N_6) amine systems containing oxime groups. In general, the ter-valent state is observed with (a) and nickel (IV) may be formed using (b). Species prepared either by chemical oxidation or by electrolysis are of sufficient stability to enable kinetic studies to be undertaken.

Results are presented on the overall reactions

$$Ni(IV)L^{2+} + 2M^{n+} \xrightarrow{[H^+]} Ni(II)(H_2L)^{2+} + 2M^{(n+1)+}$$

where M = $Fe(OH_2)_6^{2+}$, $Ni(cyclam)^{2+}$ and VO^{2+}. Data are consistent with outer-sphere mechanisms and are interpreted in terms of the Marcus theory. In each case, the rate-determining step is the formation of a nickel (III) intermediate, which reacts in a further rapid electron transfer process. Reactions are compared with those for overall two-electron reductants such as ascorbic acid and hydroquinone.

ELECTRODE-METALLOPOLYMER INTERFACES BY OXIDATIVE ELECTROPOLYMERIZATION. L.D. Margerum, C.D. Ellis, R.W. Murray, and T.J. Meyer. Kenan Labs, University of North Carolina at Chapel Hill, Chapel Hill, NC.27514

The synthesis and oxidative electropolymerization of the amine substi- monomer complexes, $(bpy)_2Ru(4\text{-}pyNH_2)_2(PF_6)_2$ (bpy= 2,2'-bipyridine, 4-$pyNH_2$= 4-aminopyridine), $(bpy)_2Ru(3\text{-}pyNH_2)_2(PF_6)_2$ (3-$pyNH_2$= 3-aminopyridine), and $(bpy)_xRu(5\text{-}phenNH_2)_{3-x}(PF_6)_2$ (5-$phenNH_2$ = 5-amino-1,10-phenanthroline) are described. The resulting polymers form fairly stable, electrochemically active films on the oxidizing electrode, which can be Pt, SnO_2 or vitreous carbon. Complexes containing only one amine substituent are difficult to polymerize, but can be copolymerized with $Ru(5\text{-}phenNH_2)_3(PF_6)_2$. Evidence is presented to suggest that the polymerization process is ligand based and affected by addition of base to the polymerization media. Stability studies carried out indicate that the films lose electrochemical activity upon repeated oxidative cycling in the presence of trace H_2O. Visible and infrared spectroscopies were employed to probe the nature of the polymer link and the results are disscussed with respect to mechanistic studies of the oxidative electropolymerization of aniline.

NEW REACTIONS INVOLVING ATOMS OR SMALL CLUSTER MOLECULES IN MATRICES AT LOW TEMPERATURES, John L. Margrave, Department of Chemistry, Rice University, Houston, TX 77251.

Atoms or small cluster molecules from all sections of the periodic table may be generated in high temperature furnaces and co-condensed at low temperatures with potential reactants, both inorganic and organic, to create an exciting new approach to synthetic chemistry. Work at Rice University over the past decade has involved studies of reactions like Li + CO, Li + C_2H_2, Li + H_2O, Si + H_2O, Ca + NH_3, Si + HF, Fe + CH_3 and some twenty five other elemental species; with these and other reactants. Low temperature adducts have been characterized by infrared, uv-visible and esr spectroscopy and gram-scale syntheses have been conducted. The chemical versatility of this synthetic approach is truly unique. One can react high-temperature species with other high-temperature species or with low-temperature species or with a mixture of such species. Already it has been possible to prepare hundreds of new molecules by this technique. For example, the interaction of silicon atoms with water has led to the identification of $Si{:}OH_2$; HSiOH; $H_2Si(OH)_2$; etc. Extension of this approach to include reactions of metal atoms with ROH, R_2O, ROR', and the many other classes of organic reactants suggests that the synthetic utilization of atoms and clusters will make major contributions to applied preparative chemistry as well as basic chemical research.

THE IDENTITY AND UV SPECTRA OF THE COMPLEXES OF Hg(II) WITH THIOCYANATE; THE COORDINATION NUMBER OF MERCURY (II). Robert E. McCoy, Department of Chemistry, Delaware State College, Dover, Delaware 19901.

From potentiometric and solubility studies Ciavatta and Grimaldi (Inorganica Chimica Acta (1970) 4, 312-318) concluded that Mercury (II) and Thiocyanate Ion form a series of complexes ending with $Hg(SCN)_4^{2-}$. It should be noted, however, that their methods would not detect higher complexes if K_5 and K_6 are no higher than 1.00. In the present work the ultraviolet spectra of sufficiently dilute solutions of Hg(II) were studied at varying concentrations of SCN^-. A peak at 1945 Å was obtained for Hg $(SCN)^+$, and a strong peak at 2220 Å was observed for $Hg(SCN)_2$. The peak for $Hg(SCN)_3^-$ has been difficult to duplicate, but appears to be well below 2400 Å, quite likely less than 2200 Å. At moderately high values of thiocyanate concentration a strong broad peak at 2410 Å is obtained; comparison with potentiometric-concentration study indicate that this belongs to $Hg(SCN)_4^{2-}$. At some elevated concentrations, shoulders (possibly due to Hg $(SCN)_5^{3-}$) are observed. In solutions of higher SCN^- concentration a peak appears at 2810 A. This is believed to correspond to $Hg(SCN)_6^{4-}$. Reasoning to support this conclusion is presented.

DETERMINATION OF PRECISE CH BOND PARAMETERS IN METHYL-METAL AND METHYL-METALLOID COMPOUNDS USING A CHD_2-SUBSTITUTION TECHNIQUE. D. C. McKean, G. P. McQuillan, Department of Chemistry, University of Aberdeen, AB9 2UE, Scotland, and D. W. Thompson, Department of Chemistry, College of William and Mary, Williamsburg, Virginia 23185.

The "isolated" CH stretching frequencies in CHD_2-substituted methyl compounds are linearly related to the CH bond length, CH bond dissociation energy and HCH angle. Using simple frequency measurements, CH bond lengths can be determined to ± 0.0005 A or better, and CH bond dissociation energies to ± 2 KJ $mole^{-1}$. Results for the series $(CHD_2)_2M$ (M = Zn, Cd, Hg) and $(CHD_2)_4$ M (M = C, Si, Ge, Sn, Pb) demonstrate a progressive increase in CH bond energy with increasing atomic number of M, and a linear, or almost linear, negative relationship between the C-H bond length and M-C bond dissociation energy. Measurements on methylphosphines enable the properties of CH bonds trans to PC, PH or phosphorus lone pairs to be individually determined. In hexamethyldialuminium, the CH bonds in the terminal methyl groups are shorter and stronger than those in bridging methyl groups. The data also show that the CH bonds in the bridging methyl groups are not all equivalent but that one of them is distinctly longer than the other two.

HYDROGEN BONDING IN SOME METAL CARBONYL - DI·2·PYRIDYLAMINE COMPLEXES. G. P. McQuillan. Chemistry Department, University of Aberdeen, Old Aberdeen AB9 2UE, Scotland.

Crystal structure determinations for the isomorphous Mo and W complexes $[M(CO)_4\cdot$ dipyam] [dipyam = $C_5H_4N)_2NH$] show that the central secondary nitrogen atom is not involved in coordination with the metal but makes a rather close contact with one of the cis-carbonyl oxygen atoms from an adjoining molecule [3·04(1)A, Mo; 2·99(2)A, W]. The calculated NH---O distances are 2·23A (Mo) and 2·19A (W) with NHO angles of 138° and 137°, suggesting a weak but tangible hydrogen-bonding interaction. Consistently with this, the NH stretching frequencies in the crystal i.r. spectra are 45-70 cm^{-1} lower than in CH_2Cl_2 solution and the B_2 carbonyl stretching modes (which are essentially vibrations of the cis-CO groups) are depressed by about 40 cm^{-1}. The chromium complex $[Cr(CO)_4\cdot$dipyam] has no NH---O distances shorter than 2·50A, and the NH and CO vibrations in the crystal spectra are less noticeably shifted than those in the other two complexes. Solubilities decrease markedly in the order Cr>Mo>W. Vibrational data for trisubstituted complexes [M(CO)$_3$·dipyam·L] (L = heterocyclic amine) indicate that many of these probably also involve hydrogen bonds at least as strong as those in the tetracarbonylmolybdenum and tungsten complexes.

SYNTHESIS OF DIMERIC LANTHANIDE AND YTTRIUM HYDRIDE COMPLEXES, $[(\eta^5\text{-}C_5H_4R)_2Ln(THF)(\mu\text{-}H)]_2$, AND THEIR REACTIVITY WITH UNSATURATED ORGANIC SUBSTRATES. James H. Meadows, and William J. Evans, Department of Chemistry, University of Chicago, Chicago, Il. 60637

The synthesis of $[(\eta^5\text{-}C_5H_4R)_2Y(THF)(\mu\text{-}H)]_2$, I, (R = H,$CH_3$), can be achieved in high yield by the hydrogenolysis of $(\eta^5\text{-}C_5H_4R)_2YCH_3$ in toluene/THF for R = H and in hexane/THF for R = CH_3. For lanthanide metals as large as terbium, analogous hydrides can be obtained similarly in high yield. The yttrium hydrides, which have been characterized by IR spectroscopy, ^1H, ^{13}C and ^{89}Y NMR spectroscopy, quantitative deuterolysis and X-ray diffraction (by J.L. Atwood and W.E. Hunter), react with a variety of unsaturated organic molecules to generate new organoyttrium complexes. I reacts with ethene and propene at room temperature to form the corresponding ethyl and propyl complexes, but the analogous reaction with 2-methylpropene fails under similar mild conditions. The reaction of 1,2-propadiene with I generates a π allyl complex. I reacts with nitriles and internal alkynes by cis addition of $(C_5H_4R)_2Y\text{-}H$ across the multiple bond. Terminal alkynes are metallated by I to form H_2 and the alkynide, $[(C_5H_4R)_2YC{\equiv}CR']_2$. The reaction of I with CO is more complex than the previous reactions and involves rapid ligand redistribution.

THE USE OF STERICALLY ENCUMBERED LIGANDS TO STABILIZE THE FE(III)-SR COORDINATION MODE: A MODEL FOR OXIDIZED RUBREDOXIN. Michelle Millar and Joe Lee, Department of Chemistry, New York University, New York, New York 10003; and Stephen A. Koch and Ronald Fikar, Department of Chemistry, SUNY Stony Brook, Stony Brook, New York 11794.

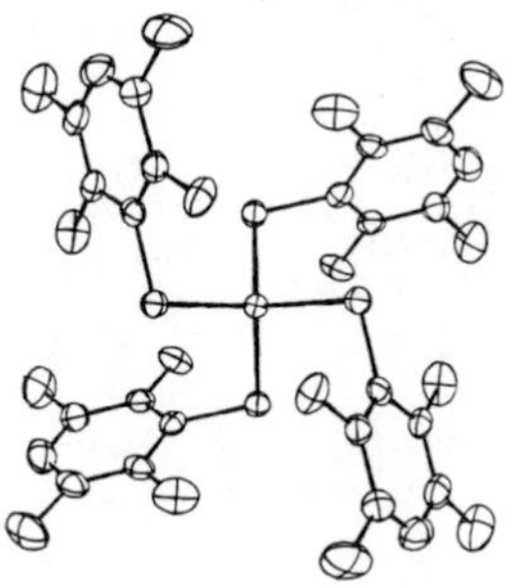

We report the preparation, structure and properties of the first example of a stable Fe(III) tetrathiolate complex containing only monodentate ligands. The reaction of four equivalents of lithium 2,3,5,6-tetramethylbenzenethiolate with $FeCl_3$ and Et_4NBr gives $(Et_4N)[Fe(SC_{10}H_{13})_4]$ (1) in high yield. An x-ray diffraction study of 1 reveals the high symmetry of the anion. Crystal site symmetry demands that the discrete anions have the rare S_4 ($\bar{4}$) point group symmetry. The FeS_4 local symmetry is therefore rigorously D_{2d}, compressed from T_d symmetry along the S_4 axis. The two equivalent S-Fe-S angles bisected by the S_4 axis are 114.4(1)°, while the remaining four angles are equal to 107.08(5)°. The Fe-S bond distance is 2.284(2) Å. The electronic and structural characteristics of 1 are compared to those of the protein, rubredoxin, one of the rare examples of a stable $[Fe(III)(SR)_4]^-$ compound.

INORGANIC APPLICATIONS OF FAST ATOM BOMBARDMENT MASS SPECTROMETRY. Jack M. Miller Department of Chemistry, Brock University, St. Catharines, Ontario, Canada, L2S 3A1

The potential of the new mass spectrometry sample introduction and ionization techique, FAST ATOM BOMBARDMENT (FAB) is discussed with examples of inorganic and bioinorganic interest. FAB spectra produce molecular or quasimolecular ions along with useable fragmentation patters (both positive and negative ion spectra) for compounds too involatile for normal inlet systems or too labile for volatilization without thermolysis. Water of hydration can sometimes be observed.

CARBON-CARBON DOUBLE BOND FORMING REACTIONS OF COORDINATED CARBON MONOXIDE.
Kenneth G. Moloy and Tobin J. Marks, Department of Chemistry, Northwestern University, Evanston, Illinois 60201.

Previous work in our laboratory has shown that the hard Lewis acid character (i.e., a high degree of coordinative unsaturation and consequent oxophilicity) of actinide organometallics may be exploited to produce exceedingly reactive carbene-like dihaptoacyls and dihaptoformyls. In order to further elucidate the mechanisms of CO coupling characteristic of these compounds, we have undertaken a study of the reactions of isonitriles in place of the isoelectronic carbon monoxide.

The reaction of isonitriles (RNC) with various thorium dihaptoacyls, $[\eta^5\text{-}C_5(CH_3)_5]_2Th(Cl)(\eta^2\text{-}COCH_2R')$, yields products resulting from a carbene-like coupling reaction. On the basis of the standard spectroscopic techniques, we assign the ketenimine structure A to these novel complexes. The potential intermediacy of the analogous and isoelectronic ketenes in the formation of enediolates (B) and enedionediolates (C) will be discussed.

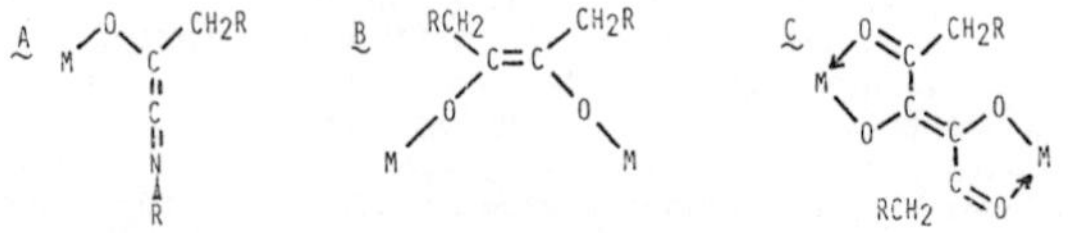

CLEAVAGE REACTIONS OF BINUCLEAR COMPLEXES OF RHODIUM(III) AND IRIDIUM(III).
Mariel M. Muir, Janio Szklaruk, and Sonia C. Fernandez. Department of Chemistry, University of Puerto Rico, Rio Piedras, Puerto Rico 00931.

Binuclear complexes of Rh(III) and Ir(III) of the type $M_2(R_3P)_4X_6$, which have two halide bridges, undergo cleavage reactions with amines and other neutral ligands. The product isolated is that expected for symmetrical cleavage, $M(R_3P)_2(A)X_3$. However, a study of the rates of reaction for several such complexes has shown that the product formed in the cleavage reaction is ionic, which results from unsymmetrical cleavage.

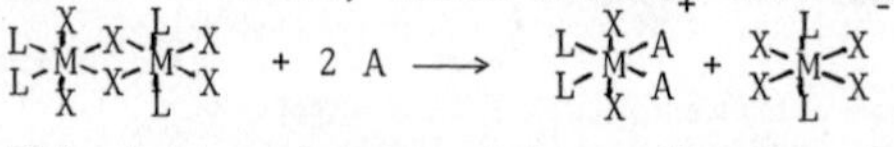

This is consistent with predictions based on the trans effect. The rate of cleavage is independent of the concentration of amine. The variation in rate with variation in the metal, the phosphine, the halide, and the solvent will be discussed.

REACTIONS OF COPPER(I) COMPLEXES WITH THE SUPEROXIDE ADDUCT OF COBALT(II) CYCLAM COMPLEX.
M. Munakata and J. F. Endicott. Wayne State University, Department of Chemistry, Detroit, Michigan 48202

The stopped flow method was used to study the oxidation of copper(I) complexes by the transient superoxide adduct, $[Co^{III}([14]aneN_4)(H_2O)(O_2^-)]$. The rate constants for the oxidation of three-coordinated copper(I) complexes, $Cu(tpy)^+$, $Cu(C_2H_4)(H_2O)_2^+$ and $CuCl_3^{2-}$ by the superoxide adduct are 5.2×10^5, 2.4×10^5, and 1.1×10^6 $M^{-1}s^{-1}$, respectively. On the other hand, the two-coordinated copper(I) complexes, $CuCl_2^-$ and $Cu(Imid)_2^+$, as well as the tetrahedral complexes, $Cu(bpy)_2^+$ and $Cu(4,4'\text{-}Me_2\text{-}bpy)_2^+$ react only very slowly, if at all. The data can be interpreted in terms of reorganizational energy associated with formation of the μ-peroxo intermediate, $Cu(II)\text{-}O_2^{2-}\text{-}Co(III)$. The μ-peroxo intermediate formed from the reaction of two-coordined copper(I) and superoxide adduct would require additional ligands to maintain the preferred four coordination geometry of Cu(II), whereas the three coordinated copper(I) complexes could readily form μ-peroxo intermediates with four-coordinated copper(II). It is more difficult to form the μ-peroxo complex of square-pyramidal copper(II) starting from tetrahedral copper(I).

THE USE OF LIGANDS AS OPTICAL STRUCTURAL PROBES (IV): A SOLUTION KERR EFFECT STUDY OF ORGANOTIN (IV) TROPOLONATES, OXINATES AND DIBENZOYLMETHANATES. W. H. Nelson and S. K. Brahma. Department of Chemistry, University of Rhode Island, Kingston, RI 02881.

Molar Kerr constants have been determined for several complexes of the type R_2SnCh_2 [R = methyl, ethyl or butyl and Ch^- = oxinate, tropolonate or dibenzoylmethanate] at 25°C in cyclohexane using a 632.8 nm laser source. As expected cis-bis(8-quinolinolato)dimethyltin (IV) is strongly negatively birefringent. The other complexes while polar are all positively birefringent. Large positive Kerr constants accompanied by nonzero permanent dipole moments in several instances point toward the predominance of distorted trans-type molecules which in several instances can best be characterized by skew-trapezoidal bipyramidal structures. Results show that the Kerr effect should be an effective means of determining the structures of complexes of the type R_2SnCh_2 in solution if the ligands Ch^- are planar and highly optically anisotropic.

A MOLECULAR ORBITAL STUDY OF THE BONDING OF THIOETHER DONORS TO COPPER, David E. Nikles, Alfred B. Anderson and F. L. Urbach, Case Western Reserve University, Department of Chemistry, Cleveland, Oh 44106.

Semi-empirical molecular orbital calculations were carried out on some thioether-containing copper(II) and copper(I) complexes to provide a theoretical basis for understanding the nature of this interaction. An initial series of model calculations, using dimethylsulfide as the prototype thioether, was undertaken to identify the thioether orbitals available for bonding to the copper ion and to understand the thioether-copper interaction. The sulfur 3d orbitals were included in these calculations to examine their ability to mix with filled copper 3d orbitals. These calculations were extended to crystallographically defined systems in order to correlate the theoretical results with experimental observations. Our molecular orbital studies provide a basis for interpreting the charge transfer spectra of thioether-containing copper(II) complexes and lend insight into the role of the sulfur 3d orbitals in stabilizing the copper(I) state.

PHOTOCHEMICAL REACTIONS OF $Re_2(CO)_{10}$ WITH OLEFINS. Philip O. Nubel and Theodore L. Brown, School of Chemical Sciences, University of Illinois, Urbana, IL 61801.

UV photolysis of $Re_2(CO)_{10}$ with excess 1-alkene in non-donor solvents results in the formation of hydridoalkenyl complexes of the general formula $(\mu\text{-H})Re_2(CO)_8(\mu\text{-}\eta^2\text{-CH=CHR})$. Spectroscopic studies indicate the alkenyl moiety to be bridging, forming a σ bond to one rhenium and a π bond to the other, with the R group trans to the Re—C σ bond. The reaction is proposed to proceed through a light-induced radical pathway, producing a 1,2-$Re_2(CO)_8(\eta^2\text{-alkene})_2$ intermediate which undergoes thermal reaction to give the observed product. Analogous reactivity is observed with cis, although not trans, isomers of internal olefins. Interestingly, photolysis of $Re_2(CO)_{10}$ with ethylene generates 1-butene alkenyl species in addition to the ethylene complex. The hydridoalkenyl compounds undergo thermal reaction with a variety of ligands at room temperature to give 1,2-$Re_2(CO)_8L_2$ complexes.

500 MHZ HIGH RESOLUTION NMR OF INORGANIC SOLIDS: TO MASS OR VASS? Eric Oldfield, Subramanian Ganapathy, Suzanne Schramm, Robert A. Kinsey, Karen Ann Smith and Michael D. Meadows. School of Chemical Sciences, University of Illinois at Urbana-Champaign, 505 S. Mathews Avenue, Urbana, Illinois 61801.

We will discuss the application of magic-angle sample spinning (MASS, at 150, 360 and 500 MHz ^1H resonance frequencies) and the new method of variable-angle sample-spinning (VASS) to the NMR of quadrupolar and spin-½ nuclei, eg. ^{11}B, ^{17}O, ^{23}Na, ^{27}Al, ^{29}Si, ^{51}V, ^{55}Mn, ^{95}Mo, ^{99}Ru, ^{103}Rh, ^{109}Ag and ^{183}W, in a variety of inorganic solids. Emphasis will be placed on aluminosilicates, especially RE-exchanged zeolites, and cluster compounds. The new technique of VASS optimizes linenarrowing of second-order quadrupole interactions--allowing study of nuclei in solids with large quadrupole coupling constants. Our results at 500 MHz emphasize the importance of very high field NMR operation for studies of quadrupolar nuclei in inorganic solids, and suggest that data obtained at even higher-field, eg. 700 MHz, will in many cases be about twice as resolved as at 500 MHz. (Work supported by NSF and NIH).

COMPLEXES OF MOLYBDENUM (VI) AND (V) WITH TRIETHYLENETETRAAMINEHEXAACETIC ACID.
N. Pariyadath. Fort Valley State College Agricultural Research Station, Box 5459, Fort Valley, GA 31030.

Triethylenetetraaminehexaacetic Acid (H_6TTHA) reacts, in aqueous medium, with sodium molybdate and molybdenum pentachloride. Product isolation, and elemental analysis led to the formulation, $Na_6Mo_2O_6(TTHA)\cdot 3H_2O$, for the colorless Mo (VI) compound, and $Na_4Mo_2O_4(TTHA)\cdot 6H_2O$ for the orange Mo (V) compound. Molar conductivity, and electronic and infrared spectra support these formulations. Spectroscopic properties of these compounds are similar to those for corresponding edta complexes.

REACTIONS OF DIMOLYBDENUM AND DITUNGSTEN HEXA-ALKOXIDES WITH POTENTIAL CARBENE SOURCES.
A.L. Ratermann, M.H. Chisholm and J.C. Huffman. Department of Chemistry and Molecular Structure Center, Indiana University, Bloomington, Indiana 47405.

Dimolybdenum hexa-alkoxides ($Mo_2(OR)_6$, R = CH_2CMe_3, $CHMe_2$) react with diazoalkanes (RR'C=N=N; R = C_6H_5, R' = C_6H_5, H) and in the presence of a nitrogen donor ligand (L = pyridine, $HNMe_2$) compounds of the general formula $Mo_2(OR)_6(RR'CN_2)_2L$ crystallize from solution. An X-ray crystal structural study of $Mo_2(O\text{-}i\text{-}Pr)_6[(C_6H_5)_2C{=}N{=}N]_2py$ shows two terminal diazoalkane ligands and three bridging alkoxide ligands in an interesting asymmmetric molecule having formally mixed valence molybdenum atoms with oxidation numbers +4½ and +5½. Ditungsten hexa-tert-butoxide reacts with diphenyldiazoalkanes to form $W_2(O\text{-}t\text{-}Bu)_6[N{=}N{=}C(C_6H_5)_2]_2$. The X-ray crystal structural analysis of this compound is under study.

Gem-dihalides ($RR'CBr_2$; R = C_6H_5, H, R' = H) and $Mo_2(OCH_2CMe_3)_6py_2$ react together and yield $Mo_2(OCH_2CMe_3)_6Br_2py$. The organic products formed are the expected products of a free radical process, (BrRR'CCRR'Br). An X-ray crystal structural analysis of $Mo_2(OCH_2CMe_3)_6Br_2py$ reveals a confacial bioctahedral geometry with one bridging and one terminal bromide ligand. The Mo-Mo bond distance of 2.53 Å indicates a metal-metal double bond consistent with an average Mo^{IV} oxidation state. Proton NMR spectroscopy indicates the molecule is fluxional at 16°C but has six inequivalent alkoxide ligands at -45°C. Further reactions of this compound are in progress.

ORGANO-TRANSITION-METAL STABILIZED ARSENIC RINGS AND CHAINS.
Arnold L. Rheingold and Patrick J. Sullivan. Department of Chemistry, University of Delaware, Newark, Delaware 19711.

Organo-transition-metal substituents exhibit the well-documented ability to stabilize otherwise unknown arrangements of main-group homonuclear catenates. In our continuing study of the reactions of As and Sb homoatomic rings and chains with cyclopentadienyl and arene metal carbonyls, several novel and unpredicted catenate structures have recently been prepared and structurally characterized by X-ray diffraction. For instance, from reactions of $[CpMo(CO)_3]_2$ with cyclopolyarsenic derivatives under varying reaction conditions, several new metal-polyarsenic structures have been isolated. Of these structures, perhaps the most extraordinary is a triple-decker sandwich, CpMo-μ_2-(η^5-<u>cyclo</u>As_5)-MoCp containing three coplanar, eclipsed five-membered rings, the middle one being a ring of five As atoms. Other products have cluster structures, e.g. $[CpMo(AsCH_3)]_4$ containing a tetrahedron of CpMo units with facially bonded $AsCH_3$, each bridging three Mo atoms. A precursor complex to the above two has also been characterized as $[CpMo(CO)_2]_2$-<u>catena</u>$(CH_3As)_5$. Additionally, structures will be reported resulting from reactions of As rings with (benzene)$Cr(CO)_3$ and $[CpFe(CO)_2]_2$. In the case of the Cr complexes, replacement of benzene occurs under thermal reaction conditions, whereas CO replacement occurs with photochemical activation.

OXIDATION AND ACIDOLYSIS OF $MoH_4(PR_3)_4$.
L. F. Rhodes, K. Folting, J. C. Huffman, and K. G. Caulton. Department of Chemistry and Molecular Structure Center, Indiana University, Bloomington, Indiana 47405.

The oxidation of $MoH_4(PMe_2Ph)_4$ by $AgBF_4$ in CH_3CN yields $[MoH_2(PMe_2Ph)_4(CH_3CN)_2](BF_4)_2$, which is characterized by spectral and X-ray crystallographic methods. Unexpectedly, silver oxidation of MoH_4-

$(PMePh_2)_4$ in CH_3CN gives $[MoH_2(PMePh_2)_3(CH_3CN)_3](BF_4)_2$ plus one mole of free phosphine. Since treatment of $MoH_4(PR_3)_4$ with $HBF_4 \cdot Et_2O$ in CH_3CN gives the same product as the silver oxidation of these two starting materials there may be some mechanistic similarities. Also, the relationship between the tetrakis(phosphino)dihydride product and the tris(phosphino)dihydride product is explored.

PERIODICITY IN THE ACID-BASE BEHAVIOR OF OXIDES AND HYDROXIDES.
Ronald L. Rich. Bluffton College, Bluffton, Ohio 45817.

Six periodic charts show the calculated aqueous solubilities, as functions primarily of pH, of oxides and hydroxides. Three of the charts relate to oxidation states (II), (III) and (IV) respectively. The other three relate to central "cations" with noble-gas electron structures, to those with mainly pseudo-noble-gas structures, and to those with "inert" electron pairs. The latter chart is partly shown here. Many graphs include two curves. One represents relatively rapid precipitations and changes of pH and does not include soluble polynuclear species. The other (or the only one) is derived from all the selected data for true equilbrium. Ionic strengths are at or near 1.0 molar, and temperatures are at or near 298 K. We observe various dependencies of acidity and basicity on oxidation state, electron structure, and position in the periodic table, along with, say, size and coordination number.

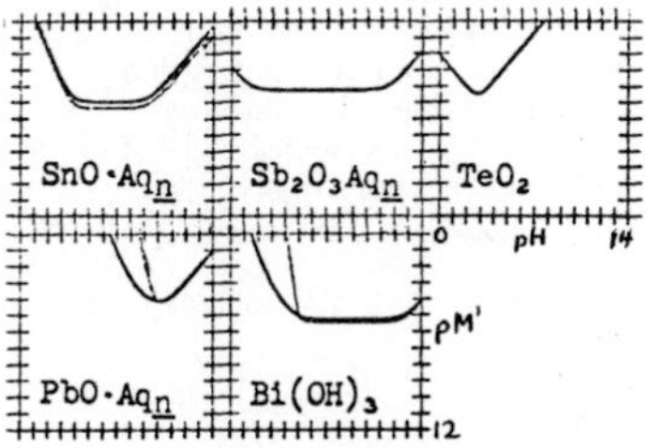

PHOTODISPROPORTIONATION OF μ-OXO-BIS (TETRAPHENYLPORPHINATO IRON(III)).
R. M. Richman and M. W. Peterson. Department of Chemistry, Carnegie-Mellon University, 4400 Fifth Avenue, Pittsburgh, PA 15213.

Irradiation into the Soret band of the title compound, $(FeTPP)_2O$, in benzene or pyridine in the presence of triphenylphosphine (PPh_3) results in the production of coordinated $Fe^{II}TPP$ and $OPPh_3$. In benzene, there is Stern-Vollmer quenching by $(PPh_3)FeTPP$. Experimental results at several concentrations of PPh_3 in both solvents are explained by a mechanism in which the primary photoprocess (quantum yield 1.0×10^{-4}) is disproportionation to the high energy oxidizing agent, FeOTPP. This mechanism represents an alternative strategy for solar energy conversion in solution that does not require a long-lived charge transfer excited state. The excited state may be short-lived and dissociative, a more common condition for transition metal complexes. The key to energy storage is that one photoproduct be a stronger oxidant than the starting compound or the other be a stronger reductant.

REDOX BEHAVIOR OF TRIS-CHELATE AND MIXED LIGAND COMPLEXES OF RUTHENIUM(II).
D. P. Rillema, G. Allen, and T. J. Meyer. Departments of Chemistry, University of North Carolina, Chapel Hill, NC 27514 and Charlotte, NC 28223.

The complexes $Ru(bpy)_3^{2+}$, $Ru(bpyrz)_3^{2+}$, $Ru(bpyrm)_3^{2+}$ (bpy = 2,2'-bipyridine, bpyrz = 2,2'-bipyrazine, bpyrm = 2,2'-bipyrimidine) and the mixed ligand complexes $Ru(bpy)_2L^{2+}$ and $Ru(bpy)L_2^{2+}$ have been synthesized (L = bpyrz and bpyrm) and their electronic and redox properties have been investigated. The absorption spectra contain intense $\pi \rightarrow \pi^*$ transitions in the 250-290 nm region and $d\pi \rightarrow \pi^*$ absorption from 400-470 nm. Polarographic half-wave potentials were 1.98V for $Ru(bpyrz)_3^{3+/2+}$, 1.72V for $Ru(bpyrz)_2(bpy)^{3+/2+}$ and 1.49V for $Ru(bpy)_2(bpyrz)^{3+/2+}$ couples. (Potentials vs. SSCE). The Ru (III-II) potentials vary linearly with the number of bpyrz or bpyrm ligands. The correlations for each series have different slopes and intersect at the $Ru(bpy)_3^{3/2+}$ potential. Reductions are ligand centered, which are sequentially reduced by one electron. The half-wave potential for the first reduction of $Ru(bpyrz)_3^{2+/+}$ is -0.68V, $Ru(bpyrz)_2(bpy)^{2+/+}$ is -0.79V, and $Ru(bpy)_2(bpyrz)^{2+/+}$ is -0.91V. A plot of $E_{1/2}$ for the Ru(III-

II) couple vs. $E_{1/2}$ for the first ligand reduction is linear for each series indicating that the oxidation-reduction processes are coupled by an intricate interplay of electron donor-electron withdrawal effects, primarily through $d\pi \rightarrow \pi^*$ interactions. The difference between the two couples represents the ground state energy gap which is found to give a linear correlation with the emission energy. This indicates that the ^{3}CT manifold from which emission occurs moves in concert with the variation in ligand field.

STEREOCHEMISTRY AND ^{113}Cd NMR SPECTROSCOPY OF Cd COMPLEXES. P.F. Rodesiler, E.A.H. Griffith and E.L. Amma,* Departments of Chemistry, University of South Carolina and Columbia College, Columbia, SC 29208.

^{113}Cd nmr with a chemical shift range of ~900 ppm has shown considerable promise as a probe of metal ion environment. Recent developments in solid state magic angle spinning ^{113}Cd nmr have further advanced the promise. A structure-chemical shift correlation appears possible and of utility as a structural tool. We have undertaken a program involving ^{113}Cd nmr in solution and in the solid state and x-ray crystallographic structure determinations. In the more shielded region, -50 to -100 ppm above the standard $Cd(H_2O)_6(ClO_4)_2$, we have found a correlation between seven coordination (7 0) and chemical shift. We have found a number of examples of pentagonal bipyramidal geometries involving benzoato ligands. In addition we have found some sulfur coordinated ^{113}Cd resonances that are much further upfield than would have been predicted by earlier correlations, +263 ppm. Furthermore we shall show examples of large (~60 ppm) ^{113}Cd resonance differences due to different oxygen ligands with the same coordination polyhedron.

*Supported by NIH, GM 27721. We are grateful for the assistance of the NSF NMR center at the University of South Carolina and Colorado State University.

THE CONVERSION OF $Co_2(CO)_8$ TO $Co_4(CO)_{12}$ ON OXIDE SUPPORT MATERIALS by R. L. Schneider, R. F. Howe and K. L. Watters, Dept. of Chemistry and Laboratory for Surface Studies, University of Wisconsin-Milwaukee, Milwaukee, Wisconsin 53201.

Dicobalt octacarbonyl, sublimed directly onto silica or zeolite surfaces, is very rapidly converted to tetracobalt dodecacarbonyl according to the reaction

$$2Co_2(CO)_8 \longrightarrow Co_4(CO)_{12} + 4CO$$

The same reaction occurs only slowly for $Co_2(CO)_8$ in solution or as a solid. The oxide surface catalyzed reaction is inhibited by low pressures (> 10 torr) of CO. A nearly stoichiometric conversion occurs on the silica surface, while zeolite supports catalyze further reaction of the $Co_4(CO)_{12}$ to species of lower carbonyl content and $Co(CO)_4^-$. The zeolite study provides an interesting example of a chemical reaction in which the reactant ($Co_2(CO)_8$) enters the zeolite supercage where it reacts to form product ($Co_4(CO)_{12}$) which is too large to exit the cage. The ir spectra of surface hydroxyl groups on silica and on HY zeolite and shifts in the ν(CO) frequencies infer a hydrogen bonding interaction of surface hydroxyl with bridging CO ligands of the cobalt carbonyls. This interaction is strongest in the zeolite case where distances between bridging CO's and surface hydroxyls match quite well. Exchange of ^{13}CO for ^{12}CO is rapid for the surface supported $Co_4(CO)_{12}$ and allows for solvent extraction of a totally exchanged form of this compound from the silica surface.

COMPLEXATION AND DEHYDRATION OF ALDOXIMES BY RU(II) Kenneth Schug and Charles P. Guengerich. Illinois Institute of Technology, Chicago, Illinois, 60616.

It was previously proposed (1) that an oxime intermediate is involved in the reaction of $(H_3N)_5RuNO^{3+}$ with the α-methylene group of ketones (leading to C-C bond scission and formation of ligand nitrile). Direct evidence for this step of the mechanism is provided by recent studies of the reaction of $(H_3N)_5RuOH_2^{2+}$ with several aldoximes:

$$(H_3N)_5RuOH_2^{\ 2+} + HO{-}N{=}C\begin{matrix}H\\R\end{matrix} \rightarrow (H_3N)_5Ru{-}N{\equiv}CR + H_2O \qquad (1)$$

The reaction proceeds as shown over a wide pH range (0.5M H^+ to 0.2M OH^-) with product yields of 75-90%. A qualitative measure of the rate of complexation is provided by experiments in which R = pyridyl; the nitrile/pyridine product ratio exceeds 9 for all three isomers. The driving force for the dehydration is attributed to the well known tendency for Ru (II) to form strong bonds to unsaturated nitrogen atoms via d→π*backbonding.

(1) K. Schug and C.P. Guengerich, Inorg. Chem. 1979, 101, 235-6

SYNTHESES, STRUCTURES AND CHEMISTRY OF CATIONIC IRON AND RUTHENIUM METALLACUMULENE COMPLEXES. John P. Selegue, Department of Chemistry, University of Kentucky, Lexington, KY 40506.

Electron-rich cyclopentadienyl metal complexes $[MX(PR_3)_2(\eta\text{-}C_5H_5)]$ (M=Fe, Ru; PR_3=PPh_3, PMe_3 or $(PR_3)_2$=dppe) react with alkyn-1-ols to give cationic allenylidene complexes via the intermediacy of hydroxyvinylidene complexes $[(Cp)(PR_3)_2M{=}C{=}CHC(OH)R_2]^+$. For example, $[RuCl(PMe_3)_2(Cp)]$ reacts smoothly with $Ph_2C(OH)C{\equiv}CH$ and NH_4PF_6 to give $[(Cp)(PMe_3)_2Ru{=}C{=}C{=}CPh_2][PF_6]$ (I) which has been characterized by an X-ray crystal structure analysis. The ruthenium-carbon bond is quite short (1.884(5)Å, with a nearly linear ruthenacumulene chain. In contrast, the reaction of $[RuCl(PPh_3)_2(Cp)]$ with $Me_2C(OH)C{\equiv}CH$ leads to the formation of $[Ru_2(\mu\text{-}C_{10}H_{12})(PPh_3)_4(Cp)_2][PF_6]_2$ (II), a dimer of the expected dimethylallenylidene complex. The ruthenium atoms are bonded to the $C_{10}H_{12}$ bridge by both vinylidene and alkylidene linkages. The structure of this remarkable molecule and reactions related to its formation will be discussed.

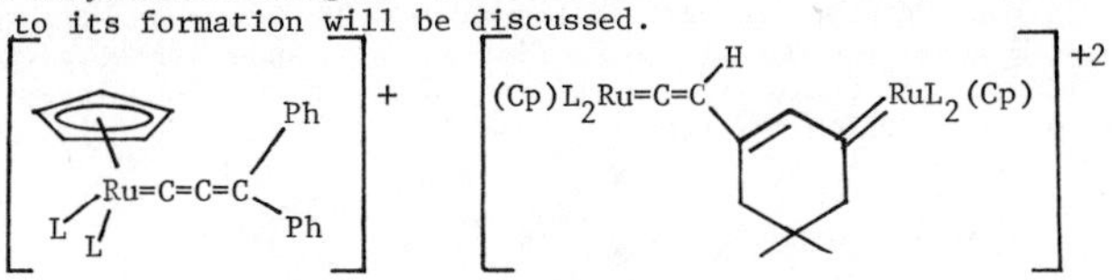

(I) (II)

FIVE-COORDINATE BENT METALLOCENES. STRUCTURE AND DYNAMICS OF BIS(η-CYCLOPENTADIENYL)-CHLORO(N,N-DIALKYLDITHIOCARBAMATO)ZIRCONIUM(IV) COMPLEXES. Michael E. Silver and Robert C. Fay, Department of Chemistry, Cornell University, Ithaca, New York 14853, and Odile Eisenstein, Department of Chemistry, University of Michigan, Ann Arbor, Michigan 48104.

The crystal and molecular structure of bis(η-cyclopentadienyl)chloro(N,N-diethyldithiocarbamato)zirconium(IV), $(\eta\text{-}C_5H_5)_2ZrCl[S_2CN(C_2H_5)_2]$, has been determined by X-ray diffraction (R_1 = 0.038; R_2 = 0.048). The complex has a five-coordinate bent metallocene structure with Zr, Cl, and the two S atoms of the bidentate dithiocarbamate ligand lying in a quasi-mirror plane that is nearly perpendicular to the plane defined by the Zr atom and the centroids of the two symmetrically attached $\eta\text{-}C_5H_5$ groups. Because of crowding in the $ZrClS_2C_{10}$ coordination group, the Zr-Cl and Zr-S bonds to the lateral coordination sites are unusually long (2.556 (2) and 2.723 (2) Å, respectively); these distances are the longest terminal Zr-Cl and Zr-S bond lengths yet observed for any structure involving Zr(IV). A 300-MHz ^1H NMR study has yielded the following kinetic data for methyl group exchange in $(\eta\text{-}C_5H_5)_2ZrCl[S_2CN(CH_3)_2]$: $k(25\ °C) = 8.6\ s^{-1}$, $\Delta G^{\neq}(25\ °C) = 16.18 \pm 0.08$ kcal/mol, $\Delta H^{\neq} = 15.9 \pm 2.0$ kcal/mol, $\Delta S^{\neq} = -1.0 \pm 6.8$ eu. Possible mechanisms for the exchange process have been analyzed on the basis of kinetic and structural data and extended Huckel M.O. calculations. A mechanism involving C$\overset{\cdots}{-}$N bond rotation is preferred.

CATIONIC FORMYL COMPLEXES OF RUTHENIUM(II), G. Smith and D.J. Cole-Hamilton, Dept. of Inorganic, Physical and Industrial Chemistry, Donnan Laboratories, University of Liverpool, P.O. Box 147, Liverpool L69 3BX, England.

Dicationic carbonyl complexes of ruthenium(II) of the form $[Ru(CO)_2(P\text{-}P)_2]^{2+}$, (P-P = $Ph_2P(CH_2)_nPPh_2$, n = 1(dppm) or 2(dppe)) have been prepared from reactions of $RuCl_2(P\text{-}P)_2$ with $AgSbF_6$ under carbon monoxide.

Treatment of $[Ru(CO)_2(P\text{-}P)_2]^{2+}$ with hydride donors at low temperature leads to the formation of monocationic formyl complexes, $[Ru(CHO)(CO)(P\text{-}P)_2]^+$ which have been isolated and fully characterized. The ^{13}C and ^2D labelled compounds have also been prepared.

At room temperature, in solution, these compounds decompose to give $[RuH(CO)(P\text{-}P)_2]^+$ and a product believed to be $[Ru(CO)(P\text{-}P)_2]^{2+}$, although for P-P = dppm, the major product is $[RuH(CO)_2(dppm)_2]^+$ with cis carbonyl groups and a monodentate dppm ligand.

These reactions, together with that of $[Ru(CHO)(CO)(dppe)_2]^+$ with CF_3COOH to give methanol and regenerate $[Ru(CO)_2(dppe)_2]^{2+}$, will be discussed.

PHOTO-OXIDATION OF COPPER (I)-AMMONIA COMPLEXES IN AQUEOUS SOLUTION. Kenneth L. Stevenson, Joseph Harber, Tamim Braish, Department of Chemistry, Indiana-Purdue University at Fort Wayne, Fort Wayne, Indiana 46805

Solutions of CuCl in aqueous ammonia exhibited electron-transfer capability when irradiated in the UV range at wavelengths shorter than about 330nm. An analysis of the absorption spectra and equilibria indicates that while the most predominant species in such a solution may be $Cu(NH_3)_2{}^+$, the photoactivity in the range of 250-330nm is probably due to absorption of light by small traces of $Cu(NH_3)_3{}^+$. The electron scavenger, N_2O was reduced by photolysis of the copper (I) complexes with the quantum yield varying with the square-root of N_2O concentration, suggesting Noyes geminate-pair scavenging kinetics for the photolytically generated electron. The highest quantum yield for Cu (I) photooxidation observed is about 0.25 at 265nm, assuming the reaction to be:

$$Cu(NH_3)_3{}^+ \xrightarrow{h\nu} Cu(NH_3)_3{}^{2+} + e^-\,aq \xrightarrow{N_2O} \text{products}$$

This occurred in 1M NH_3, 0.016M CuCl, saturated with N_2O gas. It was also observed that when $NH_4{}^+$ is added to the system instead of N_2O, an insoluble gas is evolved upon irradiation, with a quantum yield of about 0.02. This gas may be dihydrogen.

ORGANOMETALLIC SONOCHEMISTRY AND SONOCATALYSIS. Kenneth S. Suslick, Paul F. Schubert, Hsien-Hau Wang, James W. Goodale. Department of Chemistry, University of Illinois at Urbana-Champaign, Urbana, Illinois 61801.

Although the chemical effects of high intensity ultrasound have been recognized for more than 40 years, investigation of the sonochemistry of transition metal complexes has not previously been undertaken. The mechanism of sonochemistry involves cavitation within a solution: the creation, expansion, compression and dissolution of small gas vacuoles in the liquid. During this cycle, the rapid compression produces intense local heating with hot spots on the order of 1000°. We have been exploring the chemical effects of high intensity ultrasound on transition metal and organometallic complexes. A variety of unusual reactivity patterns have been observed which are not generated by either photochemical irradiation or bulk solution heating. For example, we have observed controlled multiple CO loss from metal carbonyls resulting in direct clusterification or multiple ligand substitution. The highly coordinatively-unsaturated species produced during sonolysis are extremely active and are able to induce hydrocarbon activation. In addition, such sonochemical ligand dissociation can be used to initiate homogeneous catalysis. Specifically, we have demonstrated catalytic olefin isomerization by metal carbonyls during sonication. This is the first case of "sonocatalysis" by transition metal complexes. Extensions of sonocatalysis to CO hydrogenation, hydroformylation, etc., will be discussed, and comparisons to analogous thermal and photocatalysis will be made.

EXPERIMENTAL ELECTRON DENSITY DISTRIBUTION IN 5,10,15,20-TETRAMETHYLPORPHYRINATONICKEL(II).[1] P. N. Swepston and James A. Ibers. Department of Chemistry, Northwestern University, Evanston, Illinois 60201.

An extensive ($\sin(\theta)/\lambda \leq 1.0$ Å^{-1}), low-temperature (150 K) set of X-ray crystallographic data has been used to determine the electron density distribution in the planar metalloporphyrin Ni(TMP). Experimental deformation density maps ($X-X_{H.O.}$) were calculated ($\sin(\theta)/\lambda<0.7$ $Å^{-1}$) using parameters obtained from a high-order refinement ($\sin(\theta)/\lambda\geq0.7$ $Å^{-1}$). The results indicate that unequal Ni-N bond lengths might be attributable to packing interactions between molecules. Model deformation density maps ($X_{aspherical}-X_{spherical}$) have been calculated ($\sin(\theta)/\lambda\leq1.0$ $Å^{-1}$) using parameters obtained from a multipole refinement ($\sin(\theta)/\lambda\leq1.0$ $Å^{-1}$). The accompanying figure displays a model deformation map calculated through the plane of the molecule with contour intervals of 0.10 e/$Å^3$.

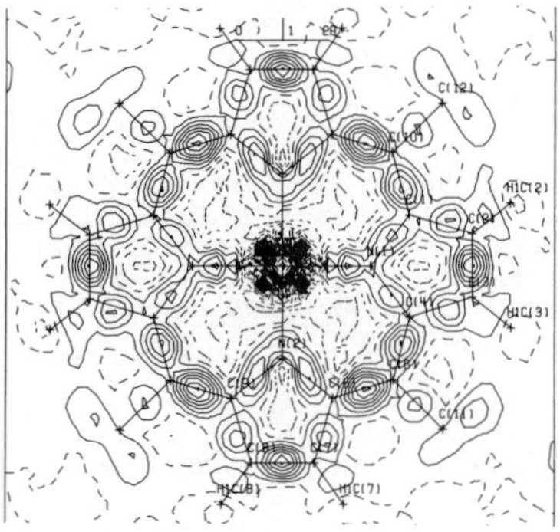

[1]Research supported by NIH (HL13157).

OXIDATION OF MONOPEROXOVANADIUM(V), VO_3^+, IN ACIDIC AQUEOUS SOLUTION.

R.C. Thompson, Department of Chemistry, University of Missouri-Columbia, Columbia, Missouri 65211.

Several distinct pathways for the oxidation of VO_3^+ in acidic solution have been found. i) A direct one-electron oxidation to produce a proposed $VO_3^{+2}\cdot$ radical cation, which may be viewed formally as a complex between oxovanadium(V) and the superoxide ion. The only fate of this intermediate appears to be an internal redox process that produces VO^{+2} and O_2. Oxidants that exhibit this mode of reaction are Co^{+3}, SO_4F^-, Ag^{+2}, and SO_4^-. All are strong oxidants, and fortunately are much more reactive toward VO_3^+ than VO^{+2}. ii) An indirect one-electron oxidation of VO_3^+ via reaction induced by VO^{+2}. Peroxomonosulfate, HSO_5^-, utilizes this pathway, and under most conditions the reaction is strictly catalyzed by VO^{+2}. Involvement of the $VO_3^{+2}\cdot$ radical cation as an intermediate is important in these systems also. iii) Indirect oxidation by reaction with the small concentrations of H_2O_2 in rapid equilibrium with VO_3^+. The aqueous Cl_2--VO_3^+ reaction proceeds solely by this reaction mode. iv) A fourth category not observed to date is a two-electron oxidation of VO_3^+ to form O_2 directly. Presumably the formation of a peroxo--oxidant complex would be required, and this path is unfavorable when the peroxide moiety is complexed to the vanadium(V) center.

Implications of these results on the redox chemistry of hydrogen peroxide and superoxide are discussed.

PHOTOCHEMICAL REACTION OF $ReH_5(PMe_2Ph)_3$ WITH CYCLIC POLYOLEFINS.

M. C. Trimarchi, J. C. Huffman, M. A. Green, and K. G. Caulton, Department of Chemistry and Molecular Structure Center, Indiana University, Bloomington, Indiana 47405.

Photolysis of hydrocarbon solutions of thermally stable ReH_5L_3 (L = PMe_2Ph) leads to dissociation of one phosphine ligand to generate a highly reactive transient species of the formula ReH_5L_2. Photolysis of ReH_5L_3 in the presence of 1,5-cyclooctadiene (1,5-COD) yields predominantly the simple diolefin complex, ReH_3L_2 (η^4-1,5 COD). Photolysis with cyclooctatetraene effects partial olefin hydrogenation to yield ReH_3L_2 (η^4-C_8H_{10}), which is thermally unstable and forms ReH_2L_2 (η^5-C_8H_{11}) upon standing. The latter complex has been structurally characterized. Similar results are obtained with cycloheptatriene, with photolytic generation of $ReH_2L_2(\eta^5$-$C_7H_9)$. Reactions of the dienyl complexes with HCl and HBF_4 will also be described.

SUBSTITUTION REACTIONS OF $Mn(CO)_3L_2\cdot$ RADICALS. Howard W. Walker, Sharon B. McCullen, and Theodore L. Brown, School of Chemical Sciences, University of Illinois, Urbana, IL 61801.

Persistent radicals of the type $Mn(CO)_3L_2\cdot$ (L = $P(n\text{-}Bu)_3$, $P(i\text{-}Bu)_3$) may be generated by photolysis of $Mn_2(CO)_8L_2$ in the presence of L. Repeated degassing of the hexane solution is required to remove the generated CO. Solutions of these radicals are stable for weeks under inert atmosphere. Reaction with CO cleanly reforms the bis-substituted dimer, equation (1).

$$2Mn(CO)_3L_2\cdot + 2CO \rightarrow Mn_2(CO)_8L_2 \quad (1)$$

The reaction is first order in $[Mn(CO)_3L_2\cdot]$ and [CO]. The second order rate constant for L = $P(n\text{-}Bu)_3$ is about two orders of magnitude faster than when L = $P(i\text{-}Bu)_3$ which has a larger cone angle. For L = $P(i\text{-}Bu)_3$, $\Delta H^{\ddagger}$ = 40kJ/mol and $\Delta S^{\ddagger}$ = -120J/mol·°K. These data are consistent with an associative displacement of L by CO, equation (2), with fast dimerization of $Mn(CO)_4L\cdot$, equation (3).

$$Mn(CO)_3L_2\cdot + CO \xrightarrow{k_1} Mn(CO)_4L\cdot + L \quad (2)$$

$$2Mn(CO)_4L\cdot \xrightarrow{fast} Mn_2(CO)_8L_2 \quad (3)$$

The rate of exchange of CO at $Mn(CO)_3L_2\cdot$ appears to be comparable with CO displacement of L.

A PHOTOELECTROCHEMICAL CELL BASED ON OXIDATIVE QUENCHING IN A METALLOPOLYMER FILM
T. David Westmoreland, Jeffrey M. Calvert, Royce W. Murray, and Thomas J. Meyer
Department of Chemistry, University of North Carolina, Chapel Hill, North Carolina 27514

Platinum electrodes coated with a film of polymer I yield cathodic photocurrent when irradiated with visible light in the presence of aqueous trisoxalato cobalt(III). These photocurrents are on the order of tens of microamps and are stable for over 1 hour. The per photon efficiency for production of current in an external circuit using 436 nm excitation is approximately 2 percent. Oxygen evolution from aqueous solution has been observed at a platinum counter-electrode.

$(2,2'\text{-bipyridine})_2Ru(PVP)_2^{2+}$

PVP is poly-4-vinylpyridine
(20 uncoordinated pyridine sites for each Ru site)

I

PREPARATION AND CHARACTERIZATION OF $HFe_4(BH_2)(CO)_{12}$. A HYDROGENATED IRON-BORIDE CLUSTER. Kwai Sam Wong, W. Robert Scheidt and Thomas P. Fehlner. Department of Chemistry, University of Notre Dame, Notre Dame, Indiana 46556.

The preparation and structural characterization of a novel iron-boron cluster containing four iron atoms and one boron is described, i.e., $HFe_4(BH_2)(CO)_{12}$. The compound consists of a $HFe_4(CO)_{12}$ butterfly fragment bridged across the wingtips by a BH_2 ligand. Both the geometrical structure and the ^{11}B NMR chemical shift suggest that this compound should be viewed as an iron-boride rather than a "normal" ferraborane. As such, it is a member of a new class of iron-boron clusters. The new compound is isoelectronic with $HFe_4(CH)(CO)_{12}$, a compound with a Fe-H-C interaction, and serves as a structural model for the interaction of CH_2 with a four atom cluster of metal atoms.

CYCLOMETALLATION OF CYCLOPROPYL PHOSPHINES WITH IRIDIUM AND PLATINUM COMPLEXES.[1] W. J. Youngs, Department of Chemistry, Northwestern University, Evanston, Illinois 60201.

The synthesis of cyclopropyl phosphines $R_2P\text{-}(CH_2)_x$  and their interaction with $[IrCl(C_8H_{14})_2]_2$ and $PtCl_2(N{\equiv}C\text{-}C_6H_5)_2$ provide several interesting examples of C-H and C-C bond activation by transition metal complexes. Depending on (1) the alkyl groups (R = t-butyl or phenyl) on the phosphorus, (2) the number of CH_2 groups between the phosphorus and the cyclopropane, and (3) the metal atom, selective cyclometallation can be observed.

[1]Research supported by the NSF (CHE80-09671).

REDUCTION OF $[Fe(CO)_3PPh_2]_2$ AND PROTONATION AND ALKYLATION OF A NEW DIANION, $[(CO)_3Fe(\mu\text{-}PPh_2)(\mu\text{-}CO)Fe(CO)_2PPh_2]^{2-}$. Yuan-Fu Yu and Andrew Wojcicki. Department of Chemistry, The Ohio State University, Columbus, Ohio 43210.

The complex $[Fe(CO)_3PPh_2]_2$ was reported by Collman to undergo a two-electron reduction reaction with Na dispersion to yield I. We find that the same result obtains when the reducing agent is Na amalgam, $LiAlH_4$, or $NaAlH_2(OCH_2CH_2OCH_3)_2$ in THF. By contrast, reduction with $LiB(C_2H_5)_3H$ or $MB[CH(CH_3)C_2H_5]_3H$ (M = Li or K), also in THF, is accompanied by cleavage of one of the Fe-P bonds to afford II, which was characterized by ^{31}P NMR and IR spectroscopy. Upon exposure to air in solution, II reverts rapidly and quantitatively to $[Fe(CO)_3PPh_2]_2$. Reaction of II with $HBF_4 \cdot (C_2H_5)_2O$ or CF_3COOH (-D) yields the yellow, air-sensitive hydride (deuteride) III (R = H or D), whereas reactions of II with CH_3I and $CH_2{=}CHCH_2I$ afford the red-brown methyl complex (III, R=CH_3) and the purple allyl complex (III, R = $CH_2CH{=}CH_2$), respectively. These air-stable products were characterized by a combination of ^1H, ^{13}C, ^{31}P NMR and IR spectroscopy.

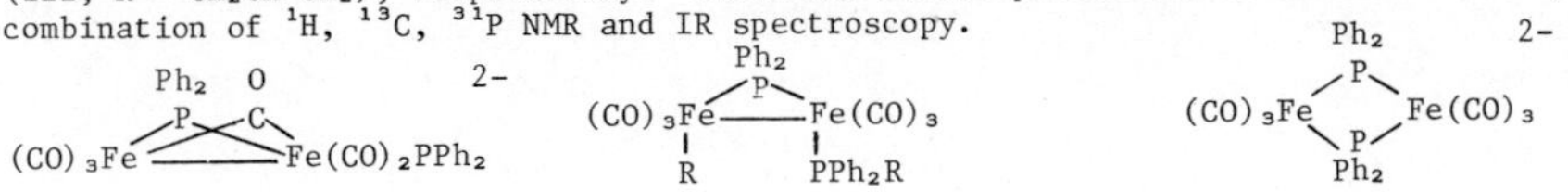

II III I

INDEX

A

B

C

D

I

J

K

L

M

N

U

V

W

X

Z

Jacket design by Kathleen Schaner
Indexing and production by Deborah Corson and Paula Bérard

Elements typeset by Service Composition Co., Baltimore, MD
Printed and bound by Maple Press Co., York, PA